Industrial Biotransformations

Edited by
A. Liese, K. Seelbach, C. Wandrey

Further Titles of Interest

Aehle, W. (Ed.)

Enzymes in Industry

Production and Applications

Second, Completely Revised Edition
2004
ISBN 3-527-29592-5

Bommarius, A. S., Riebel, B. R.

Biocatalysis

Fundamentals and Applications

2004
ISBN 3-527-30344-8

Buchholz, K., Kasche, V., Bornscheuer, U. T.

Biocatalysts and Enzyme Technology

2005
ISBN 3-527-30497-5

Jördening, H.-J., Winter, J. (Eds.)

Environmental Biotechnology

2005
ISBN 3-527-30585-8

Brakmann, S., Schwienhorst, A. (Eds.)

Evolutionary Methods in Biotechnology

Clever Tricks for Directed Evolution

2004
ISBN 3-527-30799-0

Industrial Biotransformations

Second, Completely Revised and Extended Edition

Edited by
Andreas Liese, Karsten Seelbach, Christian Wandrey

WILEY-VCH

WILEY-VCH Verlag GmbH & Co. KGaA

Editors

Prof. Dr. Andreas Liese
Technische Universität Hamburg-Harburg
Institute of Biocatalysis
Denickestr. 15
21073 Hamburg
Germany

Dr. Karsten Seelbach
Siegwerk Druckfarben AG
Alfred-Keller-Str. 55
53721 Siegburg
Germany

Prof. Dr. Christian Wandrey
Forschungszentrum Jülich
Institute of Biotechnology
52425 Jülich
Germany

First Edition 2000
Second, Completely Revised and
Extended Edition 2006

Library of Congress Card No.: applied for
British Library Cataloguing-in-Publication Data
A catalogue record for this book is available
from the British Library.

**Bibliographic information published by
Die Deutsche Bibliothek**
Die Deutsche Bibliothek lists this publication
in the Deutsche Nationalbibliografie; detailed
bibliographic data is available in the Internet at
<http://dnb.ddb.de>.

Printed in the Federal Republic of Germany.
Printed on acid-free paper.

Typesetting   Kühn & Weyh, Satz und Medien,
Freiburg
Printing   Betz Druck GmbH, Darmstadt
Bookbinding   Litges & Dopf Buchbinderei GmbH,
Heppenheim

ISBN-13:   978-3-527-31001-2
ISBN-10:   3-527-31001-0

V

Contents

Preface to the first edition

The main incentive in writing this book was to gather information on one-step biotransformations that are of industrial importance. With this collection, we want to illustrate that more enzyme-catalyzed processes have gained practical significance than their potential users are conscious of. There is still a prejudice that biotransformations are only needed in cases where classical chemical synthesis fails. Even the conviction that the respective biocatalysts are not available and, if so, then too expensive, unstable and only functional in water, still seems to be widespread. We hope that this collection of industrial biotransformations will in future influence decision-making of synthesis development in such a way that it might lead to considering the possible incorporation of a biotransformation step in a scheme of synthesis.

We therefore took great pains in explicitly describing the substrates, the catalyst, the product and as much of the reaction conditions as possible of the processes mentioned. Wherever flow schemes were available for publication or could be generated from the reaction details, this was done. Details of some process parameters are still incomplete, since such information is only sparingly available. We are nevertheless convinced that the details are sufficient to convey a feeling for the process parameters. Finally, the use of the products is described and a few process-relevant references are made.

We would go beyond the scope of this foreword, should we attempt to thank all those who were kind enough to supply us with examples. Of course, we only published openly available results (including the patent literature) or used personally conveyed results with the consent of the respective authors. We are aware of the fact that far more processes exist and that by the time the book is published, many process details will be outdated. Nonetheless, we believe that this compilation with its overview character will serve the above-mentioned purpose. This awareness could be augmented if the reader, using his or her experience, would take the trouble of filling out the printed worksheet at the end of this book with suggestions that could lead to an improvement of a given process or the incorporation of a further industrial process into the collection.

Requesting our industrial partners to make process schemes and parameters more accessible did not please them very much. Even so, we are asking our partners once again to disclose more information than they have done in the past. In

Industrial Biotransformations. Andreas Liese, Karsten Seelbach, Christian Wandrey (Eds.)
Copyright © 2006 WILEY-VCH Verlag GmbH & Co. KGaA, Weinheim
ISBN: 3-527-31001-0

many instances, far more knowledge of industrial processes has been gained than is publicly available. Our objective is to be able to make use of these "well known secrets" as well. We would like to express our gratitude to all those who supplied us with information in a progress-conducive manner. Thanks also go to those who did not reject our requests completely and at least supplied us with a photograph in compensation for the actually requested information.

The book begins with a short historical overview of industrial biotransformations. Since the process order of the compilation is in accordance with the enzyme nomenclature system, the latter is described in more detail. We also include a chapter on reaction engineering to enable an easier evaluation of the processes. The main part of the book, as you would expect, is the compilation of the industrial biotransformations. The comprehensive index will allow a facile search for substrates, enzymes and products.

We sincerely hope that this book will be of assistance in the academic as well as the industrial field, when one wants to get an insight into industrial biotransformations. We would be very thankful to receive any correction suggestions or further comments and contributions. At least we hope to experience a trigger effect that would make it worth while for the readership, the authors and the editors to have a second edition succeeding the first.

We are indebted to several coworkers for screening literature and compiling data, especially to Jürgen Haberland, Doris Hahn, Marianne Hess, Wolfgang Lanters, Monika Lauer, Christian Litterscheid, Nagaraj Rao, Durda Vasic-Racki, Murillo Villela Filho, Philomena Volkmann and Andrea Weckbecker.

We thank especially Uta Seelbach for drawing most of the figures during long nights, as well as Nagaraj Rao and the "enzyme group" (Nils Brinkmann, Lasse Greiner, Jürgen Haberland, Christoph Hoh, David Kihumbu, Stephan Laue, Thomas Stillger and Murillo Villela Filho).

And last but not least we thank our families for their support and tolerance during the time that we invested in our so called 'book project'.

Preface to the second edition

After more than five years since the first edition of "Industrial Biotransformations" many new examples have become industrially relevant, others have lost importance. Therefore we had to enlarge the chapter "Processes" by 20%. If new information about the processes of the first edition was available, this information was incorporated. All processes were checked with respect to the literature (including patent literature). We have included all the valuable corrections suggestions or further comments and contributions of many readers. This might perhaps be of great importance for the reader of the second edition. Expecting that a first edition could not be perfect, we stated in the preface to the first edition: "We would be very thankful to receive any correction suggestions or further comments and contributions. At least we hope to experience a trigger effect that would make it worthwhile for the readership, the authors and the editors to have a second edition succeeding the first." We were astonished how carefully many readers checked the information given. So the reader of the second edition will not only have an enlarged chapter "Processes", but also an updated version with – me must admit – many useful corrections. The best criticism will be given by an experienced reader. We hope very much that the "old" and the "new" readers will realize that the second edition is more than a remake of the first edition.

Since the first edition was sold out earlier than we had expected, the publisher found it scientifically – and economically – more reasonable to have a second edition than to have a reprint of the first edition. Finally after all the additional work was done we agreed with the publisher. Perhaps it is worth to be mentioned that in the meantime also the first Chinese edition appeared.

The focus of the book is still the chapter "Processes". Nevertheless all the other chapters were carefully reevaluated. In the chapter "History of Industrial Biotransformations" we included a new part "History of Biochemical Engineering".

Entirely new is the chapter "Retrosynthetic Biocatalysis". The basic idea comes from classical organic chemistry, where a complex chemical structure is reduced to building blocks, which might even be commercially available. Similarly, one can find out which easily available building blocks can be used for industrial biotransformations. We hope that the reader will find this concept useful. Especially we hope that the classical organic chemistry becomes more part of biotechnology this way.

Industrial Biotransformations. Andreas Liese, Karsten Seelbach, Christian Wandrey (Eds.)
Copyright © 2006 WILEY-VCH Verlag GmbH & Co. KGaA, Weinheim
ISBN: 3-527-31001-0

Entirely new is the chapter "Optimization of Industrial Enzymes by Molecular Engineering". The field of technical evolution of enzymes has become so important that we think it is justified to have a chapter devoted to the interesting and relevant findings in this field. There is no longer an "excuse" that there is no sufficiently stable, selective and active enzyme for a desired reaction. Technical evolution of enzymes has become similarly important as screening of enzymes from the environment. The chapter "Basics of Bioreaction Engineering" has been carefully checked and hopefully improved due to many valuable suggestions of the readers. We hope that bioreaction engineering will be understood as of equal importance for industrial biotransformations as enzyme engineering.

An additional short chapter is entitled "Quantitative Analysis of Industrial Biotransformation". Here the reader can find some quantitative information about the fact that it is a prejudice to believe that only hydrolases in water are useful for industrial biotransformations. Redox reactions and C-C-bond formations even in organic solvents or biphasic systems are also industrially relevant today.

Our original understanding of "Industrial Biotransformations" was a one step reaction of industrial relevance. This definition might become less clear in the future, because also two or three step biotransformations are or might be included. So it will become more and more difficult to distinguish an industrial biotransformation from a fermentation process. This is especially true in the age of "designer bugs", where a microorganism is first grown and than used as a more or less "non-growing catalyst" for industrial biotransformation. But we should not bother too much with definitions. Our aim was and is to show that biotransformations are of great importance in the academic and industrial fields. Since the first edition the field of "White Biotechnology" (formally known as Industrial Microbiology) has developed a lot. A quantative understanding of complex microbial systems by means of the "polyomics" techniques (genomics, proteomics, and metabolomics) has improved so much that we can expect recombinant pathways for industrial biotransformations used in a biocatalysis under non-natural conditions. We can foresee that the technical evolution not only of biocatalyst but also of bioprocesses will lead to many more industrial biotransformations. Thus, sooner or later we expect the burden/pleasure that we will have to prepare a third edition of "Industrial Biotransformations".

We would like to ask the reader again to help us with "correction suggestions or further comments and contributions". The best compensation for all the work the author of a book can get is the feeling that it is read by colleagues who understand the subject.

Last but not least we would like to mention in addition to the many coworkers who have contributed to the first edition, now the additional valuable contributions of many more who helped us to prepare the second edition, especially Karl-Heinz Drauz, Kurt Faber, Katja Goldberg, Udo Kragl, Peter Stahmann, Trevor Laird, John Villadsen, Ulrike Zimmermann. Especially we thank our families for their support during the time that we invested in the second edition of this book.

Dezember 2005

Andreas Liese
Karsten Seelbach
Christian Wandrey

List of Contributors

Dr. Arne Buchholz
DSM Nutritional Products
Feldbergstraße 8
79539 Lörrach
Germany

Dr. Thorsten Eggert
Institut für Molekulare
Enzymtechnologie
der Heinrich-Heine-Universität
Düsseldorf
im Forschungszentrum Jülich
52426 Jülich
Germany

Dr. Murillo Villela Filho
Degussa AG
Poject House ProFerm
Rodenbacher Chaussee 4
63457 Hanau-Wolfgang
Germany

Dr. Jürgen Haberland
DSM Nutritional Products
Emil-Barell-Straße 3
Bldg.50/R.209
79639 Grenzach-Wyhlen
Germany

Dipl.-Chem. Christoph Hoh
Sanofi-Aventis Deutschland GmbH
Diabel Operations
Industriepark Höchst
65926 Frankfurt am Main
Germany

Prof. Dr. Andreas Liese
Head of Institute of Biotechnology II
Technical University of Hamburg-
Harburg
Denickestr. 15
21073 Hamburg
Germany

Dipl.-Chem. Stephan Lütz
Institut für Biotechnologie
Forschungszentrum Jülich GmbH
52425 Jülich
Germany

Alan Pettman
Pfizer Limited
Walton Oaks
Dorking Road
Tadworth
Surrey, KT20 7NS
UK

Industrial Biotransformations. Andreas Liese, Karsten Seelbach, Christian Wandrey (Eds.)
Copyright © 2006 WILEY-VCH Verlag GmbH & Co. KGaA, Weinheim
ISBN: 3-527-31001-0

Prof. Dr. Durda Vasic-Racki
Faculty of Chemical Engineering
Department of Reaction Engineering
and Catalysis
University of Zagreb
Savska c.16
Zagreb HR-10000
Croatia

Dr. Nagaraj N. Rao
Rane Rao Reshamia Laboratories PVT.
LTD.
Plot 80, Sector 23. Cidco Industrial Area
Turbhe Naka
Navi Mumbai 400 705
India

Dr. Karsten Seelbach
Siegwerk Druckfarben AG
Alfred-Keller-Str. 55
53721 Siegburg
Germany

Prof. Dr. Adrie J. J. Straathof
Kluyver Laboratory for Biotechnology
Delft University of Technology
Julianalaan 136
Delft 2628 BL
The Netherlands

Dr. Junhua Tao
Pfizer Global Research & Development
La Jolla
10578 Science Center Drive
San Diego, CA 92121
USA

Prof. Dr. Christian Wandrey
Forschungszentrum Jülich
Insitute of Biotechnology
52425 Jülich
Germany

1
History of Industrial Biotransformations – Dreams and Realities

Durda Vasic-Racki

Throughout the history of mankind, microorganisms have been of enormous social and economic importance. Without even being aware of their existence, very early on in history man was using them in the production of food and beverages. The Sumerians and Babylonians were practising the brewing of beer before 6000 BC, references to wine making can be found in the Book of Genesis and the Egyptians used yeast for baking bread. However, knowledge of the production of chemicals such as alcohols and organic acids through fermentation is relatively recent and the first reports in the literature only appeared in the second half of the 19th century. Lactic acid was probably the first optically active compound to be produced industrially by fermentation. This was accomplished in the USA in 1880 [1]. In 1921, Chapman reviewed a number of early industrial fermentation processes for organic chemicals [2].

In the course of time, it was discovered that microorganisms could modify certain compounds by simple, chemically well defined reactions, which were further catalyzed by enzymes. Nowadays, these processes are called "biotransformations". The essential difference between fermentation and biotransformation is that there are several catalytic steps between the substrate and the product in a fermentation while there are only one or two in a biotransformation. The distinction is also in the fact that the chemical structures of the substrate and the product resemble one another in a biotransformation, but not necessarily in a fermentation.

1.1
From the "Flower of Vinegar" to Recombinant *E. Coli* – The History of Microbial Biotransformations

The story of microbial biotransformations is closely associated with vinegar production which dates back to around 2000 years BC.

Vinegar production is perhaps the oldest and best known example of microbial oxidation, which can illustrate some of the important developments in the field of biotransformations by living cells (Fig. 1.1).

Since ancient times, man has wanted to see things that are far smaller than can be perceived with the naked eye. In the 16th century this led to the construction of a magnifier

Industrial Biotransformations. Andreas Liese, Karsten Seelbach, Christian Wandrey (Eds.)
Copyright © 2006 WILEY-VCH Verlag GmbH & Co. KGaA, Weinheim
ISBN: 3-527-31001-0

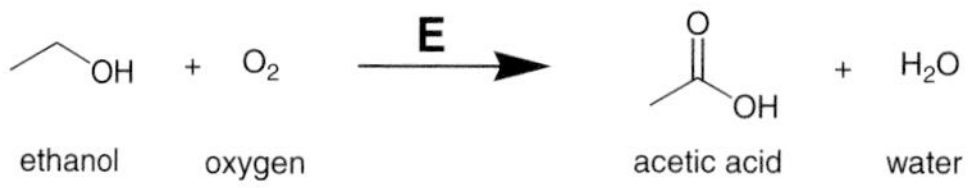

ethanol + oxygen $\xrightarrow{\text{E}}$ acetic acid + H_2O water

Fig. 1.1 Vinegar production.

consisiting of a single convex lens, and this, in turn, led eventually to the development of the microscope (Fig. 1.2). Antony von Leeuwenhoek (1632–1723) became the first person, or microscopist [3], to make and use a real microscope. He described microorganisms including bacteria, algae and protozoa in fresh water (Fig. 1.3). In fact he constructed a total of 400 microscopes during his lifetime. Subsequently, the compound microscope system was invented in the 17th century. This type of microscope, incorporating more than one lens, has made tremendous contributions to the progress of science. Using this microscope Hooke (1635–1703) discovered the fact that living things are composed of cells, and later on Pasteur, among others, discovered yeast fungus. The microscope has possibly had a greater impact on the development of knowledge than any other scientific instrument in history [4]. The discovery of new microscopic life was the starting point for experimental biology as a basis for the development of the biotransformations.

A prototype bioreactor with immobilized bacteria has been known in France since the 17th century. The oldest bioreactor to use immobilized living microorganisms, a so-called generator, was developed in 1823 [5, 6]. Even today, acetic acid is still known as "vinegar" if it is obtained by oxidative fermentation of ethanol-containing solutions by acetic acid bacteria [7].

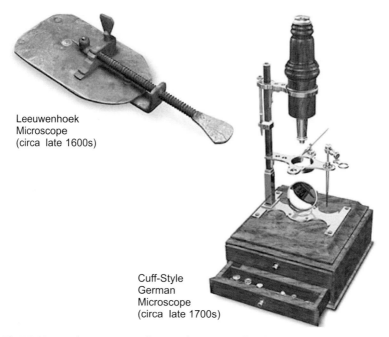

Leeuwenhoek
Microscope
(circa late 1600s)

Cuff-Style
German
Microscope
(circa late 1700s)

Fig. 1.2 Historical microscopes (photographs courtesy of Michael W. Davidson).

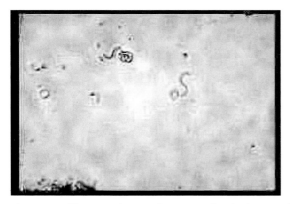

Fig. 1.3 Spiral bacteria (photograph courtesy of Michael W. Davidson).

In 1858, Pasteur [8] was the first to demonstrate the microbial resolution of tartaric acid. He performed fermentation of the ammonium salt of racemic tartaric acid, mediated by the mold *Penicillium glaucum*. The fermentation yielded (–)-tartaric acid (Fig. 1.4).

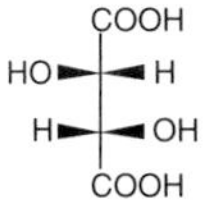

COOH
HO ◄ H
H ◄ OH
COOH

(–)-tartaric acid
(*S,S*)-tartaric acid

Fig. 1.4 Pasteur's product of the first resolution reaction.

This was also the first time that a method was used where the microorganisms degraded one enantiomer of the racemate while leaving the other untouched.

In 1862, Pasteur [9] investigated the conversion of alcohol into vinegar and concluded that pellicle, which he called "the flower of vinegar", "served as a method of transport for the oxygen in air to a multitude of organic substances".

In 1886 Brown confirmed Pasteur's findings and gave the causative agent in vinegar production the name *Bacterium xylinum*. He also found that it could oxidize propanol to propionic acid and mannitol to fructose (Fig. 1.5) [10].

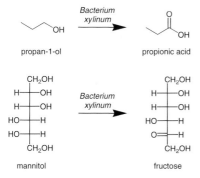

propan-1-ol → (*Bacterium xylinum*) → propionic acid

mannitol → (*Bacterium xylinum*) → fructose

Fig. 1.5 Reactions catalyzed by *Bacterium xylinum*, the vinegar biocatalyst.

In 1897 Buchner [11] reported that cell-free extracts prepared by grinding yeast cells with sand could carry out alcoholic fermentation reactions (breaking down glucose into ethanol and carbon dioxide) in the absence of living cells. This initiated the use of resting cells for biotransformations.

In 1921 Neuberg and Hirsch [12] discovered that the condensation of benzaldehyde with acetaldehyde in the presence of yeast forms optically active 1-hydroxy-1-phenyl-2-propanone (Fig. 1.6).

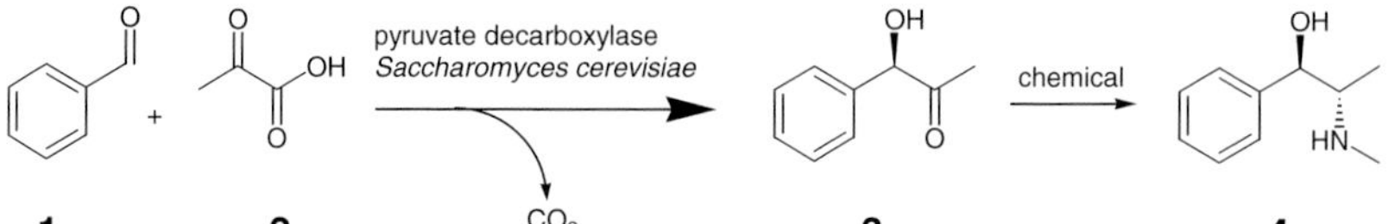

1   **2** CO_2 **3**   **4**

pyruvate decarboxylase
Saccharomyces cerevisiae

chemical

1 = benzaldehyde
2 = 2-oxo-propionic acid
3 = 1-hydroxy-1-phenylpropan-2-one
4 = 2-methylamino-1-phenylpropan-1-ol

Fig. 1.6 L-Ephedrine production.

DEUTSCHES REICH

AUSGEGEBEN AM
13. APRIL 1932

REICHSPATENTAMT

PATENTSCHRIFT

№ 548459

KLASSE **12 q** GRUPPE 32

12 q K 77. 30

Tag der Bekanntmachung über die Erteilung des Patents: 24. März 1932

Knoll A.-G. Chemische Fabriken in Ludwigshafen a. Rh.,
Dr. Gustav Hildebrandt und Dr. Wilfrid Klavehn in Mannheim

Verfahren zur Herstellung von 1-1-Phenyl-2-methylaminopropan-1-ol

Patentiert im Deutschen Reiche vom 9. April 1930 ab

Racemisches 1-Phenyl-2-methylaminopropan-1-ol kann bereits nach verschiedenen Verfahren synthetisch hergestellt werden (vgl. z.B. Nagai u. Kanao, Annalen 470 [1929], S. 157; Patentschrift 469782; Skita u. Keil, Ber. 62 [1929], S. 1142 ff.; Patentschrift 524 806).

Das racemische 1-Phenyl-2-methylaminopropan-1-ol läßt sich nach bekannten Verfahren (vgl. Nagai u. Kanao, Annalen 470 [1929], S. 157; britische Patentschrift 297 385). in seine optischen Isomeren spalten. Das bei der Spaltung entstehende l-1-Phenyl-2-methylaminopropan-1-ol ist identisch mit dem natürlichen Ephedrin und wird neuerdings therapeutisch erfolgreich verwendet.

Bisher ist jedoch kein Verfahren bekannt geworden, nach welchem l-1-Phenyl-2-methylaminopropan-1-ol auf unmittelbarem Wege dargestellt werden kann.

Es wurde nun gefunden, daß man mit guten Ausbeuten unmittelbar zum l-1-Phenyl-2-methylaminopropan-1-ol gelangt, wenn man links drehendes Phenylpropanolon (Neuberg, Biochem. Zeitschrift 115 [1921], S. 282 ff., und 128 [1922], S. 610 ff.) in Gegenwart von Methylamin der Reduktion unterwirft.

Die Bildung von l-1-Phenyl-2-methylaminopropan-1-ol war keineswegs vorauszusehen, da bekanntlich Abwandlungen von optisch aktiven Verbindungen nicht notwendig zu Verbindungen von optisch gleicher Drehungsrichtung führen müssen, sondern eben- sowohl zu solchen der entgegengesetzten Drehungsrichtung führen können (so entsteht z. B. l-Mandelsäure aus d-Benzaldehydcyanhydrin bei der Verseifung).

Ferner ist hervorzuheben, daß in der vorliegenden Erfindung eine asymmetrische Synthese vorliegt, bei welcher wiederum nicht vorauszusehen war, welche von den möglichen Konfigurationen entstehen würde. Es stand zu erwarten, daß sowohl d- oder l- oder dl-1-Phenyl-2-methylaminopropan-1-ol oder d-Pseudo-1-Phenyl-2-methylaminopropan-1-ol oder endlich ein Gemisch von mehreren dieser Komponenten entstehen würde. Daß bei der Reduktionskondensation des 1-Phenyl-propanolons mit Methylamin fast ausschließlich l-1-Phenyl-2-methylaminopropan-1-ol entsteht, war daher durchaus überraschend.

Die Erhaltung der optischen Aktivität war auch deswegen überraschend, weil auf Grund der Neubergschen Beobachtung (Biochem. Zeitschrift 128 [1922], S. 613) 1-Phenylpropanolon sich bereits in verdünnter alkalischer Lösung in kurzer Zeit racemisiert. Da bei dem Verfahren der vorliegenden Erfindung die Reduktion in alkalischer Lösung stattfindet, war vorwiegend mit der Bildung von racemischen Basen zu rechnen.

Das Verfahren stellt eine neue Methode dar, um das links drehende Phenylpropanolon in Form von l-1-Phenyl-2-methylaminopropan-1-ol nutzbringend zu verwerten.

Diese unmittelbare Synthese des l-1-Phenyl-2-methylaminopropan-1-ols hat ferner den Vorzug, daß kein therapeutisch wertloses d-1-Phenyl-2-methylaminopropan-1-ol anfällt, wie es bei den bekannten Spaltungsverfahren des Racemkörpers der Fall ist.

Beispiel 1.

120 g des durch Ätherauszug gewonnenen phenylpropanolonhaltigen Gärungsproduktes (vgl. Biochem. Zeitschrift 115 [1921], S. 282 ff.) läßt man ohne weitere Reinigung in eine Lösung von 10 g Methylamin in 500 ccm Äther und zu 20 g aktiviertem Aluminium unter Rühren im Verlaufe von 2 Stunden eintropfen. Gleichzeitig läßt man 20 bis 30 g Wasser tropfenweise zufließen. Die sofort heftig einsetzende Umsetzung wird zeitweilig durch Kühlung gemäßigt. Nach beendigter Reduktion wird der filtrierten ätherischen Lösung die entstandene optisch aktive Base mit verdünnter Säure entzogen. Die Aufarbeitung erfolgt in bekannter Weise.

Man erhält das Hydrochlorid des l-1-Phenyl-2-methylaminopropan-1-ols vom F. 214°, welches die aus der Literatur bekannte Linksdrehung zeigt. Die Ausbeute beträgt je nach Art des verwendeten Ausgangsstoffes 25 bis 45 g Hydrochlorid.

Beispiel 2

360 g des in Beispiel 1 verwendeten phenylpropanolonhaltigen Ätherauszuges werden unter vermindertem Druck destilliert. 300 g der bei 100 bis 150° unter 14 mm Druck übergehenden Fraktion werden an Gegenwart von kolloidalem Platin (70 ccm 1%ige Lösung) und 85 g 33%iger Methylaminlösung der katalytischen Reduktion unterworfen. Es ist vorteilhaft, etwas Äther zuzusetzen. Nach Beendigung der Wasserstoffaufnahme wird die ätherische Lösung mit Salzsäure ausgeschüttelt und das l-1-Phenyl-2-methylaminopropan-1-ol in bekannter Weise abgetrennt.

Das Hydrochlorid schmilzt bei 214° und zeigt die aus der Literatur bekannte Linksdrehung. Die Ausbeute an Hydrochlorid beträgt 110 g.

Beispiel 3

100 g nach Neuberg (Biochem. Zeitschrift 128 [1922], S. 611) abgetrenntes l-1-Phenylpropan-1-ol-2-on werden in 200 ccm Äther gelöst, mit 75 g 33%iger Methylaminlösung versetzt und etwa eine halbe Stunde lang geschüttelt. Hierbei findet unter Wärmeentwicklung Kondensation statt. Anschließend wird in Gegenwart von 70 ccm 1%iger kolloidaler Platinlösung mit Wasserstoff reduziert.

Die Aufarbeitung geschieht nach Beispiel 2. Das Hydrochlorid des l-1-Phenyl-2-methylaminopropan-1-ols kristallisiert aus Alkohol in derben Prismen vom F. 214 bis 216°. Der F. der freien Base liegt bei 40°.

PATENTANSPRÜCHE:

1. Verfahren zur Darstellung von l-1-Phenyl-2-methylaminopropan-1-ol, dadurch gekennzeichnet, daß man links drehendes l-Phenylpropan-1-ol-2-on mit Methylamin kondensiert und das Kondensationsprodukt gleichzeitig oder nachträglich mit Reduktionsmitteln, wie aktiviertem Aluminium in Gegenwart von Wasser oder Wasserstoff in Gegenwart eines Platinkatalysators, behandelt.

2. Verfahren nach Anspruch 1, dadurch gekennzeichnet, daß man — zur Umgehung der Reindarstellung des l-Phenylpropanolons — Destillate oder Auszüge aus l-Phenylpropanolon enthaltenden Gemischen verwendet, wie sie z. B. bei der Vergärung von Zuckern oder von zuckerhaltigen Produkten in Gegenwart von Benzaldehyd entstehen.

Fig. 1.7 Knoll's patent of 1930.

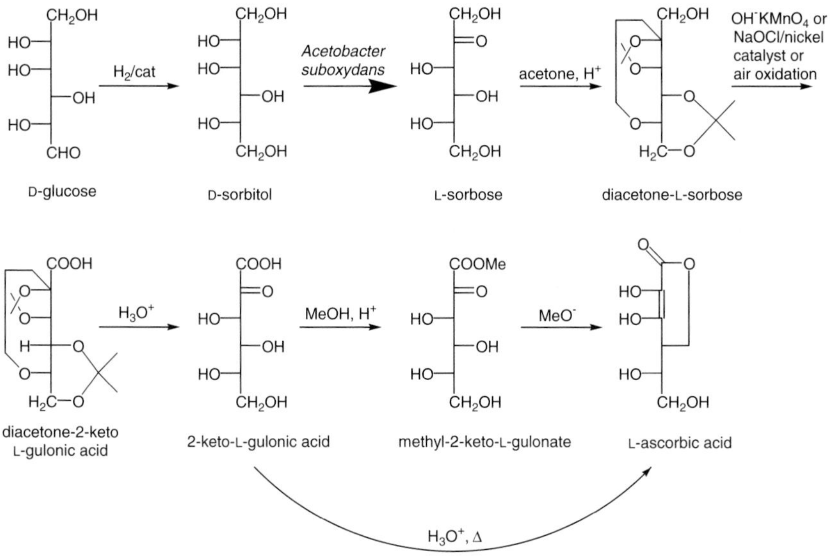

Fig. 1.8 Reichstein–Grüssner synthesis of vitamin C (L-ascorbic acid).

The compound obtained was later chemically converted into L-(–)ephedrine by Knoll AG, Ludwigshafen, Germany in 1930 (Fig. 1.7) [13].

The bacterium *Acetobacter suboxydans* was isolated in 1923 [14]. Its ability to carry out limited oxidations was utilized in a highly efficient preparation of L-sorbose from D-sorbitol (Fig. 1.8).

L-Sorbose became important in the mid-1930s as an intermediate in the Reichstein–Grüssner synthesis of L-ascorbic acid [15].

In 1953, Peterson al al. [16] reported that *Rhizopus arrhius* could convert progesterone into 11α-hydroxyprogesterone (Fig. 1.9), which was used as an intermediate in the synthesis of cortisone.

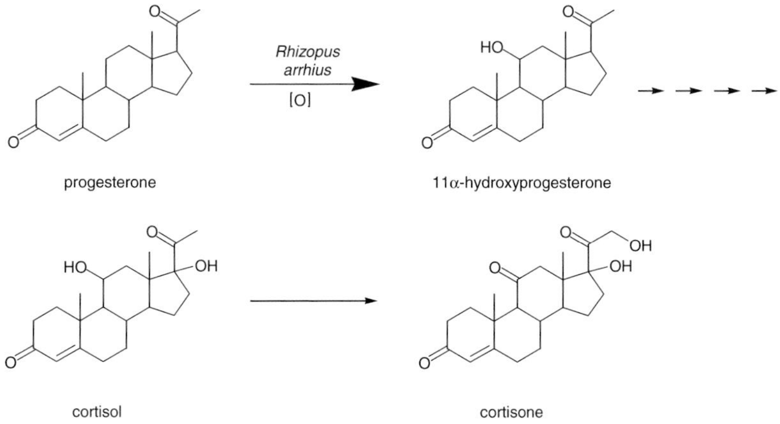

Fig. 1.9 Microbial 11α-hydroxylation of progesterone.

This microbial hydroxylation simplified and considerably improved the efficiency of the multi-step chemical synthesis of corticosteroid hormones and their derivatives. Although the chemical synthesis [17] (Fig. 1.10) from deoxycholic acid developed at Merck, Germany, was workable, it was recognized that it was complicated and uneconomical: 31 steps were necessary to obtain 1 kg of cortisone acetate from 615 kg of deoxycholic acid. The microbial 11α-hydroxylation of progesterone rapidly reduced the price of cortisone from \$200 to \$6 per gram. Further improvements have led to a current price of less than \$1 per gram [18].

In the 1950s the double helix structure and the chemical nature of RNA and DNA – the genetic code for heredity – were discovered by Watson and Crick [19]. Beadle and Tatum [20] received the Nobel Prize in 1958 for concluding that the characteristic function of the gene was to control the synthesis of a particular enzyme. They exposed the red bread mold, *Neurospora crassa*, to X-rays and studied the altered nutritional requirements of the mutants thus produced. These experiments enabled them to conclude that each gene determined the structure of a specific enzyme which, in turn, allowed a single chemical reaction to proceed. A basic hypothesis arose out of this work: one gene specifies the production of one enzyme thus the "one gene–one enzyme" hypothesis. Lederberg [21] shared the Nobel Prize with Beadle and Tatum in 1958 for his discoveries concerning genetic recombination and the organization of the genetic material of bacteria. The Beadle, Tatum and Lederberg discoveries can be regarded as milestones among the main scientific achievements of the 20th century. This led to the synthesis of recombinant DNA and gave a fillip to genetic engineering in the 1970s. In 1973, Cohen and Boyer [22] discovered recombinant DNA technology. They observed that genes from any biological species could be propagated and cloned in foreign cells by linking them to DNA molecules that posses the capacity to replicate in the intended host. Such developments quickly made the recombinant DNA technology part of industrial microbial transformations. Application of this technology for the production of small molecules began in 1983. Ensley et al. [23] reported on the construction of a strain of *E.coli* that excreted indigo, one of the oldest known dyes. They found that the entire pathway for conversion of napthalene into salicylic acid is encoded by the genes of *Pseudomonas putida*. These genes can be expressed in *E. coli*. Their results led to the unexpected finding that a subset of these genes was also responsible for the microbial production of indigo. Moreover, they have shown that indigo formation was a property of the dioxygenase enzyme system that forms *cis*-dihydrodiols from aromatic hydrocarbons. Finally, they proposed a pathway for indigo biosynthesis in a recombinant strain of *E. coli* (Fig. 1.11).

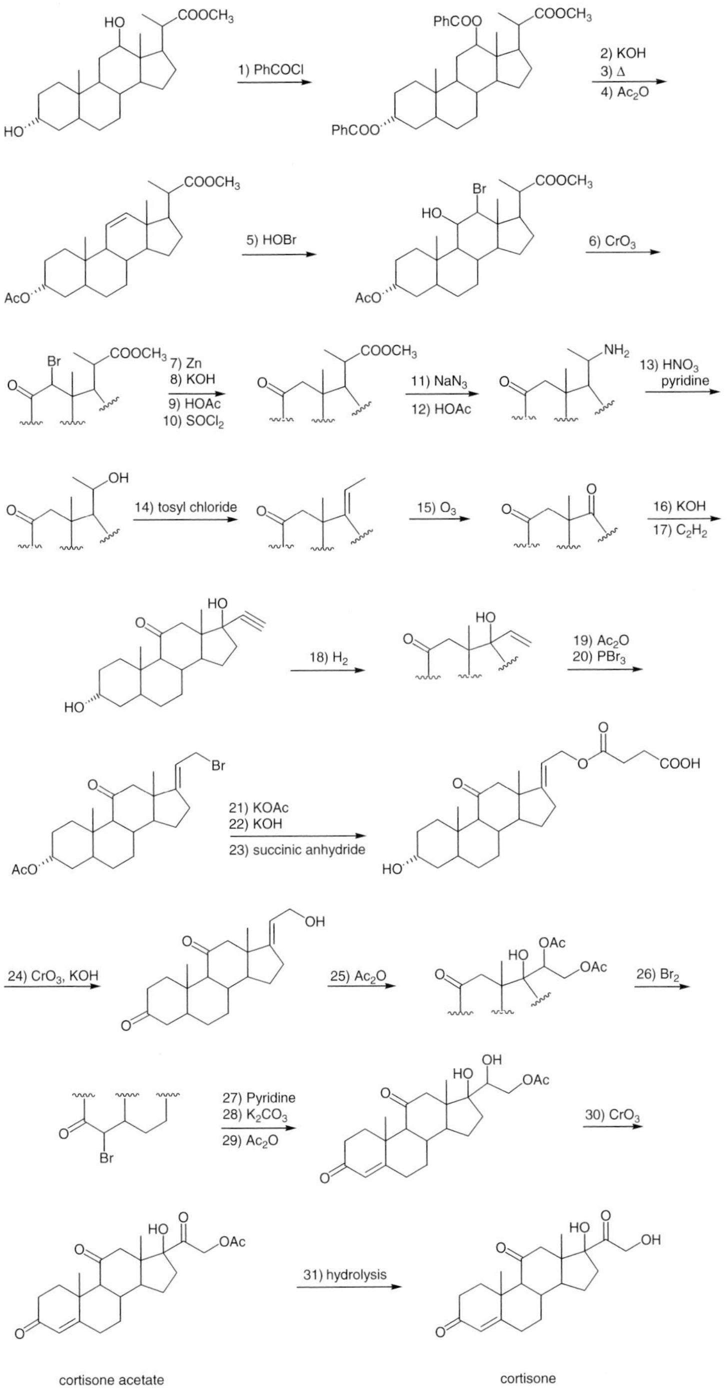

Fig. 1.10 Chemical synthesis of cortisone.

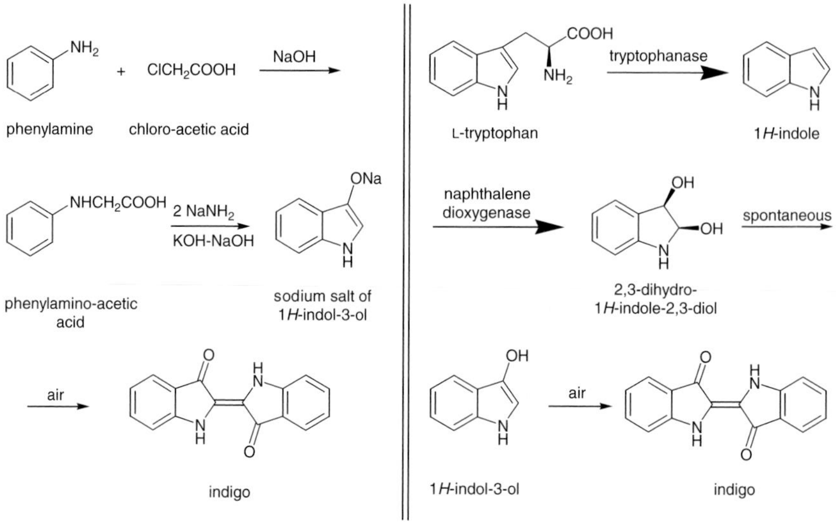

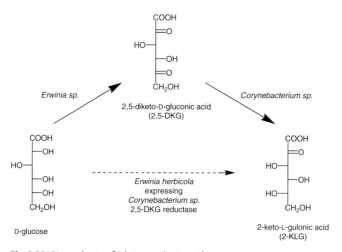

Fig. 1.11 Comparison of the chemical and biological routes to indigo.

Genencor Internationaol is developing a commercially competitive biosynthetic route to indigo using recombinant *E. coli*, which can synthesize indigo directly from glucose [24]. At the neutral pH of the fermentation the indigo precursor indoxyl yields isatin as a significant by-product. An enzyme that hydrolyzes isatin to isatic acid has been identified. After cloning and incorporating the new enzyme in the production strain, the indigo product performed equally well as the indigo produced chemically [25].

In 1984 Novozymes developed the first enzyme from a genetically modified organism for use in the starch industry – a maltogenic amylase, still marketed today under the name Maltogenase® [26].

Fig. 1.12 Biosynthesis of 2-keto-L-gulonic acid.

Anderson et al. [27] reported in 1985 on the construction of a metabolically engineered bacterial strain that was able to synthesize 2-keto-L-gulonic acid (Fig. 1.12), a key intermediate in the production of L-ascorbic acid (vitamin C).

BASF, Merck and Cerestar have built a ketogulonic acid plant in Krefeld, Germany. The operation started up in 1999. They developed a new fermentation route from sorbitol directly to ketogulonic acid [28]. This method is probably similar to that described in 1966 [29].

The vision of manufacturing L-ascorbic acid directly by fermentation without the need to isolate 2-keto-L-gulonic acid (2-KLG) has remained elusive. Nevertheless, efforts to this end are ongoing at Genencor International [30].

The Cetus Corporation (Berkeley, CA, USA) bioprocess for converting alkenes into alkene oxides emerged in 1980 [31]. This bioprocess appeared to be very interesting, because of the possibility of replacing the energy-consuming petrochemical process.

There were high hopes that the development of recombinant DNA technology would speed up technological advances. Unfortunately, there is still a great deal of work to be done on the development and application of bioprocesses before the commercial production of low-cost chemicals becomes feasible [32]. The development of some of the flagship bioprocesses of today took between 10 and 20 years: the development of the acrylamide process took 20 years and the Lonza process for L-carnitine 15 years [33]. However, today even the traditional chemical companies such as Dow Chemicals, DuPont, Degussa-Hüls AG, etc., under pressure from investors and because of technological advances, are trying to use microbial or enzymatic transformations in their production processes. This is because they need to establish whether natural feedstocks can provide more advantages than crude oil. One only needs to compare the cost of a barrel of oil to that of corn starch to see that the latter is considerably cheaper [28].

Tepha Inc. (Cambridge, MA, USA) currently produces poly-4-hydroxybutyrate (known commercially as PHA4400) (Fig. 1.13) for medical applications, using a proprietary transgenic fermentation process that has been specifically engineered to produce this homopolymer. During the fermentation process, poly-4-hydroxybutyrate accumulates inside the fermented cells as distinct granules. The polymer can be extracted in a highly pure form from the cells at the end of the fermentation process. The heart of the process is the genetically engineered *Escherichia coli* K12 microorganism, incorporating new biosynthetic pathways to produce poly-4-hydroxybutyrate [34].

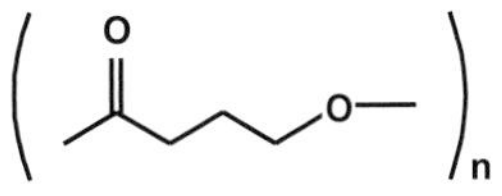

Fig. 1.13 Chemical structure of poly-4-hydroxybutyrate.

A range of polyhydroxyalkanoates with 0–24% hydroxyvalerate have been produced under the trade name of "Biopol" by Zeneca Bio Products and other manufacturers [35]. However, the polyhydroxyalkanoates production price is way above the market price of conventional plastics ($16 per kilogram for "Biopol" against $1 per kilogram for oil-derived plastics). Potentially, the production cost can be lowered by process scale-up, to around $8 per kilogram and by use of recyclable waste material as the substrates [35].

More recently, researchers from the company DSM succeeded in combining enzymatic ring opening polymerization and chemical nitroxide mediated living free radical polymerization. This genuine one-pot reaction is a method for the synthesis of block copolymers in a metal-free fashion. After proving the principle they are extending their chemo-enzymatic polymerization approach to obtain new functional polymers based on new raw materials, and to develop the technology further towards a kinetic resolution polymerization [36].

Acrylamide is one of the most important commodity chemicals with a global consumption of about 200 000 tonnes per year. It is required in the production of various polymers for use as flocculants, additives or for petroleum recovery. In conventional syntheses, copper salts are used as catalysts in the hydration of nitriles. However, this is rather disadvantageous as the preparation of the catalysts is fairly complex. In addition, it is difficult to regenerate the used catalyst and to separate and purify the acrylamide formed. Furthermore, as acrylamides are readily polymerized, their production under moderate conditions is highly desirable. In contrast to the conventional chemical process, there is no need to recover unreacted acrylonitrile in the enzymatic process, because the conversion and yield of the enzymatic hydration process are almost 100%. The removal of the copper ions from the product is no longer necessary. Overall, the enzymatic process – being carried out below 10 °C under mild reaction conditions and requiring no special energy source – proves to be simpler and more economical. The immobilized cells are used repeatedly and a very pure product is obtained. The enzymatic process, which was first implemented in 1985, is already producing about 6000 tonnes of acrylamide per year for Nitto [37, 38]. The use of a biotacalyst for the production of acrylamide may be not the first case of biotransformation being used as part of a biotechnological process in the petrochemical industry. However, it is the first example of the successful introduction of an industrial biotransformation process for the manufacture of a commodity chemical (Fig. 1.14).

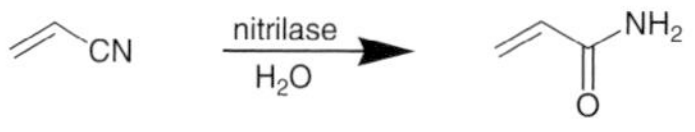

acrylonitrile   acrylamide

Fig. 1.14 Acrylamide synthesis.

Improvements to the production of 1,3-propanediol, a key component of an emerging polymer business, have been realized. By utilizing genes from natural strains that produce 1,3-propanediol from glycerol, metabolic engineering has enabled the development of a recombinant strain that utilizes the lower cost feedstock D-glucose [39].

Some representative industrial microbial transformations are listed in Table 1.1.

Tab. 1.1 Some representative industrial biotransformations catalyzed by whole cells.

Product	Biocatalyst	Operating since	Company
vinegar	bacteria	1823	various
L-2-methylamino-1-phenylpropan-1-ol	yeast	1930	Knoll AG,Germany
L-sorbose	Acetobacter suboxydans	1934	various
prednisolone	Arthrobacter simplex	1955	Shering AG, Germany
L-aspartic acid	Escherichia coli	1958	Tanabe Seiyaku Co., Japan
7-ADCA	Bacillus megaterium	1970	Asahi Chemical Industry, Japan
L-malic acid	Brevibacterium ammoniagenes	1974	Tanabe Seiyaku Co., Japan
D-p-hydroxyphenylglycine	Pseudomonas striata	1983	Kanegafuchi, Chemical Co., Japan
acrylamide	Rhodococcus sp.	1985	Nitto Chemical Ltd, Japan
D-aspartic acid and L-alanine	Pseudomonas dacunhae	1988	Tanabe Seiyaku Co., Japan
L-carnitine	Agrobacterium sp.	1993	Lonza, Czech.Rep.
2-keto-L-gulonic acid	Acetobacter sp.	1999	BASF, Merck, Cerester, Germany

1.2
From Gastric Juice to SweetzymeT – The History of Enzymatic Biotransformations

Enzymes had been in use for thousands of years before their nature became gradually understood. No one really knows when a calf stomach was used for the first time as a catalyst in the manufacture of cheese.

As early as 1783, Spallanzani showed that gastric juice secreted by cells could digest meat *in vitro*. In 1836, Schwan called the active substance pepsin [40]. The French scientist Payen [41] isolated an enzymatic complex from malt in 1833, naming it "diastase". Diastase, the enzyme that catalyzes the breakdown of starch into glucose, was the first enzyme to be discovered. In 1876, Kühne (Fig. 1.15) presented a paper to the Heidelberger Natur-Historischen und Medizinischen Verein, suggesting that such non-organized ferments should be called *e n z y m e s* [42]. At that time two terms were used: "organized ferment", such as the cell-free yeast extract from Büchner, and "unorganized ferment", such as the gastric juice secreted by cells. Today the terms "intracellular" and "extracellular" are used. Kühne also presented some interesting results from his experiments with trypsin. The word "enzyme" comes from Greek for "in yeast" or "leavened" [43].

Separat-Abdruck aus den Verhandlungen des Heidelb. Naturhist.- Med. Vereins. N. 8. I. 3. Verlag von Carl Winter's Universitätsbuchhandlung in Heidelberg.

Ueber das Verhalten verschiedener organisirter und sog. ungeformter Fermente.

Ueber das Trypsin (Enzym des Pankreas).

Von **W. Kühne.**

1876

Ueber das Verhalten verschiedener organisirter und sog. ungeformter Fermente.

Sitzung am 4. Februar 1876.

Hr. W. Kühne berichtet über das Verhalten verschiedener organisirter und sog. ungeformter Fermente. Um Missverständnissen vorzubeugen und lästige Umschreibungen zu vermeiden schlägt Vortragender vor, die ungeformten oder nicht organisirten Fermente, deren Wirkung ohne Anwesenheit von Organismen und ausserhalb derselben erfolgen kann, als *Enzyme* zu bezeichnen. — Genauer untersucht wurde besonders das Eiweiss verdauende Enzym des Pankreas, für welches, da es zugleich Spaltung der Albuminkörper veranlasst, der Name *Trypsin* gewählt wurde. Das Trypsin vom Vortr. zuerst dargestellt und zwar frei von durch dasselbe noch verdaulichen und zersetzbaren Eiweissstoffen, verdaut nur in alkalischer, neutraler, oder sehr schwach sauer reagirender Lösung. Dasselbe wird durch nicht zu kleine Mengen Salicylsäure, welche das Enzym in bedeutenden Quantitäten löst, bei 40° C. gefällt, ohne dabei seine specifische Wirksamkeit zu verlieren. Wird die Fällung in Sodalösung von 1 pCt. gelöst, so verdaut sie höchst energisch unter Bildung von Pepton, Leucin, Tyrosin u. s. w. Nur übermässiger Zusatz von Salicylsäure bis zur Bildung eines dicken Krystallbreies vernichtet die enzymotischen Eigenschaften. Dies Verhalten war kaum zu erwarten, seit Kolbe und J. Müller die hemmende, selbst vernichtende Wirkung kleiner Mengen Salicylsäure auf einige Enzyme hervorgehoben hatten. Die Beobachtungen des Vortr., der ausser dem Trypsin noch das Pepsin eingehender untersuchte, stehen jedoch mit den Angaben von J. Müller, nach welchen Salicylsäure bei einem Gehalte der

Fig. 1.15 W.F. Kühne [42].

Microorganisms synthesize numerous enzymes, each one having a specific function. Intracellular enzymes operate inside the cell in a protected and highly structured environment, while extracellular enzymes are secreted from the cell, thus working in the medium surrounding the microorganism.

The commercial usage of extracellular microbial enzymes started in the West around 1890, thanks to the Japanese entrepreneur Takamine. He settled down in the USA and started an enzyme factory based on Japanese technology. The principal product was called takadiastase. This was a mixture of amylolytic and proteolytic enzymes prepared by cultivation of *Aspergillus oryzae*. In France, Boidin and Effront developed bacterial enzymes in 1913. They found that the hay bacillus, *Bacillus subtilis,* produces an extremely heat-stable *α*-amylase when grown in still cultures on a liquid medium prepared by extraction of malt or grain [44].

In 1892, in a study of the rate of fermentation of sucrose in the presence of yeast, the British chemist Brown found that the rate seemed to be independent of the amount of sucrose present [45]. He later suggested that this result could be explained if the invertase molecules present in the yeast formed an addition complex with sucrose [46]. This was the first time that the existance of an enzyme–substrate complex had been deduced from the kinetics of an enzyme reaction [47].

As part of his studies on sugars, in 1894 Emil Fischer [48, 49] observed that the enzyme known as emulsin catalyzes the hydrolysis of *β*-methyl-D-glucoside, while the enzyme known as maltase is active towards the *α*-methyl-D-glucoside substrate (Fig. 1.16).

This led Fischer to suggest his famous "lock-and-key" theory of enzyme specificity, which he described in his own word as follows: "To use a picture, I would say that enzyme and the glucoside must fit into each other like a lock and key, in order to effect a chemical reaction on each other" [1].

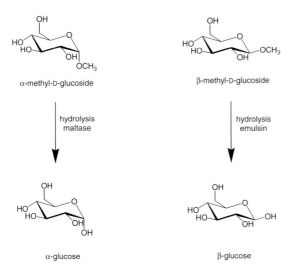

Fig. 1.16 Emil Fischer's substrates.

In 1913, the German biochemist Leonor Michaelis and his Canadian assistant Maud Leonara Menten published a theoretical consideration of enzymatic catalysis. This consideration envisaged the formation of a specific enzyme–substrate complex which further decomposed and yielded the product with the release of the enzyme. They had observed that the effect noted by Brown [45] is only observed at higher concentrations of the substrate. At lower concentrations the rate becomes proportional to the concentration of the substrate. This led to the development of the Michaelis–Menten equation to describe the typical saturation kinetics observed with purified enzymes and single substrate reactions [50]. Some years later a more general formulation of the Michaelis–Menten equation was given by Briggs and Haldane [51]. They pointed out that the Michaelis assumption that an equilibrium exists between the enzyme, substrate and enzyme–substrate complex is not always justified, and should be replaced by an assumption that the enzyme–substrate complex is not necessarily present at equilibrium but in a steady state. With the purification and crystallization of proteins in the 1920s, enzyme kinetics entered a new phase. It became possible to study the interactions between enzyme molecules and their substrate in much more detail. The British physical chemist Butler was the first to carry out kinetic studies with a pure enzyme, trypsin [52].

By 1920, about a dozen enzymes were known, none of which had been isolated [53]. Then, in 1926, Sumner [54] crystallized urease from jack bean, *Canavalia ensiformis,* and announced that it was a simple protein. He later, in 1946, received the Nobel Prize for his work with the enzyme urease, extracted from the jack bean. Urease is an enzyme that catalyzes the conversion of urea into ammonia and carbon dioxide (Fig. 1.17).

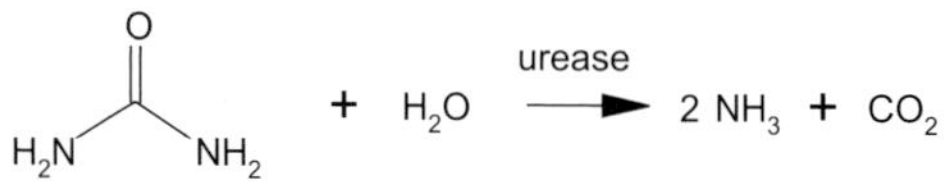

Fig. 1.17 The conversion of urea into ammonia and carbon dioxide.

Northrop and his colleagues [40] soon supported Sumner's claim that an enzyme could be a simple protein. They isolated many proteolytic enzymes, beginning with pepsin in 1930 by applying classic crystallization experiments. By the late 1940s many enzymes were available in a pure form and in sufficient amounts for investigations of their chemical structure to be carried out. Currently, more than 3000 enzymes have been catalogued [55]. The ENZYME data bank contains information related to the nomenclature of enzymes [56]. The current version contains 4309 entries. It is available through the ExPASy WWW server (http://www.expasy.org/enzyme/). Several hundred enzymes can now be obtained commercially [57].

In 1950 there was still no evidence that a given protein had a unique amino acid sequence. Lysosyme was the first enzyme to have its tertiary structure defined (Fig. 1.18), this was in 1966 with the help of X-ray crystallography [58].

However, ribonuclease A was one of the first enzymes to be prepared on a laboratory scale using organic chemistry methods. In 1969, Gutte and Merrifield synthesized its whole sequence in 11 931 steps [59].

By 1970, the complete molecular structures of several enzyme had been established and plausible reaction mechanisms could then be discussed [40].

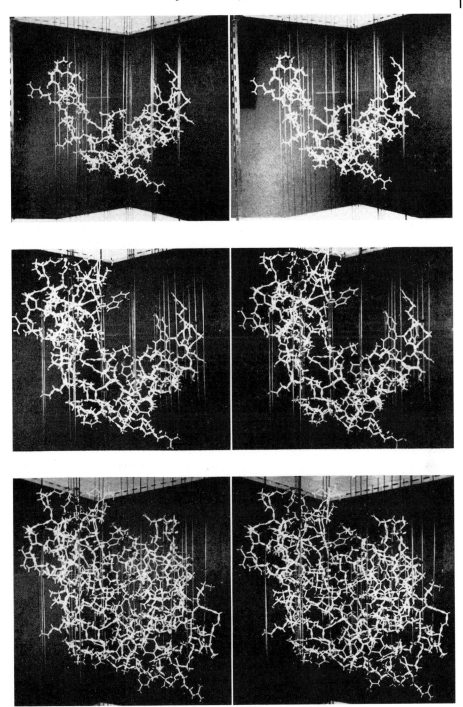

Fig. 1.18 Stereo photographs of models of part of the lysozyme molecule [58].

Hill (1897) was the first to show that the biocatalysis of hydrolytic enzymes is reversible [60].

Pottevin (1906) went further and demonstrated that crude pancreatic lipase could synthesize methyl oleate from methanol and oleic acid in a largely organic reaction mixture [61].

While the first benefits for industry from the microbiological developments had come early on, investigations with isolated enzymes had hardly any influence on the industry at that time. Consequently, industrial enzymatic biotransformations for the production of fine chemicals have a much shorter history than microbial biotransformations.

Invertase was probably the first immobilized enzyme to be used commercially for the production of Golden Syrup by Tate&Lyle during World War II, because sulfuric acid, the preferred reagent, was unavailable at that time (Fig. 1.19) [62].

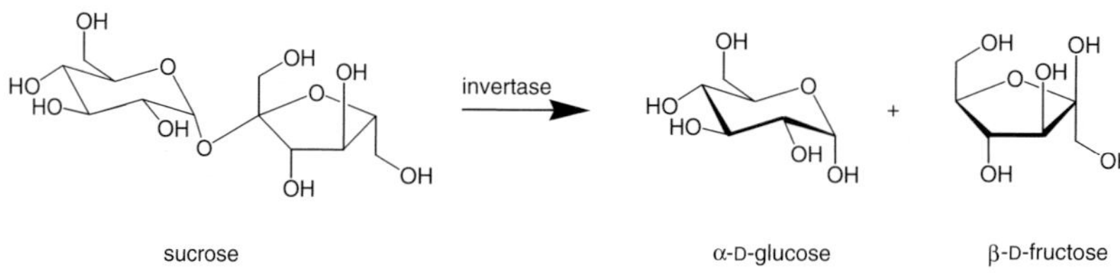

sucrose   α-D-glucose   β-D-fructose

Fig. 1.19 Inversion of sucrose by invertase.

Yeast cells were autolyzed and the autolysate clarified by adjustment to pH 4.7, followed by filtration through a calcium sulphate bed and adsorption into bone char. A bone char layer containing invertase was incorporated into the bone char bed, which was already being used for syrup decolorization. The scale of the operation was large, the bed of invertase–char being 60 cm deep in a 610 cm deep bed of char. The preparation was very stable as the limiting factor was microbial contamination or loss of decolorizing power rather than the loss of enzymatic activity. The process was cost-effective but the product did not have the flavor quality of the acid-hydrolyzed material. This is the reason why the immobilized enzyme method was abandoned once the acid became available again [63].

Industrial processes for L-amino acid production based on the batch use of soluble aminoacylase were in use by 1954. However, like many batch processes with a soluble enzyme, they had their disadvantages, such as higher labor costs, complicated product

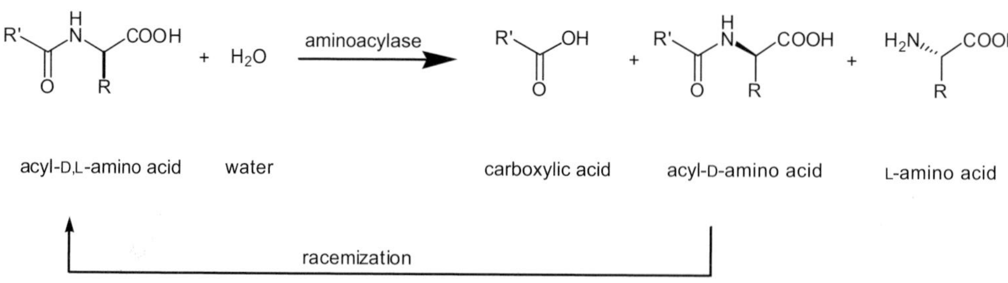

acyl-D,L-amino acid   water   carboxylic acid   acyl-D-amino acid   L-amino acid

racemization

Fig. 1.20 L-Amino acid production catalyzed by aminoacylase.

separation, low yields, high enzyme costs and non-reusability of the enzyme. During the mid-1960s the Tanabe Seiyaku Co. of Japan was trying to overcome these problems by using immobilized aminoacylases. In 1969, they started the industrial production of L-methionine by aminoacylase immobilized on DEAE–Sephadex in a packed bed reactor (Fig. 1.20). This was the first full scale industrial use of an immobilized enzyme. The most important advantages are relative simplicity and ease of control [64].

In a membrane reactor system developed at Degussa-Hüls AG in Germany in 1980 [65], native enzymes, either pure or of technical grade, are used in homogeneous solution for the large scale production of enantiomerically pure L-amino acids (Fig. 1.21).

Fig. 1.21 Enzyme membrane reactor (Degussa-Hüls AG, Germany).

A membrane reactor is particularly suitable for cofactor-dependent enzyme reactions, especially if the cofactor is regenerated by another enzyme reaction and retained by the membrane in a modified form [66]. There are several advantages to carrying out biocatalysis in membrane reactors over heterogeneous enzymatic catalysis: there are no mass transfer limitations, enzyme deactivation can be compensated for by adding a soluble enzyme and the reactors can be kept sterile more easily than with immobilized enzyme systems. The product is mostly pyrogen free (a major advantage for the production of pharmaceuticals), because the product stream passes through an ultrafiltration membrane. Scale-up of membrane reactors is simple because large units with increased surface area can be created by combining several modules.

The enzymatic isomerization of glucose to fructose (Fig. 1.22) represents the largest use of an immobilized enzyme in the manufacture of fine chemicals.

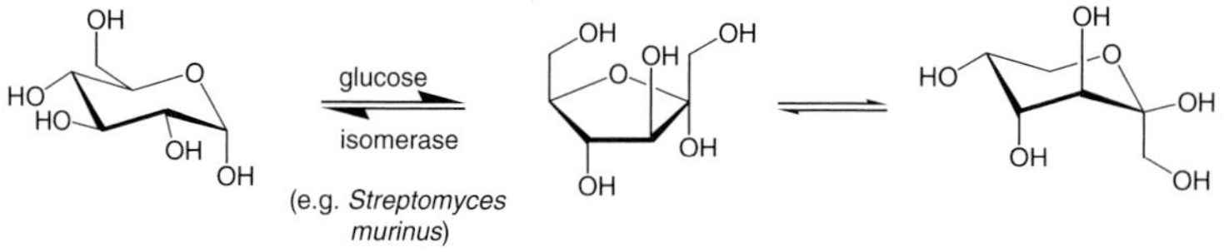

α-D-glucopyranose   α-D-fructofuranose   β-D-fructopyranose

Fig. 1.22 Isomerization of glucose to fructose.

The production of high-fructose syrup (HFCS) has grown to become a large-volume biotransformation [67]. While sucrose is sweet, fructose is approximately 1.5-times sweeter and consequently high quality invert syrups (i.e., hydrolyzed sucrose) can be produced. Invert syrups contain glucose and fructose in a 1:1 ratio. However, it took the food industry a long time to become acquainted with the potential of glucose isomerase in the production of high quality fructose syrups from glucose. Again, the Japanese were the first to employ soluble glucose isomerase to produce high quality fructose syrups, in 1966. At the beginning of 1967, the Clinton Corn Processing Company, IA, USA, was the first company to manufacture enzymatically produced fructose corn syrup [67]. The glucose-isomerase catalyzed reversible reaction gave a product containing about 42% fructose, 50% glucose and 8% other sugars. For various reasons, economic viability being the most important among them, the first commercial production of fructose syrups using glucose isomerase immobilized on a cellulose ion-exchange polymer in a packed bed reactor plant only started in 1974. It was initiated by Clinton Corn Processing [64]. In 1976, Kato was the first company in Japan to manufacture HFCS in a continuous process as opposed to a batch process. In 1984, it became the first company to isolate crystalline fructose produced in this process by using an aqueous separation technique.

The glucose isomerase Sweetzyme T, produced by Novo, Denmark, is used in the starch processing industry in the production of high fructose syrup. The key to its long life is immobilization. The enzyme is chemically bound to a carrier, making the particles too large to pass through the sieve at the bottom of isomerization columns. Sweetzyme T is packed into columns where it is used to convert glucose into fructose. The record for the longest lifetime of a column is 687 days, held by a Japanese company, Kato Kagaku in Kohwa near Nagoya. The reaction conditions are pH 7.5 and $T = 55\ °C$. Although enzyme activity is reduced at this temperature, its stability and productivity are considerably improved [68].

The engineers from Kato used to say: "The better the substrate you put in, the better the results you get out". Each column at Kato contains 1800 kg of Sweetzyme T. The column needs to be changed when the flow rate decreases to about 10% of the initial value. Sweetzyme T displays a linear decay curve under steady state operating conditions. With regard to productivity, the yield from the record-breaking column was 12 000 kg of fructose syrup (containing 42% fructose) (dry substance) per kg of Sweetzyme T. The normal column productivity was 8000–10 000 kg per kg of enzyme. Thus the 687-day record for Sweetzyme T is also a world record in the starch industry [68] (Fig. 1.23).

"Central del Latte" of Milan, Italy, was the first company to commercially hydrolyze milk lactose with immobilized lactase using SNAMprogetti technology [69]. An industrial

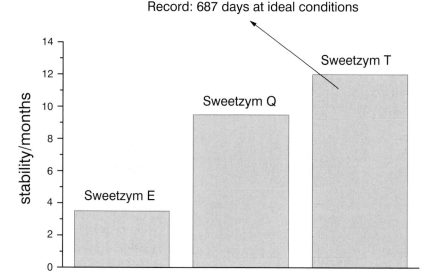

Fig. 1.23 Improved biocatalyst stability by biocatalyst engineering at NOVO.

plant with a capacity of 10 tons per day is situated in Milan. The entrapped enzyme is lactase obtained from yeast and the reaction is performed batchwise at low temperature. Lactase hydrolyzes lactose, a sugar with poor solubility properties and a relatively low degree of sweetness, to glucose and galactose (Fig. 1.24).

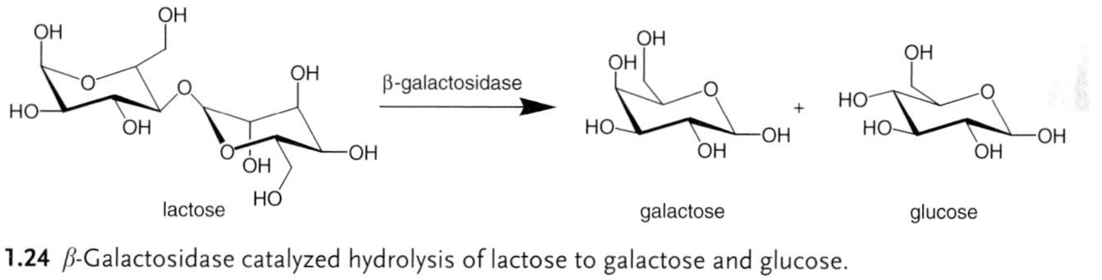

Fig. 1.24 β-Galactosidase catalyzed hydrolysis of lactose to galactose and glucose.

After the processed milk reaches the desired degree of hydrolysis of lactose, it is separated from the enzyme fibers, sterilized and sent for packing and distribution. SNAMprogetti's process facilitates the manufacture of a high-quality dietary milk at low cost. This milk has remarkable digestive tolerance, pleasant sweetness, unaltered organoleptic properties and a good shelf-life. It does not contain foreign matter. The industrial plant is shown in Fig. 1.25.

Penicillin G, present in *Penicillium notatum* and discovered by Fleming in 1929, revolutionized chemotherapy against pathogenic microorganisms. Today, β-lactam antibiotics such as penicillins and cephalosporins are used very widely. Thousands of semisynthetic β-lactam antibiotics are being synthesized to find more effective compounds. Most of

Fig. 1.25 Industrial plant for processing low-lactose milk [69].

these compounds are prepared from 6-aminopenicillanic acid (6-APA), 7-aminocephalo-sporanic acid (7-ACA) and 7-amino-desacetoxycephalosporanic acid (7-ADCA).

At present, 6-APA is mainly produced either by chemical deacylation or by enzymatic deacylation using penicillin amidase from penicillin G or V. This process, which exemplifies the best known usage of an immobilized enzyme in the pharmaceutical industry, has been used since around 1973 (Fig. 1.26). Several chemical steps are replaced by a single enzymatic reaction. Organic solvents, the use of low temperature (–40 °C) and the need for totally anhydrous conditions, which make the process difficult and expensive, were no longer necessary in the enzymatic process [70].

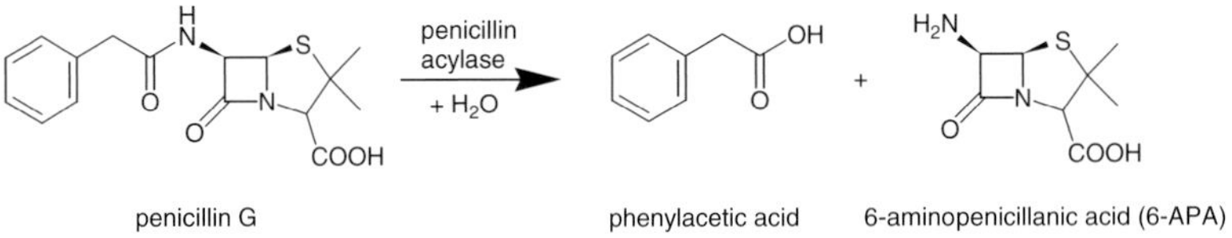

Fig. 1.26 Enzymatic synthesis of 6-aminopenicillanic acid (6-APA).

For many years enzymatic 7-ACA production was nothing but a dream. This changed in 1979, when Toyo Jozo, Japan, in collaboration with Asahi Chemical Industry, also of Japan, successfully developed the industrial production of 7-ACA by a chemoenzymatic two-step process starting from cephalosporin C (Fig. 1.27).

The chemical process requires highly purified cephalosporin C as a raw material. A number of complicated reaction steps are carried out at –40 to –60 °C, and there is a

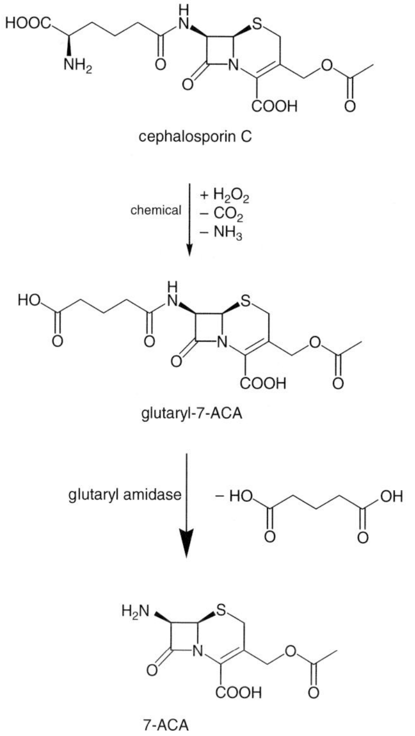

Fig. 1.27 Two-step process for production of 7-ACA from cephalosporin C.

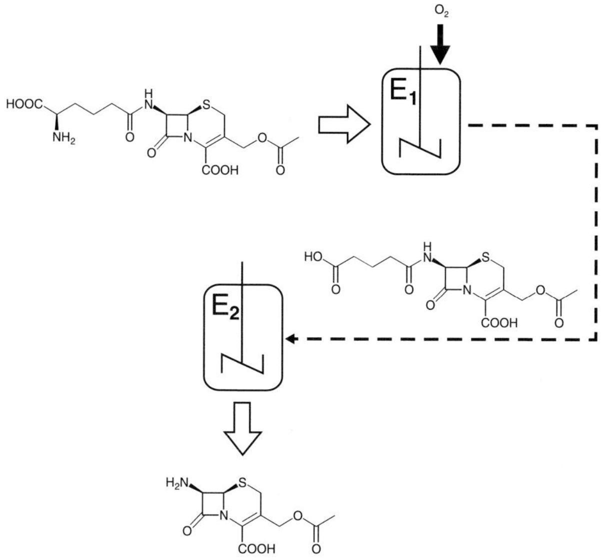

Fig. 1.28 Flow-scheme for the production of 7-ACA. Production carried out at Asahi Chemical Industry (E_1 = D-amino acid oxidase; E_2 = glutaryl amidase).

long reaction time. Furthermore, hazardous reagents, such as phosporus pentachloride, nitrosyl chloride and pyridine are used in this process and the subsequent removal of such reagents causes significant problems. Therefore, the development of an enzymatic process was a dream for a long time. In the enzymatic process, liberated glutaric acid reduces the pH and inhibits glutaryl-7-ACA amidase, the enzyme that catalyzes the deacylation of cephalosporin C. Because of this change in pH the reaction rate is decreased, necessitating strict pH control during the reaction process. For these reasons, a recircu-

Fig. 1.29 Bioreactor plant for the production of 7-ACA carried out at Asahi Chemical Industry (reprinted from Ref. [71], p. 83, courtesy of Marcel Dekker Inc.).

lating bioreactor with immobilized glutaryl-7-ACA amidase and an automatic pH controller were designed for the production of 7-ACA. The bioreactor for industrial 7-ACA production is shown in Figs. 1.28 and 1.29. The process has been in operation at Asahi Chemical Industry since 1973. It is reported that about 90 tons of 7-ACA are produced in this way annually [71].

1.3
From Wine Bottle to a State-of-the-Art Facility – The History of Biochemical Engineering

Processing of biological materials and using biological agents such as cells, enzymes or antibodies are the central features of biochemical engineering. The scope of biochemical engineering has grown from simple wine-bottle microbiology to the industrialization of the production of antibiotics, chemicals, proteins, etc. Success in biochemical engineering requires integrated knowledge of the governing biological properties and principles and of engineering methodology and strategy. The objective of biochemical engineering could be formulated as the advancement of the scientific fundamentals of engineering for a rational, sustainable and safe approach to the development of products, processes and services in the chemical, pharmaceutical and life science industries. The aim is to provide tools for the efficient development of novel know-how [72] (Figs. 1.30 and 1.31).

In the 1940s complementary developments in biochemistry, microbial genetics and engineering ushered in the era of antibiotics, which brought tremendous relief to suffering and mortality of mankind. This period marks the birth of biochemical engineering

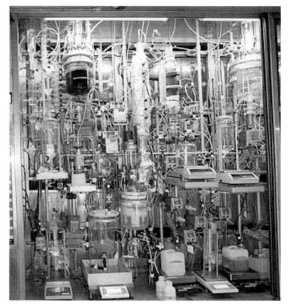

Fig. 1.30 Mini-plant at BASF (photograph courtesy of Prof. Bernhard Hauer/ZH/BASF-AG/BASF).

Fig. 1.31 Bioreactor at LONZA.

[73]. The issues of scale-up, mixing and oxygenation had never before been encountered with such urgency and criticality as that for the production of antibiotics late into and right after the Second World War. This can be marked as the genesis phase of modern biochemical engineering.

During the 1960s, the increased tempo in the cross-fertilization and integration of diverse fields of knowledge gave rise to two new interdisciplinary branches of technology: biomedical engineering, which encompasses practically all the engineering disciplines and the medical sciences, and biochemical engineering [74, 75]. Biochemical engineering is a multidisciplinary area bordering biology, chemistry and engineering sciences. As a border area, biochemical engineering plays an ever increasing role in biotechnological developments. This is demonstrated by examples ranging from substrate preparation and bioconversion to downstream processing [76] (Fig. 1.32).

Biochemical engineering has been recognized as a scientific discipline in its own right for at least three decades, but it is often misunderstood in that it is thought to be synonymous with bioprocessing development [77]. The term biochemical engineering science [78] was introduced to distinguish between the study of biochemical engineering and its application in industry [79]. Biochemical engineering science represents the fundamental research into all aspects of the interactions between engineering and other disciplines necessary to underpin the development of industrial-scale biologically based processes [78].

Bioreactor mixing and kinetics issues, the unstructured kinetic models, the emergence of whole-cell biotransformations and bioreactor instrumentation and control defined the

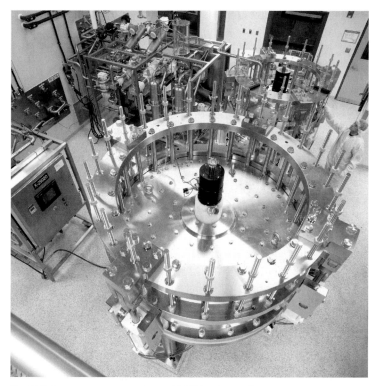

Fig. 1.32 Down-stream processing at LONZA – the purification step.

1960s. The 1970s were defined by the development of enzyme technology (including, most critically, enzyme immobilization technologies [80]), biomass engineering, single-cell protein production and bioreactor design [81], operation and control innovations. The period from about 1960 to 1980 can be defined as an evolutionary phase for biochemical engineering. The next 20 years (1980–2000) could be called the 1st step-change (growth) phase (or how-much-biology-can-an-engineer-handle phase) in biochemical engineering. As a result of this phase, biochemical engineering will never be like anything in the past ever again. The emergence of the recombinant DNA technology, the hybridoma technology, cell culture, molecular models, large-scale protein chromatography, protein and DNA sequencing, metabolic engineering and bioremediation technologies changed biochemical engineering in a drastic and profound way. New problems, new ways of thinking, the prominence and challenges of new biotechnology, everything now seems so different, the world of biochemical engineering and biology has been turned upside down [82].

The idea that understanding the secrets of cells lies in understanding the complex networks that a cell uses is now gaining acceptance. The bioprocesses suffer from a lack of robustness to disturbances of the individual steps and the responses to perturbation in a single step or in multiple steps are often detrimental to the system. Cells, however, have an amazing ability to withstand disturbances and their responses are rapid and effective.

Understanding these features and adapting them for engineered processes would enable better process control and might lead to novel methods for producing new chemicals using cell-derived catalysts [83].

Four technological advances, which had a major impact on enzymatic biotransformations, were required for the acceptance of enzymes as "alternative catalysts" in industry [84]. These advances have resulted from a greater understanding of biochemical engineering leading to changes in fermenter operation, such as controlled feeding of nutrients and improved oxygen transfer, or enabling technologies and tools, such as on-line sensors and computer simulation.

1. *The first technological advance* was the development of large-scale techniques for the release of enzymes from the interior of microorganisms [85]. Although the majority of industrial purification procedures are based on the same principles as those employed on a laboratory scale, the factors under consideration while devising industrial-scale purification regimes are somewhat different. When isolating enzymes on an industrial scale, for commercial purpose, a prime consideration has to be the cost of the production in relation to the value of the end-product. Therefore, techniques used on a laboratory scale are not always suitable for large-

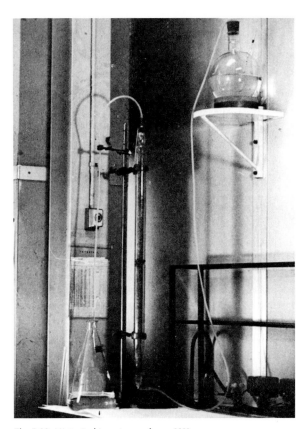

Fig. 1.33 Historical invertase column [69].

scale work [86]. Production and isolation of an intracellular microbial enzyme are fairly expensive. The cost of using a water-soluble protein as a catalyst for biotransformations can be justified only by its repeated use [87].

2. *The second technological advance* was the development of techniques for the large-scale immobilization of enzymes. As mentioned earlier, the first enzyme immobilized in the laboratory was invertase, adsorbed onto charcoal in 1916 [88]. However, only after the development of immobilization techniques on a large scale had occurred in the 1960s, were many different industrial processes using immobilized biocatalysts established. The historical invertase column that has been in operation on a laboratory scale since 1968 is shown in Fig. 1.33.

It was shown that by increasing the concentration of sucrose, the efficiency of the fiber-entrapped invertase (which hydrolyzes sucrose) can be increased. This occurred because the substrate, which is an inhibitor of the enzyme, could not reach high concentration levels inside the microcavities of the fibers due to diffusion limitations [69].

Table 1.2 lists some industrial biotransformations performed by isolated enzymes.

Tab. 1.2 Selected historical, industrial applications of isolated enzymes.

Product	Biocatalyst	Operating since	Company
L-amino acid	amino acylase	1954, 1969	Tanabe Seiyaku Co., Ltd., Japan
6-amino penicillanic acid	penicillin acylase	1973	SNAMProgetti and many others[a]
low-lactose milk	lactase	1977	Central del Latte, Milan, Italy (SNAMProgetti technology)
7-amino-cephalosporanic acid	D-amino acid oxidase	1979	Toyo Jozo and Asahi Chemical Industry, Japan

a Beecham, Squibb, Astra Lakenedal, Bayer, Gist Brocades, Pfizer, Bristol Myers, Boehringer Mannheim, Biochemie, Novo.

The first Enzyme Engineering Conference was held at Hennicker, NH, USA, in 1971. The term "immobilized enzymes" describing "enzymes physically confined at or localized in a certain region or space with retention of their catalytic activity and which can be used repeatedly and continuously" was adopted at this conference [89].

3. *The third technological advance* was the development of techniques for biocatalysis in organic media. The usage of very high proportions of organic solvents to increase the solubility of reactants was examined in 1975 for the reaction with isolated cholesterol oxidase to produce cholestenone [90]. The enzymatic synthesis was believed to be incompatible with most organic syntheses carried out in non-aqueous media. This changed after Klibanov [91] recognized in 1986 that most enzymes could function fairly well in organic solvents. Since that time various processes involving an organic phase have been established in industry (Table 1.3).

Tab. 1.3 Industrial biotransformations involving reactants that are poorly water soluble.

Process	Biocatalyst	Operating since	Company
fat interesterification	lipase	1979, 1983	Fuji Oil, Unilever
ester hydrolysis	lipase	1988	Sumitomo
transesterification	lipase	1990	Unilever
aspartame synthesis	thermolysin	1992	DSM
acylation	lipase	1996	BASF

4. *The fourth and most recent technological advance* is recombinant DNA technology. This technology is only now being widely used for biotransformation.

In general, microorganisms isolated from nature produce the desired enzyme at levels that are too low to provide a cost-efficient production process. Consequently, a modification of the organism would be highly desirable for process development. Currently, there are three principal approaches available for strain improvement. The first one, *direct evolution* [92], i.e., improvement by mutation and selection, has been successfully used in many industrial microbiological fields for many years. In 1978, Clark showed that evolution processes can be performed on a laboratory scale. Microevolution occurring in bacterial cultures grown in a chemostat gives rise to altered enzyme specificity enabling microorganisms to degrade some unusual synthetic organic compounds. Successive mutation steps could be responsible for the evolution of new enzymatic specificities. The rate of production of existing enzymes and the expression of previously dormant genes are also typically affected by this event [93]. The second method is *hybridization*. This involves modification of the cellular genetic information by transference of DNA from another strain. The third method is *recombinant DNA technology*, whereby genetic information from one strain can be manipulated *in vitro* and then inserted into the same or a different strain.

Recombinant DNA technology has dramatically changed enzyme production, because enzymes are synthesized in cells by the normal protein synthesis methods [94, 95]. The 5–10 year period required for classical enzyme development can be reduced to 1–2 years. Protein engineering, in combination with recombinant expression systems, allows new enzyme variants to be plugged in and to very quickly be at manufacturing levels [96]. Novel microbial catalysts, together with recent advances in molecular biology, offer scientists an opportunity to evolve selected genes rapidly and considerably improve bacterial biocatalysts [97]. For example, a method for the rapid generation of thermostable enzyme variants has been developed [98, 99]. This is achieved by introducing the gene coding for a given enzyme from a mesophilic organism into a thermophile. Variants that retain the enzymatic activity at higher growth temperatures of the thermophile are selected. This can be accomplished by constructing the artificial environment in which only the evolutional adaptation of the enzyme can permit cell growth. This strategy can be readily extended to the general method of screening mutant enzymes. Another example is random mutation, developed as a method for the highly efficient generation of mutant enzymes. The cloned gene coding for a given enzyme can be mutated either chemically or enzymatically *in vitro*. The mutant enzymes can be readily screened because mutant genes can be separated from intact genes. Various mutant enzymes which have changes in their properties, such as substrate specificity, thermal stability and coenzyme selectivity, have been isolated by this technique. These methods do not require predictive strategies, unlike, for example, site-directed mutagenesis. It is hoped that in the course of time that enzymes can be made which are excellent catalysts, fulfilling all the requirements for industrial use. This research field is referred to as *biocatalyst engineering* [32].

In parallel with developments in genetic engineering have come improvements in biochemical engineering that have yielded commercial benefits in reactor and fermenter design and operation, improved control techniques and downstream separation. These have resulted in more rapid delivery of new products [100] to the marketplace.

1.4
Advantages of Biotransformations Over Classical Chemistry Enzymes are proteins, things of beauty and a joy forever [40]

Biocatalysis is a relatively green technology. Enzyme reactions can be carried out in water at ambient temperature and neutral pH, without the need for high pressure and extreme conditions, thereby saving energy normally required for processing. Biocatalysis has proven to be a useful supplementary technology for the chemical industry, allowing, in some cases, reactions that are not easily conducted by classical organic chemistry or, in other cases, allowing reactions that can replace several chemical steps. Today, highly chemo-, regio- and stereoselective biotransformations can simplify manufacturing processes and make them even more economically attractive and environmentally acceptable [101].

Both new discoveries and incrementalism describe how the industrial enzyme business changed during 1996. Enzymes have competed well with chemical methods for resolution but not with synthesis. Ibuprofen, phenylethylamine and acrylamide are commonly cited as compounds prepared using enzyme-based chiral processes. There is also an unconfirmed suspicion that the fat substitute Olestra, because of some of its structural features, may require enzymatic steps in its synthesis. The outlook for industrial enzymes is positive. The suppliers have extensive portfolios of promising new enzymes in their product pipelines. The range of customers considering the utilization of enzymes as replacements for conventional chemical methods, appears to be growing. New niche applications continue to be discovered in otherwise mature segments [102]. It appears that enzyme-based processes are gradually replacing conventional chemical-based methods, e.g., the use of enzymes as catalysts provides a totally new method of polymer synthesis; most of these polymers are otherwise very difficult to synthesize by conventional chemical catalysts [36]. Finally, the latest literature on enzymology suggests that other biocatalysts will add to future sales, both in established and in new markets. The enzyme "nitrogenase", which converts dinitrogen into ammonia, a basic chemical compound, has been discovered recently [103]. Dream reactions of organic chemists might come true in the future, with the use of biocatalysts where functional or chiral groups are introduced into molecules by utilizing H_2, O_2 or CO_2. Recently Aresta reported on a carboxylase enzyme that utilizes CO_2 in the synthesis of 4-hydroxybenzoic acid starting from phenyl phosphate [104]. The process of carbon dioxide fixation can be carried out successfully with a pure stream of carbon dioxide in bioreactor [105]. In order to make the bioprocess feasible the enzyme D-Ribulose-1,5-bisphosphate carboxylase/oxygenase (Rubisco) was recently immobilized [106]. Also, the novel trickling spray reactor employing immobilized carbonic anhydrase, which enables concentration of CO_2 from the emission stream was developed [107] Carbonic anhydrase is one of the fastest enzymes that make fast mass transfer from gas phase to aqueous phase. The single one-step chemical process shown in Fig. 1.34 has the possibility of retaining carbon in the bound form for a very short period of time, therefore the biocatalytic fixation process, just as the basic process in plants, could be the answer to atmospheric pollution. A large number of commercially useful substances, which could be made from fixed carbon, also require the bioprocess of carbon dioxide fixation.

$$CaO \ + \ CO_2 \ \longrightarrow \ CaCO_3$$

Fig. 1.34 The single one-step chemical process for the fixation of carbon dioxide.

Although the production of D-amino acids is currently of great interest, there has been no known industrial manufacture of D-amino acids except for D-*p*-hydroxyphenylglycine and D-phenylglycine. At present chemical methods are not suitable for the large-scale production of D-amino acids due to the low yield and high cost. Most L-amino acids are efficiently manufactured by fermentation, but D-amino acids are rarely produced by fermentation, apart from a few exceptions, because it is difficult to obtain high optical purity and productivity. Enzymatic methods are most plausible for the industrial manufacture of D-amino acids with respect to the optical purity and productivity. D-Amino acids such as D-*p*-hydroxy phenylglycine and D-phenylglycine are produced from D,L-hydantoins. From an industrial point of view, the availability of cheap starting materials and the development of suitable biocatalysts are most important. The number of substrates that are available on an industrial scale is limited. Based on these criteria, synthetic intermediates of D,L-amino acids and L-amino acids produced by biotransformations would be the most important starting materials for the production of D-amino acids. The enzymatic production of D-amino acids is classified into three categories based on the starting materials [108]:

1. D,L-amino acids (D-amino acylase);
2. synthetic intermediates (D,L-hydantoin:D-hydantoinhydrolase, D,L-amino acid amides:D-amidase);
3. prochiral substrates (α-keto acids, L-amino acids; D-transaminase and amino acid racemase).

The fed batch process [109] used in the production of L-DOPA, giving a final product concentration of 110 g L^{-1}, has many advantages over the classical chemical process, such as a single reaction step, water as the only reaction by-product, no need for optical separation, a shorter production cycle of three days, simple down-stream processing and process sustainability. L-DOPA is a metabolic precursor of dopamine, a very important drug in the treatment of Parkinson's disease.

It is difficult to assess directly the true commercial value of biocatalysis, because the real value of the products made using the biocatalysts must be taken into account. Of course, its major advantages lies in stereoselective reactions. A good example of the technological power and commercial potential is the aforementioned stereoselective hydroxylation of steroids.

In comparison with fermentation processes, fewer side-products are formed in enzymatic biotransformations, complex expensive fermenters are not required, aeration, agitation and sterility need not necessarily be maintained and the substrate is not diverted into the formation of a *de novo* cellular biomass [70]. Isolated biocatalysts are especially useful if the reaction they catalyze is about to be completed, if they are resistant to product inhibition, and if they are active in the presence of low concentrations of the substrate (such as in detoxification reactions where pollutants are present in the waste stream). "One-pot" multi-enzyme reactions are much more feasible than a combined use of several chemical catalysts or reagents, especially as the latter often have be used in reactors made of special resistant materials that can tolerate extreme conditions, such as the use of concentrated acids under elevated temperatures and pressures [70].

Silicatein is an enzyme that has purified the glassy skeletal elements of a marine sponge. It was previously shown to be capable of catalyzing and structurally directing the hydrolysis and polycondensation of silicon alkoxides to yield silica and silsesquioxanes at low temperature and pressure and neutral pH, and of catalyzing and templating the hydrolysis and subsequent polycondensation of a water-stable alkoxide-like conjugate of titanium to form titanium dioxide. Although biocatalysis has been used for a long time to produce compounds and materials based on carbon, these observations are the first to have extended this approach to valuable inorganic materials such as titanium dioxide [110].

It is no longer the case that biotransformations are relevant only to high added-value products such as pharmaceuticals. Bulk chemicals, including polymers [35, 36], may incorporate biotransformations, such as the conversion of methane into methanol (Chevron Research and Technology and Maxygen), the conversion of sugars into 3-hydroxypropionic acid (Cargill Inc., USA) [111] or the dehalogenation step in Dow's alkene oxide process. The next generation of bioprocesses will target large volume chemicals and polymers and will compete directly with petroleum-based products. Biotransformations are becoming competitive with conventional routes, but industry experts believe that further improvements in enzymatic catalysis and fermentation engineering may be required before many companies are prepared to announce large-scale bioprocessing plants for world-wide production. Bioprocessing proponents envisage a future in which micro-organisms are replaced by purified enzymes, synthetic cells or crop plants [112].

Today, both the academic and the industrial community see biocatalysis as a highly promising area of research, especially in the development of sustainable technologies for the production of chemicals and for the more selective and complex active ingredients in pharmaceuticals and agrochemicals [113–116].

References

1 Sheldon, R.A. 1993, *Chirotechnology*, Marcel Dekker, New York, (p.) 105.

2 Turner, M.K. 1998 (Perspectives in Biotransformations), in *Biotechnology*, (vol. 8), eds. Rehm,H.-J., Reed, G., *Biotransformations I*, ed. Kelly, D.R., Wiley-VCH, Weinheim, (p.) 9.

3 Ford, J.B. 1995, (First Steps in Experimental Microscopy. Leeuwenhoek as a Practical Scientist), *Microscope* 43, 47–57.

4 www.inventors.com/rd/results/ rdq_history_of_microscope/www.cas. muohio.edu/~mbi- ws/microscopes/ history.html/search4it.html

5 Mitchel, C.A. **1916**, *Vinegar: Its Manufacture and Examination*, Griffin, London.

6 Mori, A. **1993**, (Vinegar Production in a Fluidized Bed Reactor with Immobilized Bacteria), in *Industrial Application of Immobilized Biocatalysts*, eds. Tanaka, A., Tosa, T., Kobayashi, T., Marcel Dekker, New York, (pp.) 291–313.

7 Ebner, H.,Sellmer, S., Follmann, H. **1996**, (Acetic Acid), in *Biotechnology*, (vol. 6), eds. Rehm, H.J., Reed, G., Pühler, A., Stadler, P., *Products of Primary Metabolism*, ed. Roehr, M., VCH, Weinheim, (p.) 383.

8 Pasteur, L. **1858**, (Mémoire sur la fermentation de l'acide tartrique), *C.R. Acad. Sci. (Paris)* 46, 615–618.

9 Pasteur, L. **1862**, (Suite a une précédente communication sur les mycodermes; Nouveau procédé industriel de fabrication du vinaigre), *Compt. Rend.* 55, 28–32.

10 Sebek, O.K. **1982**, (Notes on the Historical Development of Microbial Transformations), in: *Microbial Transformations of Bioreactive Compounds* (Vol. 1), ed. Rosazza, J. P., CRC-Press, Boca Raton, FL, 2–6.

11 Buchner, E. **1897**, (Alkoholische Garung ohne Hefezellen), *Ber. Dtsch. Chem. Ges.* 30 (117), 1110.

12 Neuberg, C., Hirsch, J. **1921**, [Über ein Kohlenstoffketten knüpfendes Ferment (Carboligase)], *Biochem. Z.* 115, 282–310.

13 Hildebrandt, G., Klavehn, W. **1930**, (Verfahren zur Herstellung von L-1-Phenyl-2-methylaminopropan-1-ol), Knoll AG Chemische Fabriken in Ludwigshafen, *Ger. Pat.* 548 459.

14 Kluyver, A.J., de Leeuw, F.J. **1924**, (*Acetobacter suboxydans*, een merkwaardige azijnbacterie), *Tijdschr. Verg. Geneesk.* 10, 170.

15 Reichstein, T., Grüssner, H. **1934**, [Eine ergiebige Synthese der L-Ascorbinsäure (C-Vitamin)], *Helv. Chim. Acta* 17, 311–328.

16 Peterson, D.H., Murray, H.C., Epstein, S.H., Reineke, L.M., Weintraub, A., Meister, P.D., Leigh, H.M. **1952**, (Microbiological Oxygenation of Steroids. I. Introduction of Oxygen at Carbon-11 of Progesterone), *J. Am. Chem. Soc.* 74, 5933–5936.

17 Sarett, L.H. **1946**, (Partial Synthesis of Pregnene-4-triol-17(β),20(β),21-dione-3,11 and Pregnene-4-diol-17(β),21-trione-3,11,20 monoacetate), *J. Biol. Chem.* 162, 601–631.

18 Sebek, O.K., Perlman D. **1979**, (Microbial Transformation of Steroids and Sterols), in *Microbial Technology*, (vol. 1), 2nd edn, Academic Press, New York, (pp.) 484, 488.

19 Watson, J.D., Crick, F.H.C. **1953**, (Molecular Structure of Nucleic Acid: A Structure of Deoxyribose Nucleic Acid), *Nature* 171, 737–738.

20 Beadle, G.W., Tatum, E.L. **1941**, (Genetic Control of Biochemical Reactions in *Neurospora*), *Proc. Natl. Acad. Sci. USA* 27, 499–506.

21 Tatum, E.L., Lederberg, J. **1947**, (Gene Recombination in the Bacterium *Escherichia coli*), *J. Bact.* 53, 673–684.

22 Cohen, N.S., Boyer, W.H. **1973**, (Construction of Biologically Functional Bacterial Plasmids *In Vitro*), *Proc. Natl. Acad. Sci. USA* 70, 3240–3244.

23 Ensley, D.B., Ratzkin, J.B., Osslund, D.T., Simon, J.M. **1983**, (Expression of Naphthalene Oxidation Genes in *Escherichia coli* Results in the Biosynthesis of Indigo), *Science* 222, 167–169.

24 Wick, C.B. **1995**, (Genencor International Takes a Green Route to Blue Dye), *Gen. Eng. News* January 15 (2), 1, 22.

25 Weyler, W., Dodge, T.C., Lauff, J.J., Wendt, D.J. **1999**, *US Patent* 5866396.

26 www.novozymes.com; *The History of Enzymes.*

27 Anderson, S., Berman-Marks, C., Lazarus, R., Miller, J., Stafford, K., Seymour, J., Light, D., Rastetter, W., Estell, D. **1985**, (Production of 2-Keto-L-Gulonate, an Intermediate in L-Ascorbate Synthesis, by Genetically Modified *Erwinia herbicola*), *Science* 230, 144–149.

28 McCoy, M. **1998**, (Chemical Makers Try Biotech Paths), *Chem. Eng.* 76 (25), 13–19.

29 Motizuki, K., Kanzaki, T., Okazaki, H., Yoshino, H., Nara, M., Isono, M., Nakanishi, I., Sasajima, K., **1966**, (Method for Producing 2-Keto-L-Gulonic Acid), *US Patent* 3,234,105.

30 Chotani, G., Dodge, T., Hsu, A., Kumar, M., LaDuca, R., Trimbur, D., Weyler, W., Sanford, K. **2000**, (The Commercial Production of Chemicals Using Pathway Engineering), *Biochim. Biophys. Acta* 1543, 434–455.

31 Neidleman, S.L. **1980**, (Use of Enzymes as Catalysts for Alkene Oxide Production), *Hydrocarbon Process.* 59 (11) 135–138.

32 Ikemi, M., **1994**, (Industrial Chemicals: Enzymatic Transformation by Recombinant Microbes), *Bioproc. Technol.* 19, 797–813.

33 Bommarius, A.S., Riebel, B.R. **2004**, *Biocatalysis*, Wiley-VCH Verlag, Weinheim, (p.) 5.

34 Martin, D.P., Williams, S.F., **2003**, (Medical Applications of Poly-4-hydroxybutyrate: a Strong Flexible Absorbable Biomaterial), *Biochem. Eng. J.* 16, 97–105.

35 Salehizadeh, H., Loosdrecht, M.C.M. **2004**, (Production of Polyhydroxyalkanoates by Mixed Culture: Recent Trends and Biotechnological Importance), *Biotechnol. Adv.* 22, 261–279.

36 Heise, A. **2003**, Biocatalysis – a New Pathway to Polymeric Materials? *Annual DPI Meeting*, November 27.

37 Nagasawa T., Yamada H., **1989**, (Microbial Transformations of Nitriles), *TIBTECH* 7, 153–158.

38 Kobayashi, M., Nagasawa, T., Yamada, H. **1992**, (Enzymatic Synthesis of Acrylamide: a Success Story Not Yet Over), *TIBTECH* 10, 402–408.

39 Nakamura, C.E., Whited, G.M. **2003**, (Metabolic Engineering for the Microbial Production of 1,3-Propanediol), *Curr. Opin. Biotechnol.* 14, 454–459.

40 Perham, R.N. **1976**, (The Protein Chemistry of Enzymes, in "Enzymes: One Hundred Years", ed. Gutfreund, H.), *FEBS Lett.* 62 Suppl. E20–E28.

41 Payen, A., **1833**, (Memoire sur la Diastase; les principaux Produits de ses Réactions et leurs applicants aux arts industriels), *J. Ann. Chem. Phys.* 73–93.

42 Kühne, W. **1876**, *Über das Verhalten verschiedener organisirter und sog. ungeformter Fermente. Über das Trypsin (Enzym des Pankreas)*, Verhandlungen des Heidelb. Naturhist. Med. Vereins. N.S.I3, Verlag von Carl Winter's, Universitäatsbuchandlung in Heidelberg.

43 Gutfreund, H. **1976**, (Wilhelm Friedrich Kühne; An Appreciation, in "Enzymes: One Hundred Years" ed. Gutfreund, H.), *FEBS Lett.* 62 Suppl. E1–E12.

44 Aunstrup, K. **1979**, (Production, Isolation and Economics of Extracellular Enzymes), in *Appl. Biochem. Bioeng.*, (vol. 2), *Enzyme Technology*, eds. Wingard, L.B., Katchalski-Katzir, E., Goldstein, L., Academic Press, New York, (pp.) 27–69.

45 Brown, A.J. **1892**, (Influence of Oxygen and Concentration on Alcohol Fermentation), *J. Chem. Soc.* 61, 369–385.

46 Brown, A.J. **1902**, (Enzyme Action), *J. Chem. Soc.* 81, 373–386.

47 Laidler, K.J. **1997**, (A Brief History of Enzyme Kinetics), in *New Beer in an Old Bottle: Eduard Buchner and the Growth of Biochemical Knowledge*, ed. Cornish-Bowden, A., Universitat de Valencia, Valencia, Spain, (pp.) 127–133.

48 Fischer, E. **1894**, (Einfluss der Configuration auf die Wirkung der Enzyme), *Ber. Dtsch. Chem. Ges.* 27, 2985–2993.

49 Fischer, E. **1894**, (Synthesen in der Zuckergruppe II), *Ber. Dtsch. Chem. Ges.* 27, 3189–3232.

50 Michaelis, L., Menten, M.L. **1913**, (Die Kinetik der Invertinwirkung), *Biochem. Z.* 49, 333–369.

51 Briggs, G.E, Haldane, J.B.S. **1925**, (A Note on the Kinetics of Enzyme Action), *Biochem. J.*, 19, 338–339.

52 Butler, J.A.V. **1941**, (Molecular Kinetics of Trypsin Action), *J. Am. Chem. Soc.* 63, 2971–2974.

53 Schoffers, E., Golebiowski, A., Johnson, C.R. **1996**, (Enantioselective Synthesis Through Enzymatic Asymmetrization), *Tetrahedron* 52, 3769–3826.

54 Sumner, J.B. **1926**, (The Isolation and Crystallization of the Enzyme Urease), *J. Biol. Chem.* 69, 435–441.

55 *Enzyme Nomenclature* **1992**, Academic Press, New.York.

56 Bairoch, A. **1999**, (The ENZYME Data Bank in 1999), *Nucleic Acids Res.*, Jan 1, 27 (1), 310–311.

57 Godfrey, T., West, S. (eds.) **1996**, (Data Index 4, Alphabetica Listing of Industrial Enzymes and Source), in *Industrial Enzymology*, 2nd edn, Stockton Press, New York, (pp.) 583–588.

58 Phillips, D.C. **1967**, (The Hen-Egg-White Lysozyme Molecule), *Proc. Natl. Acad. Sci. USA* 57, 484–495.

59 Gutte,B., Merrifield, R.B. **1969**, (The Total Synthesis of an Enzyme with Ribonuclease A Activity), *J. Am. Chem. Soc.* 91, 501–502.

60 Hill, A.C. **1897**, (Reversible Zymohydrolysis), *J. Chem. Soc.* 73, 634–658.

61 Pottevin, H. **1906**, (Actions diastasiques réversibles. Formation et dédoublement des ethers-sels sous l'influence des diastases du pancréas), *Ann. Inst. Pasteur,* 20, 901–923.

62 Cheetham, P.S.J. **1995**, (The Applications of Enzymes in Industry), in *Handbook of Enzyme Biotechnology*, ed. Wiseman, A., Ellis Horwood, London, (pp.) 420.

63 Chaplin, M.F., Bucke, C. **1990**, *Enzyme Technology*, Cambridge University Press, Cambridge, (pp.) 190.

64 Trevan, M.D., **1980**, *Immobilized Enzymes. An Introduction and Applications in Biotechnology*, John Wiley, New York, (p.) 71.

65 Bommarius, A.S., Drauz, K., Groeger, U., Wandrey, C. **1992**, (Membrane Bioreactors for the Production of Enantiomerically Pure *a*-Amino Acids), in *Chirality in Industry*, eds. Collins, A.N., Sheldrake, G.N., Crosby, J., John Wiley, New York, (pp.) 372–397.

66 Kragl, U., Vasic-Racki, D., Wandrey, C. **1993**, (Continuous Processes with Soluble Enzymes), *Ind. J. Chem.* 32B, 103–117.

67 Antrim, R.L., Colilla, W., Schnyder, B.J. **1979**, (Glucose Isomerase Production of High-Fructose Syrups), in *Appl. Biochem. Bioeng.*, (vol 2), *Enzyme Technology*, eds. Wingard, L.B., Katchalski-Katzir, E., Goldstein, L., Academic Press, New York, (pp.) 97–207.

68 **1998** (687 days is the record for Sweetzyme T), *BioTimes*®, Novozymes, *March 1* (www.novozyme.com).

69 Marconi, W., Morisi, F., **1979**, (Industrial Application of Fiber-Entrapped Enzymes), in *Appl. Biochem. Bioeng.*, (vol. 2), *Enzyme Technology*, eds. Wingard, L.B., Katchalski-Katzir, E., Goldstein, L., Academic Press, New York, (pp.) 219–258.

70 Cheetham, P.S.J. **1995**, (The Application of Enzymes in Industry), in *Handbook of Enzyme Biotechnology*, ed. Wiseman, A., Ellis Horwood, London, (pp.) 493-498.

71 Matsumoto, K. **1993**, (Production of 6-APA, 7-ACA, and 7-ADCA by Immobilized Penicillin and Cephalosporin Amidases), in *Industrial Application of Immobilized Biocatalysts*, eds. Tanaka, A., Tosa, T., Kobayashi, T., Marcel Dekker, New York, (pp.) 67–88.

72 von Stockar, U., Valentinotti, S., Marison, I., Cannizzaro, C., Herwig, C. **2003**, (Know-how and Know-why in Biochemical Engineering), *Biotechnol. Adv.* 21, 417–430.

73 Bailey, J.E., Ollis, D.F. **1986**, *Biochemical Engineering Fundamentals*, 2nd edn, McGraw-Hill, New York, (p.) 1.

74 Aiba, S., Humphrey, A.E, Millis, N. F., **1965**, *Biochemical Engineering*, Academic Press, New York.

75 Wang, D.I.C., Humphrey A.E., **1969**, (Biochemical Engineering), *Chem. Eng.* Dec. 15, 108–120.

76 Wandrey, C. **1991**, (The Growing Importance of Biochemical Engineering), *Proceedings Biochemical Engineering-Stuttgart*, eds.

Reuss, M., Chimiel, H., Gilles, F.D., Knack-muss, H.J., Gustav Fischer, Stuttgart, (pp.) 43–67.

77 Zlokarnik, M., **1990**, (Trends and Needs in Bioprocess Engineering), *Chem. Eng. Prog.* 62–67.

78 Lilly, M.D. **1997**, (The Development of Biochemical Engineering Science in Europe), *J. Biotechnol.* 59, 11–18.

79 Gram, A. **1997**, (Biochemical Engineering and Industry), *J. Biotechnol.* 59, 19–23.

80 Katalski-Katzir, E. **1993**, (Immobilized Enzymes–Learning from Past Successes and Failures), *TIBTECH* 11, 471–478.

81 Lilly, M.D., Dunnill, P. **1971**, (Biochemical Reactors), *Process Biochem.* August, 29–32.

82 Papoutsakis, E.T. **2003**, (Murray Moo-Young: the Gentleman of Biochemical Engineering), *Biotechnol. Adv.* 21, 381–382.

83 Chakrabarti, S., Bhattacharya, S., Bhattacharya, S.K. **2003**, (Biochemical Engineering: Cues from Cells), *Trends Biotechnol.* 21, 204–209.

84 Lilly, M.D. **1994**, (Advances in Biotransformation Processes), *Chem. Eng. Sci.* 49, 151–159.

85 Hetherington, P.J., Follows, M., Dunnill, P., Lilly, M.D. **1971**, [Release of Protein from Baker's Yeast (*Saccharomyces cerevisiae*) by Disruption in an Industrial Homogeniser], *Trans. Inst. Chem. Engrs.* 49, 142–148.

86 Atkinson, T., Scawen, M.D., Hammond, P.M. **1987**, (Large Scale Industrial Techniques of Enzyme Recovery), in *Biotechnology*, eds. Rehm, H.J., Reed, G., (vol. 7a), *Enzyme Technology*, ed. Kennedy, J.F., VCH Verlagsgesellschaft, Weinheim, (pp.) 279–323.

87 Kennedy, J.F., Cabral, J.M.S. **1987**, (Enzyme Immobilization), in *Biotechnology*, eds. Rehm, H.J., Reed, G., (vol. 7a), *Enzyme Technology*, ed. Kennedy, J.F., VCH Verlagsgesellschaft, Weinheim, (pp.) 347–404.

88 Nelson, J.M., Griffin E.G. **1916**, (Adsorption of Invertase), *J. Am. Chem. Soc.* 38, 1109–1115.

89 Powel, L.W. **1996**, (Immobilized Enzymes), in *Industrial Enzymology*, eds. Godfrey, T., West, S., 2nd edn, Stockton Press, New York, (p.) 267.

90 Buckland, B.C., Dunnill, P., Lilly, M.D. **1975**, (The Enzymatic Transformation of Water-insoluble Reactants in Nonaqueous Solvents. Conversion of Cholesterol to Cholest-4-ene-3-one by a *Nocardia* sp.), *Biotechnol. Bioeng.*, 17, 815–826.

91 Klibanov, A.M. **1986**, (Enzymes that Work in Organic Solvents), *CHEMTECH* 16, 354–359.

92 Arnold, F.H., Morre, J.C. **1997**, (Optimizing Industrial Enzymes by Directed Evolution), in *New Enzymes for Organic Synthesis,* (vol. 58), *Adv. Biochem. Eng. Biotechnol.*, Springer, Berlin, (pp.) 2–14.

93 Borriss, R. **1987**, (Biotechnology of Enzymes), in *Biotechnology*, (vol 7a), eds. Rehm, H.J., Reed, G., *Enzyme Technology*, ed. Kennedy, J.F., VCH Verlagsgesellschaft, Weinheim (pp.) 35–62.

94 Gerhartz, W. **1990**, in *Enzymes in Industry*, ed. Gerhartz, W., VCH Verlagsgesellschaft, Weinheim, (p.) 11.

95 Clarke P.H., **1976**, (Genes and Enzymes), in *Enzymes: One Hundred Years*, ed. Gutfreund H., *FEBS Lett.* 62 Suppl., E37–E46

96 Hodgson, J. **1994**, (The Changing Bulk Biocatalyst Market), *Bio/Technology* 12 (August), 789–790.

97 Wacket L.P. **1997**, (Bacterial Biocatalysis: Stealing a Page from Nature's Book), *Nature Biotechnol.* 15, 415–416.

98 Matsumura, M., Aiba, S. **1985**, (Screening for Thermostable Mutant of Kanamycin Nucleotidyltransferase by the Use of a Transformation System for a Thermophile, *Bacillus stearothermophilus*), *J. Biol. Chem.* 260, 15298–15303.

99 Liao, H., McKenzie, T., Hageman, R. **1986**, (Isolation of a Thermostable Enzyme Variant by Cloning and Selection in a Thermophile), *Proc. Natl. Acad. Sci. USA*, 83, 576–580.

100 Straathof, A.J.J., Panke, S., Schmid, A. **2002**, (The Production of Fine Chemicals by Biotransformations), *Curr. Opin. Biotechnol.* 13, 548–556.

101 Petersen, M., Kiener, A. **1999**, (Biocatalysis. Preparation and Functionalization of *N*-Heterocycles), *Green Chem.* 1, 99–106.

102 Rawls, R.L. **1998**, (Breaking Up is Hard to Do), *Chem. Eng. News* 76 (25), 29–34.

103 Wrotnowski C. **1996**, (Unexpected Niche Applications for Industrial Enzymes Drives Market Growth), *Gen. Eng. News* (February 1) 14, 30.

104 Yagasaki M., Ozaki A. **1998, (**Industrial Biotransformations for the Production of D-
Amino Acids), *J. Mol. Cat. B: Enzymatic* 4,
1–11.

105 Bhattacharya, S., Chakrabarti, S.,
Bhattacharya, S.K. **2002,** (Bioprocess for
Recyclable CO_2 Fixation: A General Description), in *Recent Research Developments in Biotechnology and Bioengineering*, eds. Bhattacharya, S.K., Mal, T., Chakrabarti, S.,
Research Signpost India, Kerala, 109–120.

106 Chakrabarti, S., Bhattacharya, S.,
Bhattacharya, S.K. **2003,** (Immobilization of
D-Ribulose-1,5-bisphosphate. A Step
Toward Carbon Dioxide Fixation Biopro-
cess), *Biotechnol. Bioeng.* 81, 705–711.

107 Bhattacharya, S., Nayak, A., Schiavone, M.,
Bhattacharya, S.K., **2004,** (Solubilization
and Concentration of Carbon Dioxide:
Novel Spray Reactors with Immobilized Carbonic Anhydrase), *Biotechnol. Bioeng.* 86,
37–46.

108 Enie, H., Nakazawa, E., Tsuchida, T., Name-
rikawa, T., Kumagai, H. **1996,** *Japan Bioind.
Lett.* 13 (1), 2–4.

109 Ghisalba, O. **2000,** (Biocatalysed Reactions),
in *New Trends in Synthetic Medicinal Chemistry*, ed. Gualtieri, F., Wiley-VCH, Weinheim,
(pp.) 175–214.

110 Sumerel, J.L., Yang, W., Kisailus, D.,
Weaver, J.C., Choi, J.H., Morse, D.E. **2003,**
(Biocatalytically Templated Synthesis of
Titanium Dioxide), *Chem. Mater.* 15,
4804–4809.

111 Gokarn, R., Selifenova, O., Jessen, H., Gort,
S., Liao, H., Anderson, J., Gwegorryn, S.,
Cameron, D. **2003,** (3-Hydroxypropionic
Acid: A Building Block for Bio-based Poly-
mers), *Abstract of Papers of the American
Chemical Society,* 226:238-POLY, Part 2.

112 *The Application of Biotechnology to Industrial
Sustainability,* **2001,** OECD publication, (p.)
22.

113 Schmid, A., Dordick, J.S., Hauer, B., Kiener,
A., Wubbolts, M., Witholt, B. **2001,** (Industrial Biocatalysis Today and Tomorrow), *Nature* 409, 258–268.

114 Schoemaker, H.E., Mink, D.,
Wubbolts, M.G. **2003,** (Dispelling the Myths
– Biocatalysis in Industrial Synthesis),
Science 299, 1694–1697.

115 Shaw, N.M., Robins, K.T., Kiener, A. **2003,**
(Lonza: 20 Years of Biotransformations),
Adv. Synth. Catal. 345, 425–435.

116 Schmid, A., Hollmann, F., Park, J.B., Büh-
ler, B. **2002,** (The Use of Enzymes in the
Chemical Industry in Europe), *Curr. Opin.
Biotechnol.* 13, 359–366.

2
The Enzyme Classification

Christoph Hoh and Murillo Villela Filho

2.1
Enzyme Nomenclature

In early stages of studies on biochemistry there were no guidelines for naming enzymes. The denomination of a newly discovered enzyme was given arbitrarily by individual workers. This practice proved to be inappropriate. Occasionally two different enzymes had the same name while in other cases two different names were given to the same enzyme. Furthermore, denominations emerged which provided no clues about the catalyzed reaction (e.g., catalases, or pH 5 enzyme).

With the great progress achieved in the area of biochemistry in the 1950s, a large number of enzymes could be isolated and characterized. By this time it had become evident that it was necessary to regulate the enzyme nomenclature. Thus, the International Union of Biochemistry and Molecular Biology (IUBMB), formerly the International Union of Biochemistry (IUB), in consultation with the International Union for Pure and Applied Chemistry (IUPAC), set up an Enzyme Commission to be in charge of guiding the naming and establishing a systematic classification for enzymes. In 1961, the report of the commission was published. The proposed classification was used to name 712 enzymes. This work has been widely used as a guideline for enzyme nomenclature in scientific journals and textbooks ever since. It has been periodically updated, new entries have been included or old ones deleted, while some other enzymes have been reclassified. The sixth complete edition of the *Enzyme Nomenclature* (1992) contains 3196 enzymes [1]. Ten supplements to the *Enzyme Nomenclature* with various additions and corrections have been published to date, signaling the constantly increasing number of new enzyme entries [2–7]. The five latest supplements (2000–2004) are only available online [7]. An updated documentation of the classified enzymes is available on the ENZYME data bank server [8, 9].

The *Enzyme Nomenclature* suggests two names for each enzyme: a **recommended name**, that is convenient for everyday use, and a **systematic name**, used to minimize ambiguity. Both names are based on the nature of the catalyzed reaction. The recommended name is often the former trivial name, which is sometimes changed slightly to prevent misinterpretation. The systematic name also includes the substrates involved. This taxonomy leads to the classification of enzymes into six main classes (Table 2.1).

Industrial Biotransformations. Andreas Liese, Karsten Seelbach, Christian Wandrey (Eds.)
Copyright © 2006 WILEY-VCH Verlag GmbH & Co. KGaA, Weinheim
ISBN: 3-527-31001-0

Tab. 2.1 The main enzyme classes.

Enzyme class	Catalyzed reaction
1. Oxidoreductases	oxidation–reduction reactions
2. Transferases	transfer of functional groups
3. Hydrolases	hydrolysis reactions
4. Lyases	group elimination (forming double bonds)
5. Isomerases	isomerization reactions
6. Ligases	bond formation coupled with a triphospate cleavage

As the systematic name may be very extensive and awkward to use, the Enzyme Commission (EC) has also developed a numerical system based on the same criteria, which can be used together with the recommended name to specify a particular enzyme. According to this system, each enzyme is assigned a four-digit EC number (Table 2.2). The first digit denotes the main class, which specifies the catalyzed reaction type. These are divided into subclasses, according to the nature of the substrate, the type of the transferred functional group or the nature of the specific bond involved in the catalyzed reaction. These subclasses are designated by the second digit. The third digit reflects a further division of the subclasses according to the substrate or co-substrate, giving rise to the sub-subclasses. In the fourth digit a serial number is used to complete the enzyme identification.

Tab. 2.2 Constitution of the four-digit EC number.

EC number: EC (i).(ii).(iii).(iv)

(i)	the main class, denotes the type of catalyzed reaction
(ii)	subclass, indicates the substrate type, the type of transferred functional group or the nature of one specific bond involved in the catalyzed reaction
(iii)	sub-subclass, expresses the nature of the substrate or co-substrate
(iv)	an arbitrary serial number

As an example, aminoacylase (*N*-acyl-L-amino-acid amidohydrolase, as it is known according to systematic nomenclature), an enzyme used in the industrial production of L-methionine, has the classification number EC 3.5.1.14 (see the process on page 403). The first number (i = 3) indicates that this enzyme belongs to the class of hydrolases. The second number (ii = 5) shows that a carbon–nitrogen bond is hydrolyzed and the third number (iii = 1) denotes that the substrate is a linear amide. The serial number (iv = 14) is required for the full classification of the enzyme.

As the biological source of an enzyme is not included in its classification, it is important to mention this together with the enzyme number in order to provide full identification. So the enzyme used in the production of "acrylamide" should be described as "nitrile hydratase (EC 4.2.1.84) from "*Rhodococcus rhodochorous*".

An important aspect associated with the application of enzyme nomenclature is the direction in which a catalyzed reaction is written for classification purposes. To make the classification more transparent the direction should be the same for all enzymes of a given class, even if this direction has not been demonstrated for all enzymes of this class. Many examples of the use of this convention can be found in the class of oxidoreductases.

A further implication of this system is that full classification of an enzyme is impossible if the catalyzed reaction is not clear. Complete classification of the enzymes is only dependent on the natural substrates. Non-natural substrates are not considered in the classification of the biocatalyst.

Finally, it is important to emphasize that the advantages of the enzyme classification is not just limited to enzyme nomenclature in biochemistry. It is also very beneficial to organic preparative chemists because it facilitates the choice of enzymes for synthetic applications. As the classification of enzymes is based on the catalyzed reactions it helps chemists to find an appropriate biocatalyst for a given synthetic task. To date an analogous nomenclature for chemical catalysts has not been set up. In this context it must be stressed that the trade names of several commercially available enzyme preparations do not correlate with the enzyme nomenclature, e.g., "*Novozyme 435*™" (Novozymes) for *C. antarctica* lipase B, a triacylglycerol lipase (EC 3.1.1.3) or "*PGA-450*™" (Roche Diagnostics) for a carrier-fixed Penicillin amidase (EC 3.5.1.11). This could complicate the rapid selection of an appropriate biocatalyst.

The number of enzymes existing in nature is estimated to be as high as 25 000 [10]. One of the essential areas of biochemistry and related sciences is trying to find and identify them. The scientist isolating and characterizing a new enzyme is free to report the discovery of that "new" biocatalyst to the Nomenclature Committee of the IUBMB and may create a new systematic name for this enzyme. An appropriate form to draw the attention of the editor of *Enzyme Nomenclature* to enzymes and other catalytic entities missing from this list is available online [11].

2.2
Enzyme Classes

The following section of this chapter aims at giving a compact overview of the six main enzyme classes and their subclasses. As the industrial bioprocesses and biotransformations illustrated in the following chapters of the book are divided according to the enzymes involved and their classes, this short survey should provide the reader with the most important information on these enzyme classes.

The six main enzyme classes are resumed separately by giving a general reaction equation for every enzyme subclass according to *Enzyme Nomenclature*. The reaction equations are illustrated in a very general manner indicating only the most important attributes of the catalyzed reactions. The authors would like to emphasize that no attempt has been made to provide a complete summary of the reactions catalyzed by the enzymes listed in *Enzyme Nomenclature*. The reaction schemes have been elaborated to give reaction equations being as general and clear as possible and as detailed as necessary.

An important point that needs to be considered in this context concerns the enzymes classified as EC (i).99 or EC (i).(ii).99. These enzymes are either very substrate specific and therefore cannot be classified in the already existing enzyme subclasses (or sub-subclasses) or a substrate of these enzymes has not been fully identified yet.

For instance, in the enzyme main class EC 5.(ii).(iii).(iv) (isomerases), the EC number 5.99 only describes "other isomerases" that cannot be classified within the remaining existing subclasses EC 5.1 to EC 5.5. It is important to point out that the enzymes classified with a 99-digit have not been considered in the reaction equations unless stated explicitly. The catalyzed reactions of these enzymes differ significantly from those of the other enzymes in the same main division.

The following short comments on the generalized reaction schemes should help the reader to understand the illustrated enzyme catalyzed reactions.

1. Each main enzyme class is introduced by a short paragraph giving a general impression of the respective enzymes.
2. By generalizing nearly all catalyzed reactions of one enzyme subclass to only one or a few reaction equations, some details of the single reactions had to be neglected, e.g., specification of the cofactor, reaction conditions (pH, temperature), electric charge or stoichiometry. Also, correct protonation of the substrates and products depending on the pH value of each reaction mixture have not been taken into consideration. In addition, the enzyme itself does not appear in the reaction schemes of this chapter.
3. If the catalyzed reaction leads to a defined equilibrium, only one direction of this reaction is considered according to its direction in *Enzyme Nomenclature*. As a consequence, no equilibrium arrows are used in any reaction scheme of this chapter.
4. Enzymes of a given subclass may show some frequently appearing common properties or a very worthwhile uniqueness. These qualities are taken into account by additional comments below the reaction schemes.

2.2.1
EC 1 Oxidoreductases

The enzymes of this first main division catalyze oxidoreduction reactions, which means that all these enzymes act on substrates through the transfer of electrons. In the majority of cases the substrate that is oxidized is regarded as a hydrogen donor. Various cofactors or coenzymes serve as acceptor molecules. The systematic name is based on *donor:acceptor oxidoreductase*.

Whenever possible the nomination as a *dehydrogenase* is recommended. Alternatively, the term *reductase* can be used. If molecular oxygen (O_2) is the acceptor, the enzymes may be named as *oxidases*.

EC 1.1 Acting on CH–OH group of donors

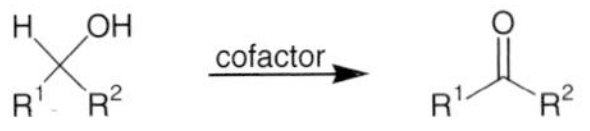

R^1 = hydrogen, organic residue
R^2 = hydrogen, organic residue, alcoxy residue

The sub-subclasses are defined by the type of cofactor.

EC 1.2 Acting on the aldehyde or oxo group of donors

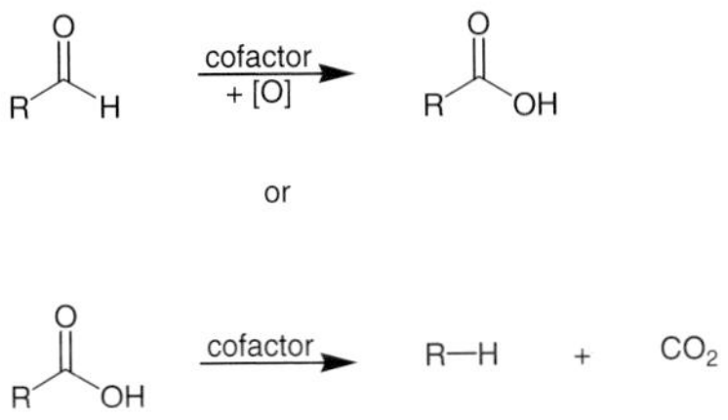

or

R = hydrogen, organic residue

Analogous with the first depicted reaction, the aldehyde can be oxidized to the respective thioester with coenzyme A (CoA). In the case of oxidation of carboxylic acids, the organic product is not necessarily bound to hydrogen as suggested in the figure. It can also be bound to the cofactor. The sub-subclasses are classified according to the cofactor.

EC 1.3 Acting on the CH–CH group of donors

$R^{1,2,3,4}$ = hydrogen, organic residue

In some cases the residues can also contain heteroatoms, e.g., dehydrogenation of *trans*-1,2-dihydroxycyclohexa-3,5-diene to 1,2-dihydroxybenzene (catechol). Further classification is based on the cofactor.

EC 1.4   Acting on the CH–NH₂ group of donors

$$\underset{R^1 \quad R^2}{\overset{H \quad NH_2}{\bigtimes}} \xrightarrow{\text{cofactor}} \underset{R^1 \quad R^2}{\overset{NH}{\bigparallel}} \xrightarrow{\text{--- H}_2O} \underset{R^1 \quad R^2}{\overset{O}{\bigparallel}} + \quad NH_3$$

$R^{1,2}$ = hydrogen, organic residue

In most cases the imine formed is hydrolyzed to give an oxo-group and ammonia (deaminating). The division into sub-subclasses depends on the cofactor.

EC 1.5   Acting on the CH–NH group of donors

$$\underset{R^1 \quad R^2}{\overset{H \quad NHR^3}{\bigtimes}} \xrightarrow{\text{cofactor}} \underset{R^1 \quad R^2}{\overset{NR^3}{\bigparallel}} \xrightarrow{\text{--- H}_2O} \underset{R^1 \quad R^2}{\overset{O}{\bigparallel}} + \quad R^3NH_2$$

$R^{1,2}$ = hydrogen, organic residue
R^3 = organic residue

In some cases the primary product of the enzymatic reaction can be hydrolyzed. The further classification is based on the cofactors.

EC 1.6   Acting on NAD(P)H

$$\boxed{NAD(P)}\!-\!H \quad + \quad A \quad \longrightarrow \quad \boxed{NAD(P)}^+ \quad + \quad A\!-\!H$$

A = acceptor

In general, enzymes that use NAD(P)H as the reducing agent are classified according to the substrate of the reverse reaction. Only enzymes that require some other redox carrier as acceptors to oxidize NAD(P)H are classified in this subclass. Further division depends on the redox carrier used.

EC 1.7   Acting on other nitrogen compounds as donors

$$N_{red}R_3 \xrightarrow{\text{cofactor}} N_{ox}R_3$$

R = hydrogen, organic residue, oxygen

The enzymes that catalyze the oxidation of ammonia to nitrite and the oxidation of nitrite to nitrate belong to this subclass. The subdivision is based on the cofactor.

EC 1.8 Acting on sulfur group of donors

S_{red} $\xrightarrow{\text{cofactor}}$ S_{ox}

S_{red} = sulfide, sulfite, thiosulfate, thiol, etc.

S_{ox} = sulfite, sulfate, tetrathionate, disulfite, etc.

The substrates may be either organic or inorganic sulfur compounds. The nature of the cofactor defines the further classification.

EC 1.9 Acting on a heme group of donors

heme—Fe^{2+} $\xrightarrow{\text{cofactor}}$ heme—Fe^{3+}

The sub-subclasses depend again on the cofactor.

EC 1.10 Acting on diphenols and related substances as donors

X_d = OH, NH_2
X_a = O, NH

The aromatic ring may be substituted; ascorbates are also substrates for this subclass. The primary product can undergo further reaction. The subdivision into four sub-subclasses depends on the cofactor.

EC 1.11 Acting on a peroxide as acceptor

H_2O_2 + D_{red} $\longrightarrow$ H_2O + D_{ox}

D = donor

The single sub-subclass contains the peroxidases.

EC 1.12 Acting on hydrogen as donor

$$H_2 \quad + \quad A^+ \quad \longrightarrow \quad H^+ \quad + \quad A-H$$

Sub-subclass 1.12.1 contains enzymes using NAD^+ and $NADP^+$ as cofactors. Other hydrogenases are classified under 1.12.99. Enzymes using iron–sulfur compounds as the cofactor are listed under 1.18.

EC 1.13 Acting on single donors with incorporation of molecular oxygen

$$A \quad + \quad O_2 \quad \longrightarrow \quad AO_{(2)}$$

If two oxygen atoms are incorporated, the enzyme belongs to the sub-subclass 1.13.11 and if only one atom of oxygen is used the enzyme is classified as 1.13.12. All other cases are classified under 1.13.99.

EC 1.14 Acting on paired donors with incorporation of molecular oxygen

$$A \quad + \quad O_2 \quad \xrightarrow{\text{cofactor}} \quad AO_{(2)}$$

The classification into sub-subclasses depends on whether both oxygen atoms or just one are bound to the substrate. The difference between subclass 1.13 is the requirement of a cofactor.

EC 1.15 Acting on superoxide radicals as acceptor

$$O_2^{\bullet -} \quad + \quad O_2^{\bullet -} \quad + \quad H^+ \quad \longrightarrow \quad \tfrac{3}{2}O_2 \quad + \quad H_2O$$

The only enzyme classified into this subclass is the superoxide dismutase.

EC 1.16 Oxidizing metal ions

$$M^{m+} \quad \xrightarrow{\text{cofactor}} \quad M^{n+}$$

$m \geq 0$
$n > m$

The two sub-subclasses are divided according to the cofactor.

EC 1.17   Acting on CH$_2$ groups

The origin of the oxidizing oxygen is either molecular oxygen or water.

EC 1.18   Acting on reduced ferredoxin as donor

ferredoxin$_{red}$ $\xrightarrow{\text{cofactor}}$ ferredoxin$_{ox}$

EC 1.19   With dinitrogen as acceptor

N$_2$ $\xrightarrow{\text{cofactor}}$ NH$_3$

The only enzyme classified in this subclass is nitrogenase.

EC 1.20   Acting on phosphorus or arsenic in donors
The nature of the substrate may differ greatly within the sub-subclasses EC 1.20.1 and 1.20.4. The sub-subclasses EC 1.20.2 and EC 1.20.3 do not have any entries as yet.

EC 1.20.1   Acting on phosphorus or arsenic in donors, with NAD(P)$^+$ as acceptor

HO–P(=O)(H)–OH   +   NAD(P)$^+$ +   H$_2$O $\longrightarrow$ HO–P(=O)(OH)–OH   +   NAD(P)H

At present no entries for reactions including arsenic in donors are listed.

EC 1.20.4   Acting on phosphorus or arsenic in donors, with disulfide as acceptor

HO–As(OH)(X)   +   –S–S–   +   H$_2$O $\longrightarrow$ HO–As(=O)(OH)–X   +   –SH HS–

X = OH or methyl group

At present no entries for reactions including phosphorus in donors are listed.

EC 1.21   Acting on X–H and Y–H to form an X–Y bond

$$X{-}H \quad + \quad Y{-}H \quad + \quad 2\,A \quad \longrightarrow \quad X{-}Y \quad + \quad 2\,\,H{-}A$$

A = hydrogen acceptor

The hydrogen acceptor is either oxygen or a disulfide.

2.2.2
EC 2 Transferases

The transferases are enzymes that transfer a chemical group from one compound (generally regarded as the donor) to another compound (generally regarded as the acceptor). Of all biological reactions, this class of biocatalysts is one of the most common [12]. To avoid any confusion, the following reaction schemes of the subclasses all show the same pattern: the donor molecule is always the first of the substrates shown, the acceptor is the second one. Where possible, some detailed information is given on the acceptor, but also the general denomination of A = acceptor has been chosen in three cases.

In general, the systematic names of these biocatalysts are formed according to the scheme *donor:acceptor group-transferase*. In many cases, the donor is a cofactor (coenzyme) carrying the often activated chemical group to be transferred.

EC 2.1   Transferring one-carbon groups

A = acceptor
R = organic residue
Ⓒ = methyl-, hydroxymethyl-, formyl-, carboxyl-, carbamoyl- and amidino-groups

EC 2.2   Transferring aldehyde or ketone residues

R^1 = hydrogen or methyl residue
R^2 = methyl residue or polyol chain
R^3 = hydrogen or polyol chain

Three of the only four enzymes in this subclass depend on thiamin-diphosphate as a cofactor. The catalyzed reactions can be regarded as an aldol addition. Some enzymes also accept hydroxypyruvate as a donor to form CO_2 and the resulting addition product.

EC 2.3 Acyltransferases

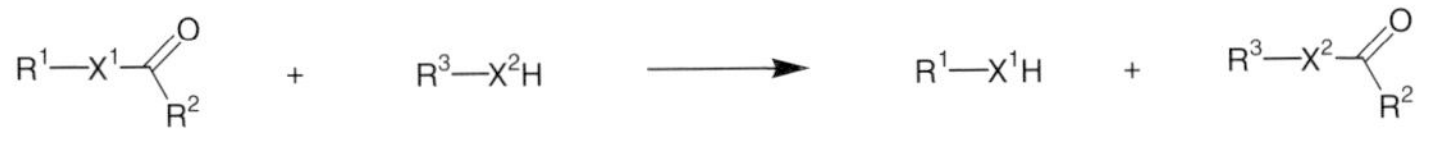

R^1-X^1 ... R^2 ... O + R^3-X^2H $\longrightarrow$ R^1-X^1H + R^3-X^2 ... R^2 ... O

X^1 = S, O, NH
X^2 = S, O, NH, CH_2
R^1 = hydrogen, alkyl-, aryl- or monophosphate residue
R^2 = hydrogen, alkyl- or aryl-residue
R^3 = alkyl-, aryl-, acyl- or monophosphate residue, aryl-NH

Transferred acyl-groups are often activated as coenzyme A (CoA) conjugates.

EC 2.4 Glycosyltransferases

OH ... HO— ... HO ... OH ... $-X^1R^1$ + R^2-X^2H $\longrightarrow$ R^1-X^1H + OH ... HO— ... HO ... OH ... $-X^2R^2$

X^1 = O, PO_4^{3-}
X^2 = O, NH
R^1 = hydrogen, hexosyl, pentosyl, oligosaccharide, monophosphate
R^2 = hexosyl, pentosyl, oligosacharide, monophosphate, organic residue with OH- or NH_2-groups
X^1R^1 = nucleoside di- or monophosphates (e.g. UDP, ADP, GDP or CMP), purine

This enzyme subclass is subdivided into the hexosyl- (sub-subclass 2.4.1) and pentosyl-transferases (sub-subclass 2.4.2). Although a hexosyl transfer is illustrated in the figure, this general scheme is intended to describe both enzyme sub-subclasses.

EC 2.5   Transferring alkyl or aryl groups, other than methyl groups

$$\text{X—R} \quad + \quad \text{A} \quad \longrightarrow \quad \text{X} \quad + \quad \text{A—R}$$

A = acceptor
X = OH, NH, SR, SO_4^-, mono-, di- or triphosphate
R = organic residue other than a methyl group

EC 2.6   Transferring nitrogenous groups

If NX = NH_2, then --- is a single bond.
If NX = NOH, then --- is a double bond.
R^1 = hydrogen, carboxy or methyl residue
R^2 = organic residue
R^3 = hydrogen, carboxy or hydroxymethyl residue
R^4 = organic residue

Pyridoxal-phosphate is the cofactor that appears most frequently for these enzymes. For NX = NH_2 the substrates are often a-amino acids and 2-oxo acids.

EC 2.7   Transferring phosphorus-containing groups

$$\text{R}^1\text{—}\textcircled{P} \quad + \quad \text{R}^2\text{—X} \quad \longrightarrow \quad \text{R}^1 \quad + \quad \text{R}^2\text{—X—}\textcircled{P}$$

or

$$\text{R}^1\text{—}\textcircled{P} \quad + \quad \text{R}^2\text{—X} \quad \longrightarrow \quad \text{R}^1\text{—X—R}^2 \quad + \quad \textcircled{P}$$

X = OH, COOH, NH_2, PO_4^{2-}
R^1 = hydrogen, NDP, NMP, adenosine, monosaccharide residue, acyl residue, polyphosphate, histidine, syn-glycerol, organic residues carrying more functional groups
R^2 = hydrogen, monosaccharide residue, nucleosides, nucleotides, organic residues carrying more functional groups, proteins, polyphosphate
$\textcircled{P}$ = mono- or diphosphates

The enzymes transferring a phosphate residue from an ATP molecule to an acceptor are known as kinases. The enzyme EC 2.7.2.2 (carbamate kinase) transfers a phosphate residue from an ATP molecule on CO_2 or NH_3 to form carbamoyl phosphate.

EC 2.8 Transferring sulfur-containing groups

A = acceptor, e.g. cyanide, phenols, alcohols, carboxylic acids, amino acids, amines, saccharides
R = sulfur atom, (phosphorous-) organic residue
Ⓢ = sulfur atom, SO_3^{2-}, SH, CoA

EC 2.9 Transferring selenium-containing groups

The only enzyme classified under this subclass is L-seryl-tRNA (Sec) selenium transferase.

2.2.3
EC 3 Hydrolases

This third main class of enzymes plays the most important role in today's industrial enzymatic processes. Hydrolases catalyze the hydrolytic cleavage of C–O, C–N, C–C and some other bonds, including P–O bonds in phosphates. The applications of these enzymes are very diverse: the most well-known examples are the hydrolysis of polysaccharides, nitriles (see the processes on pages 317 and 320), proteins or the esterification of fatty acids (see the process on page 297). Most of these industrial enzymes are used in processing-type reactions to degrade proteins, carbohydrates and lipids in detergent formulations and in the food industry.

Interestingly, all hydrolytic enzymes could be classified as transferases, as every hydrolysis reaction can be regarded as the transfer of a specific chemical group to a water molecule. However, because of the ubiquity and importance of water in natural processes, these biocatalysts are classified as hydrolases rather than as transferases.

The term *hydrolase* is included in every systematic name. The recommendation for the naming of these enzymes is the formation of a name that includes the name of the substrate and the suffix -*ase*. It is understood that the name of a substrate with this suffix implies a hydrolytic enzyme.

EC 3.1 Acting on ester bonds
The nature of the substrate can differ greatly, as shown in the three examples.

EC 3.1.1 Carboxylic ester hydrolase

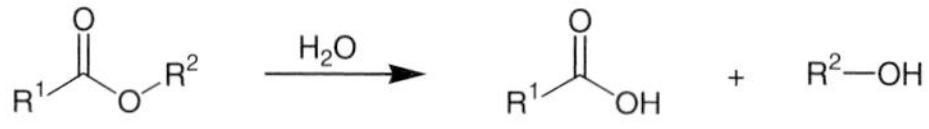

R¹—C(=O)—O—R² $\xrightarrow{\text{H}_2\text{O}}$ R¹—C(=O)—OH + R²—OH

R¹ = hydrogen, organic residue
R² = organic residue

EC 3.1.2 Thiolester hydrolase

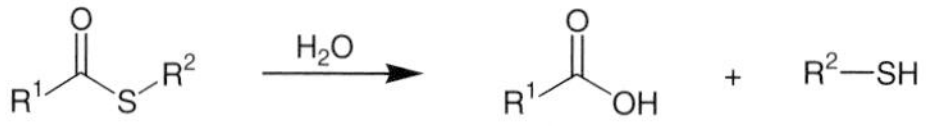

R¹—C(=O)—S—R² $\xrightarrow{\text{H}_2\text{O}}$ R¹—C(=O)—OH + R²—SH

R¹ = hydrogen, organic residue
R² = organic residue

EC 3.1.3 Phosphohydrolase ("phosphatase")

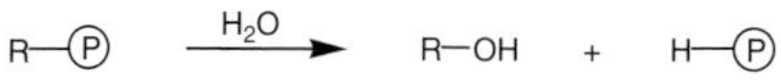

R—Ⓟ $\xrightarrow{\text{H}_2\text{O}}$ R—OH + H—Ⓟ

Ⓟ = monophosphate
R = organic residue

EC 3.2 Glycosidases

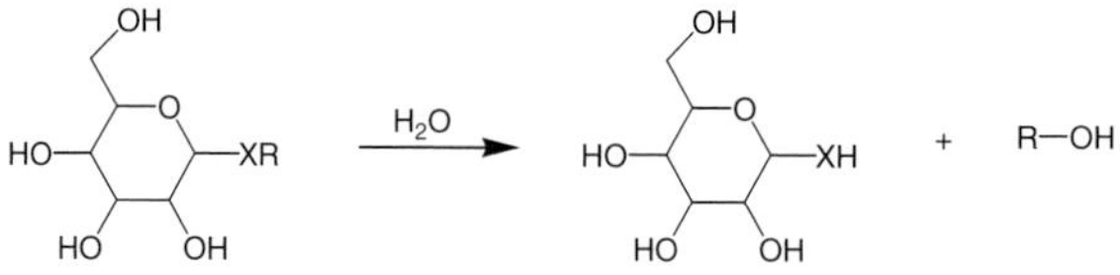

X = O, N or S
R = organic residue

The illustration shows the hydrolysis of a hexose derivative although pentose derivatives are also accepted as substrates.

EC 3.3   Acting on ether bonds

$$R^1—X—R^2 \quad \xrightarrow{H_2O} \quad R^1—X—H \quad + \quad R^2—OH$$

X = O or S
$R^{1,2}$ = organic residue

EC 3.4   Acting on peptide bonds

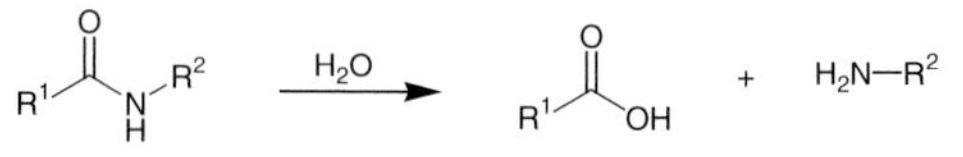

$R^{1,2}$ = part of amino acids or proteins

EC 3.5   Acting on carbon–nitrogen bonds other than peptide bonds

R^1 = organic residue
R^2 = hydrogen or organic residue

For some nitriles a similar reaction takes place. The enzyme involved is known as nitri-lase (EC 3.5.5.)

R = aromatic, heterocyclic and certain unsaturated aliphatic residues

EC 3.6 Acting on acid anhydrides

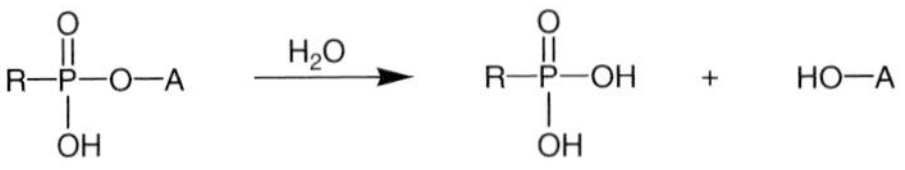

R—P(=O)(OH)—O—A $\xrightarrow{\text{H}_2\text{O}}$ R—P(=O)(OH)—OH + HO—A

A = phosphate, organic phosphate, sulfate
R = organic residue, hydroxy group

EC 3.7 Acting on carbon–carbon bonds

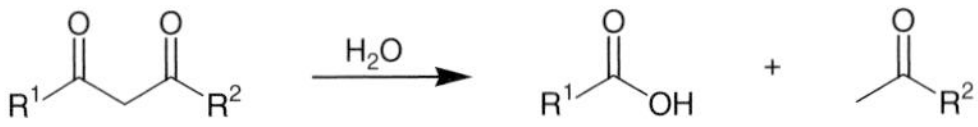

$\xrightarrow{\text{H}_2\text{O}}$

$R^{1,2}$ = organic residue, hydroxy group

There is only one sub-subclass.

EC 3.8 Acting on halide bonds

R_3C—X $\xrightarrow{\text{H}_2\text{O}}$ R_3C—OH + HX

X = halogen
R = hydrogen, organic residue, hydroxy group

EC 3.9 Acting on phosphorus–nitrogen bonds

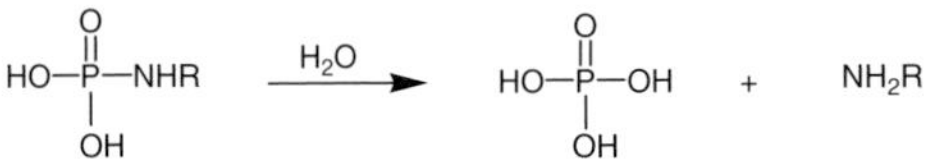

HO—P(=O)(OH)—NHR $\xrightarrow{\text{H}_2\text{O}}$ HO—P(=O)(OH)—OH + NH_2R

R = organic residue

The only enzyme classified under this subclass is phosphoamidase.

EC 3.10   Acting on sulfur–nitrogen bonds

Ⓢ—NR$_2$ $\xrightarrow{\text{H}_2\text{O}}$ Ⓢ—OH   +   HNR$_2$

Ⓢ = sulfon group
R = organic residue

There is only one subdivision of this subclass.

EC 3.11   Acting on carbon–phosphorus bonds

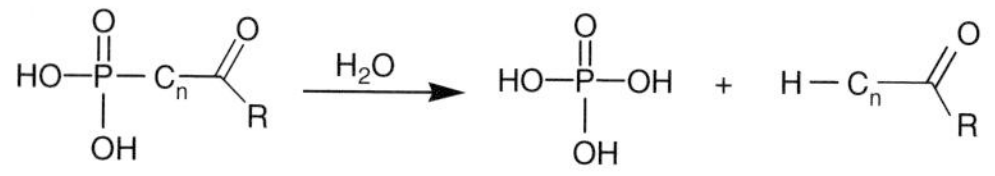

HO—P(=O)(OH)—C$_n$(=O)(R) $\xrightarrow{\text{H}_2\text{O}}$ HO—P(=O)(OH)—OH   +   H—C$_n$(=O)(R)

R = CH$_3$, OH
n = 0, 1

If n = 0, the product is an aldehyde.

EC 3.12   Acting on sulfur–sulfur bonds

O=S(O$^-$)(=O)—S—S(O$^-$)(=O)=O $\xrightarrow{\text{H}_2\text{O}}$ O=S(O$^-$)(=O)—OH   +   HS—S(O$^-$)(=O)=O

The only enzyme classified under this subclass is trithionate hydrolase.

EC 3.13   Acting on carbon–sulfur bonds

O=S(=O)(OH)—R $\xrightarrow{\text{H}_2\text{O}}$ HO—S(=O)(=O)—OH   +   HO—R

R = organic residue

The only enzyme classified under this subclass is UDP-sulfoquinovose synthase.

2.2.4
EC 4 Lyases

From a commercial perspective, these enzymes are an attractive group of catalysts as demonstrated by their use in many industrial processes (see Chapter 4). The reactions catalyzed are the cleavage of C–C, C–O, C–N and some other bonds. It is important to mention that this bond cleavage is different from hydrolysis, often leaving unsaturated products with double bonds that may be subject to further reactions. In industrial processes these enzymes are most commonly used in the synthetic mode, meaning that the reverse reaction – addition of a molecule to an unsaturated substrate – is of interest. To shift the equilibrium these reactions are conducted at very high substrate concentrations, which results in very high conversions into the desired products. For instance, a specific type of lyase, the *phenylalanine ammonia lyase* (EC 4.3.1.5), catalyzes the formation of an asymmetric C–N bond yielding the L-amino acid dihydroxy-L-phenylalanine (L-DOPA). This amino acid is produced on a ton scale and with very high optical purities (see the process on page 460).

Systematic denomination of these enzymes should follow the pattern *substrate group-lyase*. To avoid any confusion the hyphen should not be omitted, e.g., the term *hydro-lyase* should be used instead of *hydrolyase*, which actually looks fairly similar to *hydrolase*.

In the recommended names, terms such as *decarboxylase*, *aldolase* or *dehydratase* (describing the elimination of CO_2, an aldehyde or water) are used. If the reverse reaction is much more important, or it is the only one known, the term *synthase* may be used.

EC 4.1   Carbon–carbon lyases

$R^{1,2,3,4,5}$ = hydrogen, organic residue

If the substrate is a carboxylic acid, one of the products will be carbon dioxide. If the substrate is an aldehyde, carbon monoxide could be a product.

EC 4.2   Carbon–oxygen lyases

$R^{1,2,3,4,5}$ = hydrogen, organic residue

A further addition of water to the product could lead to an oxo acid. This is the case for some amino acids, where ammonia is then eliminated.

EC 4.3 Carbon–nitrogen lyases

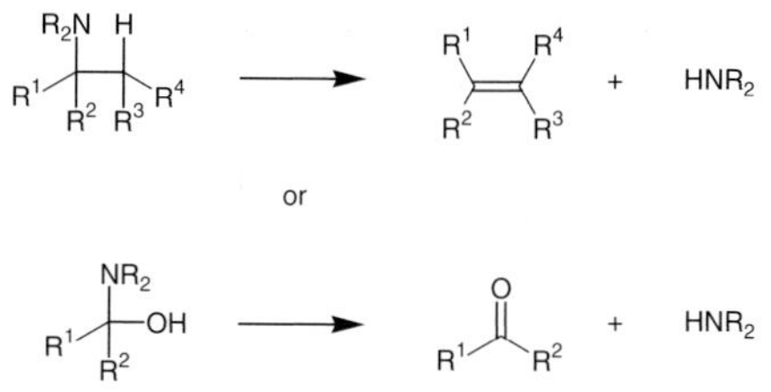

R = organic residue

The resulting double bond can change its position in order to deliver a more stable product, for instance in the case of keto–enol tautomerism. The product can also undergo a further reaction.

EC 4.4 Carbon–sulfur lyases

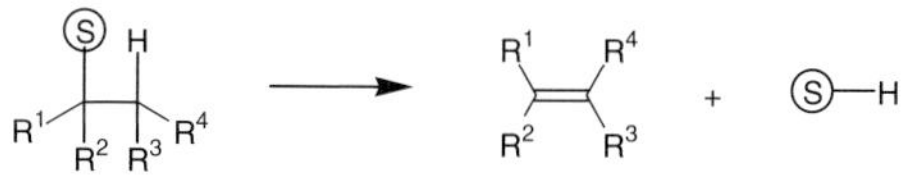

Ⓢ = SH, (di)substituted sulfide, sulfur-oxide, SeH
R = organic residue

According to *Enzyme Nomenclature* the carbon–selenium lyase also belongs to this subclass. Similarly to other lyases, further reactions can occur on the product. In the case of disubstituted sulfides, there is no hydrogen bonded to the sulfur in the product.

EC 4.5 Carbon–halide lyases

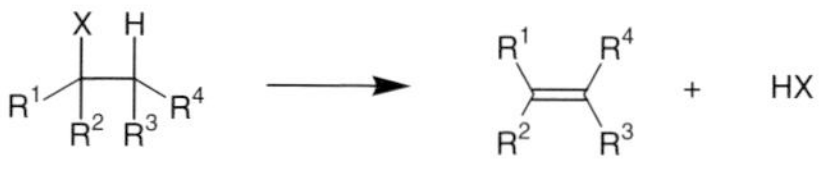

X = halogen
R = organic residue

The primary product can also undergo further reaction. In the case of dihalosubstituted methane the sequential reaction will lead to the aldehyde. Amino compounds can react through elimination of ammonia to give oxo compounds. If thioglycolate is a cofactor, a sulfur–carbon bond will replace the halogen–carbon one.

EC 4.6   Phosphorus–oxygen lyase

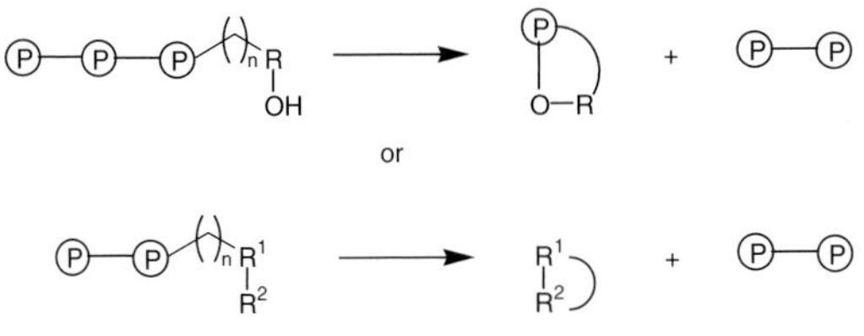

P = monophosphate
R = organic residue

With the exception of EC 4.6.1.4 all enzymes of this subclass lead to a cyclic product.

2.2.5
EC 5 Isomerases

This enzyme class only represents a small number of enzymes, but nevertheless one of them plays a major role in today industry. This enzyme, known as *xylose isomerase* (EC 5.3.1.5), catalyzes the conversion of D-glucose into D-fructose, which is necessary in the production of high-fructose corn syrup (HFCS) (see the process on page 507). This syrup is a substitute for sucrose and is used by the food and beverage industries as a natural sweetener.

In general, the isomerases catalyze geometric or structural changes within one single molecule. Depending on the type of isomerism, these enzymes are known as *epimerases, racemases,* cis–trans-*isomerases, tautomerases* or *mutases.*

EC 5.1   Racemases and epimerases

X = NH_2, NHR, NR_2, OH, CH_3, COOH
$R^{1,2}$ = organic residue

EC 5.2 *cis–trans*-Isomerases

X = C or N

$R^{1,2,3,4}$ = organic residue

If X = N the substrate is an oxime. In this case R^4 represents the single electron pair.

EC 5.3 Intramolecular oxidoreductases

General scheme for the subclasses 5.3.1–5.3.4:

$R^{1,2}$ = hydrogen, organic residue

General scheme for the subclass 5.3.99:

$R^{1,2}$ = hydrogen, organic residue

For these enzymes the centers of oxidation and reduction in the substrate need not be adjacent.

To avoid misunderstandings the sub-subclasses 5.3.1–5.3.4 are presented separately.

EC 5.3.1 Interconverting aldoses and ketoses

$R^{1,2}$ = hydrogen, organic residue

If R^1 is a hydrogen atom, then R^2 is any organic residue and *vice versa*.

EC 5.3.2 Interconverting keto–enol-groups

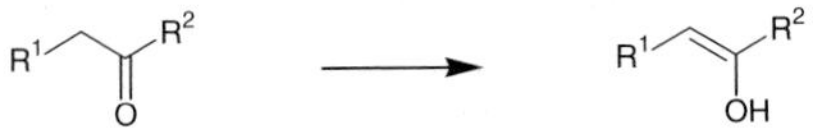

$R^{1,2}$ = hydrogen, organic residue

EC 5.3.3 Transposing C=C bonds

$R^{1,2,3,4,5}$ = organic residue

EC 5.3.4 Transposing S–S bonds

cysteine¹—SH + cysteine²—S—S—cysteine³ ⟶ cysteine¹—S—S—cysteine² + HS—cysteine³

The cysteine residues are parts of proteins.

EC 5.4   Intramolecular transferases (mutases)

This enzyme subclass can be divided in two groups.

The enzymes belonging to 5.4.1 and 5.4.2 catalyze the transfer of a functional group from one oxygen atom to another oxygen atom of the same molecule.

TG = transferred groups are acyl or orthophosphate groups
$R^{1,2}$ = organic residue
n = 0 or 4

The enzymes classified as 5.4.3 catalyze the transfer of a whole amino group from one carbon atom of a molecule to a neighboring atom of the same molecule.

$R^{1,2}$ = organic residue

EC 5.5   Intramolecular lyases

X = O, CH_2
$R^{1,2}$ = organic residue

2.2.6
EC 6 Ligases

In contrast to all other five enzyme classes this last main division in the *Enzyme Nomenclature* is the only one where no member is used for the production of any fine chemicals in an industrial process. Nevertheless, these biocatalysts play a major role in genetic engineering or genetic diagnostics, as specific enzymes in this class that are known as DNA ligases catalyze the formation of C–O bonds in DNA synthesis. This reaction is essential in genetic engineering sciences, allowing the connection of two DNA strings to give a single string.

To generalize, ligases are enzymes that catalyze a bond formation between two molecules. This reaction is always coupled with the hydrolysis of a pyrophosphate bond in ATP or a similar triphosphate. The bonds formed are, e.g., C–O, C–S and C–N bonds.

The systematic names should be formed by the system *X:Y ligase.*

EC 6.1 Forming carbon–oxygen bonds

$\textcircled{P}$ = diphosphate
R = organic residue

The tRNA-hydroxy group is the 2′- or 3′-hydroxy group of the 3′-terminal nucleoside.

EC 6.2 Forming carbon–sulfur bonds

$\textcircled{P}$ = diphosphate
R   = organic residue
NTP = nucleotide triphosphate (ATP, GTP)
NMP = nucleotide monophosphate (AMP, GMP)

The thiol group is the terminal group of the coenzyme A (CoA) molecule.

EC 6.3 Forming carbon–nitrogen bonds

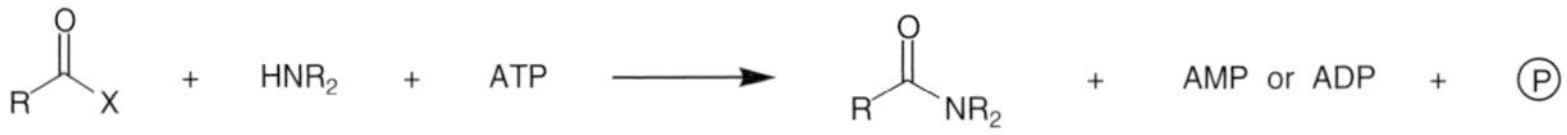

$\textcircled{P}$ = monophosphate, diphosphate
X = OH, H, COOH
R = hydrogen, organic residue

CO_2 is the substrate for the enzyme EC 6.3.3.3. There are exceptions to this reaction pattern, such as ligase EC 6.3.4.1, which catalyzes the following reaction.

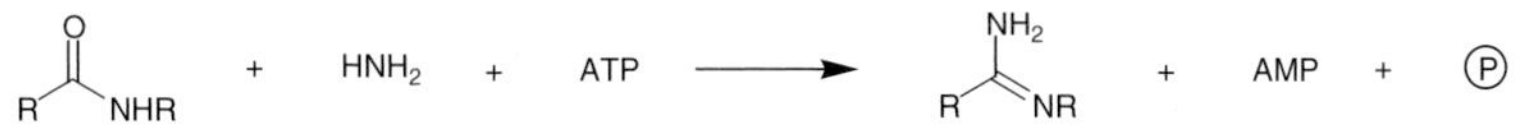

Ⓟ = diphosphate

EC 6.4   Forming carbon–carbon bonds

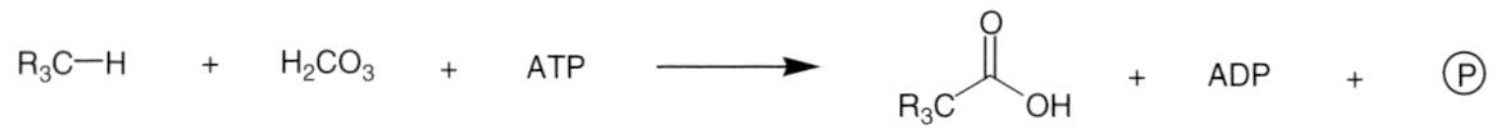

Ⓟ = monophosphate
R   = hydrogen, organic residue

EC 6.5   Forming phosphoric ester bonds

Ⓟ   = monophosphate
PPᵢ = diphosphate

This subclass contains the repair enzymes for DNA. The enzyme EC 6.5.1.2 uses NAD^+. Enzymes acting on RNA form cyclic products.

EC 6.6   Forming nitrogen-metal bonds

M = metal ion
L = chelate ligand
[ML] = chelate complex
Ⓟ= monophosphate

The metal ion is either Mg^{2+} or Co^{2+} and is coordinated by nitrogen atoms of the chelate ligand.

References

1 International Union of Biochemistry and Molecular Biology **1992**, *Enzyme Nomenclature*, Academic Press, San Diego.

2 Supplement 1: Nomenclature Committee of the International Union of Biochemistry and Molecular Biology (NC-IUBMB), **1994**, *Eur. J. Biochem.* 223, 1–5.

3 Supplement 2: Nomenclature Committee of the International Union of Biochemistry and Molecular Biology (NC-IUBMB) **1995**, *Eur. J. Biochem.* 232, 1–6.

4 Supplement 3: Nomenclature Committee of the International Union of Biochemistry and Molecular Biology (NC-IUBMB) **1996**, *Eur. J. Biochem.* 237, 1–5.

5 Supplement 4: Nomenclature Committee of the International Union of Biochemistry and Molecular Biology (NC-IUBMB) **1997**, *Eur. J. Biochem.* 250, 1–6.

6 Supplement 5: Nomenclature Committee of the International Union of Biochemistry and Molecular Biology (NC-IUBMB) **1999**, *Eur. J. Biochem.* 264, 610–650.

7 Available through http://www.chem. qmw.ac.uk/iubmb/enzyme/supplements

8 Appel, R.D., Bairoch, A., Hochstrasser, D.F. **1994**, (A New Generation of Information Tools for Biologists: the Example of the ExPASy WWW Server), *Trends Biochem. Sci.* 19, 258–260.

9 Bairoch, A. **1999**, (The ENZYME Data Bank in 1999), *Nucleic Acids Res.* 27, 310–311 (available through http://www. expasy.ch/enzyme/).

10 Kindel, S. **1981**, *Technology* 1, 62.

11 http://www.expasy.ch/enzyme/enz_new_form.html

12 Ager, D.J. **1999**, *Handbook of Chiral Chemicals*, Marcel Dekker, New York, Basel.

13 Straathof, A.J.J., Panke, S., Schmid, A. **2002**, (The Production of Fine Chemicals by Biotransformations), *Curr. Opin. Biotechnol.* 13, 548–556.

3
Retrosynthetic Biocatalysis
Junhua Tao, Alan Pettman, and Andreas Liese

In this chapter the classical chemical retrosynthetic approach is shown, limited to biocatalytic transformations. References, in most cases, lead to standard textbooks on biocatalysis, where detailed information and further literature citations to the respective biocatalytic step can be found. Enantioselectivity can be expected when chiral substrates or products are involved in biotransformations. Stereochemistry will be depicted if a dynamic resolution process can be applied. Two examples of retrosynthetic analysis to incorporate an enzymatic step in the chemical synthesis of chiral molecules are included.

3.1
Alkanes

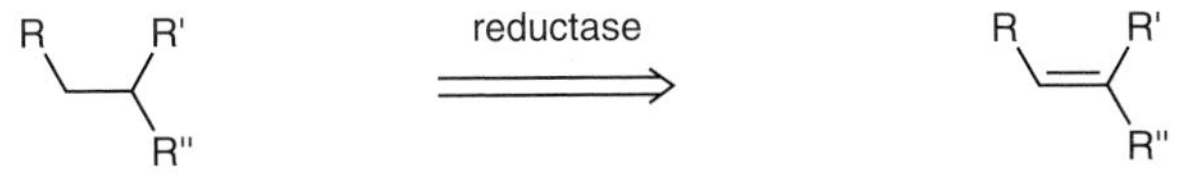

R, R', R" = one of substituents electron withdrawing

Alkanes can be prepared from the corresponding electron deficient alkenes by whole cell catalyzed enzymatic reductions [1]. The transformation requires at least one electron-withdrawing substituent attached to the C=C bond.

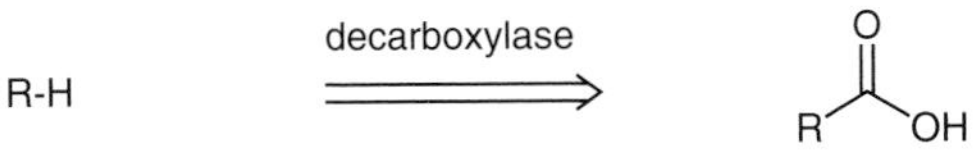

Alternatively, enzymatic decarboxylation can also lead to alkanes [2]. There are only a limited number of enzymes available to date, and the substrate-tolerance has not been fully explored.

Industrial Biotransformations. Andreas Liese, Karsten Seelbach, Christian Wandrey (Eds.)
Copyright © 2006 WILEY-VCH Verlag GmbH & Co. KGaA, Weinheim
ISBN: 3-527-31001-0

3.2
Alkenes

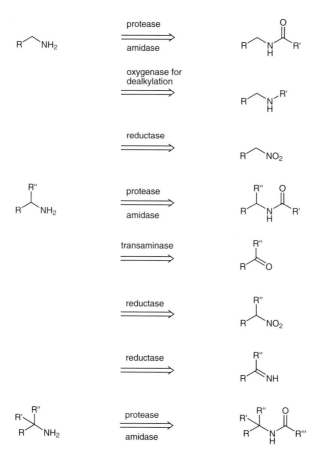

X = OH, NH₂, SH, halides

Alkenes can be formed by enzymatic dehydration or elimination of water or alcohols, amines, thiols or HX using C–X lyases [3].

3.3
Amines

Amines can be prepared from a variety of precursors. The most common pathway is from amides through enzymatic hydrolysis [4]. In this route, hydrolysis of amides possessing an amino moiety on a tertiary carbon atom is very difficult using standard proteases due to steric hindrance.

Transaminases have been demonstrated, at scale, to convert ketones into amines in one pot [5]. Amines can also be prepared by reduction of nitro, imine, azide or diazo compounds using whole cell biotransformations [6].

Chemoenzymatic dynamic kinetic resolution can be used for the preparation of enantiomerically enriched amines. Here either Pd- or Ru-catalysts have been reported for the regeneration of racemic amines or ketoximes [7].

Pd-, Ru- catalysts

R_1, R_2 = organic substituents

3.4
Alcohols

Alcohols can be prepared from a variety of precursors, for example, through hydrolysis of carboxyl esters, sulfate esters, phosphate esters [8]. Hydrolysis of esters composed of *tert*-alcohols can be very difficult by standard carboxyl ester hydrolases.

The reversal reactions are esterification, sulfonylation (sulfotransferases) and phosphorylation (catalyzed by kinases) [9]. Carbonyl-reduction of aldehyde or ketone precursors can also be carried out on a practical scale [10].

As for hydroxylation of alkanes, most reactions are carried out under whole cell conditions, although some recombinant P450s or monooxygenases are available commercially [11].

Several approaches have been developed for the dynamic resolution of enzyme-catalyzed resolution of alcohols. For example, Ru-catalysts have been used to racemize the wrong enantiomeric allylic alcohol [12].

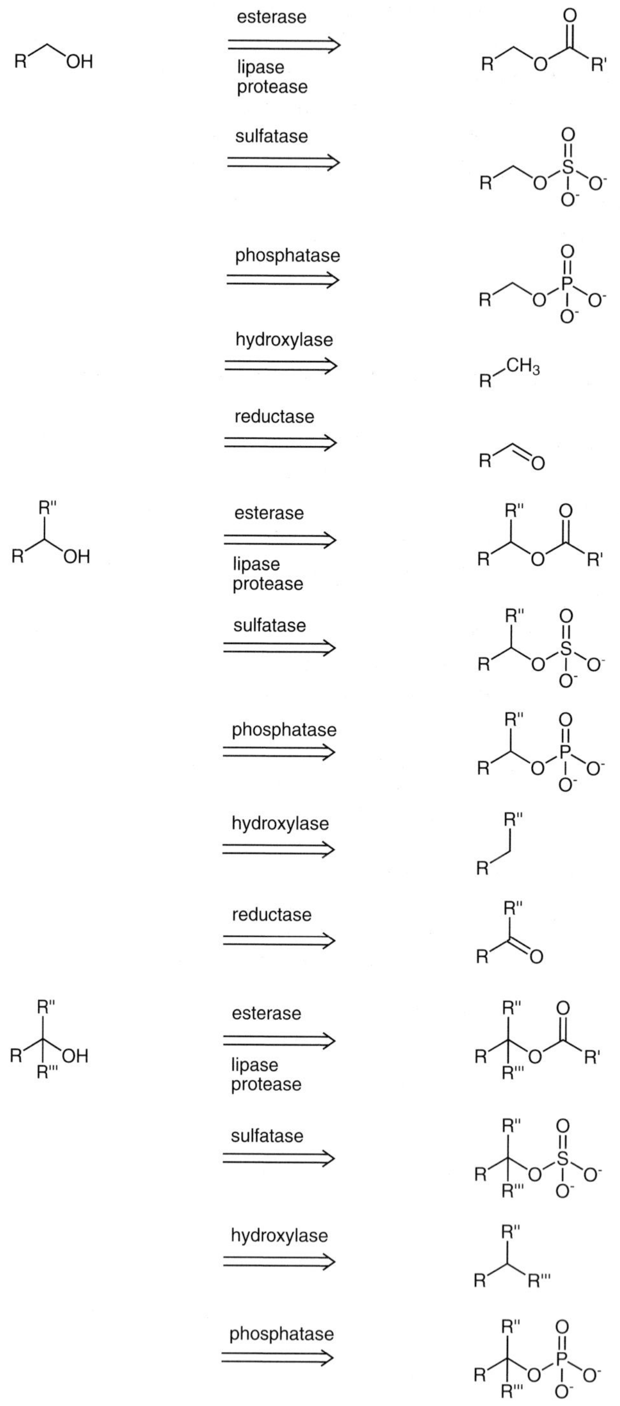
esterase
lipase
protease
sulfatase
phosphatase
hydroxylase
reductase
R
R'
R''
R'''
OH
O
S
P
O-
CH3

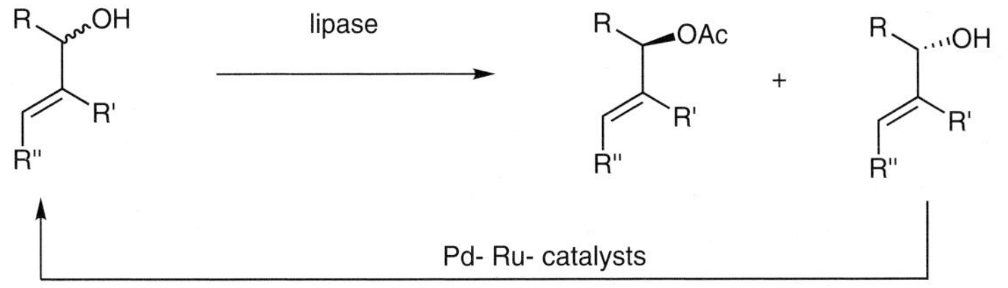

R, R', R" = alkyl, aryl substituents

Similarly, dynamic resolution of non-functionalized *sec*-alcohols can be achieved by coupling a lipase-catalyzed acyl-transfer reaction to *in situ* racemization via ketone intermediates catalyzed by transition metal catalysts. The stereoselectivity is determined by the enzymes and a variety of functionalized alcohols can be used as substrates [13].

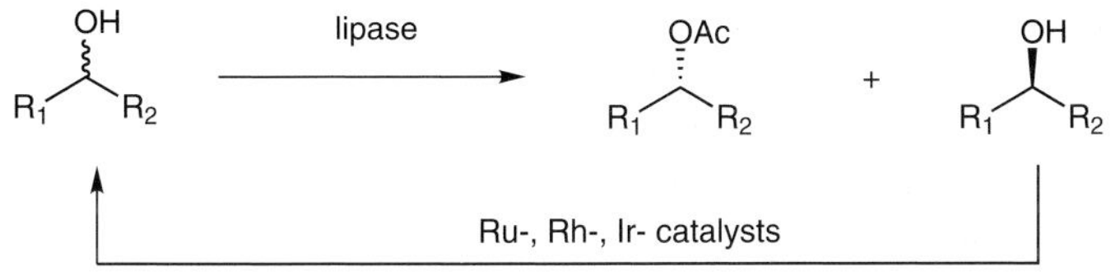

R_1 = alkyll, aryl
R_2= OR (organic substituents), COOR, N_3, CN, X (halogens), allylic, protected aldehydes, phosphonates,

3.5
Aldehydes

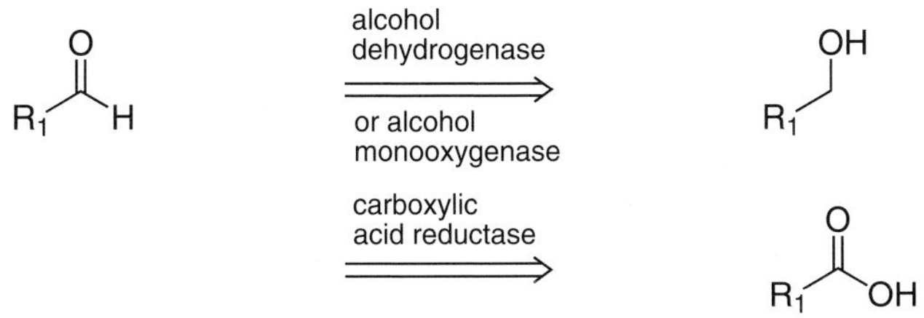

Aldehydes can be prepared from alcohols by oxidation using alcohol dehydrogenases, alcohol oxidases or monooxygenases [14], or by enzymatic reduction from carboxylic acids [15].

3.6
Ketones

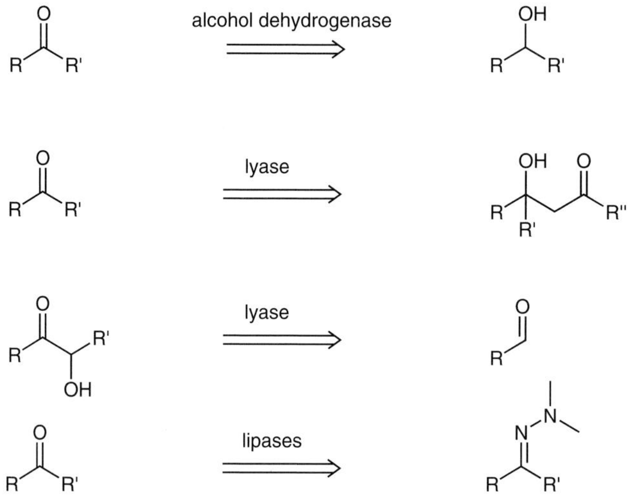

Ketones can be prepared from alcohols under cofactor-dependent oxidation [16]. Studies on *sec*-alcohol oxidases have not been well reported in the literature. Some specific ketones can be formed by cleavage of *a*-hydroxyl ketones (acyloins) [17]. Lipase-catalyzed hydrolysis of dimethylhydrazones serves as a mild method for the deprotection of ketones [18].

3.7
Epoxides

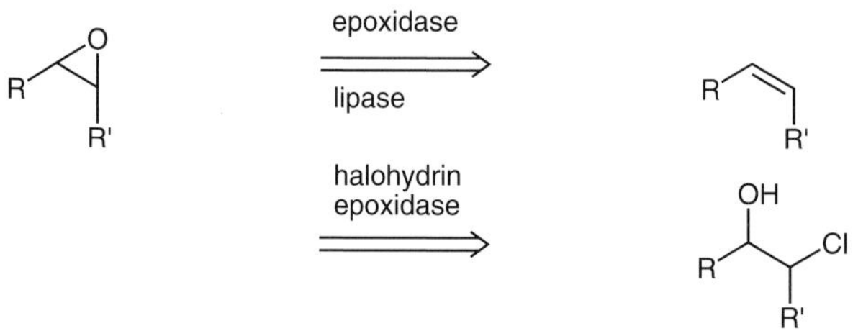

Epoxides can be obtained by enzymatic oxidation of alkenes or dehydrohalogenation of chloro-alcohols using halohydrin epoxidases [19]. Alternatively, epoxidation of alkenes can be catalyzed indirectly by lipases with *in situ* formation of a peroxo acid from H_2O_2.

3.8
Diols

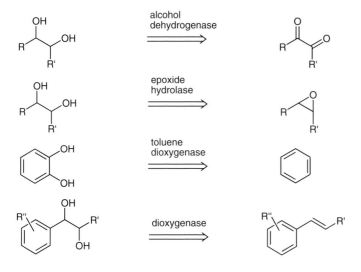

Diols can be prepared from diketones by reduction and epoxides by enzymatic hydrolysis. Mono-, di-, and tri-substituted diols can be obtained in this fashion [20]. Alternatively, they can be prepared by dioxygenation of double bonds.

The dynamic resolution of 1,2-, 1,3-diols has been reported using lipase–Ru-catalysts [21].

R = alkyl, R' = Tr

3.9
Carboxylic Acids

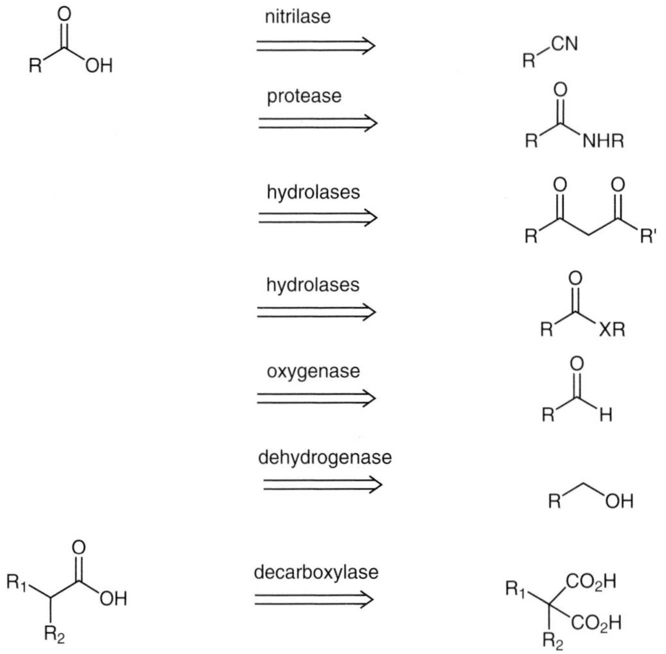

X = O, S
R' = organic substitutent or OH

With the exception of oxygenases, most other approaches described above show a broad substrate spectrum [22]. Some of these enzymes are also available commercially. Hydrolases can also work on C–C bonds, as shown above [23].

While in many cases the wrong enantiomer has to be separated and epimerized for recycling, dynamic resolution can be applied at higher pH when the α-proton is sufficiently acidic, as shown in the example [24].

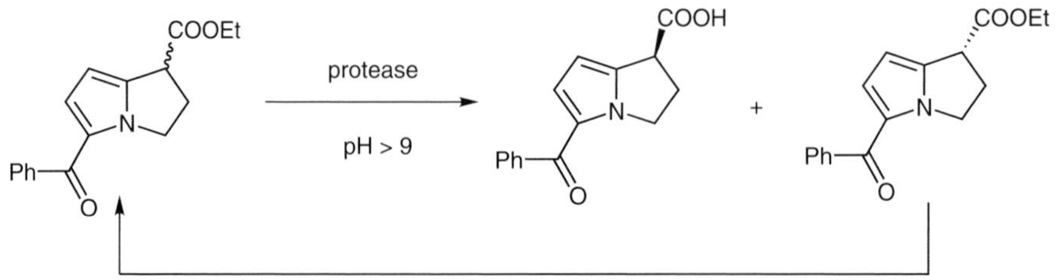

Other approches include the use of thioesters to increase the acidity of the α-proton [25], or aryl acetic esters [26].

A reversible Michael reaction can also be used to design dynamic resolution [27].

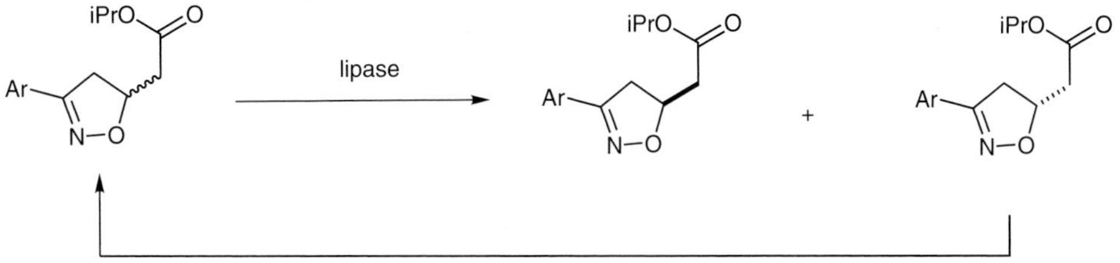

The internal low stability of hemiacetols has been explored for the dynamic resolution of 6-acetyloxy-2H-pyran-3(6H)-one [28].

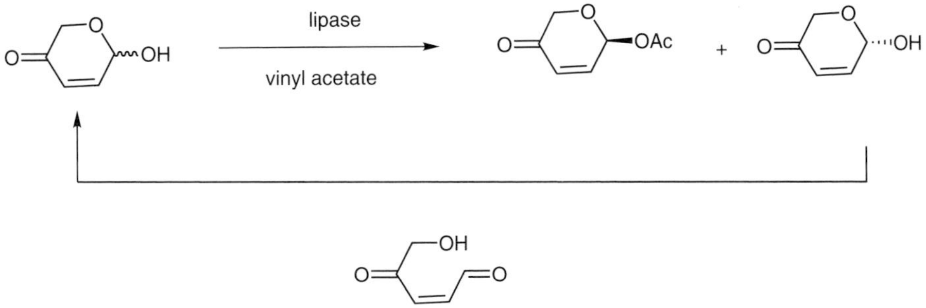

Microbial deracemization has been reported for 2-methyl carboxylic acids via a CoA intermediate [29]. The use of thioesters is known to facilitate racemization of the α-chiral center [30].

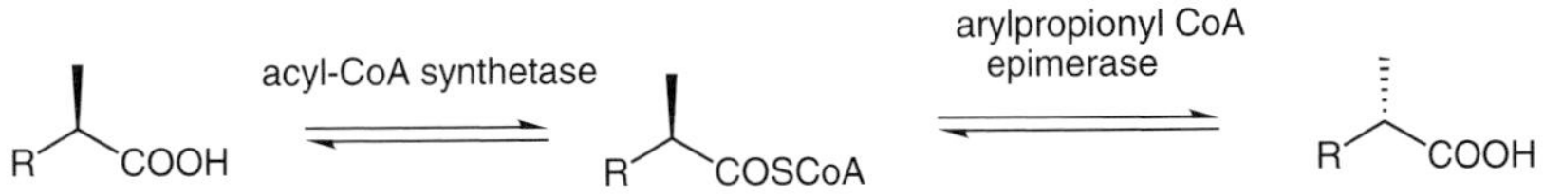

β-Ketohydrolases can convert diketones into keto acids [31].

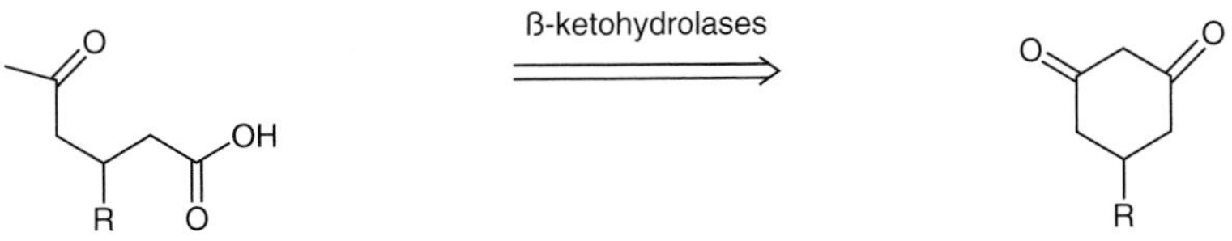

3.10
Esters

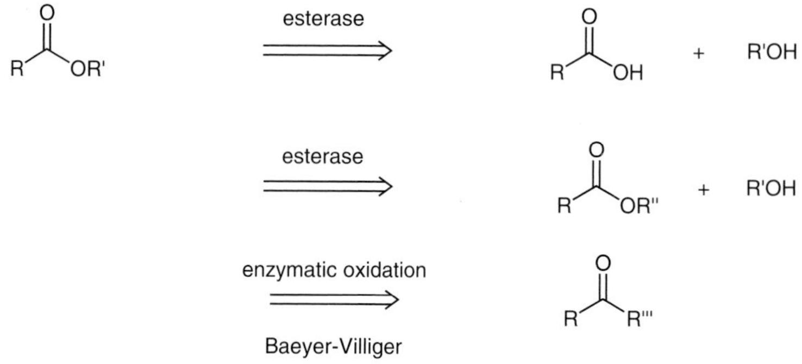

R–C(=O)–OR' →(esterase)→ R–C(=O)–OH + R'OH

→(esterase)→ R–C(=O)–OR" + R'OH

→(enzymatic oxidation)→ R–C(=O)–R'''

Baeyer-Villiger

The most common method to prepare esters is through enzymatic hydrolysis, esterification or transesterification. In addition to esterases, other hydrolases including proteases, lipases and amidases can also catalyze the reaction. Most of these enzymes are available commercially, some of them in large quantities [17].

3.11
Amides

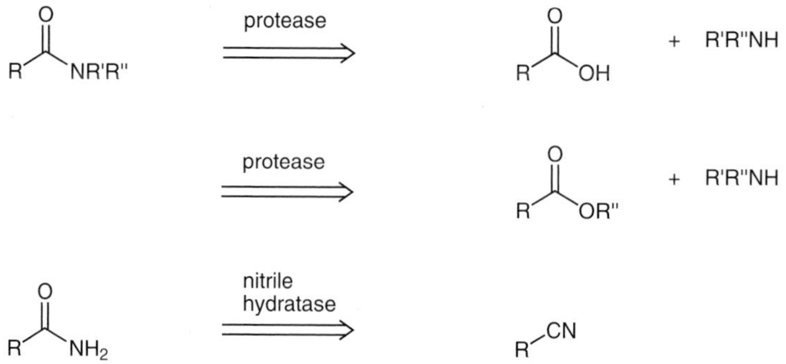

R–C(=O)–NR'R" →(protease)→ R–C(=O)–OH + R'R"NH

→(protease)→ R–C(=O)–OR" + R'R"NH

R–C(=O)–NH₂ →(nitrile hydratase)→ R–CN

The most common method to prepare amides is through aminolysis of esters, amide bond formation ("peptide coupling") between a carboxylic acid and an amine. Alternatively, specific amides can be formed by enzymatic acylation from activated esters (enol esters, such as vinyl acetate, haloethyl esters) or anhydrides. Most of these enzymes are available commercially [20].

3.12
Imines

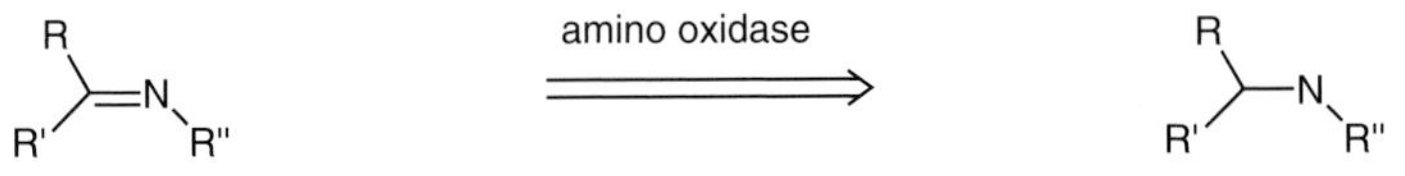

Imines can be prepared from amines by amino oxidases [32].

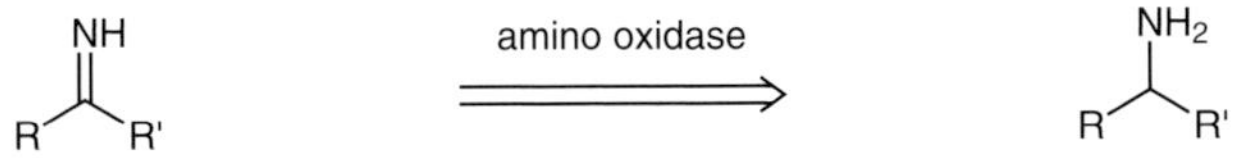

Imines can also be obtained from amines by enzymatic oxidation using amino acid oxidases or amino oxidases. A variety of D-amino acid oxidases are particularly useful for this purpose [33].

3.13
Amino Acids

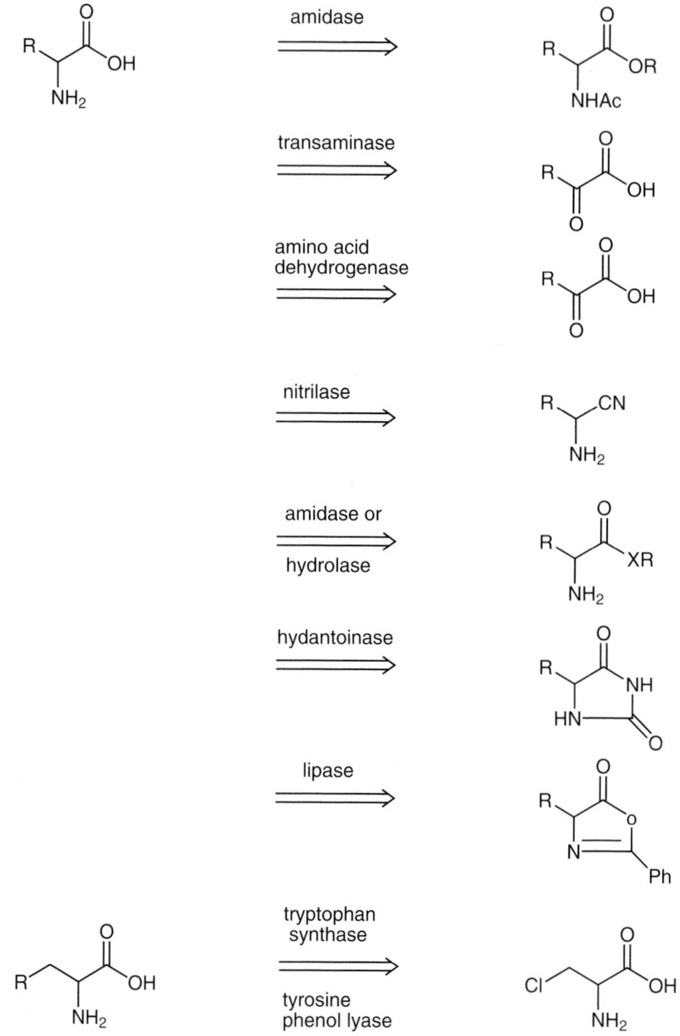

X = O, NH

α-Amino acids can be synthesized from amides or esters by enzymatic hydrolysis, α-keto acids by enzymatic transamination [34] or reductive amination, hydrolysis of hydantoins or amino nitriles by nitrilases [35].

In general, most of these approaches have been demonstrated on a practical scale and many of the enzymes are commercially available. A variety of amino acids can also be prepared enzymatically from β-chloroalanine [29].

Several dynamic resolution approaches have been reported for the synthesis of amino acids. For example, the wrong enantiomeric amino ester can be epimerized by forming a Schiff-base with pyridoxal-5-phosphate or some other aldehydes [36].

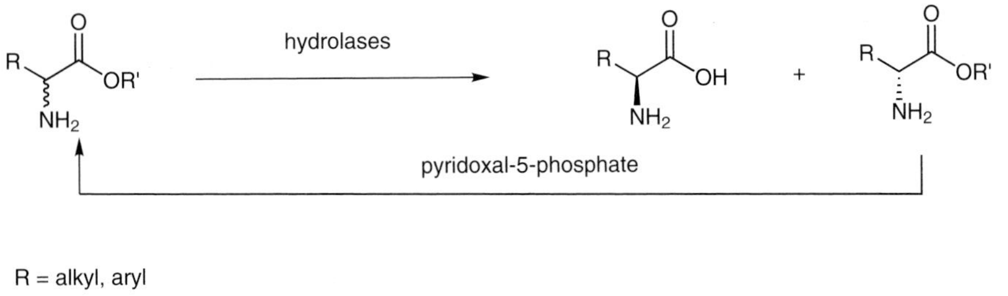

R = alkyl, aryl

Lipase-catalyzed dynamic resolution of oxazolin-5-ones can also be accomplished successfully, as the α-proton is sufficiently acidic for *in situ* racemization. In this particular case, both enantiomers can be produced depending on the lipases used [37].

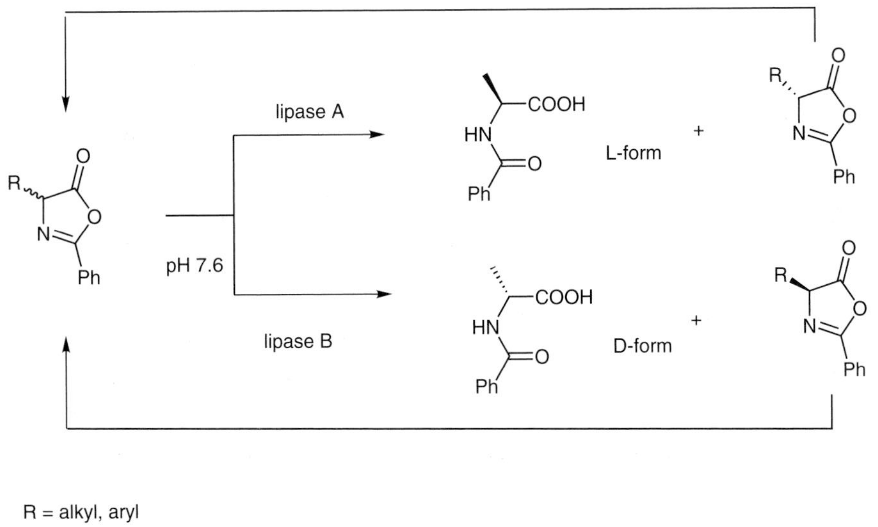

R = alkyl, aryl

Similar approaches have been developed for hydantoins and thiazoline derivatives [38]. The use of a hydantoin racemase for dynamic resolution has also been reported [39].

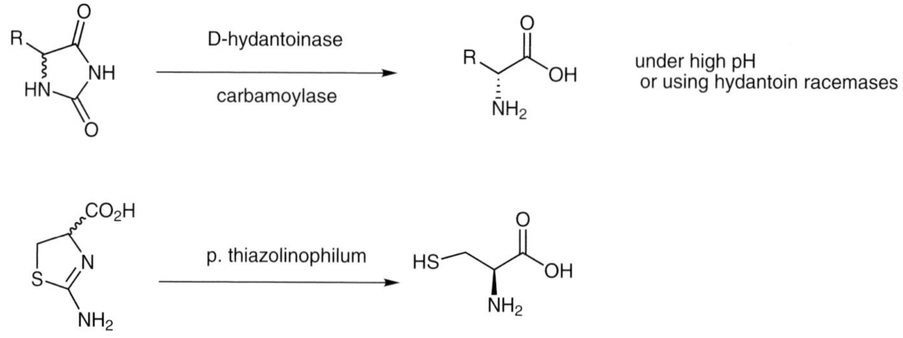

R = alkyl or aryl

More recently, stereoinversion of chiral amines or amino acids has been reported using amino acid or amino oxidases, which will allow the accumulation of the desired enantiomer with yields of up to 100% [40].

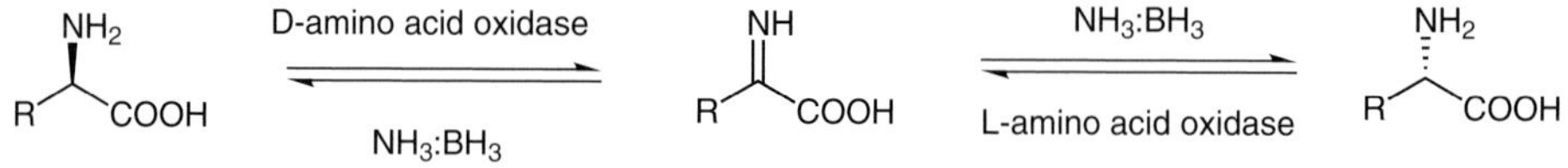

R = alkyl, aryl

3.14
Hydroxy Acids

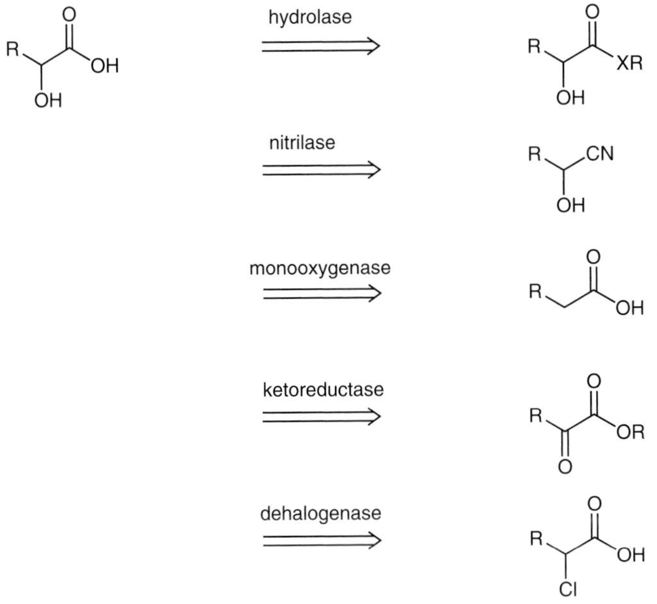

X = O, NH

α-Hydroxy acids can be prepared from amides or esters by enzymatic hydrolysis [17], α-keto acids by enzymatic reduction, cyanohydrins by nitrilases, or α-chloro acids by dehalogenation [41, 42]. Most of these methods have been demonstrated on practical scales and many of the enzymes are commercially available. α-Hydroxy acids can also be transformed from carboxylic acids via oxidation [43], or α-chloral acids by dehalogenation [17]. Both of these two approaches have been demonstrated on a large scale.

Dynamic resolutions of α-hydroxy esters have also been reported [44].

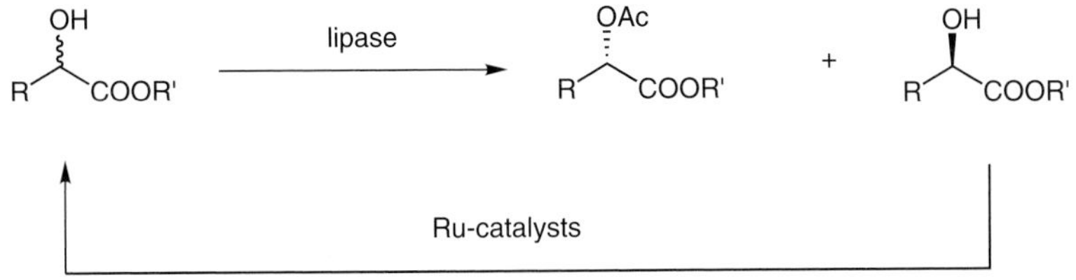

R = alkyl, aryl

When cyanohydrins are used, dynamic resolution can be achieved due to the intrinsic instability of the substrate itself [45].

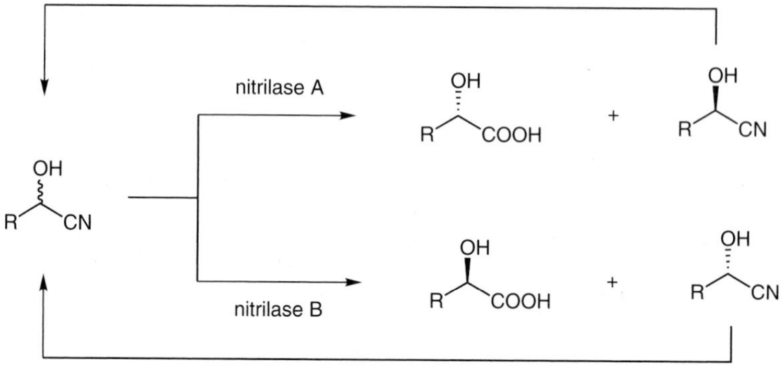

R = alkyl, aryl

Polyfunctionalized α-hydroxy acids such as those listed in the scheme can be prepared by enzymatic Michael type additions to electron deficient alkenes. Both processes have been commercialized [46].

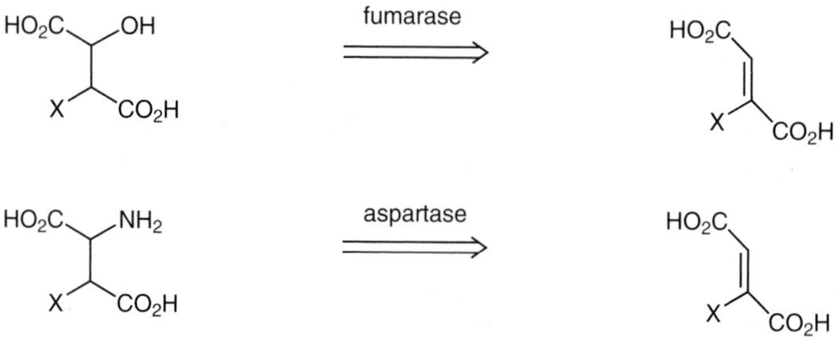

X = H, Cl, alkyl

3.15
a-Hydroxy Ketones

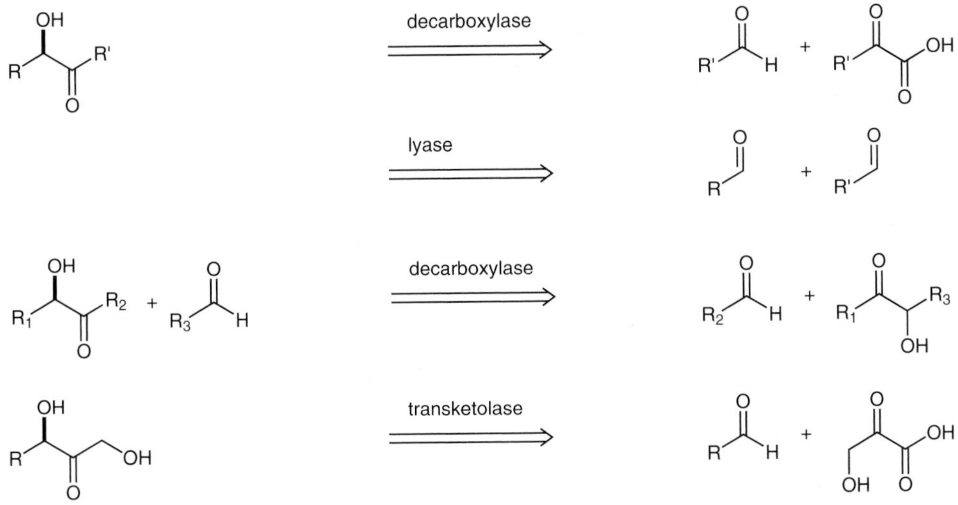

acyloin reaction

The classic method to prepare some of these compounds is to use decarboxylative acyloin reactions (a type of aldolase reaction) [47].

The dynamic resolution of 2-hydroxy ketones has been reported using whole cell systems where two different alcohol dehydrogenases with complementary enantioselectivity are involved [48]. For this particular example, the enantioselectivity is pH dependent.

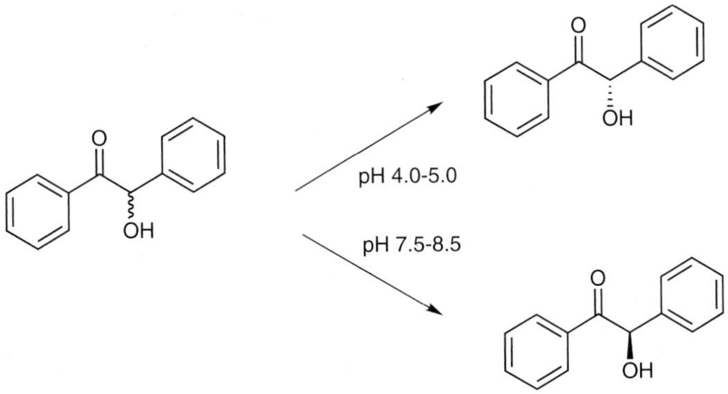

3.16

β-Hydroxy Aldehydes, Ketones or Carboxylic Acids

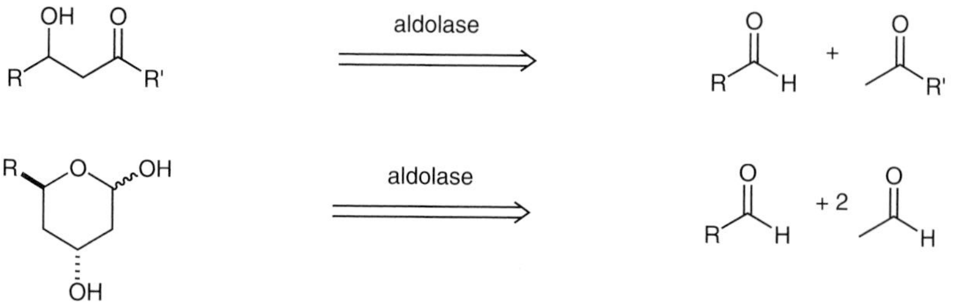

R' = H, Me, F

These compounds can be produced via enzymatic aldol reactions. Some of the enzymes are available commercially. α-Hydroxy ketones can also be obtained by transferring aldehyde or ketone residues using thiamine-diphosphate dependent enzymes.

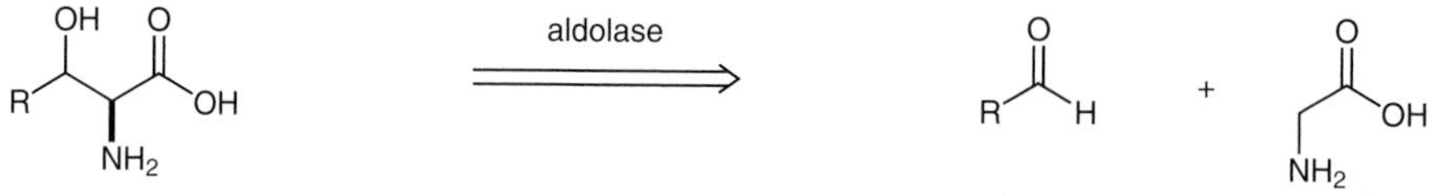

Glycine dependent aldolases are known to catalyze the production of α-amino β-hydroxyl acids from glycine and aldehydes [49].

β-Hydroxy esters can be prepared from keto precursors. With this approach, the reduction can be both enantioselective and diastereoselective as a result of rapid *in situ* racemization of the α-chiral center. The *syn-* and *anti-*selectivity can be as high as 100:1. The alcohol moiety can be *S-* or *R-*selective depending on the reductases [50].

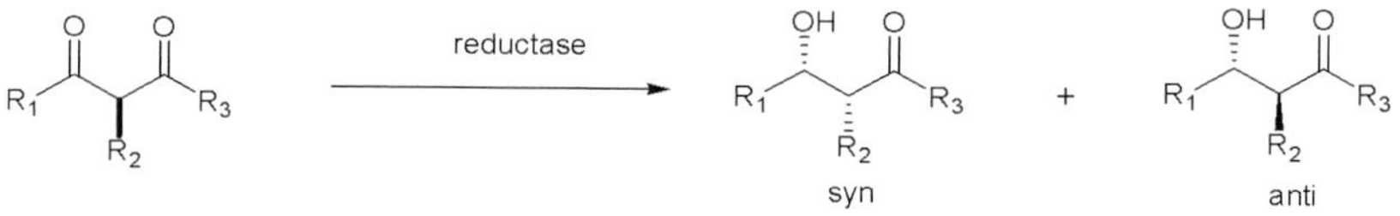

R_1, R_2, R_3 = alkyl, aryl

Ru-catalyzed dynamic resolution has also been extended to the resolution of hydroxy acids, diols and hydroxy aldehydes [51].

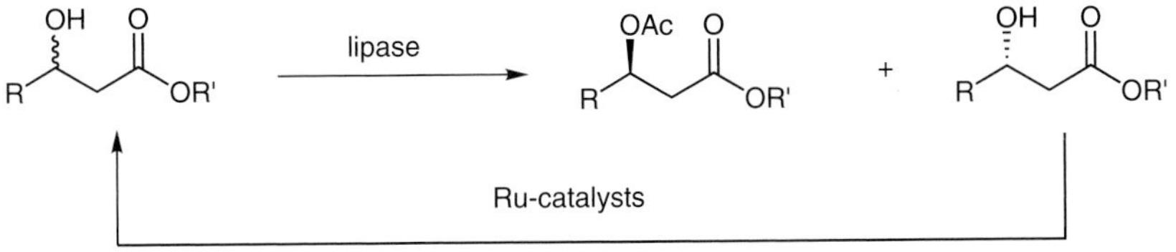

R = alkyl, aryl

3.17
Cyanohydrins, Hemithioacetals and Hemiaminals

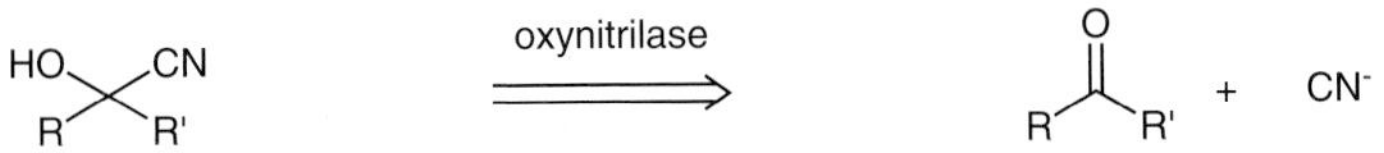

Many (R)- or (S)-selective hydroxynitrile lyases (formerly also denoted as "oxynitri-lases") are available for the synthesis of cyanohydrins, which can then be converted into a variety of synthetically useful chiral intermediates [52].

Racemic cyanohydrins can be converted into one enantiomer with high optical purity in a dynamic resolution fashion due to the intrinsic instability of the respective molecule. The same principle can also be applied to hemithioacetals [53] and N-acyl hemiaminals [54].

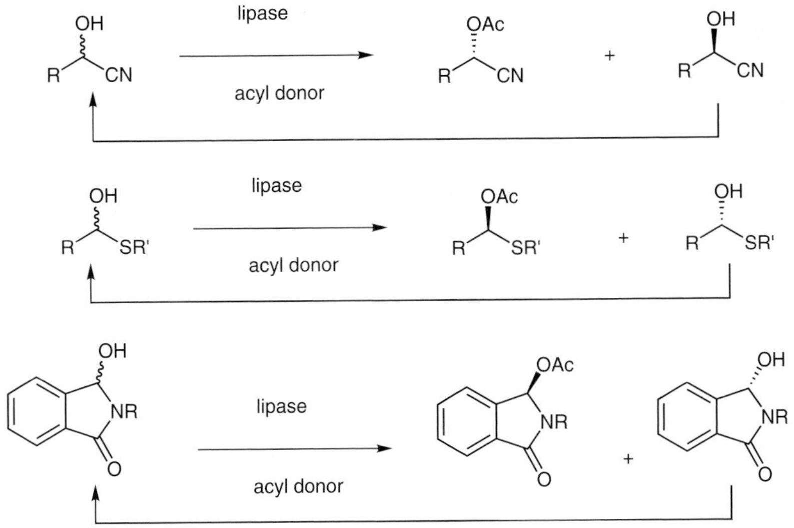

R, R' = organic substituents

3.18
Sulfoxides and Sulfones

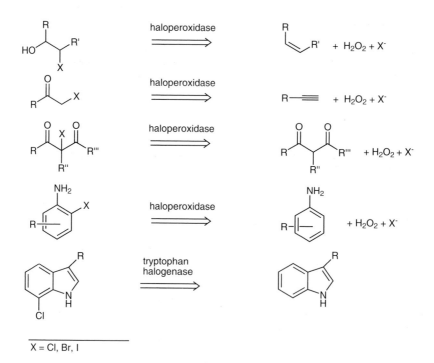

Sulfoxides can be prepared from organic sulfur compounds by oxidation using monooxygenases or chloroperoxidases; some of these are commercially available. Sulfones can be produced from sulfoxides on slower further oxidation. When whole cells are used, a mixture of two could be produced. The reversal of these types of reactions can also take place under reductive conditions [56].

3.19
Halides

Halogenations can be performed with electron-rich aromatics (amino or phenol derivatives) using haloperoxidases or halogenases [56]. Alternatively, activated (acidic) C–H
bonds can also undergo halogenation.

3.20
Aromatics (Ring Functionalization)

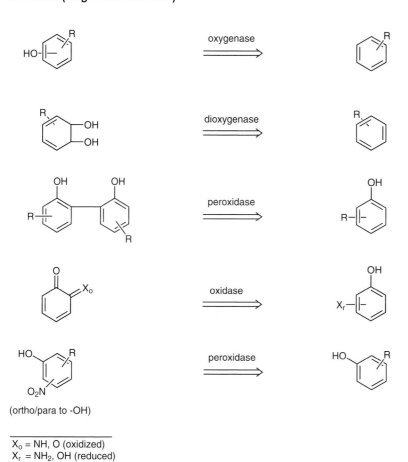

X_o = NH, O (oxidized)
X_r = NH$_2$, OH (reduced)

Aromatics can be monooxygenated or dioxygenated using whole cell biotransformations. Practical processes have been demonstrated for these conversions. This is the common pathway for microorganisms to detoxify aromatics. Biaryl compounds can be prepared from monomers by oxidative cross-coupling. Halogenations can be performed on
electron-rich aromatics (amino or phenol derivatives) using haloperoxidases [57].

3.21
Cyclic Compounds

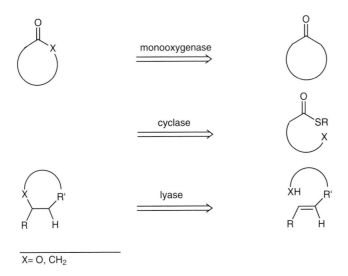

X= O, CH₂

Lactones can be formed from ketones under enzymatic Baeyer–Villiger reaction conditions [58], or enzymatic macrocyclization, commonly seen in the biosynthesis of natural products, where the hetero atom X can be oxygen, nitrogen [59].

Intramolecular lyases can give similar products [60].

3.22
Carbohydrates

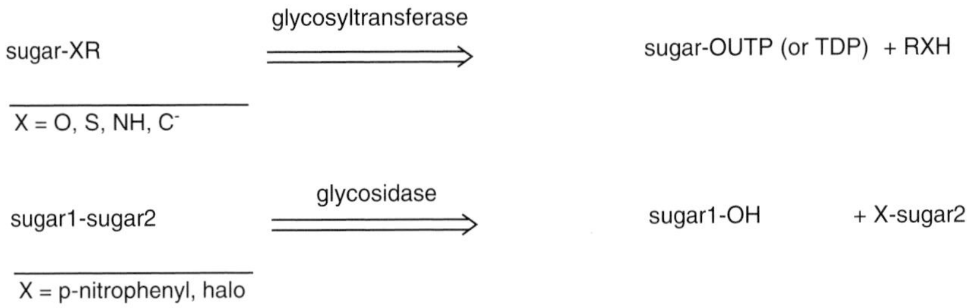

                    glycosyltransferase
sugar-XR   ═══════════════⟹   sugar-OUTP (or TDP) + RXH

X = O, S, NH, C⁻

                    glycosidase
sugar1-sugar2   ═══════════════⟹   sugar1-OH   + X-sugar2

X = p-nitrophenyl, halo

There are many glycosyltransferases available for the synthesis of carbohydrates. The nucleophiles can be alcohol (the most common type), thiol, amine or even carbanions. The reversal hydrolytic reactions are catalyzed by glycosidases [61]. When activated sugars are used, disaccharides can be prepared on a large scale by glycosidases.

3.23
Peroxides

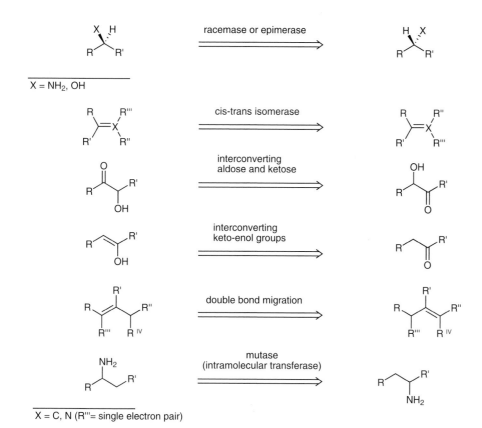

Enzymes are capable of catalyzing peroxidation using O_2 or H_2O_2, as shown above, usually in the allylic position [62].

3.24
Isomers

There are many isomerases in nature, which catalyze geometric, structural or stereochemical changes. As yet, most have not been fully explored for chemical transformations [63].

3.25
Examples of Retrosynthetic Biotransformations

3.25.1
Example 1

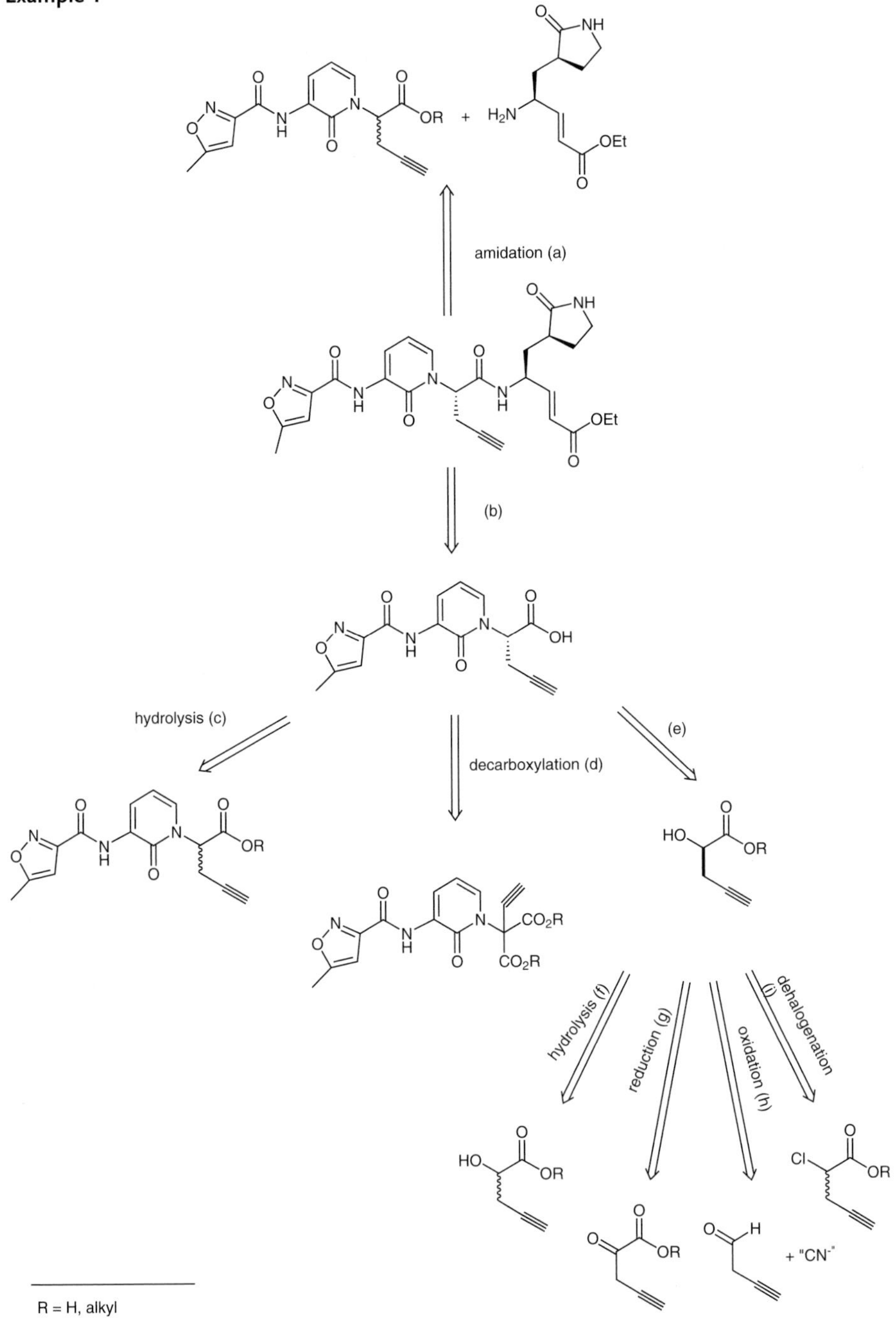

amidation (a)

(b)

hydrolysis (c)

decarboxylation (d)

(e)

hydrolysis (f)

reduction (g)

oxidation (h)

dehalogenation (i)

+ "CN⁻"

R = H, alkyl

Chemoenzymatic strategies can be used to retrosynthetically dissect the target molecule. Firstly, enzymatic amidation can bring an ester and amine together, thus a racemic ester can be used to obtain high optical purity under kinetic resolution [method (a)]. Alternatively, the enantiomerically pure carboxylic acid could be the next target [method (b)], which can be obtained under enzymatic hydrolysis of an ester [method (c)], enzymatic decarboxylation of a manolate derivative [method (d)] or chemically synthesized from an optically pure a-hydroxy ester [method (e)]. The 2-hydroxy ester can be transformed from a variety of simpler starting materials, including its corresponding racemic esters [method (f)], 2-ketoesters or acids [method (g)], aldehyde [method (h)] and 2-chloroesters [method (i)], to name just a few.

3.25.2
Example 2

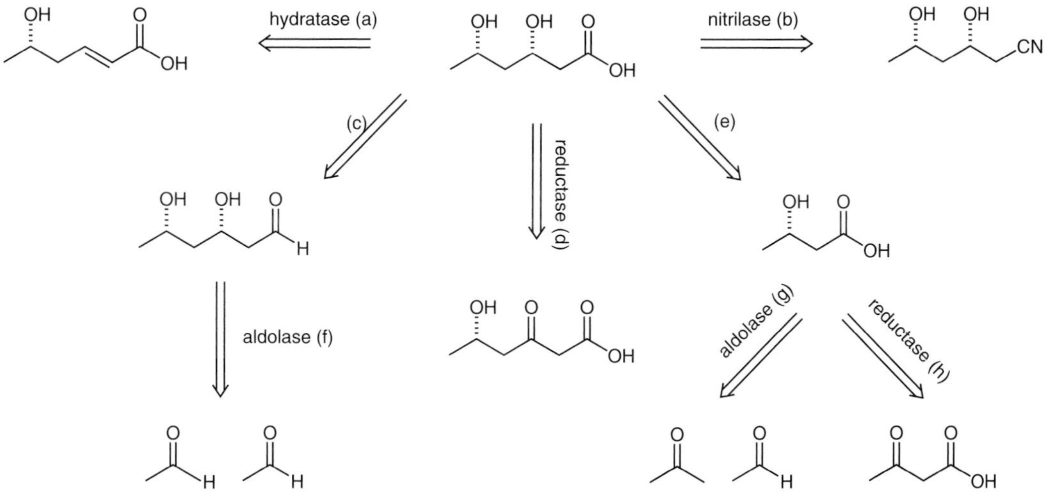

In this retrosynthetic analysis, method (a) involves enzymatic hydration and method (b) uses nitrilase catalyzed hydrolysis of a nitrile. The target molecule can also be prepared from its aldehyde precursor using oxidases [this aldehyde can be synthesized from acetone and acetaldehyde through an enzymatic aldol reaction under method (f)] [method (c)], 3-keto acid by reductases [method (d)] or chemically from a simpler precursor, 3-hydroxy acid [method (e)], which can be converted from readily available starting materials using aldolases [method (g)] or ketoreductase [method (h)].

References

1 Faber, K. 2004, *Biotransformations in Organic Chemistry*, Springer-Verlag, Berlin, (pp.) 205–212.

2 Ohta, H., Sugai, T. 2000, (Enzyme-mediated Decarboxylation Reactions in Organic Synthesis), *Stereoselective Biocatalysis*, ed. Patel, P.N., Marcel Dekker, New York, (pp.) 467–526.

3 Ohta, H., Sugai, T. 1998, (Enzymes in Organic Syntheses: A Novel Elimination Reaction in Andrographolide Triacetate), *Nat. Prod. Lett.* 12 (1), 1–4.

4 a) Faber, K. 2004, *Biotransfromations in Organic Chemistry*, Springer-Verlag, Berlin, pp. 52–63; b) Bornscheuer, U.T., Kazlauskas, R.J. 1999, (Survey of Enantio-selective Protease- and Amidase-Catalyzed Reactions), in *Hydrolases in Organic Synthesis*, Wiley-VCH, Weinheim, (pp.) 179–193.

5 Rozzell, J.D., Bommarius, A.S. 2002, (Trans-amination), in *Enzyme Catalysis on Organic Synthesis*, eds. Drauz, K., Waldmann, H., Wiley-VCH, Weinheim, (pp.) 875–893.

6 Nakamura, K., Matsuda, T. 2002, (Reduction Reactions), in *Enzyme Catalysis on Organic Synthesis*, eds. Drauz, K., Waldmann, H., Wiley-VCH, Weinheim, (pp.) 1035–1036.

7 a) Choi, Y.K., Kim, M.J., Ahn, Y., Kim, M.J. 2001, (Lipase/Palladium-Catalyzed Asymmetric Transformations of Ketoximes to Optically Active Amines), *Org. Lett.* 3, 4099; b) Pàmies, O., Éll, A.H., Samec, J.S.M., Hermanns, N., Bäckvall, J.E. 2002, (An Efficient and Mild Ruthenium-catalyzed Racemization of Amines: Application to the Synthesis of Enantiomerically Pure Amines), *Tetrahedron Lett.* 43, 4699–4702.

8 Faber, K. 2004, *Biotransformations in Organic Chemistry*, Springer-Verlag, Berlin, (pp.) 63–134.

9 Bornscheuer, U.T., Kazlauskas, R.J. 1999, (Choosing Reaction Media: Water and Organic Solvents), in *Hydrolases in Organic Synthesis*, Wiley-VCH, Weinheim, (pp.) 39–63.

10 a) Nakamura, K., Matsuda, T. 2002, (Reduction Reactions), in *Enzyme Catalysis on Organic Synthesis*, (vol. 3), eds. Drauz, K., Waldmann, H., Wiley-VCH, Weinheim, (pp.) 991–1034; b) Klibanov, A.M. 2003, (Asymmetric Enzymic Oxidoreductions in Organic Solvents), *Curr. Opin. Biotechnol.* 14 (4), 427–431.

11 Flitsch, S., Grogan, G., Ashcroft, D. 2002, (Oxidation Reactions), in *Enzyme Catalysis on Organic Synthesis*, eds. Drauz, K., Waldmann, H., Wiley-VCH, Weinheim, (pp.) 1065–1099.

12 a) Allen, J.V., Williams, J.M.J. 1996, (Dynamic Kinetic Resolution with Enzyme and Palladium Combinations), *Tetrahedron Lett.* 37 (11), 1859–1862; b) Lee, D., Huh, E.A., Kim, M.-J., Jung, H.M., Koh, J.H. Park, J. 2000, (Dynamic Kinetic Resolution of Allylic Acohols Mediated by Ruthenium- and Lipase-based Catalysts), *Org. Lett.* 2, 2377–2379.

13 Pàmies, O., Bäckvall, J.-E. 2003, *Chem. Rev.* 103, 3247.

14 Kroutil, W., Mang, H., Edegger, K., Faber, K. 2004, (Biocatalytic Oxidation of Primary and Secondary Alcohols), *Org. Bioorg. Chem.* (346), 125–142

15 He, A., Li, T., Daniels, L., Fotheringham, I., Rosazza, J.P.N. 2004, (Nocardia sp. Carboxylic Acid Reductase: Cloning, Expression, and Characterization of a New Aldehyde Oxidoreductase Family), *Appl. Environ. Microbiol.* (70), 1874–1881.

16 Schmid, A., Hollmann, F., Bühler, B. 2002, (Oxidation of Aldehydes), in *Enzyme Catalysis on Organic Synthesis*, eds. Drauz, K., Waldmann, H., Wiley-VCH, Weinheim, (pp.) 1194–1201.

17 Hoh, C., Villela, M.F. 2000, (Enzyme Classification), in *Industrial Biotransformations*, eds. Liese, A., Seelbach, K., Wandrey, C., Wiley-VCH, Weinheim, (pp.) 48–49.

18 Mino, T., Matsuda, T., Hiramatsu, D., Yamashita, M. 2000, (Deprotection of Ketone Dimethylhydrazones Using Lipases), *Tetrahedron Lett.* 41, 1461–1463.

19 Flitsch, S., Grogan, G., Ashcroft, D. 2002, (Oxidation Reactions), in *Enzyme Catalysis on Organic Synthesis*, eds. Drauz, K., Waldmann, H., Wiley-VCH, Weinheim, (pp.) 1084–1099.

20 Faber, K., Orru, R.V.A. 2002, (Hydrolysis of Epoxides), in *Enzyme Catalysis on Organic Synthesis*, (vol. 2), eds. Drauz, K., Waldmann, H., Wiley-VCH, Weinheim, (pp.) 579–604.

21 Kim, M.-J., Choi, Y.K., Choi, M.Y., Kim, M.J., Park, J. 2001, (Lipase/Ruthenium-Catalyzed Dynamic Kinetic Resolution of Hydroxy Acids, Diols, and Hydroxy Aldehydes Protected with a Bulky Group), *J. Org. Chem.* 66, 4736–4738.

22 Faber, K. 2004, *Biotransformations in Organic Chemistry*, (vol. 5), Springer-Verlag, Berlin, (pp.) 63–122.

23 Hoh, C., Villela, M.F. 2000, (Enzyme Classification), in *Industrial Biotransformations*, eds. Liese, A., Seelbach, K., Wandrey, C., Wiley-VCH, Weinheim, (p.) 46.

24 Fülling, G., Sih, C.J. 1987, (Enzymatic Second-Order Asymmetric Hydrolysis of Ketorolac Esters: *In Situ* Racemization), *J. Am. Chem. Soc.* 109, 2845–2845.

25 a) Chang, C.-S., Tsai, S.-W., Kuo, J. 1999, *Biotechnol. Bioeng.* 64 (1), 121; b) Drueckhammer, D.G., Um, P.-J. 1998, (Dynamic Enzymatic Resolution of Thioesters), *J. Am. Chem. Soc.* 120, 5605–5610.

26 Williams, J.M.J., Dinh, P.M., Harris, W. 1999, (Selective Racemisation of Esters: Relevance to Enzymatic Hydrolysis Reactions), *Tetrahedron Lett.* 40, 749–752.

27 Pesti, J.A., Yin, J., Zhang, L.-h., Anzalone, L. 2001, (Reversible Michael Reaction-Enzymatic Hydrolysis: A New Variant of Dynamic Resolution), *J. Am. Chem. Soc.* 123, 11075–11076.

28 van den Heuvel, M., Cuiper, A.D., van der Deen, H., Kellogg, R.M., Feringa, B.L. 1997, Optically Active 6-Acetyloxy-2H-pyran-3(6H)-one Obtained by Lipase Catalyzed Transesterification and Esterification), *Tetrahedron Lett.* 38, 1655–1658.

29 Kato, D., Mitsuda, S., Ohta, H. 2002, (Microbial Deracemization of {alpha}-Substituted Carboxylic Acids), *Org. Lett.* 4, 371–373.

30 Um, P.-J., Drueckhammer, D.G. 1998, (Dynamic Enzymatic Resolution of Thioesters), *J. Am. Chem. Soc.* 120, 5605–5610.

31 Chai, W., Sakamaki, H., Kitanaka, S., Saito, M., Horiuchi, C.A. 2003, (Biotransformation of Cycloalkanediones by Caragana chamlagu), *Bull. Chem. Soc. Jpn.* 76 (1), 177–182.

32 Flitsch, S., Grogan, G., Ashcroft, D. 2002, (Oxidation Reactions), in *Enzyme Catalysis on Organic Synthesis*, (vol. 3), eds. Drauz, K., Waldmann, H., Wiley-VCH, Weinheim, (pp.) 1256–1260.

33 Turner, N.J. 2004, (Enzyme Catalysed Deracemisation and Dynamic Kinetic Resolution Reactions), *Curr. Opin. Chem. Biol.* 8, 114–119.

34 Schulze, B. 2002, (Hydrolysis and Formation of C–N Bonds), in *Enzyme Catalysis on Organic Synthesis*, (vol. 2), eds. Drauz, K., Waldmann, H., Wiley-VCH, Weinheim, (pp.) 878–892.

35 a) Yamada, H., Shimizu, S., Yoneda, K. 1980, (Synthesis of D-Amino Acids Using Hydantoinase of Microorganism, Production of p-D-Hydroxyphenylglycine), *Hakko to Kogyo* 38 (10), 937–46; b) Pietzsch, M. Syldatk, C., 2002, (Hydrolysis and Formation of Hydantoins), in *Enzyme Catalysis on Organic Synthesis*, (vol. 2), eds. Drauz, K., Waldmann, H., Wiley-VCH, Weinheim, (pp.) 761–796; c) Martinkova, L., Kren, V., 2002, (Nitrile- and Amide-converting Microbial Enzymes: Stereo-, Regio- and Chemoselectivity), *Biocatal. Biotransform.* 20 (2), 73–93.

36 Chen, S.T., Huang, W.H., Wang, K.T.J. 1994, (Resolution of Amino Acids in a Mixture of 2-Methyl-2-propanol/Water (19:1) Catalyzed by Alcalase via *In Situ* Racemization of One Antipode Mediated by Pyridoxal 5-Phosphate), *Org. Chem.* 59, 7580–7581.

37 Crich, J.Z., Brieva, R., Marquart, P., Gu, R.L., Flemming, S., Sih, C.J. 1993, (Enzymic Asymmetric Synthesis of Alpha-amino Acids. Enantioselective Cleavage of 4-Substituted Oxazolin-5-ones and Thiazolin-5-ones), *J. Org. Chem.* 58, 3252–3258.

38 a) Drauz, K., Kottenhahn, M., Makryaleas, K., Klenk, H., Bernd, M. 1991, (Chemoenzymatic Syntheses of {omega}-Ureido D-Amino Acids), *Angew. Chem., Int. Ed. Engl.* 30, 712–714; b) Sano, K., Mitsugi, K. 1978, (Enzymatic Production of l-Cysteine from d,l-2-Amino-Δ^2-thiazoline-4-carboxylic Acid by Pseudomonas thiazolinophilum: Optimal Conditions for the Enzyme Formation and Enzymatic Reaction), *Agric. Biol. Chem.* 42, 2315–2321.

39 May, O., Verseck, S., Bommarius, S., Drauz, K. 2002, (Development of Dynamic Kinetic Resolution Processes for Biocatalytic Production of Natural and Nonnatural l-Amino Acids), *Org. Proc. Res. Dev.* 6, 452–457.

40 Carr, R., Alexeeva, M., Enright, A., Eve, T.S.C., Dawson, M.J., Turner, N.J. 2003,

(Directed Evolution of an Amine Oxidase Possessing both Broad Substrate Specificity and High Enantioselectivity), *Angew. Chem., Int. Ed. Engl.* 42, 4807–4810.

41 Roberts, St.M. 2001, (Preparative Biotransformations), *J. Chem. Soc., Perkin Trans.* 1 (13), 1475–1499.

42 Faber, K. 2004, *Biotransformations in Organic Chemistry*, Springer-Verlag, Berlin, (pp.) 330–331.

43 Faber, K. 2004, *Biotransformations in Organic Chemistry*, Springer-Verlag, Berlin, (pp.) 251–252.

44 Huerta, F.F., Laxmi, Y.R.S., Bäckvall, J.-E. 2000, (Dynamic Kinetic Resolution of {alpha}-Hydroxy Acid Esters), *Org. Lett.* 2 (4), 1037–1040.

45 Yamamoto, K., Oishi, K., Fujimatsu, I., Komatsu, I. 1991, Production of *R*-(–)-mandelic Acid from Mandelonitrile by *Alcaligenes faecalis* ATCC 8750), *Appl. Environ. Microbiol.* 57, 3028–3032.

46 Faber, K. 2004, *Biotransformations in Organic Chemistry*, Springer-Verlag, Berlin, (pp.) 302–305.

47 a) Ward, O.P., Singh, A. 2000, (Enzymatic Asymmetric Synthesis by Decarboxylases), *Curr. Opin. Biotechnol.* 11 (6), 520–526; b) Seoane, G. 2000, (Enzymatic C–C Bond-forming Reactions in Organic Synthesis), *Curr. Org. Chem.* 4 (3), 283–304.

48 Demir, A.S., Hamamci, H., Sesenoglu, O., Neslihanoglu, R., Asikoglu, B., Capanoglu, D. 2002, (Fungal Deracemization of Benzoin), *Tetrahedron Lett.* 43, 6447–6449.

49 Wong, C.-H. 2002, (Formation of C–C Bonds), in *Enzyme Catalysis on Organic Synthesis*, (vol. 2), eds. Drauz, K., Waldmann, H., Wiley-VCH, Weinheim, (pp.) 931–966.

50 a) Nakamura, K., Miyai, T., Nozaki, K., Ushio, K., Ohno, A. 1986, (Diastereo- and Enantio-selective Reduction of 2-Methyl-3-oxobutanoate by Bakers' Yeast), *Tetrahedron Lett.* 27, 3155–3156; for more examples see Ward, R. 1995, (Dynamic Kinetic Resolution), *Tetrahedron Asymmetry* 6, 1475–1490.

51 a) Kim, M.-J., Choi, Y.K., Choi, M.Y., Kim, M.J., Park, J., 2001, (Lipase/Ruthenium-Catalyzed Dynamic Kinetic Resolution of Hydroxy Acids, Diols, and Hydroxy Aldehydes Protected with a Bulky Group UND: Enantioselective Synthesis of {beta}-Hydroxy Acid Derivatives via a One-Pot

Aldol Reaction-Dynamic Kinetic Resolution), *J. Org. Chem.* 66, 4736; b) Huerta, F.F., Bäckvall, J.-E. 2001, (Enantioselective Synthesis of ß-Hydroxy Acid Derivatives via a One-Pot Aldol Reaction-Dynamic Kinetic Resolution), *Org. Lett.* 3, 1209–1212.

52 Osprian, I., Fechter, M.H., Griengl, H. 2003, (Biocatalytic Hydrolysis of Cyanohydrins: An Efficient Approach to Enantiopure β-Hydroxy Carboxylic Acids), *J. Mol. Catal. B: Enzymatic* 24–25, 89–98.

53 a) Inagaki, M., Hiratake, J., Nishioka, T., Oda, J. 1992, (One-pot Synthesis of Optically Active Cyanohydrin Acetates from Aldehydes via Lipase-catalyzed Kinetic Resolution Coupled with *In Situ* Formation and Racemization of Cyanohydrins), *J. Org. Chem.* 57, 5643–5649; b) Brand, S., Jones, M.F., Rayner, C.M. 1995, (The First Examples of Dynamic Kinetic Resolution by Enantioselective Acetylation of Hemithioacetals: An Efficient Synthesis of Homochiral-Acetoxysulfides), *Tetrahedron Lett.* 36, 8493–8496.

54 Sharfuddin, M., Narumi, A., Iwai, Y., Keiko, M., Yamada, S., Kakuchi, T., Kaga, H. 2003, (Lipase-catalyzed Dynamic Kinetic Resolution of Hemiaminals), *Tetrahedron Asymmetry* 14, 1581–1585.

55 Faber, K. 2004, *Biotransformations in Organic Chemistry*, Springer-Verlag, Berlin, (pp.) 240–242.

56 Faber, K. 2004, *Biotransformations in Organic Chemistry*, Springer-Verlag, Berlin, (pp.) 322–327.

57 a) Faber, K. 2004, *Biotransformations in Organic Chemistry*, Springer-Verlag, Berlin, (pp.) 253–256; b) Boll, M., Fuchs, G., Heider, J. 2002, (Anaerobic Oxidation of Aromatic Compounds and Hydrocarbons), *Curr. Opin. Chem. Biol.* 6 (5), 604–611.

58 Flitsch, S., Grogan, G., Ashcroft, D. 2002, (Oxidation Reactions), in *Enzyme Catalysis on Organic Synthesis*, eds. Drauz, K., Waldmann, H., Wiley-VCH, Weinheim, (pp.) 1213–1241.

59 Kohli, R.M., Walsh, C.T. 2003, (Enzymology of Acyl Chain Macrocyclization in Natural Product Biosynthesis), *Chem. Commun.* (3), 297–307.

60 Hoh, C., Villela, M.F. 2000, (Enzyme Classification), in *Industrial Biotransformations*,

eds. Liese, A., Seelbach, K., Wandrey, C.,
Wiley-VCH, Weinheim. (p.) 54.

61 Crout, D.H.G., Vic, G. 1998, (Glycosidases
and Glycosyl Transferases in Glycoside and
Oligosaccharide Synthesis), *Curr. Opin.
Chem. Biol.* 2 (1), 98–111.

62 Faber, K. 2004, *Biotransformations in Organic
Chemistry*, Springer-Verlag, Berlin, (pp.)
249–252.

63 Hoh, C., Villela, M.F. 2000, (Enzyme Classification), in *Industrial Biotransformations*,
eds. Liese, A., Seelbach, K., Wandrey, C,
Wiley-VCH., Weinheim, (pp.) 51–53.

4
Optimization of Industrial Enzymes by Molecular Engineering

Thorsten Eggert

4.1
Introduction

The biocatalyst assisted conversion of educts into desired products has developed over a long period of time alongside the cultural history of human society. As described in Chapter 1, fermentation was discovered as a process for the production of alcohol about 5000 years ago. Also, without having any knowledge of the existence of microorganisms or even enzymes, the Egyptian civilization used yeast for baking bread, a technique which later became known as whole cell biocatalysis. Nowadays, biocatalysis using whole cells, crude cell extracts or purified enzymes has achieved a position of steadily increasing importance for the biotechnological production of food additives, agrochemicals, cosmetics and flavors, and, in particular, for pharmaceuticals. The rapidly increasing market for these compounds has resulted in a growing demand to identify biocatalysts with novel and specific properties. Hence, there are extensive programs aimed at collecting novel microorganisms, plants or animals from all over the world to use them as sources for the identification of novel enzymes. However, current estimates indicate that more than 99% of the prokaryotes present in natural environments such as soil, water, sediments, or plant surfaces cannot be readily cultured in the laboratory by standard techniques and therefore remain inaccessible for biotechnological applications [1]. The so-called metagenome approach can overcome the cultivation problem by direct isolation and cloning of environmental deoxyribonucleic acid (eDNA) resulting in metagenome libraries, which represent the genomes of all microorganisms present in a given sample independent of their culturability [2].

Nevertheless, today's enzymes are the product of biological evolution, which has taken several millions of years. They usually catalyze a given reaction with high specificity and enantioselectivity; however, as they are adjusted perfectly to their physiological role in an aqueous-based environment, their activity and stability are often far removed from what organic chemists require. This is true for the stability of enzymes in organic solvents and particularly for the enantioselectivity of reactions yielding industrially important compounds [3, 4].

The aim of this chapter is to give the reader an overview of the requirements and of the possibilities of overcoming the addressed drawbacks of biocatalysts through the use

Industrial Biotransformations. Andreas Liese, Karsten Seelbach, Christian Wandrey (Eds.)
Copyright © 2006 WILEY-VCH Verlag GmbH & Co. KGaA, Weinheim
ISBN: 3-527-31001-0

of molecular biotechnology techniques. This is an ongoing revolution in biological sciences, which began with the discovery of the structure of deoxyribonucleic acid (DNA) by Watson and Crick in 1953 [5]. Other important developments followed, such as the isolation of specific restriction enzymes by Smith and coworkers in 1970 [6, 7] and the polymerase chain reaction (PCR) discovered by Mullis et al. in 1986 [8], which is summarized in Table 4.1.

Tab. 4.1 Milestones in molecular biotechnology.

Year	Scientific milestone	Researcher(s)
1953	determination of molecular structure of deoxyribonucleic acid (DNA)	F.H.C. Crick and J.D. Watson (Nobel Prize in Medicine awarded in 1962)
1958	first crystal structure of a protein	J.C. Kendrew (Nobel Prize in Chemistry awarded in 1962)
1961	discovery of genetic code, three nucleotide bases, or one codon, specify one amino acid	S. Brenner and F.H.C. Crick
1970	isolation of the first DNA restriction enzyme	H.O. Smith (Nobel Prize in Medicine awarded in 1978)
1972	creation of recombinant DNA using restriction enzymes and T4 ligase	S. Cohen and H. Boyer
1977	techniques developed to determine DNA sequences	F. Sanger and W. Gilbert (Nobel Prize in Chemistry awarded in 1980)
1977	recombinant production of a human protein (somatostatin) in *E. coli*	Genentech (founded by Cohen and Boyer)
1978	recombinant production of human insulin in *E. coli*	Genentech
1983	development of polymerase chain reaction (PCR) to specifically amplify target DNA sequences	K.B. Mullis (Nobel Prize in Chemistry awarded in 1993)
1983	first full gene sequence of a virus is completed (Bacteriophage lambda)	D.L. Daniels and coworkers
1995	first full gene sequence of a living organism is completed (*Haemophilus influenzae*)	R.D. Fleischmann and coworkers
2000	first draft of human genome sequence is completed	Human genome project and Celera Genomics

4.2
Learning from Nature

Nature itself appears to provide a solution to the apparent dilemma of the frequent incompatibilities of enzyme properties and organic chemistry: natural evolution produces enzyme variants by mutation and subsequently tests activity by selecting the "fittest" variant. This process can be mimicked in a test-tube using modern molecular biological methods of site directed or random mutagenesis and subsequent analysis of the enzyme activity. This collection of (molecular) biological methods has been termed "molecular engineering", which provides a powerful tool for the development of biocatalysts with novel properties.

Before beginning an enzyme optimization by means of molecular engineering, three requirements are necessary: (a) the enzyme coding gene must be known and isolated from gene-libraries or amplified directly from genomic DNA using PCR; (b) the identified gene must be cloned into an appropriate expression vector for production of the enzymatically active biocatalyst; and (c) the analysis of enzymatic activity must be established to identify changes in biocatalyst properties. In the age of complete genome [9–13] and environmental genome sequencing [14] projects, the availability of gene sequences do not usually constitute a major drawback to molecular engineering approaches. However, having the gene in ones hands does not as a consequence mean that the catalytically active enzyme is actually available. Therefore, one essential requirement is the availability of powerful expression systems based on homologous or heterologous microbial hosts, as described in the next section.

4.3
Enzyme Production Using Bacterial Expression Hosts

An efficient bacterial overexpression system consists of a vector harboring the gene (or genes) of interest under the control of a promoter that might be regulated by *trans*-acting elements (so-called regulatory proteins) leading to a modulated gene expression in the prokaryotic host cell. These expression vectors are naturally occurring extra-chromosomal DNA, so-called plasmids, usually carrying resistance genes enabling the cells to survive in toxic environments or to cope with antibiotic attacks from other microorganisms. One of the first and best-studied "general purpose" cloning vectors is pBR322[1].

This plasmid, isolated by Bolivar and Rodriguez et al. [15], consists of 4361 base pairs. As shown in Fig. 4.1(A), pBR322 contains two antibiotic resistance genes (ampicillin and tetracycline), an origin of replication for *Escherichia coli* as the host cell and unique

---

1) Plasmid names usually start with a "p", which is the abbreviation for plasmid. The subsequent letters and/or numbers are abbreviations chosen by the scientist who created or isolated the plasmid. In the case of pBR322, the letters represent the initials from the molecular biology scientists Bolivar and Rodriguez, whereas the "322" is a serial number.

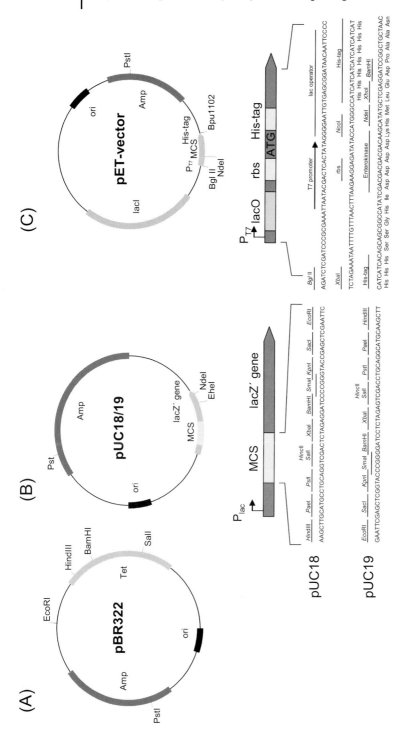

Fig. 4.1 Genetic map of cloning and expression vectors. (A) Cloning vector pBR322 [15], (B) pUC-vectors [16] and (C) pET-vectors (Novagen, Madison, WI, USA) are given in a schematical overview. All plasmids contain typical expression vector elements such as ampicillin (Amp) or tetracycline (Tet) resistance genes, an origin of replication (ori) and unique restriction endonuclease recognition sites (e.g., *HindIII*, *EcoRI* or *BamHI*). In pUC- and pET-vectors many unique restriction sites are located in multiple cloning sites (MCS) enabling convenient cloning of the target gene. For a tightly controlled overexpression, the pET-vector series contain the T7 promoter (P_{T7}), the lactose operator (lacO) and the ribosome binding site (rbs) in front of the ATG-start codon.

restriction sites for gene cloning (*Pst*I, *Eco*RI, *Hind*III, *Bam*HI, *Sal*I[2]), which represent the minimal requirements of a bacterial cloning vector (Table 4.2). Because of the time-consuming selection procedure of foreign DNA containing plasmids and the low expression efficiency of pBR322, other systems were developed to overcome these limitations.

Tab. 4.2 Characteristic elements of plasmids used for overexpression of target genes.

Element	Function
origin of replication	amplification of the plasmid within the host cell; strain specific
selection marker	usually antibiotic resistant genes to apply selection pressure for plasmid containing cells
fusion-tags	additional amino acids at the *N*- or *C*-terminus for affinity purification (e.g., His-tag, Strep-tag)
promoter	DNA element for expression control of the target gene
ribosome binding site	the ribosome binding site, also called Shine–Dalgarno sequence, is necessary for efficient translation of mRNA into the amino acid sequence
multiple cloning site	DNA sequence downstream of the promoter having unique restriction sites for cloning

For example, the plasmids pUC18 and pUC19[16] [Fig. 4.1(B)] are versatile vectors consisting of a 54 bp multiple cloning site containing 14 unique restriction sites (*Eco*RI, *Sac*I, *Kpn*I, *Xma*I, *Sma*I, *Bam*HI, *Xba*I, *Sal*I, *Hinc*II, *Acc*I, *Bsp*I, *Pst*I, *Sph*I, *Hind*III), an ampicillin resistance marker, an origin of replication derived from pBR322 and a portion of the *lacZ*[3] gene from *E. coli*. The pUC-vectors allow easy identification of recombinant plasmids by blue/white color screening on indicator agar plates containing 5-bromo-4-chloro-3-indolyl-*β*-D-galactopyranoside (X-gal). Bacterial colonies carrying vectors without foreign DNA show blue color on these indicator plates, whereas cells with recombinant vectors remain white, because of the disruption of the LacZ-*α*-peptide encoded by the pUC-vectors.

The expression vectors used for enzyme production today are genetically reconstructed plasmids, having elements from different sources to fulfill the requirements summarized in Table 4.2. One of the best known bacterial overexpression systems developed for *E. coli* host strains is the pET-system commercialized by Novagen (Madison, WI, USA) with the expression of genes controlled by a strong promoter and a DNA-dependent RNA polymerase derived from the bacteriophage T7 [Fig. 4.1(C)].

2) Abbreviations of restriction enzymes are derived from the name of the source organism (e.g., enzyme *Pst*I is isolated from *Providencia stuarti* whereas *Eco*RI is an enzyme from *Escherichia coli*.

3) Abbreviation of the *E. coli β*-lactamase gene used as a molecular marker

However, a considerable number of enzymes cannot be expressed using one of the *E. coli* systems, because their production requires accessory cellular functions such as essential cofactors, post-translational modifications or unique chaperones for folding. Therefore, alternative expression hosts such as *Bacillus* species, Pseudomonads, or yeast strains using expression vectors different from *E. coli* are necessary [17, 18]. Very often the expression vectors are constructed as shuttle vectors containing more than one origin of replication for convenient cloning in *E. coli* and subsequent expression in the final host. Furthermore, so-called broad-host-range plasmids are available for cloning and expression in a wide range of microorganisms [19–22].

4.4
Improvements to Enzymes by Molecular Engineering Techniques

Two different strategies, namely *rational design* and *directed evolution*, represent the "state-of-the-art" technologies in molecular enzyme engineering. Both techniques depend on enzyme modification at the DNA level by introducing mutations into the gene; however, in both cases different additional information about the enzyme or molecular methods must be available. These additional requirements, as summarized in Table 4.3, favor or exclude one or the other strategy. To help the reader evaluate which strategy might be the best with respect to the laboratory equipment available and knowledge of the target enzyme, both strategies, *rational design* and *directed evolution*, are described in more detail in the next section, and the pros and cons of both approaches are discussed.

Tab. 4.3 Rational design versus directed evolution; the additional requirements and necessary molecular methods of both strategies are given to evaluate the pros and cons of both *in vitro* strategies for molecular enzyme engineering.

Experimental	Rational design	Directed evolution
additional requirements	• 3D-structure, or structure model • knowledge of the reaction mechanism • modeling of enzyme–substrate interaction	• high-throughput screening or selection technology to analyze enzyme properties
molecular methods	• site-directed mutagenesis	• Random mutagenesis (point mutations or recombination)

4.4.1
Rational Enzyme Design

In 1958 the first three-dimensional structure of a protein, namely myoglobin, was reported by Kendrew and coworkers [23, 24]. Kendrew won the Nobel Prize four years later in 1962, together with Perutz, for their studies on the structures of globular proteins. Since that time the number of novel protein and peptide structures solved by X-ray diffraction and, more recently, by NMR spectroscopy has reached 30 800 (May 2005); as a consequence thereof, our knowledge of enzyme architecture and functionality has been improved considerably. From 1971 onwards, these biological macromolecular structures have been deposited in the Protein Data Bank (PDB) at Brookhaven National Laboratories [25]. All deposited structures are available via the internet (http://www.rcsb.org/pdb/).

In addition, methods in the field of recombinant DNA are becoming more and more "state-of-the-art" technology and widely used in the natural sciences (Table 4.1). In particular, the polymerase chain reaction (PCR) developed by Mullis (Nobel Prize in 1993) simplifies the amplification of enzyme coding genes significantly and also the introduction of site directed mutations [8, 26].

Using the novel techniques of recombinant DNA and the structural knowledge of the phage T4 lysozyme, in 1988 Matsumura and coworkers published their ground breaking results on enzyme stabilization by molecular engineering. In site-directed mutagenesis experiments, two, four or six amino acid residues, spatially close to each other on the surface of the natural enzyme, were changed into cysteine residues. Consequently, the variant enzymes contained one, two or three disulfide bonds, hampering the thermal unfolding of the native enzyme structure. The triple-disulfide variant unfolds at a temperature 23.4 °C higher than the wild-type lysozyme [27–29].

Many examples using the rational design approach to increase enzyme stability (temperature, pH, organic solvents) or specific activity followed [30]. Successful examples of rational enzyme design to improve or invert enantioselectivity are relatively rare. However, Pleiss and coworkers have reported an improvement in the enantioselective hydrolysis of linalyl acetate by *B. subtilis* *p*-nitrobenzyl esterase variants, as predicted by computer simulations. Furthermore, an inverted enantiopreference using 2-phenyl-3-butin-2-yl-acetate as the model substrate was achieved [31]. Other successful examples of improved enantioselectivities by rational protein engineering have been presented by the group working with Hult (Department of Biotechnology, Royal Institute of Technology, Stockholm, Sweden) on *Candida antarctica* lipase B (CALB) [32–35] and the group with Raushel (Department of Chemistry, Texas A&M University, USA) on phosphotriesterase from *Pseudomonas diminuta* [36, 37]. Here at least two variants were created showing a million-fold difference in enantioselectivity towards the substrate ethyl phenyl *p*-nitrophenyl phosphate [37].

Nevertheless, there are still fundamental problems when applying rational enzyme design: (a) the three-dimensional structure of the enzyme and (b) ideas with respect to the molecular functions of certain amino acid side chains must be available when rational design has to be applied. An alternative might be the development of a reliable structural model based on related enzymes. (c) In general it is not possible to predict exactly the final structure of a variant enzyme using computer simulations; however, the methods are continuously improving, and include theoretical methods using combined quantum mechanical and molecular mechanical calculations (QM/MM) [38]; however, in all simulations the effect of the enzyme dynamics are neglected. (d) Furthermore, solid state structures derived from crystallography could be different from protein structures in solution.

4.4.2
Directed Evolution

Owing to the major difficulties encountered with rational protein design when creating bio-catalysts that perform better towards interesting non-natural substrates, over the last decade scientific researchers have established a collection of methods termed "directed" or "*in vitro*" evolution. In this way they have provided a powerful tool for the development of biocatalysts with novel properties without requiring knowledge of the enzyme structures or catalytic mechanisms (Table 4.4). This strategy mimics natural evolution in a test-tube; however, it reduces the time scale from millions of years to several months or even weeks.

The general strategy for isolating enzymes with novel properties by directed evolution is outlined in Fig. 4.2. Molecular diversity is created by random mutagenesis and/or recombi-nation of a target gene or a set of related genes (i.e., gene family). A powerful (over)ex-pression system is required to express the variant proteins at a level high enough to allow for screening and/or selection of better variants. In many cases, secretion of the protein of interest into the bacterial culture supernatant greatly facilitates screening, in particular when microtiter plates are used. As soon as enzyme variants with improved properties are identified the corresponding genes are used to parent the next round of evolution.

Tab. 4.4 Molecular methods for directed evolution.

Method	Reference
1. *random point mutations*	
saturation mutagenesis	90
error-prone polymerase chain reaction (epPCR)	43–46
2. *insertion and deletion*	
random insertion/deletion mutagenesis (RID)	91
random deletions and repeats	92
3. in vitro *recombination (homology-dependent)*	
DNA shuffling	59, 60, 93
family shuffling	94
staggered extension process (StEP)	67
random priming recombination (RPR)	95
heteroduplex recombination	96
ssDNA family shuffling	97, 98
degenerate oligonucleotide gene shuffling (DOGS)	99
random chimeragenesis on transient templates (RACHITT)	100
mutagenic and unidirectional reassembly (MURA)	101
assembly of designed oligonucleotides (ADO)	68
multiplex-PCR-based recombination (MUPREC)	126
4. in vitro *recombination (homology independent)*	
incremental truncation for the creation of hybrid enzymes (ITCHY)	71, 102
sequence homology independent protein recombination (SHIPREC)	103
combination of ITCHY and DNA shuffling (SCRATCHY)	73, 104

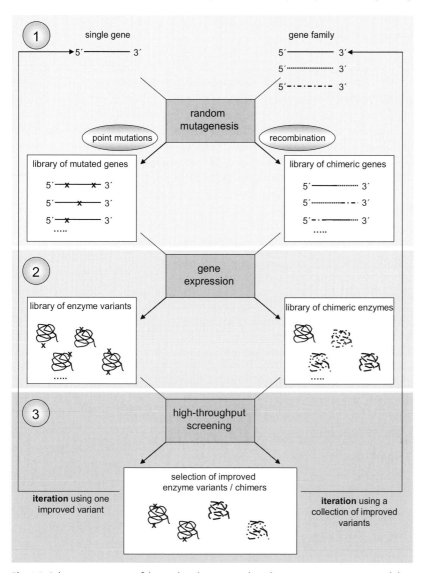

Fig. 4.2 Schematic overview of directed evolution experiments. (1) Variant gene libraries are generated by *in vitro* random mutagenesis using non-recombinative (introducing point mutations) or recombinative methods. (2) Gene libraries are cloned into expression vectors and the corresponding biocatalyst libraries are produced *in vivo* by microbial host strains. (3) Biocatalysts showing the desired properties are identified by high-throughput screening or selection systems.

4.4.3
Random Mutagenesis Methods

The simple concept of laboratory evolution – random mutagenesis and screening or selection of the "fittest" – should be planned carefully by the experimenter right from the beginning of the project. The first problem the scientist is confronted with is implementing the huge sequence space of proteins: using the 20 building blocks of proteins, the different amino acids encoded by the DNA, an enormous number of novel combinations can be generated. For example, an enzyme consisting of 300 amino acid residues can theoretically exist in 20^{300} possible linear combinations of all 20 amino acids at each position. This infinite number of combinations cannot be generated by any one scientist – at least nor by nature itself, because the mass of such a library would exceed the mass of the whole universe – which makes *de novo* enzyme design by completely random approaches impossible. Therefore, two different strategies using good starting points in the "fitness landscape" of protein-sequence space have been developed: *random point mutations* and in vitro *recombination*.

The insertion of random point mutations into DNA has been used to generate mutant strains since industrial biocatalysis began. UV and X-ray radiation or mutagenic chemicals such as nitrous acid, formic acid or hydrazine have been used to generate production strains in industry as a way of so-called "classical strain improvement": random mutagenesis of the whole genome and subsequent screening for better performing variants. Today, this is still an option as a way of improving production strains without creating so-called genetically manipulated organisms (GMOs). Many successful examples have been published or patented; however, the accumulation of deleterious mutations in the genome of the target organism makes the process difficult and unpredictable. Therefore, the Biotech Company Maxygen (Redwood City, CA, USA) and its subsidiary Codexis (Redwood City, CA, USA) have recently introduced a new strategy to speed up classical strain improvement using *whole genome shuffling*. Here, the genomes of mutants created by classical strain improvement are recombined by protoplast fusions, creating a pool of recombinants, which is screened for better production strains. This technique has proved to be useful for the rapid improvement in the production of the macrolite antibiotic tylosin from *Streptomyces fradiae* [39] as well as changing the pH tolerance of a *Lactobacillus* strain of commercial interest, so that it will grow under acidic conditions [40].

Nevertheless, most directed evolution strategies use *in vitro* mutagenesis methods applied to defined target genes. According to this, the most commonly used method is *error prone PCR (epPCR)*, a technique that is as simple as regular PCR. The *Taq*-polymerase[4], an enzyme without *in vitro* proof-reading activity, is used to amplify the target gene under "suboptimal" conditions. This increases the naturally occurring error rate from $0.1–2 \times 10^{-4}$ [41, 42] to $1–5 \times 10^{-3}$ base substitutions by: (a) increasing the concentration of $MgCl_2$, (b) addition of $MnCl_2$, (c) using an unbalanced concentration of nucleotides, or (d) using a mixture of triphosphate nucleoside analogues [43–47]. In addition, companies such as Statagene (La Jolla, CA, USA) or BD Biosciences – Clontech (Palo Alto, CA,

---

4) Abbreviation for the hyperthermophilic bacterium *Thermus aquaticus*

USA) offer *ready-to-use* error prone PCR kits using low fidelity polymerases and opti-mized buffer systems.

The diversity of an enzyme library generated by epPCR is usually calculated by corre-lating the base pair substitutions introduced per gene with the amino acid exchanges introduced per enzyme molecule, i.e., an average of 1–2 base pair substitutions usually results in one amino acid exchange. The overall size of a variant library can subsequently be calculated by a combinatorial algorithm, as shown in Table 4.5. This algorithm is based on the assumption that all 19 remaining amino acids can be introduced with the same probability at a single position ($E = 19$). Unfortunately, this high diversity is not created by epPCR, because an event with two or even three base pair exchanges per codon is highly unlikely. Under optimal conditions, one nucleotide of a given codon will be exchanged thereby leading to just 9 (instead of 64 possible) different codons encoding 4–7 (instead of 20) different amino acids. An estimation of amino acid exchanges that could be introduced by epPCR into four different lipase genes, two originating from *B. subtilis* and two from Pseudomonads, revealed that the actual number of variants was only 22.5 and 19.5% of the theoretical number, respectively [48]. Despite these drawbacks many examples are known from the literature showing improvements in enzyme proper-ties, such as (thermo-, pH- or solvent-) stability [49–51], specific activity [52] or enantios-electivity [53, 54] by using epPCR in the first round of directed evolution.

Tab. 4.5 Theoretical number of enzyme variants in a library obtained for an enzyme consisting of 181 amino acids (e.g., lipase A from *Bacillus subtilis*) with 1–5 amino acid exchanges per molecule.

Number of amino acid exchanges (M)	Number of variants[a] (N)
1	3 439
2	5 880 690
3	6 666 742 230
4	5 636 730 555 465
5	3 791 264 971 605 760

a Values calculated with $E = 19$ using the algorithm:

$$N = \frac{E^M X!}{(X-M)!M!}$$

N = number of variants at maximum size of diversity
E = number of amino acids exchanged per position
M = total number of amino acid exchanges per enzyme molecule
X = number of amino acids per enzyme molecule

Site specific saturation mutagenesis is a method that can overcome the drawback of error prone PCR mentioned above. This method generates all 19 amino acid exchanges virtually, at a given position, by using customized degenerated oligonucleotides[5], introducing all possible codons. These degenerated oligonucleotides are usually incorporated into the target gene by PCR reactions. This can be achieved by applying mega-primer PCR or overlap extension PCR protocols [55, 56]. This technique was initially established to find ideal amino acid exchanges at hot spot positions identified by other methods such as error prone PCR. This approach is presently been extended to complete gene sequences, thereby generating a complete saturation mutagenesis library containing all single mutants calculated by the formulae given in Table 4.5 [54, 57, 58].

When evolving an enzyme by iterative rounds of random mutagenesis and screening, the best performing variant identified in each generation constitutes the starting point for the next round of mutagenesis (Fig. 4.2), thus many useful variants are discarded. Furthermore, it is not obvious whether the selected variant will have the potential for further improvement in the next generation or not. Deleterious mutations could be accumulated, leading to a dead end for the *in vitro* evolution. The second evolution strategy based on *in vitro* recombination techniques helps to overcome this problem. By recombining a pool of better performing variants, libraries of chimeric genes can be produced, providing the possibility of removing deleterious mutations and combining beneficial ones.

DNA shuffling was the first method described for *in vitro* recombination, and it immediately proved to be a valuable tool for directed evolution of biocatalysts [59, 60]. Homologous DNA sequences carrying mutations are recombined *in vitro* in a process consisting of random gene fragmentation using DNase I and subsequent reassembly in a self-priming chain extension method catalyzed by DNA polymerase. Recombination occurs by template switching: a fragment originating from one gene anneals to a fragment from another gene.

DNA shuffling has been applied in a wide variety of experiments to improve single gene encoded enzymes [61, 62] as well as operon encoded multi-gene pathways [63]. Furthermore, DNA shuffling has been used to improve viruses in terms of stability and processing yields as well as changes in specificity (also known as tropism) for better application in human gene therapy [64–66].

In addition to DNA shuffling, other efficient recombinative methods to generate variant libraries followed including the PCR-based *staggered extension process (StEP)* [67] or the *assembly of designed oligonucleotides (ADO)* [68]. Furthermore, many closely related methods or "updates" have been described, as summarized in two recently published review articles [69, 70] (see also Table 4.4).

All *in vitro* recombination methods described so far require relatively high levels of DNA homology in the target sequences, otherwise the recombination events will only occur in regions of homology or they will not occur at all. A different approach to create libraries of fused gene fragments independent of sequence homology has been developed, which was termed *incremental truncation for the creation of hybrid enzymes (ITCHY)* [71]. Two parental genes are digested with exonuclease III in a tightly controlled manner to generate truncated gene libraries with progressive 1 base pair deletions. The truncated

---

5) The genetic code is said to be degenerated
   because more than one nucleotide triplet codes
   for the same amino acid.

5'-fragments of the one gene and the truncated 3'-fragments of the other gene are fused to yield a library of chimeric sequences, which are expressed and screened or selected for improved enzyme activity. This method allows the creation of functional fusions of genes from overlapping amino- or carboxy-terminal gene fragments independent of DNA sequence homology. Limitations include the fusion of only two genes per experiment and the creation of just a single crossover between two fragments, thereby restricting diversity of the libraries [72]. Therefore, an improved ITCHY method has been described, which consists of the sequential combination of classical ITCHY and DNA shuffling, termed *SCRATCHY*, creating multiple crossovers independent of the homology of DNA sequences [73].

4.5
Identification of Improved Enzyme Variants

By using directed evolution to improve the properties of biocatalysts, large libraries of enzyme variants can be generated, as described previously. Finding the desired variant is not always trivial. Therefore, high-throughput screening (HTS) or selection systems must be available to assay the biocatalytic activity from 10^4 to 10^7 individual variants simultaneously.

Genetic selection is by far the most elegant and powerful way to identify the "one in a million". However, a microbial system must be established in which the catalytic activity provides a growth advantage. Often the substrate of interest is provided as a sole carbon or nitrogen source, whereby hydrolysis of the compound enables the cells to grow. Other selection systems are based on *in vivo* or *in vitro* display technologies: the most popular one being phage display, originally developed by Smith [74]. Here, the members of a variant enzyme library are displayed on the surface of the filamentous phage fd[6] as a fusion to the *N*-terminus of the minor coat protein[7], thereby physically linking the phenotype and the genotype of the biocatalyst. Enzymes showing the desired binding properties can be selected from a pool of randomly mutagenized variants. Successful examples of phage display selection have been reported to identify enzyme variants with improved biophysical properties and/or enhanced catalytic activities [75–78]. Furthermore, promising preliminary results show covalent and selective binding of phage-bound lipase to a chiral phosphonate inhibitor [79, 80]; however, the selection of enzyme variants showing improved enantioselectivities using chiral suicide inhibitors remains to be demonstrated.

Unfortunately, in the majority of cases growth selection or phage display is not practical, making clone separation and individual assaying necessary. This can be performed in agar plates using indicator media, such as tributyrin for detection of esterolytic and lipolytic activity [Fig. 4.3(A)] or in microtiter plates using high-throughput screening (HTS) assays. The screening capacity of these systems can be extended by using automated technology and standard procedures operated by pipeting robots [Fig. 3(B)]. Depending on the enzyme property, a wide range of spectrophotometric assays using chromogenic/fluorogenic substrates or substrate analogues have been developed [81].

6) fibrious DNA phage
7) Also referred to as the gene-3-protein (g3p)

(A)

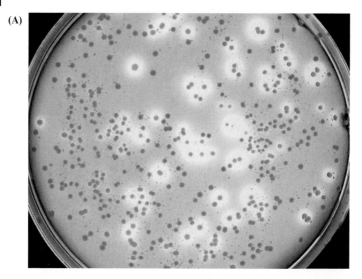

(B)

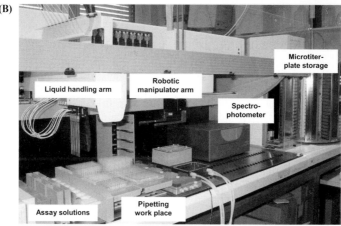

Fig. 4.3 High-throughput screening systems. (A) Screening for estero-lytic and lipolytic activity using tributyrin indicator agar-plates. Activity towards the substrate is indicated by clear halos surrounding the bac-terial colonies. (B) Spectrophotometric assay with robotics-based high-throughput screening.

These assays are mainly based on three different strategies: (a) chromogenic or fluoro-genic substrates, (b) staining of the product or (c) the use of indicator dyes. In Table 4.6 a range of well established screening methods is presented. For a comprehensive sum-mary of "state-of-the-art" high-throughput screening technology recent review articles [81–83] and books on assay development [84, 85] can be recommended. Furthermore, sophisticated HTS systems have been developed mainly by the group working with Reetz (Max-Planck Institut für Kohlenforschung, Mülheim an der Ruhr, Germany) for the identification of enantioselective biocatalysts [86, 87]. Although during recent years a vast number of novel high-throughput screening assays have been developed, in the context of biocatalyst improvement, the screening is still considered to be the bottleneck in enzyme evolution. Therefore, miniaturization and automation of variant library screening must be continued in the future.

Tab. 4.6 Spectrophotometric assays for high-throughput screening (HTS) of biocatalyst libraries.

Method[a] and coloring substance	Enzyme	Reference
1. *chromogenic or fluorogenic substrates*		
p-nitrophenol	lipase, esterase, protease, monooxygenase	105, 106
umbelliferone	lipase, esterase, protease, phosphatase, epoxidhydrolase,	107, 108
	transaldolase, transketolase	109, 110
4-(*p*-nitrobenzyl)pyridine (NBP)	Epoxidhydrolase	111
9,11,13,15-octadecatetraenoic acid ester (parinaric acid ester)	lipase	112
resorufin ester (e.g., 1,2-*O*-dilauryl-*rac*-glycero-3-glutaric acid-resorufin ester)	lipase	113
2. *product-staining or -conversion*		
4-nitro-7-chloro-benzo-2-oxa-1,3-diazole (NBD-Cl)	amidase	114
4-hydrazino-7-nitro-2,1,3-benzoxadiazole (NBD-H)	lipase	115
o-phthaldialdehyde-2-mercaptoethanol	nitrilase	116
NAD(P)H accumulation by alcohol dehydrogenase activity	lipase, esterase	117
3. *product detection using indicator dyes*		
pH indicators (e.g., Bromothymol Blue, Phenol Red)	lipase, esterase, amidase, haloalkane dehalogenase	118–120
2,3,5-triphenyltetrazolium chloride (Tetrazolium Red)	pyruvate decarboxylase, benzoylformate decarboxylase	121, 122
fuchsin	epoxidhydrolase	123
6-methoxy-*N*-(3-sulfopropyl) quinolinium (SPQ)	dehalogenase	124

a This list of methods gives an overview of widely used screening assays but it is not exhaustive. Further information can be found in recent reviews [81, 83, 125].

4.6
Conclusions and Future Prospects

In this chapter strategies and methods for molecular enzyme engineering have been presented and their pros and cons have been discussed. For further information three recently published books are recommended [84, 88, 89] as they give detailed insights into (random-) mutagenesis and high-throughput screening methods.

Although directed evolution works even without any knowledge of an enzymes' structure or reaction mechanism, knowledge of the three-dimensional protein structure can significantly speed up a directed evolution approach because the size of the sequence space to be sampled can be narrowed down. Eventually, a structure- *and* theory-assisted method of molecular enzyme engineering using directed evolution will facilitate the creation of better performing biocatalysts for industrial applications.

Acknowledgments

The author thanks Prof. Karl-Erich Jaeger and P.D. Martina Pohl (Institute for Molecular Enzyme Technology, University of Düsseldorf, Germany) for critical reading of the manuscript and valuable advice.

References

1 Amann, R.I., Ludwig, W., Schleifer, K.H. 1995, (Phylogenetic Identification and *In Situ* Detection of Individual Microbial-cells Without Cultivation), *Microbiol. Rev.* 59, 143–169.

2 Handelsman, J., Rondon, M.R., Brady, S.F., Clardy, J., R. Goodman, M. 1998, (Molecular Biological Access to the Chemistry of Unknown Soil Microbes: A New Frontier for Natural Products), *Chem. Biol.* 5, 245–249.

3 Patel, R.N. 2003, (Microbial/Enzymatic Synthesis of Chiral Pharmaceutical Intermediates), *Curr. Opin. Drug Discov.* 6, 902–920.

4 Straathof, A.J.J., Panke, S., Schmid, A. 2002, (The Production of Fine Chemicals by Biotransformations), *Curr. Opin. Biotechnol.* 13, 548–556.

5 Watson, J.D., Crick, F.H.C. 1953, (Molecular Structure of Nucleic Acids – a Structure for Deoxyribose Nucleic Acid), *Nature* 171, 737–738.

6 Smith, H.O., Wilcox, K.W. 1970, (A Restriction Enzyme from *Hemophilus influenzae*. 1. Purification and General Properties), *J. Mol. Biol.* 51, 379–391.

7 Kelly, T.J., Smith, H.O. 1970, (A Restriction Enzyme from *Hemophilus influenzae*. 2. Base Sequence of Recognition Site. *J. Mol. Biol.* 51, 393–409.

8 Mullis, K., Faloona, F., Scharf, S., Saiki, R., Horn, G., Erlich, H. 1986, (Specific Enzymatic Amplification of DNA *In Vitro* – the Polymerase Chain-reaction), *Cold Spring Harbor Symposia on Quantitative Biology* 51, 263–273.

9 Kunst, F., Ogasawara, N., Moszer, I., Albertini, A.M., Alloni, G., Azevedo, V., Bertero, M.G., Bessieres, P., Bolotin, A., Borchert, S., Borriss, R., Boursier, L., Brans, A., Braun, M., Brignell, S.C., Bron, S., Brouillet, S., Bruschi, C.V., Caldwell, B., Capuano, V., Carter, N.M., Choi, S.K., Codani, J.J., Connerton, I.F., Cummings, N.J., Daniel, R.A., Denizot, F., Devine, K.M., Dusterhoft, A., Ehrlich, S.D., Emmerson, P.T., Entian, K.D., Errington, J., Fabret, C., Ferrari, E., Foulger, D., Fritz, C., Fujita, M., Fujita, Y., Fuma, S., Galizzi, A.,

Galleron, N., Ghim, S.Y., Glaser, P., Goffeau, A., Golightly, E.J., Grandi, G., Guiseppi, G., Guy, B.J., Haga, K., Haiech, J., Harwood, C.R., Henaut, A., Hilbert, H., Holsappel, S., Hosono, S., Hullo, M.F., Itaya, M., Jones, L., Joris, B., Karamata, D., Kasahara, Y., KlaerrBlanchard, M., Klein, C., Kobayashi, Y., Koetter, P., Koningstein, G., Krogh, S., Kumano, M., Kurita, K., Lapidus, A., Lardinois, S., Lauber, J., Lazarevic, V., Lee, S.M., Levine, A., Liu, H., Masuda, S., Mauel, C., Medigue, C., Medina, N., Mellado, R.P., Mizuno, M., Moestl, D., Nakai, S., Noback, M., Noone, D., Oreilly, M., Ogawa, K., Ogiwara, A., Oudega, B., Park, S.H., Parro, V., Pohl, T.M., Portetelle, D., Porwollik, S., Prescott, A.M., Presecan, E., Pujic, P., Purnelle, B., Rapoport, G., Rey, M., Reynolds, S., Rieger, M., Rivolta, C., Rocha, E., Roche, B., Rose, M., Sadaie, Y., Sato, T., Scanlan, E., Schleich, S., Schroeter, R., Scoffone, J., Sekiguchi, F., Sekowska, A., Seror, S.J., Serror, P., Shin, B.S., Soldo, B., Sorokin, A., Tacconi, E., Takagi, T., Takahashi, H., Takemaru, K., Takeuchi, M., Tamakoshi, A., Tanaka, T., Terpstra, P., Tognoni, A., Tosato, V., Uchiyama, S., Vandenbol, M., Vannier, F., Vassarotti, A., Viari, A., Wambutt, R., Wedler, E., Wedler, H., Weitzenegger, T., Winters, P., Wipat, A., Yamamoto, H., Yamane, K., Yasumoto, K., Yata, K., Yoshida, K., Yoshikawa, H.F., Zumstein, E., Yoshikawa, H., Danchin, A. 1997, (The Complete Genome Sequence of the Gram-positive Bacterium *Bacillus subtilis*) *Nature* 390, 249–256.

10 Istrail, S., Sutton, G.G., Florea, L., Halpern, A.L., Mobarry, C.M., Lippert, R., Walenz, B., Shatkay, H., Dew, I., Miller, J.R., Flanigan, M.J., Edwards, N.J., Bolanos, R., Fasulo, D., Halldorsson, B.V., Hannenhalli, S., Turner, R., Yooseph, S., Lu, F., Nusskern, D.R., Shue, B.C., Zheng, X.Q.H., Zhong, F., Delcher, A.L., Huson, D.H., Kravitz, S.A., Mouchard, L., Reinert, K., Remington, K.A., Clark, A.G., Waterman, M.S., Eichler, E.E., Adams, M.D., Hunkapiller, M.W., Myers, E.W., Venter, J.C. 2004, (Whole-genome Shotgun Assembly and Comparison of Human Genome Assemblies), *Proc. Natl. Acad. Sci. USA* 101, 1916–1921.

11 Fleischmann, R.D., Adams, M.D., White, O., Clayton, R.A., Kirkness, E.F., Kerlavage, A.R., Bult, C.J., Tomb, J.F., Dougherty, B.A., Merrick, J.M., McKenney, K., Sutton, G., Fitzhugh, W., Fields, C., Gocayne, J.D., Scott, J., Shirley, R., Liu, L.I., Glodek, A., Kelley, J.M., Weidman, J.F., Phillips, C.A., Spriggs, T., Hedblom, E., Cotton, M.D., Utterback, T.R., Hanna, M.C., Nguyen, D.T., Saudek, D.M., Brandon, R.C., Fine, L.D., Fritchman, J.L., Fuhrmann, J.L., Geoghagen, N.S.M., Gnehm, C.L., McDonald, L.A., Small, K.V., Fraser, C.M., Smith, H.O., Venter, J.C., 1995, (Whole-genome Random Sequencing and Assembly of *Haemophilus influenzae* Rd.) *Science* 269, 496–512.

12 Bult, C.J., White, O., Olsen, G.J., Zhou, L.X., Fleischmann, R.D., Sutton, G.G., Blake, J.A., Fitzgerald, L.M., Clayton, R.A., Gocayne, J.D., Kerlavage, A.R., Dougherty, B.A., Tomb, J.F., Adams, M.D., Reich, C.I., Overbeek, R., Kirkness, E.F., Weinstock, K.G., Merrick, J.M., Glodek, A., Scott, J.L., Geoghagen, N.S.M., Weidman, J.F., Fuhrmann, J.L., Nguyen, D., Utterback, T.R., Kelley, J.M., Peterson, J.D., Sadow, P.W., Hanna, M.C., Cotton, M.D., Roberts, K.M., Hurst, M.A., Kaine, B.P., Borodovsky, M., Klenk, H.P., Fraser, C.M., Smith, H.O., Woese, C.R., Venter, J.C. 1996, (Complete Genome Sequence of the Methanogenic Archaeon, *Methanococcus jannaschii*), *Science* 273, 1058–1073.

13 Blattner, F.R., Plunkett, G., Bloch, C.A., Perna, N.T., Burland, V., Riley, M., ColladoVides, J., Glasner, J.D., Rode, C.K., Mayhew, G.F., Gregor, J., Davis, N.W., Kirkpatrick, H.A., Goeden, M.A., Rose, D.J., Mau, B., Shao, Y. 1997, (The Complete Genome Sequence of *Escherichia coli* K-12), *Science* 277, 1453–1474.

14 Venter, J.C., Remington, K., Heidelberg, J.F., Halpern, A.L., Rusch, D., Eisen, J.A., Wu, D.Y., Paulsen, I., Nelson, K.E., Nelson, W., Fouts, D.E., Levy, S., Knap, A.H., Lomas, M.W., Nealson, K., White, O., Peterson, J., Hoffman, J., Parsons, R., Baden-Tillson, H., Pfannkoch, C., Rogers, Y.H., Smith, H.O. 2004, (Environmental Genome Shotgun Sequencing of the Sargasso Sea), *Science* 304, 66–74.

15 Bolivar, F., Rodriguez, R.L., Greene, P.J., Betlach, M.C., Heyneker, H.L., Boyer, H.W.,

Crosa, J.H., Falkow, S. 1977, (Construction and Characterization of New Cloning Vehicles. 2. Multipurpose Cloning System), *Gene* 2, 95–113.

16 Yanisch-Perron, C., Vieira, J., Messing, J. 1985, (Improved M13 Phage Cloning Vectors and Host Strains – Nucleotide-sequences of the M13mp18 and pUC19-Vectors), *Gene* 33, 103–119.

17 Jaeger, K.E., Rosenau, F. 2004, (Overexpression and Secretion of *Pseudomonas* Lipases), in *Pseudomonas – Biosynthesis of Macromolecules and Molecular Metabolism*, (vol. 3), ed. Ramos, J.-L., Kluwer Academic/Plenum Publishers, New York, (pp.) 491–508.

18 Rosenau, F., Jaeger, K.E., 2004, (Overexpression and Secretion of Biocatalysts in *Pseudomonas*), in *Enzyme Functionality: Design, Engineering, and Screening*, ed. Svendsen, A., Marcel Dekker, New York, (pp.) 617–631.

19 Makrides, S.C. 1996, (Strategies for Achieving High-level Expression of Genes in *Escherichia coli*), *Microbiol. Rev.* 60, 512–538.

20 Baneyx, F. 1999, (Recombinant Protein Expression in *Escherichia coli*), *Curr. Opin. Biotechnol.* 10, 411–421.

21 Watson, A.A., Alm, R.A., Mattick, J.S. 1996, (Construction of Improved Vectors for Protein Production in *Pseudomonas aeruginosa*), *Gene* 172, 163–164.

22 Kovach, M.E., Phillips, R.W., Elzer, P.H., Roop, R.M., Peterson, K.M. 1994, (pBBR1MCS - a Broad-host-range Cloning Vector), *Biotechniques* 16, 800–802.

23 Kendrew, J.C., Dickerson, R.E., Strandberg, B.E., Hart, R.G., Davies, D.R., Phillips, D.C., Shore, V.C. 1960, (Structure of Myoglobin – 3-Dimensional Fourier Synthesis at 2 Å Resolution), *Nature* 185, 422–427.

24 Kendrew, J.C., Bodo, G., Dintzis, H.M., Parrish, R.G., Wyckoff, H., Phillips, D.C. 1958, (3-Dimensional Model of the Myoglobin Molecule Obtained by X-ray Analysis), *Nature* 181, 662–666.

25 Westbrook, J., Feng, Z.K., Chen, L., Yang, H.W., Berman, H.M. 2003, (The Protein Data Bank and Structural Genomics), *Nucleic Acids Res.* 31, 489–491.

26 Saiki, R.K., Gelfand, D.H., Stoffel, S., Scharf, S.J., Higuchi, R., Horn, G.T., Mullis, K.B., Erlich, H.A. 1988, (Primer-directed Enzymatic Amplification of DNA

with a Thermostable DNA-polymerase), *Science* 239, 487–491.

27 Matsumura, M., Becktel, W.J., Levitt, M., Matthews, B.W. 1989, (Stabilization of Phage-T4 Lysozyme by Engineered Disulfide Bonds), *Proc. Natl. Acad. Sci. USA* 86, 6562–6566.

28 Matsumura, M., Becktel, W.J., Matthews, B.W. 1988, (Hydrophobic Stabilization in T4 Lysozyme Determined Directly by Multiple Substitutions of Ile-3), *Nature* 334, 406–410.

29 Matsumura, M., Signor, G., Matthews, B.W. 1989, (Substantial Increase of Protein Stability by Multiple Disulfide Bonds), *Nature* 342, 291–293.

30 Bornscheuer, U.T., Pohl, M. 2001, (Improved Biocatalysts by Directed Evolution and Rational Protein Design), *Curr. Opin. Chem. Biol.* 5, 137–143.

31 Henke, E., Bornscheuer, U.T., Schmid, R.D., Pleiss, J. 2003, (A Molecular Mechanism of Enantiorecognition of Tertiary Alcohols by Carboxylesterases), *Chembiochem.* 4, 485–493.

32 Rotticci, D., Haeffner, F., Orrenius, C., Norin, T., Hult, K. 1998, (Molecular Recognition of sec-Alcohol Enantiomers by *Candida antarctica* Lipase B), *J. Mol. Catal. B-Enzymatic* 5, 267–272.

33 Magnusson, A., Hult, K., Holmquist, M. 2001, (Creation of an Enantioselective Hydrolase by Engineered Substrate-assisted Catalysis), *J. Am. Chem. Soc.* 123, 4354–4355.

34 Orrenius, C., Haeffner, F., Rotticci, D., Ohrner, N., Norin, T., Hult, K. 1998, (Chiral Recognition of Alcohol Enantiomers in Acyl Transfer Reactions Catalysed by *Candida antarctica* Lipase B), *Biocatal. Biotrans.* 16, 1–15.

35 Rotticci, D., Rotticci-Mulder, J.C., Denman, S., Norin, T., Hult, K. 2001, (Improved Enantioselectivity of a Lipase by Rational Protein Engineering), *Chembiochem.* 2, 766–770.

36 Chen-Goodspeed, M., Sogorb, M.A., Wu, F.Y., Hong, S.B., Raushel, F.M. 2001, (Structural Determinants of the Substrate and Stereochemical Specificity of Phosphotriesterase), *Biochemistry* 40, 1325–1331.

37 Chen-Goodspeed, M., Sogorb, M.A., Wu, F.Y., Raushel, F.M. 2001, (Enhance-

ment, Relaxation, and Reversal of the Stereoselectivity for Phosphotriesterase by Rational Evolution of Active Site Residues), *Biochemistry* 40, 1332–1339.

38 Schoneboom, J.C., Lin, H., Reuter, N., Thiel, W., Cohen, S., Ogliaro, F., Shaik, S. 2002, (The Elusive Oxidant Species of Cytochrome P450 Enzymes: Characterization by Combined Quantum Mechanical/Molecular Mechanical (QM/MM) Calculations), *J. Am. Chem. Soc.* 124, 8142–8151.

39 Zhang, Y.X., Perry, K., Vinci, V.A., Powell, K., Stemmer, W.P.C., del Cardayre, S.B. 2002, (Genome Shuffling Leads to Rapid Phenotypic Improvement in Bacteria), *Nature* 415, 644–646.

40 Patnaik, R., Louie, S., Gavrilovic, V., Perry, K., Stemmer, W.P.C., Ryan, C.M., del Cardayre, S. 2002, (Genome Shuffling of *Lactobacillus* for Improved Acid Tolerance), *Nat. Biotechnol.* 20, 707–712.

41 Eckert, K.A., Kunkel, T.A. 1990, (High Fidelity DNA-synthesis by the *Thermus aquaticus* DNA-polymerase), *Nucleic Acids Res.* 18, 3739–3744.

42 Tindall, K.R., Kunkel, T.A. 1988, (Fidelity of DNA-synthesis by the *Thermus aquaticus* DNA-polymerase), *Biochemistry* 27, 6008–6013.

43 Cadwell, R.C., Joyce, G.F. 1992, Randomization of Genes by PCR Mutagenesis), *PCR Methods Appl.* 2, 28–33.

44 Cadwell, R.C., Joyce, G.F. 1995, (Mutagenic PCR), in *PCR Primer: a Laboratory Manual*, eds. Dieffenbach, C.H., Dveksler, G.S., CSHL Press, Cold Spring Harbor, (p.) 583.

45 Zaccolo, M., Williams, D.M., Brown, D.M., Gherardi, E. 1996, (An Approach to Random Mutagenesis of DNA Using Mixtures of Triphosphate Derivatives of Nucleoside Analogues), *J. Mol. Biol.* 255, 589–603.

46 Zhou, Y.H., Zhang, X.P., Ebright, R.H. 1991, (Random Mutagenesis of Gene-sized DNA-molecules by use of PCR with Taq DNA-polymerase), *Nucleic Acids Res.* 19, 6052–6052.

47 Jaeger, K.E., Eggert, T., Eipper, A., Reetz, M.T. 2001, (Directed Evolution and the Creation of Enantioselective Biocatalysts), *Appl. Microbiol. Biotechnol.* 55, 519–530.

48 Eggert, T., Reetz, M.T., Jaeger, K.E. 2004, (Directed Evolution by Random Mutagenesis: a Critical Evaluation), in *Enzyme Functionality: Design, Engineering, and Screening*, ed. Svendsen, A., Marcel Dekker, New York, 375–390.

49 Bessler, C., Schmitt, J., Maurer, K.H., Schmid, R.D. 2003, (Directed Evolution of a Bacterial Alpha-amylase: Toward Enhanced pH-performance and Higher Specific Activity), *Protein Sci.* 12, 2141–2149.

50 Chen, K., Arnold, F.H. 1993, (Tuning the Activity of an Enzyme for Unusual Environments: Sequential Random Mutagenesis of Subtilisin E for Catalysis in Dimethylformamide), *Proc. Natl. Acad. Sci. USA* 90, 5618–5622.

51 Giver, L., Gershenson, A., Freskgard, P.O., Arnold, F.H. 1998, (Directed Evolution of a Thermostable Esterase), *Proc. Natl. Acad. Sci. USA* 95, 12809–12813.

52 Meyer, A., Schmid, A., Held, M., Westphal, A.H., Rothlisberger, M., Kohler, H.P., van Berkel, W.J., Witholt, B. 2002, (Changing the Substrate Reactivity of 2-Hydroxybiphenyl 3-Monooxygenase from *Pseudomonas azelaica* HBP1 by Directed Evolution), *J. Biol. Chem.* 277, 5575–5582.

53 Liebeton, K., Zonta, A., Schimossek, K., Nardini, M., Lang, D., Dijkstra, B.W., Reetz, M.T., Jaeger, K.E. 2000, (Directed Evolution of an Enantioselective Lipase), *Chem. Biol.* 7, 709–718.

54 Funke, S.A., Eipper, A., Reetz, M.T., Otte, N., Thiel, W., Van Pouderoyen, G., Dijkstra, B.W., Jaeger, K.E., Eggert, T. 2003, (Directed Evolution of an Enantioselective *Bacillus subtilis* Lipase), *Biocatal. Biotrans.* 21, 67–73.

55 Barettino, D., Feigenbutz, M., Valcarcel, R., Stunnenberg, H.G. 1994, (Improved Method for PCR-mediated Site-directed Mutagenesis), *Nucleic Acids Res.* 22, 541–542.

56 Urban, A., Neukirchen, S., Jaeger, K.E. 1997, (A Rapid and Efficient Method for Site-directed Mutagenesis Using One-step Overlap Extension PCR), *Nucleic Acids Res.* 25, 2227–2228.

57 Gray, K.A., Richardson, T.H., Kretz, K., Short, J.M., Bartnek, F., Knowles, R., Kan, L., Swanson, P.E., Robertson, D.E. 2001, (Rapid Evolution of Reversible Denaturation and Elevated Melting Temperature in a Microbial Haloalkane Dehalogenase), *Adv. Synth. Catal.* 343, 607–617.

58 DeSantis, G., Wong, K., Farwell, B., Chatman, K., Zhu, Z.L., Tomlinson, G., Huang, H.J., Tan, X.Q., Bibbs, L., Chen, P., Kretz, K., Burk, M.J. 2003, (Creation of a Productive, Highly Enantioselective Nitrilase Through Gene Site Saturation Mutagenesis (GSSM)), *J. Am. Chem. Soc.* 125, 11476–11477.

59 Stemmer, W.P.C. 1994, (DNA Shuffling by Random Fragmentation and Reassembly – *In Vitro* Recombination for Molecular Evolution), *Proc. Natl. Acad. Sci. USA* 91, 10747–10751.

60 Stemmer, W.P.C. 1994, (Rapid Evolution of a Protein *In Vitro* by DNA Shuffling), *Nature* 370, 389–391.

61 Farinas, E.T., Bulter, T., Arnold, F.H. 2001, (Directed Enzyme Evolution), *Curr. Opin. Biotechnol.* 12, 545–551.

62 Minshull, J., Stemmer, W.P.C. 1999, (Protein Evolution by Molecular Breeding) *Curr. Opin. Chem. Biol.* 3, 284–290.

63 Crameri, A., Dawes, G., Rodriguez, E., Silver, S., Stemmer, W.P.C. 1997, (Molecular Evolution of an Arsenate Detoxification Pathway DNA Shuffling. *Nat. Biotechnol.* 15, 436–438.

64 Stemmer, W.P.C., Soong, N.W. 1999, (Molecular Breeding of Viruses for Targeting and Other Clinical Properties), *Tumor Target* 4, 59–62.

65 Powell, S.K., Kaloss, M.A., Pinkstaff, A., McKee, R., Burimski, I., Pensiero, M., Otto, E., Stemmer, W.P.C., Soong, N.W. 2000, (Breeding of Retroviruses by DNA Shuffling for Improved Stability and Processing Yields), *Nat. Biotechnol.* 18, 1279–1282.

66 Soong, N.W., Nomura, L., Pekrun, K., Reed, M., Sheppard, L., Dawes, G., Stemmer, W.P.C. 2000, (Molecular Breeding of Viruses), *Nat. Genet.* 25, 436–439.

67 Zhao, H., Giver, L., Shao, Z., Affholter, J.A., Arnold, F.H. 1998, (Molecular Evolution by Staggered Extension Process (StEP) *In Vitro* Recombination), *Nat. Biotechnol.* 16, 258–261.

68 Zha, D.X., Eipper, A., Reetz, M.T. 2003, (Assembly of Designed Oligonucleotides as an Efficient Method for Gene Recombination: A New Tool in Directed Evolution), *Chembiochem.* 4, 34–39.

69 Lutz, S., Patrick, W.M. 2004, (Novel Methods for Directed Evolution of Enzymes: Quality, not Quantity), *Curr. Opin. Biotechnol.* 15, 291–297.

70 Neylon, C. 2004, (Chemical and Biochemical Strategies for the Randomization of Protein Encoding DNA Sequences: Library Construction Methods for Directed Evolution), *Nucleic Acids Res.* 32, 1448–1459.

71 Ostermeier, M., Nixon, A.E., Shim, J.H., Benkovic, S.J. 1999, (Combinatorial Protein Engineering by Incremental Truncation), *Proc. Natl. Acad. Sci. USA* 96, 3562–3567.

72 Lutz, S., Benkovic, S.J. 2000, (Homology-independent Protein Engineering), *Curr. Opin. Biotechnol.* 11, 319–324.

73 Lutz, S., Ostermeier, M., Moore, G.L., Maranas, C.D., Benkovic, S.J. 2001, (Creating Multiple-crossover DNA Libraries Independent of Sequence Identity), *Proc. Natl. Acad. Sci. USA* 98, 11248–11253.

74 Smith, G.P. 1985, (Filamentous Fusion Phage – Novel Expression Vectors that Display Cloned Antigens on the Virion Surface), *Science* 228, 1315–1317.

75 Fernandez-Gacio, A., Uguen, M., Fastrez, J. 2003, (Phage Display as a Tool for the Directed Evolution of Enzymes), *Trends Biotechnol.* 21, 408–414.

76 Lin, H.N., Cornish, V.W. 2002, (Screening and Selection Methods for Large-scale Analysis of Protein Function), *Angew. Chem., Int. Ed. Engl.* 41, 4403–4425.

77 Sieber, V., Plückthun, A., Schmid, F.X. 1998, (Selecting Proteins with Improved Stability by a Phage-based Method), *Nat. Biotechnol.* 16, 955–960.

78 Verhaert, R.M.D., Beekwilder, J., Olsthoorn, R., van Duin, J., Quax, W.J. 2002, (Phage Display Selects for Amylases with Improved Low pH Starch-binding), *J Biotechnol.* 96, 103–118.

79 Dröge, M.J., Rüggeberg, C.J., van der Sloot, A.M., Schimmel, J., Dijkstra, D.S., Verhaert, R.M.D., Reetz, M.T., Quax, W.J. 2003, (Binding of Phage Displayed *Bacillus subtilis* Lipase A to a Phosphonate Suicide Inhibitor), *J. Biotechnol.* 101, 19–28.

80 Reetz, M.T., Rüggeberg, C.J., Dröge, M.J., Quax, W.J. 2002, (Immobilization of Chiral Enzyme Inhibitors on Solid Supports by Amide-forming Coupling and Olefin Metathesis), *Tetrahedron* 58, 8465–8473.

81 Goddard, J.-P., Reymond, J.-L. 2004, (Enzyme Assays for High-throughput Screening), *Curr. Opin. Biotechnol.* 15, 314–322.

82 Reymond, J.L., Wahler, D., 2002, (Substrate Arrays as Enzyme Fingerprinting Tools), *Chembiochem.* 3, 701–708.

83 Wahler, D., Reymond, J.L. 2001, (High-throughput Screening for Biocatalysts), *Curr. Opin. Biotechnol.* 12, 535–544.

84 Arnold, F.H., Georgiou, G. 2003, *Directed Enzyme Evolution: Screening and Selection Methods*, Humana Press, Totowa, New Jersey.

85 Eisenthal, R., Danson, M. 2002, *Enzyme Assays: A Practical Approach*, Oxford University Press, Oxford.

86 Reetz, M.T. 2003, (An Overview of High-throughput Screening Systems for Enantioselective Enzymatic Transformations), in *Directed Enzyme Evolution: Screening and Selection Methods*, eds. Arnold, F.H., Georgiou, G., (vol. 230), Humana Press, Totowa, New Jersey, (pp.) 59–282.

87 Reetz, M.T. 2002, (New Methods for the High-throughput Screening of Enantioselective Catalysts and Biocatalysts), *Angew. Chem., Int. Ed. Engl.* 41, 1335–1338.

88 Svendsen, A. 2004, *Enzyme Functionality: Design, Engineering, and Screening*, Marcel Dekker, New York.

89 Arnold, F.H., Georgiou, G. 2003, *Directed Evolution Library Creation: Methods and Protocols*, Humana Press, Totowa, New Jersey.

90 Hughes, M.D., Nagel, D.A., Santos, A.F., Sutherland, A.J., Hine, A.V. 2003, (Removing the Redundancy from Randomised Gene Libraries), *J. Mol. Biol.* 331, 973–979.

91 Murakami, H., Hohsaka, T., Sisido, M. 2002, (Random Insertion and Deletion of Arbitrary Number of Bases for Codon-based Random Mutation of DNAs), *Nat. Biotechnol.* 20, 76–81.

92 Pikkemaat, M.G., Janssen, D.B. 2002, (Generating Segmental Mutations in Haloalkane Dehalogenase: A Novel Part in the Directed Evolution Toolbox), *Nucleic Acids Res.* 30, e35.

93 Zhao, H., Arnold, F.H. 1997, (Optimization of DNA Shuffling for High Fidelity Recombination), *Nucleic Acids Res.* 25, 1307–1308.

94 Crameri, A., Raillard, S.A., Bermudez, E., Stemmer, W.P.C. 1998, (DNA Shuffling of a Family of Genes from Diverse Species Accelerates Directed Evolution), *Nature* 391, 288–291.

95 Shao, Z., Zhao, H., Giver, L., Arnold, F.H. 1998, (Random-priming *In Vitro* Recombination: an Effective Tool for Directed Evolution), *Nucleic Acids Res.* 26, 681–683.

96 Volkov, A.A., Shao, Z., Arnold, F.H. 1999, (Recombination and Chimeragenesis by *In Vitro* Heteroduplex Formation and *In Vivo* Repair), *Nucleic Acids Res.* 27, e18.

97 Kikuchi, M., Ohnishi, K., Harayama, S. 1999, (Novel Family Shuffling Methods for the *In Vitro* Evolution of Enzymes), *Gene* 236, 159–167.

98 Kikuchi, M., Ohnishi, K., Harayama, S. 2000, (An Effective Family Shuffling Method using Single-stranded DNA), *Gene* 243, 133–137.

99 Gibbs, M.D., Nevalainen, K.M., Bergquist, P.L. 2001, (Degenerate Oligonucleotide Gene Shuffling (DOGS): A Method for Enhancing the Frequency of Recombination with Family Shuffling), *Gene* 271, 13–20.

100 Coco, W.M., Levinson, W.E., Crist, M.J., Hektor, H.J., Darzins, A., Pienkos, P.T., Squires, C.H., Monticello, D.J. 2001, (DNA Shuffling Method for Generating Highly Recombined Genes and Evolved Enzymes), *Nat. Biotechnol.* 19, 354–359.

101 Song, J.K., Chung, B., Oh, Y.H., Rhee, J.S. 2002, (Construction of DNA-shuffled and Incrementally Truncated Libraries by a Mutagenic and Unidirectional Reassembly Method: Changing from a Substrate Specificity of Phospholipase to that of Lipase), *Appl. Environ. Microbiol.* 68, 6146–6151.

102 Ostermeier, M., Shim, J.H., Benkovic, S.J. 1999, (A Combinatorial Approach to Hybrid Enzymes Independent of DNA Homology), *Nat. Biotechnol.* 17, 1205–1209.

103 Sieber, V., Martinez, C.A., Arnold, F.H. 2001, (Libraries of Hybrid Proteins from Distantly Related Sequences), *Nat. Biotechnol.* 19, 456–460.

104 Kawarasaki, Y., Griswold, K.E., Stevenson, J.D., Selzer, T., Benkovic, S.J., Iverson, B.L., Georgiou, G. 2003, (Enhanced Crossover SCRATCHY: Construction and High-throughput Screening of a Combinatorial Library Containing Multiple Non-homologous Crossovers), *Nucleic Acids Res.* 31, e126.

105 Reetz, M.T., Zonta, A., Schimossek, K., Liebeton, K., Jaeger, K.E. 1997, (Creation of Enantioselective Biocatalysts for Organic

Chemistry by *In Vitro* Evolution), *Angew. Chem., Int. Ed. Engl.* 36, 2830–2832.

106 Farinas, E.T. 2003, (Colorimetric Screen for Aliphatic Hydroxylation by Cytochrome P450 using *p*-Nitrophenyl-substituted Alkanes), in *Directed Enzyme Evolution: Screening and Selection Methods*, (vol. 230), eds. Arnold, F.H., Georgiou, G., Humana Press, Totowa, New Jersey, (pp.) 149–155.

107 Badalassi, F., Wahler, D., Klein, G., Crotti, P., Reymond, J.L. 2000, (A Versatile Periodate-coupled Fluorogenic Assay for Hydrolytic Enzymes), *Angew. Chem., Int. Ed. Engl.* 39, 4067–4070.

108 Leroy, E., Bensel, N., Reymond, J.L. 2003, (Fluorogenic Cyanohydrin Esters as Chiral Probes for Esterase and Lipase Activity), *Adv. Synth. Catal.* 345, 859–865.

109 Sevestre, A., Helaine, V., Guyot, G., Martin, C., Hecquet, L. 2003, (A Fluorogenic Assay for Transketolase from *Saccharomyces cerevisiae*), *Tetrahedron Lett.* 44, 827–830.

110 Gonzalez-Garcia, E., Helaine, V., Klein, G., Schuermann, M., Sprenger, G.A., Fessner, W.D., Reymond, J.L. 2003, (Fluorogenic Stereochemical Probes for Transaldolases), *Chem. Eur. J.* 9, 893–899.

111 Zocher, F., Enzelberger, M.M., Bornscheuer, U.T., Hauer, B., Schmid, R.D. 1999, (A Colorimetric Assay Suitable for Screening Epoxide Hydrolase Activity), *Anal. Chim. Acta* 391, 345–351.

112 Beisson, F., Ferte, N., Nari, J., Noat, G., Arondel, V., Verger, R. 1999, (Use of Naturally Fluorescent Triacylglycerols from Parinari glaberrimum to Detect Low Lipase Activities from Arabidopsis thaliana Seedlings), *J. Lipid Res.* 40, 2313–2321.

113 Beisson, F., Tiss, A., Riviere, C., Verger, R. 2000, (Methods for Lipase Detection and Assay: A Critical Review), *Eur. J. Lipid Sci. Technol.* 102, 133–153.

114 Henke, E., Bornscheuer, U.T. 2003, (Fluorophoric Assay for the High-throughput Determination of Amidase Activity), *Anal. Chem.* 75, 255–260.

115 Konarzycka-Bessler, M., Bornscheuer, U.T. 2003, (A High-throughput-screening Method for Determining the Synthetic Activity of Hydrolases), *Angew. Chem., Int. Ed. Engl.* 42, 1418–1420.

116 Banerjee, A., Sharma, R., Banerjee, U.C. 2003, (A Rapid and Sensitive Fluorometric Assay Method for the Determination of Nitrilase Activity), *Biotech. Appl. Biochem.* 37, 289–293.

117 Li, Z., Bütikofer, L., Witholt, B. 2004, (High-throughput Measurement of the Enantiomeric Excess of Chiral Alcohols by using Two Enzymes), *Angew. Chem., Int. Ed. Engl.* 43, 1698–1702.

118 Moris-Varas, F., Shah, A., Aikens, J., Nadkarni, N.P., Rozzell, J.D., Demirjian, D.C. 1999, (Visualization of Enzyme-catalyzed Reactions using pH Indicators: Rapid Screening of Hydrolase Libraries and Estimation of the Enantioselectivity), *Bioorg. Med. Chem.* 7, 2183–2188.

119 Janes, L.E., Löwendahl, A.C., Kazlauskas, R.J. 1998, (Quantitative Screening of Hydrolase Libraries using pH Indicators: Identifying Active and Enantioselective Hydrolases), *Chem. Eur. J.* 4, 2324–2331.

120 Zhao. H. 2003, (A pH-indicator-based Screen for Hydrolytic Haloalkane Dehalogenase), in *Directed Enzyme Evolution: Screening and Selection Methods*, (vol. 230), eds. Arnold, F.H., Georgiou, G., Humana Press, Totowa, New Jersey, (pp.) 213–221.

121 Breuer, M., Pohl, M., Hauer, B., Lingen, B. 2002, (High-throughput Assay of (*R*)-phenylacetylcarbinol Synthesized by Pyruvate Decarboxylase), *Anal. Bioanal. Chem.* 374, 1069–1073.

122 Lingen, B., Kolter-Jung, D., Dünkelmann, P., Feldmann, R., Grotzinger, J., Pohl, M., Müller, M. 2003, (Alteration of the Substrate Specificity of Benzoylformate Decarboxylase from *Pseudomonas putida* by Directed Evolution), *Chembiochem.* 4, 721–726.

123 Doderer, K., Lutz-Wahl, S., Hauer, B., Schmid, R.D. 2003, (Spectrophotometric Assay for Epoxide Hydrolase Activity Toward any Epoxide), *Anal. Biochem.* 321, 131–134.

124 Marchesi, J.R. 2003, (A Microplate Fluorimetric Assay for Measuring Dehalogenase Activity), *J. Microbiol. Meth.* 55, 325–329.

125 Wahler, D., Reymond, J.L. 2001, (Novel Methods for Biocatalyst Screening), *Curr. Opin. Chem. Biol.* 5, 152–158.

126 Eggert, T., Funke, S.A., Rao, O.M., Acharya, P., Krumm, H., Reetz, M.T., Jaeger, K.-E. 2005 (Multiplex-PCR-based recombination as a novel high fidelity method for directed evolution), *Chembiochem* 6, 1–6.

5
Basics of Bioreaction Engineering

Nagaraj N. Rao, Stephan Lütz, Karsten Seelbach, and Andreas Liese

A prerequisite for any successful industrial process is rational design and development, taking into consideration both the technical and economic aspects. The goal of rational process development is to maximize selectivity, yield, chemical and enantiomeric purity as well as profitability, while trying to minimize consumption of the catalyst, energy, raw materials and solvents. To this end, detailed information about the characteristics of the biocatalyst, of the biotransformation reaction taking place and of the bioreactor in which this occurs is necessary (Fig. 5.1).

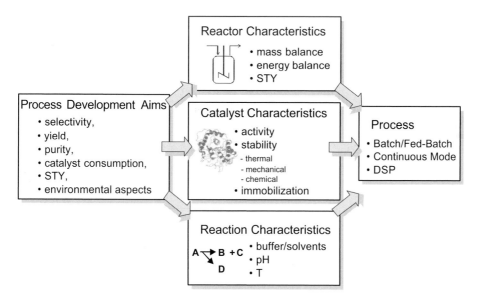

Fig. 5.1 Rational process design.

In this chapter, some fundamental aspects of reaction engineering, including definitions, kinetics, basic reactor types, immobilization and scaling-up are discussed from the point of view of understanding the data presented in Chapter 6: Biotransformations on an Industrial Scale.

Industrial Biotransformations. Andreas Liese, Karsten Seelbach, Christian Wandrey (Eds.)
Copyright © 2006 WILEY-VCH Verlag GmbH & Co. KGaA, Weinheim
ISBN: 3-527-31001-0

5.1
Definitions

5.1.1
Process Definitions

5.1.1.1 Conversion

The conversion is the number of converted molecules per number of starting molecules:

$$X_s = \frac{n_{s_0} - n_s}{n_{s_0}} \qquad (1)$$

where X_s = conversion of substrate s (–); n_{s0} = amount of substrate s at the start of the reaction (mol); n_s = amount of substrate s at the end of the reaction (mol).

Good conversion makes the recycling of unconverted reactant solution unnecessary and also minimizes reactor volumes. However, long reaction times or high catalyst concentrations may be needed to achieve this. This can lead to undesired subsequent reactions of the product as well as the formation of by-products.

5.1.1.2 Yield
The yield is the number of synthesized molecules of product p per number of starting molecules:

$$Y_p = \frac{n_p - n_{p_0}}{n_{s_0}} \cdot \frac{|v_s|}{|v_p|} \qquad (2)$$

where Y_p = yield of product p (–); n_{p0} = amount of product p at the start of the reaction (mol); n_p = amount of product p at the end of the reaction (mol); v_s = stoichiometric factor for substrate s (–); and v_p = stoichiometric factor for product p (–).

In combination with the conversion or the selectivity values, the yield describes how many product molecules are synthesized in relation to the starting number of substrate molecules. The described yield is based on analytical results. In practice, the isolated yield is usually given, because this describes the actual amount of product obtained after down stream processing (DSP). The isolated yield does not help in understanding single reaction steps and developing correct kinetic models. If an entire process is considered, the overall yield can be calculated by multiplication of the yields of all the single steps. The yield becomes the key economic factor if the unreacted substrate cannot be recovered during down stream processing.

5.1.1.3 Selectivity
The selectivity, sometimes also known as integral selectivity, is the number of molecules synthesized per number of molecules *converted* and is described by the formula:

$$\sigma_p = \frac{n_p - n_{p_0}}{n_{s_0} - n_s} \cdot \frac{|v_s|}{|v_p|} \qquad (3)$$

where σ_p = selectivity to component p (–); n_{s0} = amount of substrate s at the start of the reaction (mol); n_s = amount of substrate s at the end of the reaction (mol); n_{p0} = amount of product p at the start of the reaction (mol); n_p = amount of product p at the end of the reaction (mol); v_s = stoichiometric factor for substrate s (–); and v_p = stoichiometric factor for product p (–).

The selectivity describes the extent to which the product molecules are synthesized in relation to the substrate molecules converted. The selectivity must be as close to unity as possible, in order to avoid wastage of the starting material. If the unreacted substrate can be recovered during down stream processing and recycled into the reactor, the selectivity becomes a very crucial factor in the economic operation of the plant.

If only a very short reaction course is considered, the selectivity assumes a differential form. This is interesting for gaining information on the synthesis of by-products at every step of the conversion. This information is important to estimate whether a premature halting of the reaction would be more useful in improving the overall yield of the process.

The combination of conversion, yield and selectivity leads to the equation:

$$\eta = \sigma \cdot X \tag{4}$$

5.1.1.4 Enantiomeric Excess

The enantiomeric excess (ee) is the difference between the numbers of both enantiomers per sum of the enantiomers:

$$ee_R = \frac{n_R - n_S}{n_R + n_S} \tag{5}$$

where ee_R = enantiomeric excess of the (R)-enantiomer (–); n_R = amount of the (R)-enantiomer (mol); and n_S = amount of the (S)-enantiomer (mol).

The enantiomeric excess describes the enantiomeric purity of an optically active molecule. Minor differences in the spatial sequence of the binding partners of one central atom (particularly in the active site region) can lead to dramatic differences in chemical behavior, biological pathways and (substrate) recognition. It is absolutely essential to choose biotransformation processes that lead to products with high enantiomeric excess. This is because the pharmacological activities of the enantiomers might not be identical. Each enantiomer could even have a completely different pharmacological activity profile, including serious side effects in certain cases. The enantiomers often have different organoleptic properties, such as taste, flavor and odor. Many pharmaceuticals and agrochemicals, which were previously sold as racemates, are now sold as single enantiomer products ("racemic switches"). The US FDA (Food and Drug Administration) makes it mandatory for drug companies to carry out clinical trials of the individual enantiomers before selection of the correct enantiomer as the active pharmaceutical ingredient [1]. The use of enantiomerically pure drugs instead of racemates avoids the intake of inactive (or even toxic) compounds. Similarly, this applies to the use of enantiomerically pure agrochemicals.

5.1.1.5 Turnover Number

The turnover number (*tn*) is the number of synthesized molecules per number of catalyst molecules used:

$$tn = \frac{n_p}{n_{cat}} \cdot \frac{1}{|v_p|} \tag{6}$$

where *tn* = turnover number (–); n_p = amount of product *p* at the end of the reaction (mol); v_p = stoichiometric factor for the product *p* (–); and n_{cat} = amount of catalyst (mol).

The turnover number is a measure of the efficiency of a catalyst. Kinases, dehydrogenases and aminotransferases have turnover numbers of the order of 10^3. Enzymes such as carbonic anhydrase and superoxide dismutase have turnover numbers of the order of 10^6 [2]. Clearly the turnover numbers have to be high for commercial applications of biotransformations. Particularly, while using expensive catalysts, the *tn* should be as high as possible to reduce the final production cost.

It is very important to name the defined reaction parameters in combination with the *tn* to make the values comparable. Instead of the *tn*, the deactivation rate or half-life may also be given.

The turnover number can also be given for cofactors/coenzymes.

The turnover number is more frequently used, in addition to the turnover frequency and the selectivity, in order to evaluate a catalyst in homogeneous (chemical) catalysis. In biocatalysis, possibly because the molar mass has to be taken into account to obtain a dimensionless number, this parameter is used less frequently.

5.1.1.6 Turnover Frequency

The turnover frequency is a mass-independent quantity for describing the activity of the biocatalyst and is defined as:

$$tof = \frac{n_s}{t \cdot n_{cat}} \tag{7}$$

where *tof* = turnover frequency (s^{-1}); n_s = moles of converted starting material *s* (mol, μmol); *t* = time for conversion (s, min); and n_{cat} = moles of active sites (mol, μmol).

The turnover frequency allows an evaluation of the performance between different catalyst systems, irrespective of whether they are biological or non-biological. Its threshold is at the value of one event per second per active site.

According to this definition, the *tof* can be determined only if the number of active sites is known. For an enzymatic reaction obeying Michaelis–Menten kinetics (*vide infra*), the turnover frequency is given by:

$$tof = \frac{1}{k_{cat}} \tag{8}$$

Biological catalysts have high turnover frequencies compared with chemical catalysts. However, when substrate/product solubility, stability and molecular mass of the biocatalyst are taken into consideration, this advantage is lost. As an example, the epoxidation and sulfidation catalyst Mn-Salen has a *tof* of 3 h^{-1}, while its enzymatic counterpart chloroperoxidase (CPO) has a *tof* of 4500 h^{-1}. However, the molecular masses are 635 and 42 000 g mol^{-1}, respectively [3, 4].

5.1.1.7 Enzyme Activity

The enzyme activity is defined as the reaction rate per mass of catalyst (protein):

$$V = \frac{\partial n_s}{\partial t \cdot m_{cat}} \tag{9}$$

where V = maximum activity of the enzyme under defined conditions (katal kg^{-1}, U mg^{-1}); ∂n_s = differential amount of converted substrate (mol, µmol); ∂t = differential time for conversion (s, min); and m_{cat} = mass of catalyst (kg, mg).

The SI unit for enzyme activity is the katal (kat = mol s^{-1}, 1 kat = 6×10^7 U), which results in very low values, so that µ-, n- or pkat are often used. A more practical unit of measurement of enzyme activity is the Unit (1 U = 1 µmol min^{-1}). It is very important that the activity be mentioned along with the substrate used and with all the necessary reaction conditions, such as temperature, buffer salts, pH value and so on. The concentration of a protein solution can be determined using indirect photometric tests (e.g., Bradford [5]) or analytical methods (e.g., electrophoresis). Only when the enzyme contains a photometrically active component, e.g., a heme protein, can the concentration be readily determined by direct photometric absorption. The activity can also be given per unit of reaction volume (U mL^{-1}), if the enzyme concentration cannot be determined due to a lack of information on its molecular mass or due to the inability to analyze the mass of the solubilized enzyme.

5.1.1.8 Deactivation Rate

The deactivation rate is defined as the loss of catalyst activity per unit of time:

$$V_1 = V_0 \cdot e^{-k_{deact} \cdot (t_1 - t_0)} \tag{10}$$

where k_{deact} = deactivation rate (min^{-1}, h^{-1}, d^{-1}); V_0 = enzyme activity at the start of the measurement (U mg^{-1}); V_1 = enzyme activity at the end of the measurement (U mg^{-1}); t_0 = time at the start of the measurement (min, h, d); and t_1 = time at the end of the measurement (min, h, d).

The deactivation rate expresses the stability of a catalyst.

5.1.1.9 Half-life

The half-life is defined as the time in which the activity is halved:

$$V_1 = V_0 \cdot e^{-k_{deact} \cdot (t_1 - t_0)} \tag{11}$$

$$V_2 = V_0 \cdot e^{-k_{deact} \cdot (t_2 - t_0)} \tag{12}$$

$$V_1 = \frac{1}{2} \cdot V_2 \tag{13}$$

$$\Rightarrow t_{1/2} = \frac{\ln(2)}{k_{deact}} \tag{14}$$

where $t_{1/2}$ = half-life of the catalyst (min, h, d); V_x = enzyme activity at time t_x (U mg^{-1}); t_x = time of measurement (min, h, d); and k_{deact} = deactivation rate (min^{-1}, h^{-1}, d^{-1}).

The half-life expresses the stability of a catalyst. The activity usually shows a typical exponential decay. Therefore the half-life can be calculated and it gives the extent of the catalyst deactivation independent of considered time differences.

5.1.1.10 Catalyst Consumption

The biocatalyst consumption (*bc*) is defined as the mass of catalyst consumed per mass of synthesized product:

$$bc = \frac{m_{cat}}{m_p} \tag{15}$$

where bc = *biocatalyst* consumption (g kg^{-1}); m_{cat} = mass of catalyst used to give the synthesized mass of product (g); and m_p = mass of synthesized product (g).

If an expensive catalyst is used it is clear that the biocatalyst consumption should be as low as possible. Pharmaceutical products are sometimes of such high value that in discontinuous reactions the catalyst is often discarded without recycling. As the catalyst stability can change with conversion due to the formation of deactivating by-products, it is interesting to look at the differential catalyst consumption to find the optimal conversion for terminating the reaction and for separating the reaction solution from the catalyst.

5.1.1.11 Biomass Concentration

The *cell dry weight* (cdw, g L^{-1}) is the most widely used unit for the analysis of the density of bacteria and fungi. It corresponds generally to about 20–25% of the cell wet weight. The cell dry weight is used when a balance between the substrates and products has to be calculated.

The *optical density* is a fast and relatively easy method to determine the cell density. It should always be given with reference to the wavelength at which it is measured. Optical density can also be measured on-line during the biotransformation process.

5.1.1.12 Residence Time

The residence time (τ) is defined as the quotient of reactor volume and feed rate:

$$\tau = \frac{V_R}{F} \tag{16}$$

where τ = residence time or reaction time (h); V_R = reactor volume (L); and F = feed rate (L h^{-1}).

The residence time describes the average time a molecule is in the reactor. As the residence times of different molecules are not the same, the average residence time is usually used. Diffusion effects and non-ideal stirring in a continuously operated stirred tank reactor (CSTR) or back mixing in plug-flow reactors result in a broad distribution of single residence times. To give a detailed simulation of the process this distribution has to be taken into account. For example, one substrate molecule could leave the reactor directly after it has been fed into the reactor or it could stay in the reaction system forever. Therefore, the selectivity can be strongly influenced by a broad distribution. The dilution rate in fermentations is the inverse residence time.

5.1.1.13 Space–Time Yield

The space–time yield (STY) is the mass of product synthesized per reactor volume and time. It is also known as the *volumetric productivity* or the *reactor productivity.*

$$STY = \frac{m_p}{\tau \cdot V_R} \tag{17}$$

where STY = space–time yield (g L^{-1} d^{-1}); m_p = mass of synthesized product (g); τ = residence time or reaction time (d); and V_R = reactor volume (L).

A low STY can be caused by low substrate solubility or poor catalyst reactivity and leads to low product concentration or poor reaction rates. This, in turn, leads to more complex down stream processing steps or large reactor volumes, respectively. Commercial processes usually account for an STY of more than 500 g L^{-1} d^{-1}. However, in cases where the product is of high value, an STY of even 100 g L^{-1} d^{-1} may be attractive.

5.1.2
Definitions of Unit Operations

5.1.2.1 Filtration

Filtration is a process of separating solid particles from liquids (or liquid droplets from gases) with the help of a filter medium. Filtration is used either for clarification of liquids (such as the separation of the biomass from the reaction solution) or for the recovery of solids.

In *cross-flow filtration*, the suspension flows at a high speed, tangential to the filter surface, thereby avoiding the formation of a cake. The rate of flow of the liquid through the filter medium is slow. In biotechnology, membranes are applied as filters and are classified according to the molecular weight of the compound they retain (molecular weight cut off, MWCO) [6, 7].

In *microfiltration*, particles with a diameter of 0.1–10 μm are separated from a solvent or other components with molecular masses less than 10^5 to 10^6 g mol^{-1}. The separation occurs due to a sieving effect and the particles are separated on the basis of their dimensions. The hydrostatic pressure difference can be in the region of 10–500 kPa. In *ultrafiltration* (UF), the components to be retained by the membrane are molecules or small particles with a size less than 0.1 μm in diameter. The hydrostatic pressure of the feed solution is negligible. UF membranes are usually asymmetrical in structure and the pores have a diameter of 1–10 nm. In *nanofiltration*, the cut-off size is about 0.001 µm, so nanofiltration can remove larger molecules, particulates and divalent species. In *reverse osmosis*, the cut-off size is about 0.0001 µm. Particles, macromolecules and low molecular mass compounds such as salts and sugars can be separated from a solvent, usually water. The feed solution often has high osmotic pressure and this must be overcome by the hydrostatic pressure that is applied as the driving force. Thus, microfiltration, ultrafiltration, nanofiltration and reverse osmosis differ from each other in the size of the particles being separated (Fig. 5.2).

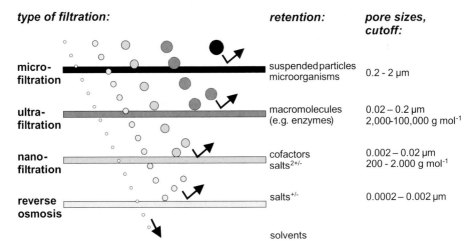

type of filtration: **retention:**   **pore sizes, cutoff:**

micro-filtration   suspended particles microorganisms   0.2 - 2 µm

ultra-filtration   macromolecules (e.g. enzymes)   0.02 – 0.2 µm, 2,000-100,000 g mol^{-1}

nano-filtration   cofactors, salts$^{2+/-}$ 0.002 – 0.02 µm, 200 - 2.000 g mol^{-1}

reverse osmosis   salts$^{+/-}$ 0.0002 – 0.002 µm

solvents

Fig. 5.2 Classification of membrane filtration processes.

Apart from being used as separation devices, membranes can also be used to decouple the residence time of the biocatalyst and reactants, which leads to continuously operated reactors [8–10].

5.1.2.2 Crystallization
Crystallization is a heat- and mass-transfer process that is strongly influenced by fluid and particle mechanics. Nucleation and growth kinetics are the key processes in this unit of operation. Even small traces of impurities may influence the process dramatically.

5.1.2.3 Precipitation
A very rapid process of crystallization could be considered to be precipitation; however, precipitation is often an irreversible process. Many precipitates are virtually insoluble substances produced by a chemical reaction. On the other hand, crystallized products can often be redissolved when the original conditions of the solvent, temperature and concentration are restored.

In *iso-electric precipitation*, advantage is taken of the fact that at the pH at which the protein or a zwitter-ion bears zero net charge, its solubility is at a minimum. The value of the pH at which this occurs is called the iso-electric point. Most proteins have an iso-electric point of less than 7. *Iso-electric focusing* is an electrophoretic separation method by which amphoteric compounds are fractionated according to their iso-electric point along a continuous pH gradient.

5.1.2.4 Dialysis and Electrodialysis

In dialysis, one or more solutes are transferred from one solution ("feed") to another ("dialysate") through a membrane down their concentration gradient. When pressure is employed in the separation, in addition to the concentration gradient, the process is called *pervaporation*.

In *electrodialysis*, the separation of components of an ionic solution occurs in a cell consisting of a series of anion- and cation-exchange membranes arranged alternately between an anode and a cathode, to form individual electrodialysis cells. During the process of electrodialysis, there is an increase in the ion concentration of one type of ion in one type of compartment and is accompanied by a simultaneous decrease in the concentration in the other type of compartment.

5.1.2.5 Extraction

Liquid–liquid extraction is a classical process used to purify bioproducts. Both aqueous and organic extraction media are employed. Extraction using aqueous two-phase polymer systems (PEG, dextran) provides an excellent method of protein separation [11, 12].

5.1.2.6 Adsorption

Adsorption is a common operation employed for the purification of biological products. Activated carbon or ion-exchange resins can be used. In adsorption, material accumulates on the surface of a solid adsorbent that has a multiplicity of pores of different sizes.

In *affinity adsorption*, proteins can be absorbed biospecifically on the basis of their interaction with a complimentary tertiary structure. In ion-exchange adsorption, the adsorbents have ionic groups with easily dissociable counter ions.

5.1.2.7 Chromatography

Chromatography covers a wide variety of methods of separation and purification of molecules. It is based on the differences in equilibrium constants for the components of a mixture placed in a biphasic system. Ion-exchange chromatography, affinity chromatography, gel filtration chromatography and hydrophobic interaction chromatography (HIC) are used in protein purification in decreasing order of importance. Reversed phase chromatography is primarily of interest in the separation of peptides and small molecules.

5.2
Biocatalyst Kinetics

5.2.1
Types of Biocatalysts

A biocatalyst is always described as a whole cell or an enzyme. In the first case we are looking at a mini-reactor with all the necessary cofactors and sequences of enzymes concentrated in one cell. In the second case, the main catalytic unit is isolated and purified. In both cases optimization is possible. Furthermore, multistep biosynthetic pathways can be changed to prevent degradation of the desired product or to produce precursors not normally prioritized in the usual pathway. All these changes for the whole cell lead to an optimized mini-plant. The optimization of the main catalyst is comparable to catalyst development inside a reaction system.

Whole cells can be bacteria, fungi, plant cells or animal cells. They are subdivided into the two groups: prokaryotic cells and eukaryotic cells.

5.2.1.1 **Prokaryotic Cells**

Prokaryotic cells are the "lowest microorganisms" and do not possess a true nucleus. The nuclear material is contained in the cytoplasm of the cell. They reproduce by cell division, are relatively small in size (0.2–10 µm) and exist as single cells or as mycelia. When designing bioreactors, an adequate supply of nutrients as well as oxygen to the bioreactor must be assured, as the cells, e.g., bacteria, grow rapidly. Parameters such as pH, oxygen feed rate and temperature in the bioreactor must be optimized. Perhaps the most widely used prokaryotic microorganism in industrial biotransformations is *Escherichia coli*, which is a native to the human intestinal flora.

5.2.1.2 **Eukaryotic Cells**

Eukaryotic cells are higher microorganisms and have a true nucleus bounded by a nuclear membrane. They reproduce by an indirect cell division method called mitosis, in which the two daughter nuclei normally receive identical compliments of the number of chromosomes characteristic of the somatic cells of the species. They are larger in size (5–30 µm) and have a complex structure. When eukaryotic cells are used as biocatalysts, high or low mechanical stress must be avoided by using large stirrers at slow speed and by eliminating dead zones in the fermenter. *Saccharomyces cerevisiae* and *Zymomonas mobilis* represent the most important eukaryotic cells used in industrial biotransformations.

5.2.2
Enzyme Structure

An enzyme is an accumulation of one or more polypeptide chains in the form of a protein. It is unique in being capable of accelerating or producing by catalytic action a transformation in a substrate, for which it is often specific. The three-dimensional structure of an enzyme is determined on different levels [13, 14].

- *Primary structure*: sequence of connected amino acids of a protein chain.
- *Secondary structure*: hydrogen bonds from the particular type of R–N–H–O=C–R are responsible for the formation of the secondary structure, the helix or the β-sheet, of one protein chain.
- *Tertiary structure*: hydrogen and disulfide bonds, and ionic and hydrophobic forces lead to the tertiary structure, the folded protein chain.
- *Quaternary structure*: if several protein chains are combined in the form of subunits, the quaternary structure is formed. It is not covalent bonds, but molecular interactions occurring in the secondary, tertiary and quaternary structures that are responsible for the formation of the well-functioning catalytic system.

5.2.3
Kinetics

In this section the fundamentals of enzyme kinetics will be discussed in brief. For a detailed description of enzyme kinetics and a discussion of the different kinetic models please refer to the literature [15–17].

The determination of the kinetic parameters can be carried out in two different ways: either by measurement of the initial reaction rate under different reaction conditions or by batch experiments. In both cases a kinetic model has to exist to describe the reaction rate as a function of the concentrations of the different reaction components. The two methods differ in the number of variable components. In the case of the initial determination of the reaction rate, only the concentration of one compound is altered, whereby all others are constant. On the contrary, in the case of batch reactions, the time course of all concentrations of all (!) components is measured. Therefore, all mass balances [see Eqs. (31–35)] are required for the determination of the kinetic parameters that form a system of coupled differential equations. The values of the kinetic parameters are determined by fitting the kinetic equations to the measured data by non-linear regression (Fig. 5.3). In the case of batch experiments, this is supplemented by numerical integration of the reaction rate equations. An appropriate test of the kinetic model and the kinetic parameters is the simulation of the time-courses of batch reactor experiments with different starting concentrations of the substrate. These are then compared with the actual batch experiments.

The fundamental description of enzyme kinetics dates back to Michaelis and Menten [18]. In 1913, in their theory on enzyme catalysis they postulated the existence of an enzyme–substrate (ES) complex that is formed in a reversible reaction from the substrate (S) and enzyme (E).

$$E \; + \; S \; \underset{k_{-1}}{\overset{k_1}{\rightleftharpoons}} \; ES \; \overset{k_2}{\longrightarrow} \; E \; + \; P$$

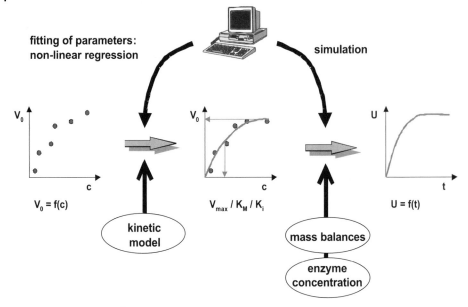

Fig. 5.3 Determination of kinetic parameters.

The rate limiting step is the dissociation of the ES complex $(k_{-1} \gg k_2)$. The reaction rate is proportional to the rapid, "preceding equilibrium". As a consequence of the latter assumptions the following reaction rate equation is derived [Eq. (18)]:

$$v = \frac{v_{max} \cdot [S]}{K + [S]} \qquad \text{with}: \quad K = K_S = \frac{k_{-1}}{k_1} \tag{18}$$

where v = reaction rate (U mg^{-1}); V_{max} = maximum reaction rate (U mg^{-1}); K = dissociation constant of ES complex (mM); k_x = reaction rate constant of reaction step x (min^{-1}); and $[S]$ = the substrate concentration (mM).

Here K is identical to the dissociation constant K_S of the ES complex. Briggs and Haldane extended this theory in 1925 [19]. They substituted the assumption of the "rapid equilibrium" by a "steady state assumption". This means that after starting the reaction an almost steady state level of the ES complex is established in a very short time. The concentration of the ES complex is constant with time $(d[ES]/dt = 0)$. In this assumption, the constant K has to be increased by k_2, resulting in the Michaelis–Menten constant K_M.

$$K = K_M = \frac{k_{-1} + k_2}{k_1} \tag{19}$$

where K_M = Michaelis–Menten constant (mM).

The Michaelis–Menten constant does not now describe the dissociation, but rather it is a kinetic constant. It denotes the special substrate concentration where half of the maximal activity is reached. As the Michaelis–Menten constant approaches the dissociation constant K_S of the ES complex, it is valuable for estimating individual reaction kinetics.

K_M values usually range from 10 to 0.01 mM. A low K_M value implies a high affinity between the enzyme and the substrate. The function $v = f(S)$ is shown in Fig. 5.4:

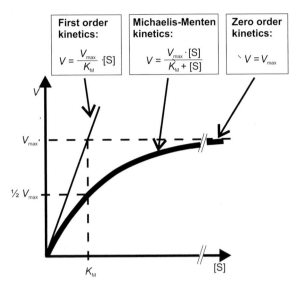

First order kinetics:	Michaelis-Menten kinetics:	Zero order kinetics:
$V = \dfrac{V_{max}}{K_M} \cdot [S]$	$V = \dfrac{V_{max} \cdot [S]}{K_M + [S]}$	$V = V_{max}$

Fig. 5.4 Typical Michaelis–Menten curve.

Here, two borderline cases have to be discussed. If the substrate concentration $[S]$ is a long way below the K_M value, a linear first-order kinetic results. The active sites of the enzyme are almost all free and the substrate concentration is rate limiting. If the substrate concentration is high, all active sites are saturated and zero-order kinetics results. By the Michaelis–Menten equation given above, a one-substrate reaction can be described. If a two-substrate reaction is to be addressed, two reaction rate terms are connected by multiplication. For a simple two-substrate reaction of $A + B$ the double substrate kinetics for the forward reaction are:

$$v_{forward} = \frac{V_{max}^{forward} \cdot [A] \cdot [B]}{(K_{MA} + [A]) \cdot (K_{MB} + [B])} \tag{20}$$

A corresponding equation for the reverse reaction can also be set up. The resulting total reaction rate equals the difference between the forward and reverse reactions.

$$v = \frac{d[P]}{dt} = v_{forward} - v_{reverse} \tag{21}$$

The easiest way to describe a double substrate reaction is by using the kinetics already described [Eqs. (20 and 21)] derived from the single substrate Michaelis–Menten kinetics [Eq. (18)]. The disadvantages of this approach are:
• no information about the mechanism is included;
• forward and reverse reactions are addressed as two totally independent reactions;
• no information about the equilibrium is included.

However, as opposed to these disadvantages there are also significant advantages of the Michaelis–Menten kinetics:

- over broad ranges, real reactors can be described by this simple type of kinetics;
- kinetic parameters are independent of the definition of reaction direction;
- all parameters possess a graphical meaning.

A mechanistically correct description of the total reaction is only possible with a more complex model, e.g., ordered bi–bi, random bi–bi, ping–pong, and so on [16–20]. In these models all equilibria leading to the formation of transition states are described individually. The single kinetic parameters no longer have a descriptive meaning. The advantage of these mechanistic models is the possibility of having an exact description of the individual equilibria.

5.2.3.1 Inhibition

The activity of a biocatalyst can be reduced by binding with substances that do not lead to active $[ES]$ complexes. These substances can bind both reversibly and irreversibly to the enzyme and are called inhibitors. Reversible binding inhibitors form, as the term implies, enzyme–inhibitor complexes in a reversible manner. Some forms of inhibition will be discussed briefly, but more details can be found in the literature [15–17].

Competitive inhibitors bind to the substrate binding site of the enzyme, thereby competing with the substrate molecules for the same binding site of the protein. The presence of high substrate concentrations suppresses the binding of the inhibitors, and *vice versa*. The V_{max} value of the enzyme remains unaffected; however, the K_M value increases.

The rate law for a competitive inhibitor is given in Eq. (22).

$$v = V_{max} \frac{[S]}{K_M \left(1 + \frac{[I]}{K_I}\right) + [S]} \tag{22}$$

where $[I]$ = concentration of inhibitor I (mM); and K_I = inhibition constant (mM).

A typical method of deducing the form of inhibition is by plotting the experimental data in reciprocal and thus linearized graphs. One way of visualizing the inhibition is a Lineweaver–Burk plot ($1/v$ versus $1/[S]$), which gives straight lines that intersect at the same point on the ordinate axis, the slopes of which are greater as the inhibitor concentration increases, for the case of a competitive inhibitor. The corresponding relationship is given in Eq. (23):

$$\frac{1}{v_i} = \frac{\left(1 + \frac{I}{K_I}\right) K_M}{V_{max}} \left(\frac{1}{[S]}\right) + \frac{1}{V_{max}} \tag{23}$$

It is predominantly linearization that has been used to estimate the kinetic parameters from the slope and the intercept. However today non-linear regression via mathematical computer tools is easy to use and is recommended for calculating the kinetic parameters [21].

The product P of a biotransformation is often a competitive inhibitor leading to *product inhibition*. The corresponding Eq. (24) is obtained by inserting the product concentration into the inhibition term.

$$v = V_{max} \frac{[S]}{K_M \left(1 + \frac{[P]}{K_I}\right) + [S]} \tag{24}$$

Non-competitive inhibitors bind only to the ES complex, leading to ESI complexes. The binding of the inhibitor, which does not need to resemble the substrate, can be assumed to cause a structural distortion, rendering the enzyme less active [Eq. (25)].

$$v = V_{max} \frac{[S]}{K_M + \left(1 + \frac{[I]}{K_I}\right) \cdot [S]} \tag{25}$$

The linearization leads to Eq. (26):

$$\frac{1}{v_i} = \left(\frac{K_M}{V_{max}}\right) \frac{1}{[S]} + \frac{\left(1 + \frac{I}{K_I}\right)}{V_{max}} \tag{26}$$

A special case of non-competitive inhibition is the *substrate surplus inhibition*, in which an ESS complex is formed. Using the substrate concentration as the inhibitor concentration in Eq. (25) gives Eq. (27):

$$v = V_{max} \frac{[S]}{K_M + [S] + \frac{[S]^2}{K_I}} \tag{27}$$

Non-competitive inhibitors can bind to the ES complex and to the free enzyme molecule. It is therefore also referred to as *mixed inhibition*. The corresponding equation is Eq. (28):

$$v = V_{max} \frac{[S]}{\left(1 + \frac{[I]}{K_I}\right)(K_M + [S])} \tag{28}$$

The Lineweaver–Burk form is given in Eq. (29):

$$\frac{1}{v_i} = \left[\frac{\left(1 + \frac{I}{K_I}\right) K_M}{V_{max}}\right] \left(\frac{1}{[S]}\right) + \frac{\left(1 + \frac{I}{K_I}\right)}{V_{max}} \tag{29}$$

The inhibition schemes and equations are summarized in Table 5.1.

Tab. 5.1 Inhibitor schemes and equations.

Inhibition type	Scheme	Equation
competitive	E + S $\rightleftharpoons$ ES $\longrightarrow$ E + P + I $\updownarrow$ EI	$v = V_{max} \dfrac{[S]}{K_M \left(1 + \dfrac{[I]}{K_I}\right) + [S]}$
product	see above, $I = P$	$v = V_{max} \dfrac{[S]}{K_M \left(1 + \dfrac{[P]}{K_I}\right) + [S]}$
noncompetitive	E + S $\rightleftharpoons$ ES $\longrightarrow$ E + P + I $\updownarrow$ ESI	$v = V_{max} \dfrac{[S]}{K_M + \left(1 + \dfrac{[I]}{K_I}\right) \cdot [S]}$
substrate surplus	see above, $I = S$	$v = V_{max} \dfrac{[S]}{K_M + [S] + \dfrac{[S]^2}{K_I}}$
non-competitve/mixed	E + S $\rightleftharpoons$ ES $\longrightarrow$ E + P + + I I $\updownarrow$ $\updownarrow$ EI + S $\rightleftharpoons$ ESI	$v = V_{max} \dfrac{[S]}{\left(1 + \dfrac{[I]}{K_I}\right)(K_M + [S])}$

5.3
Basic Reactor Types and their Modes of Operation

While designing and selecting reactors for biotransformations, certain characteristic features of the biocatalyst have to be considered.

In whole cell biotransformations, materials are processed in each active microbial cell, so that the main function of the bioreactor should be to provide and maintain the optimal conditions for the cells to perform the biotransformation. The performance of the biocatalysts depends on concentration levels and physical requirements (such as salts and appropriate temperature, respectively). Microorganisms can adapt the structure and activity of their enzymes to the process condition, unlike isolated enzymes.

The microbial mass can increase significantly as the biotransformation progresses, leading to a change in rheological behavior. Also, metabolic products of the cells may influence the performance of the biocatalyst.

- Microorganisms are often sensitive to strong shear stress.
- Bioreactors generally have to function under sterile conditions to avoid microbial contamination, so they must be designed for easy sterilization.
- In the case with both enzymes and whole cells as biocatalysts, the substrate and/or the product may inhibit or deactivate the biocatalyst, necessitating special reactor layouts.
- Biotransformations with enzymes are usually carried out in a single (aqueous) or in two (aqueous/organic) phases, whereas the whole cells generally catalyze in gas–liquid–solid systems. In this instance the liquid phase is usually aqueous.
- Foam formation is undesirable but often unavoidable in the case of whole cell biotransformations, where most processes are aerobic. Due consideration must be given to this aspect while designing or selecting a bioreactor.

Only the three basic types of reactors are presented here. All others are variations or deductions thereof:

- sirred-tank reactors (STRs) or batch reactors
- continuously operated stirred-tank reactors (CSTRs)
- plug-flow reactors (PFRs).

In contrast to the stirred tank reactor which is operated batchwise, the last two are operated continuously. By knowing the main characteristics of these fundamental reactors and some of their variations, it is possible to choose the appropriate reactor for a specific application [22]. This is especially important when dealing with a kinetically or thermodynamically limited system. In the following only the basic terms are explained. For further reading, the reader is referred to particular textbooks [23–32].

The *stirred tank reactor* is operated in a non-stationary manner (Fig. 5.5). Assuming ideal mixing, as a function of time, the concentration is the same in every volume element. With advancing conversion the substrate concentration decreases and the product concentration increases.

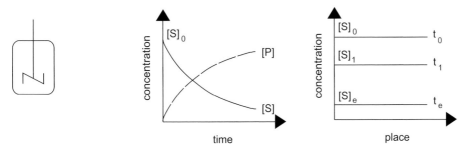

Fig. 5.5 Concentration–time and concentration–place profile for a stirred tank reactor.

This reactor type is widely used on an industrial scale. A variation is operation in a *repetitive batch* or *fed batch* mode. Repetitive batch means that the catalyst is separated after complete conversion by filtration or even decantation. New substrate solution is added and the reaction is started again. Fed batch means that one reaction compound, in most cases the substrate, is fed to the reactor during the conversion.

The *continuous stirred tank reactor* (CSTR) works under product outflow conditions, meaning that the concentrations in every volume element are the same as those at the outlet of the reactor (Fig. 5.6).

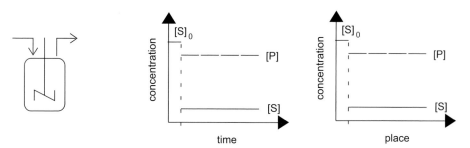

Fig. 5.6 Concentration–time and concentration–place profile for a continuously operated stirred tank reactor.

If the steady state is reached, the concentrations are independent of time and place. The conversion is controlled by the catalyst concentration and the residence time.

One very common application of the CSTR is a *cascade* of n CSTRs (Fig. 5.7).

With an increasing number of reaction vessels (n) the cascade approximates the plug flow reactor. The product concentration increases stepwise from vessel to vessel.

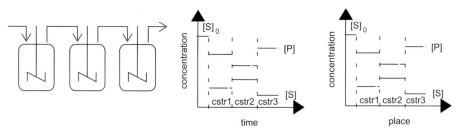

Fig. 5.7 Concentration–time and concentration–place profile for a cascade of continuously operated stirred tank reactors.

In a *plug flow reactor* the product concentration increases slowly over the length of the reactor (Fig. 5.8). Therefore, the average reaction rate is faster than in a continuously operated stirred tank reactor. In each single volume element in the reactor the concentration is constant in the steady state. In other words, the dimension of time is exchanged for the dimension of place in comparison with a stirred tank reactor.

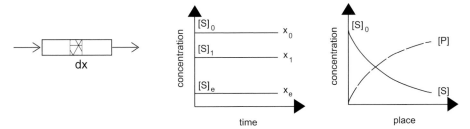

Fig. 5.8 Concentration–time and concentration–place profile for a plug-flow reactor.

If a reaction is limited by a surplus of substrate or product inhibition the choice of the correct reactor is important to yield a high reaction rate.

Any reaction exhibiting substrate surplus inhibition should not be carried out in a batch reactor set-up, as this results in longer reaction times. The high substrate concentration at start-up lowers the reaction velocity. Here, a continuously operated stirred tank reactor is preferred. By establishing a high conversion in the steady state a low substrate concentration is achieved. In addition, the use of a fed batch results in a small substrate concentration.

If product inhibition occurs, either a stirred tank reactor, a plug flow reactor or a cascade of n continuously operated stirred tank reactors should be chosen. In all these reactors the product concentration increases over time. Alternatively, a differential reactor with integrated product separation can be used.

5.3.1
Mass and Energy Balances

The principle of conservation of matter and of energy forms the basis for any reactor calculations [26, 31]. A material balance can be set up for any molecular species taking part in the reaction. Thus, for a component X, such a balance over a system may be formulated as:

(accumulation of mass of X in the system) = (mass of X into the system) –
(mass of X out of the system) + (mass of X produced by the reaction) (30)

The energy balance states the following:

(rate of accumulation of energy) = (rate of energy in) – (rate of energy out) +
(rate of energy production) (31)

Looking at one volume element, a change in energy or mass can only occur by reaction or a difference between the input and output. This is illustrated in Fig. 5.9.

The performance of the different reactor types with respect to one reaction can be simulated mathematically. This is also the verification of the kinetic model of the reaction, as it should describe the course of the concentration for each compound with only a small

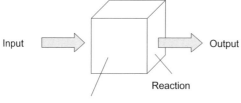

Change in Concentration

Fig. 5.9 Influence on the mass-balance in a volume element.

error. The main part of the simulation model is the coupled system of differential equations of the first order, which are the mass balances of all reactants and products. The change in the concentration of one compound in time and in a volume element (= "accumulation") is the sum of convection, reaction and diffusion.

$$\text{accumulation} = \text{convection} + \text{reaction} + \text{diffusion} \tag{32}$$

The convection term describes the change in the concentration of one compound in the reactor as the difference between the influx into the reactor and the efflux. The reaction term describes, by using the kinetic model, the change in the concentration of one compound as a result of the reaction. The reaction velocity v is the sum of the individual reaction velocities describing the consumption of a substrate or formation of a product. Diffusion is only given in the case where no ideal mixing is stated. Depending on the reactor type chosen, the mass balance can be simplified, stating ideal mixing.

5.3.1.1 Mass Balance in Stirred Tank Reactor

The mass balance of each compound is defined by the reaction rate only, as no fluid enters or leaves the reactor. At a defined time the concentrations are the same in every volume element (diffusion = 0). There is no influx or efflux of substrate or products for a single volume element in time (convection = 0).

The mass balance is simplified to:

$$-\frac{d[S]}{dt} = v \tag{32}$$

The time t that is necessary to reach a desired conversion X can be determined by integrating the reciprocal rate equation from zero to the desired conversion X.

$$dt = -\frac{d[S]}{v} = \frac{[S]_0 \cdot dX}{v} \quad \Rightarrow \quad t = [S]_0 \cdot \int_0^x \frac{1}{v} \cdot dX \tag{33}$$

5.3.1.2 Mass Balance in a Plug-flow Reactor

The change in reaction rate within a unit volume passing the reactor length is equivalent to a change corresponding to the residence time within the reactor. Diffusion is neglected in an ideal plug-flow reactor (diffusion = 0) and all the concentrations will not change with time in the steady state (accumulation = 0). Just by exchanging t for τ, Eq. (33) can be also used for the plug-flow reactor to determine the residence time necessary to reach a desired conversion:

$$\tau = [S]_0 \cdot \int_0^x \frac{1}{v} \cdot dX \tag{34}$$

5.3.1.3 Mass Balance in a Continuously Operated Stirred Tank Reactor

The concentration of substrate S within the reactor is affected by convection as well as by reaction. There is no diffusion between different volume elements (diffusion = 0) and in the steady state the concentrations will not change with time (accumulation = 0). The mass balance is simplified to:

$$0 = -\frac{d[S]}{dt} = \frac{[S]_0 - [S]}{\tau} + v_S \tag{35}$$

The residence time that is necessary to reach a desired conversion can be determined by Eq. (36):

$$\tau = [S]_0 \cdot \frac{1}{v} \cdot dX \tag{36}$$

5.4
Biocatalyst Recycling and Recovery

To facilitate downstream processing or increase the turnover number of the catalyst, it is often beneficial to remove the catalyst from the reaction solution and recycle it. Several approaches for the recycling and recovery of the catalyst are feasible (Fig. 5.10).

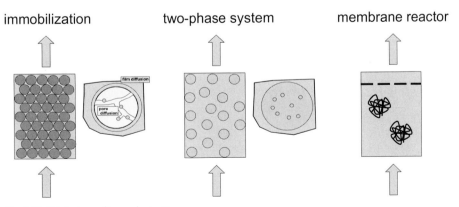

Fig. 5.10 Catalyst recycling and retention.

One of the most important techniques is immobilization [33], mostly onto a heterogeneous support. Nevertheless, immobilization is not generally required, as a large number of industrially applied processes use non-immobilized biocatalysts [34]. By using a second liquid phase (e.g., organic solvent, ionic liquid, PEG/dextran) the biocatalyst can be immobilized in one phase homogeneously. Filtration can also be used to decouple the residence times of the catalyst and reactants, as indicated in Section 5.1.2.1.

The immobilization techniques for biocatalysts discussed here are those often used on an industrial scale to reduce the cost of the catalyst and to increase the stability.

If a good biocatalyst is found for a specific reaction, one possibility for further improvement of its properties is by immobilization. The best way of immobilizing the biocatalyst has to be determined by experiment. This is dependent on the nature of the reaction, the stability of the biocatalyst, the possibilities for the immobilization of the biocatalyst and the activity of the immobilized biocatalyst. No straightforward plan for testing immobilization is known, but at least the biocatalyst structure should be taken into consideration. Often mathematical black-box optimization methods such as the genetic algorithm are applied for the optimization of the immobilization conditions and to yield maximum activity.

The main *advantages* of immobilization are:
- easy separation of biocatalyst
- reduced costs of down stream processing
- possibility of biocatalyst recycling

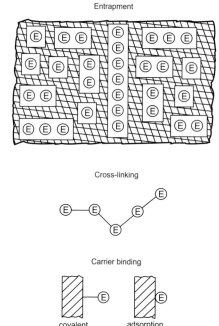

Fig. 5.11 Common immobilization methods.

- better stability, especially towards organic solvents and heat
- use of fixed bed reactors
- easier realization of continuous production.

The main *disadvantages* of immobilization are:
- loss of absolute activity due to the immobilization process
- lower activity of immobilized biocatalyst compared with non-immobilized biocatalyst as used in the processes with membrane filtration
- additional costs for carrier or immobilization matrix and immobilization procedure
- carrier or matrix cannot be recycled
- diffusion limitations lowering reaction rates.

In spite of these disadvantages, immobilization has become an indispensable aspect of industrial biotransformations. The most common methods for the immobilization are entrapment in matrices, cross-linking and covalent binding (Fig. 5.11).

5.4.1
Entrapment

The biocatalyst can be entrapped in natural or synthetic gel matrices. A very simple method is the entrapment in sodium alginate, a natural polysaccharide. The water soluble alginate is mixed with the biocatalyst solution and dropped into a calcium chloride solution in which water-insoluble alginate beads are formed (Fig. 5.12).

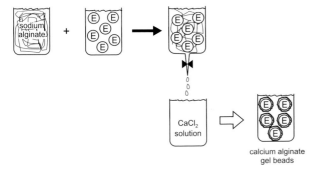

Fig. 5.12 Alginate gel formation.

Another naturally occurring polysaccharide widely used for immobilization is κ-carrageenan. In a manner similar to the alginate method, a mixture of κ-carrageenan in saline and the biocatalyst solution (or suspension) is dropped into a solution of a gelling reagent, such as potassium chloride. Ammonium, calcium and aluminum cations also serve as good gelling reagents. The gel can be hardened by glutaraldehyde, hexamethylenediamine or other cross-linking reagents, often enhancing biocatalyst stability (Fig. 5.13).

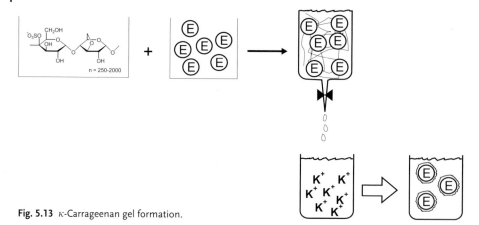

Fig. 5.13 κ-Carrageenan gel formation.

An often-used synthetic immobilization method employs polyacrylamide gel. The biocatalyst, a monomer (acrylamide) and a cross-linker (e.g., *N,N'*-methylenebisacrylamide) are mixed and polymerized by starting the reaction with an initiator (e.g., potassium persulfate) in the presence of a stimulator (e.g., 3-dimethylamino propionitrile) (Fig. 5.14).

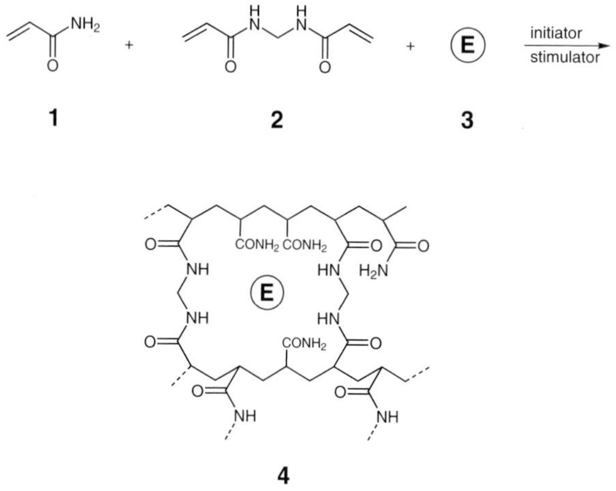

1 = acrylamide
2 = N,N'-methylenebisacrylamide
3 = enzyme
4 = polyacrylamide gel

Fig. 5.14 Polyacrylamide gel formation.

5.4.2
Cross-linking

The most popular cross-linking reagent is glutaraldehyde, although other bi- or multi-functional reagents can be used instead. The single biocatalyst (protein molecule or cell) is cross-linked to insoluble macromolecules or cell pellets (Fig. 5.15).

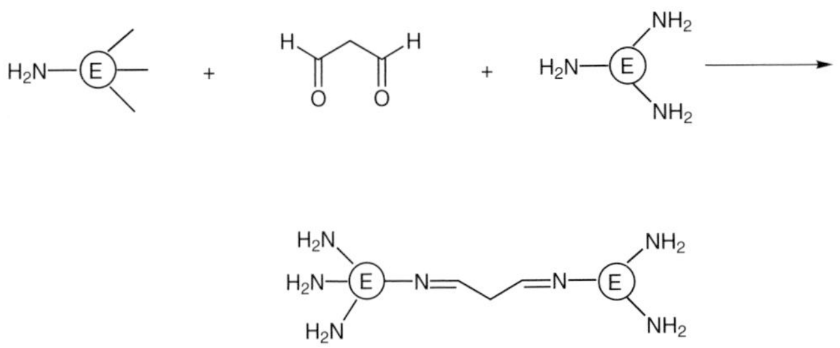

Fig. 5.15 Cross-linking with glutaraldehyde.

5.4.3
Covalent Binding

The biocatalyst, in this case usually an isolated enzyme, can be attached to a carrier by a reaction of the amino or acid groups of the proteins. Generally, amino acid residues, which are not involved in the active site or in the substrate-binding site of the enzyme, can be used for covalent binding with carriers. Usually the carrier is a polymer (polysaccharide, polysiloxane, polyacrylamide, etc.) bearing hydroxy groups or amino groups on its surface. To combine the enzyme with these groups, different activation methods can be applied. These activated carriers are commercially available. Examples are spacers with epoxy groups, which are activated by cyanogen bromide or other activating groups such as acid azide, leading to a spontaneous reaction with the amino group of the biocatalyst (Fig. 5.16).

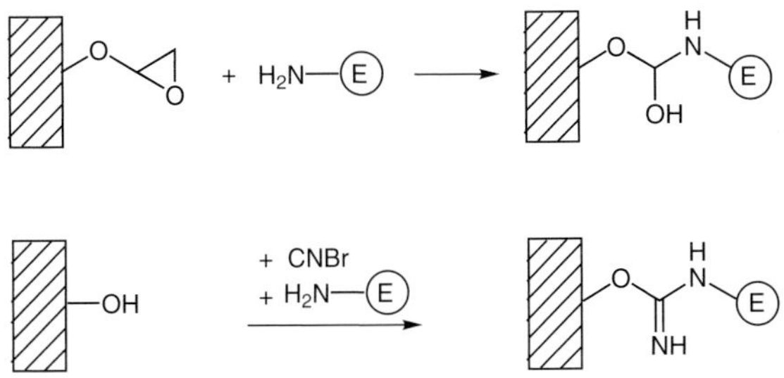

Fig. 5.16 Two examples of carrier-coupling using the amino group of an enzyme.

If the acid group is to be coupled to a carrier, the enzyme also has to be activated, e.g., by reaction with a diimide. The activated carrier and enzyme can now be coupled (Fig. 5.17).

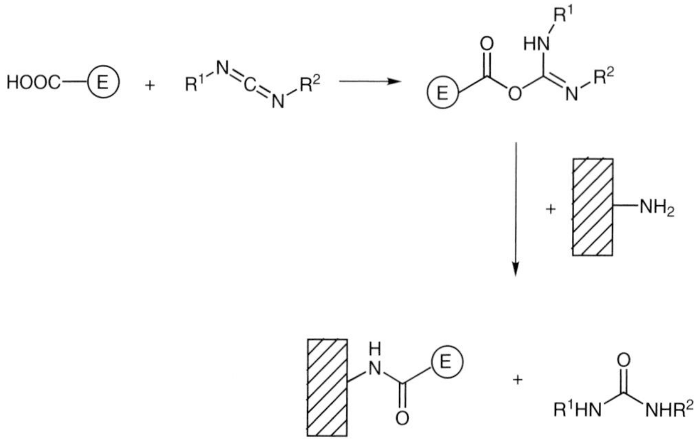

Fig. 5.17 Example of carrier-coupling using the acid group of an enzyme.

5.4.4
Membrane Filtration

An alternative approach to immobilization for an easy biocatalyst recovery is the use of a filtration unit. For the separation of suspended whole cells or solubilized enzymes, membranes are often applied as filters. For a classification of membranes refer to Section 5.1.2.1 on filtration (Fig. 5.2).

Microfiltration membranes are applied for the separation of whole cells (for examples see the processes on pages 219, 350, 487). Due to the lower molecular weight and physical size of enzymes, ultrafiltration membranes have to be used to retain them (for examples see the processes on pages 157, 179, 203).

5.5
Reaction Parameters

The reaction parameters have to be optimized for the reaction with respect to high space–time yields and high stability of the biocatalyst, which implies low production costs.

Parameters that can be varied include:
• pH
• temperature
• solvents
• buffer salts
• cofactors

- immobilization methods
- substrate and product concentrations
- addition of antioxidants or stabilizers
- reactor material or coating
- physical treatment (stirring, pumping, gas–liquid phases, etc.).

During optimization of the space–time yield it is necessary to consider the catalyst costs. In particular, the combination of stability and activity has to be considered. Sometimes it is desirable to work at very low temperatures with low reaction rates, which have to be compensated for by a large amount of biocatalyst if the turnover number has thus to be increased. In other cases the turnover number will have a lower priority because as much product as possible has to be synthesized. No pragmatic rule exists for the best strategy to adopt to optimize the reaction conditions. Empirical methods based on statistical methods have good chances of being successful (e.g., genetic algorithms).The most important improvements can be made by finding and constructing an optimal catalyst.

5.6
Scale-up of Bioreactors

One of the primary goals of bioprocess engineering is to achieve large-scale production levels by scaling-up processes that were originally developed in the laboratory on a bench-scale. It implies an increase in the volume handled in a production cycle, or in other words, a decrease in the surface to volume ratio. This fact prohibits a direct scaling-up of a process, because mass and energy transport limitations arise.

Scaling-up involves scientific, engineering and economic considerations, in addition to correct judgments and experience. It also results in the introduction of larger apparatus, whose effects on the process must be analyzed and optimized. The possible major effects on physical and biotransformation parameters as a result of scaling-up are summarized in Table 5.2.

5.7
Recent Developments and Trends

New challenges in the reaction engineering of biotransformations can be foreseen in various areas. In this section we will briefly discuss recent developments and techniques which have already been applied successfully on a laboratory scale but have not found widespread technical application.

In the field of redox biotransformations [36, 37] a great deal of effort has been spent on establishing methods for efficient cofactor regeneration. Recent research has elucidated the potential of using cheap redox equivalents such as molecular hydrogen with hydrogenase enzymes [38, 39] or electrons in indirect electrochemical cofactor reduction for the production and regeneration of reduced [40, 41] and oxidized cofactors [42, 43]. Even enzymes that are not cofactor dependent can be combined with electrochemical steps, e.g., for the synthesis of a cosubstrate [44].

Tab. 5.2 Issues involved in scaling-up [35].

Physical property/unit	Considerations during scale-up
1. *mixing*	
1.1 mechanical agitation	higher stirring speeds for efficient mixing
	change in impeller design/multiple impellers
	change in ratio of reactor diameter to height
	change in reactor type
	top-drive or bottom-drive mixing
1.2 aeration	change in sparger design
	change in ratio of volumetric air flow rate to cross-sectional area of reactor
	foaming – anti-foam agents or mechanical foam separators
	increase in superficial gas velocity for constant aeration rate – limited by flooding
2. *heat transfer*	efficient removal of heat generated by reaction itself and by broth agitation
	with increase in reactor size, ratio of surface area available for heat transfer to reactor volume can decrease at constant aspect ratio
	vigorous mixing ensures better heat transfer but has its own limitations
	lowering temperature of cooling water if viable
	external or internal cooling; double-jacket, full pipe or half-pipe cooling
	Material of construction
3. *power consumption*	changes in power requirement for mechanical agitation, aeration and circulation
	affected by rheology in bioreactor
4. *rheology*	changes due to increase in biomass concentration
	changes due to formation of mycelial networks in cases with filamentous fungi as biocatalysts
	synthesis of extracellular polysaccharides
	baffle design and number
	changes due to heat transfer rates
5. *biotransformation*	biocatalyst immobilization method
	effect of shear stress and concentration gradients
	changes in microorganism metabolism
	changes in age profiles of microorganism
	changes in regulatory phenomena due to changes in extracellular environment
	ease of sterilization
	changes in down-stream processing techniques and/or capacities

Owing to the need for stabilized biocatalysts, a large amount of effort is directed at improving established immobilization techniques and to develop new ones. A technique of great interest is the use of cross-linked enzyme crystals (CLEC) and cross-linked enzyme aggregates (CLEA). For the preparation of CLEAs, protein preparations of even technical purity can be used [45].

For substrates and products of poor solubility, aqueous–organic mixtures are being used as reaction media for the biotransformation. The two-phase systems have recently attracted new attention, as it has been pointed out that not only the $\log P$ [46], but also structural elements of the solvents have to be considered to guide the choice of the solvent [47].

For compounds with low water solubility the use of new solvent systems such as ionic liquids [48] or supercritical fluids [49–51] (e.g., carbon dioxide) in industrial biocatalysis can be foreseen. Owing to the water-free environment in ionic liquids even enzymatic condensation reactions can be carried out [52]. Alternative techniques for the separation of the products from the new reaction medium have to be developed: while distillation is feasible for volatile products, nanofiltration can be used for nonvolatile compounds [53]. Instead of using organic solvents, a two-phase system of water and ionic liquids can also be used in biocatalysis both with isolated enzymes [54] or whole cells [55].

One of the most important tasks for biocatalysis that has to be considered in the early stages of industrial development is that of reducing the time-to-market. Thus, in the future some problems in the process development could frequently be solved *in silico* (reactor design, process strategy, kinetic simulations). Scale-up and scale-down operations will be facilitated by mini- and microreactor technology and parallelization. Isolated enzymes and whole cells will give rise to new synthetic routes increasing the use of renewable resources. Using modern shuffling techniques, directed evolution will give rise to improved enzymes with new biocatalytic properties [56, 57] (see also Chapter 4).

References

1 Stinson, S.C. **1997**, (FDA may Confer New Status on Enantiomers), *Chem. Eng. News* 75, 28–29.

2 Silverman, R. **2000**, *The Organic Chemistry of Enzyme Catalyzed Reactions*, Academic Press, San Diego.

3 Palucki, M., Hanson, P., Jacobsen, E.N. **1992**, (Asymmetric Oxidation of Sulfides with H_2O_2 Catalyzed by (Salen)Mn(III) Complexes), *Tetrahedron Lett.* 33, 7111–7114.

4 van Deurzen, M.P.J., van Rantwijk, F., Sheldon, R.A. **1997**, (Selective Oxidations Catalyzed by Peroxidases), *Tetrahedron* 53, 13183–13220.

5 Bradford, M.M. **1976**, (Rapid and Sensitive Method for Quantitation of Microgram Quantities of Protein Utilizing Principle of Protein–Dye Binding), *Anal. Biochem.* 72, 248–254.

6 Mulder, M. **1996**, *Basic Principles of Membrane Technology*, Kluwer Academic, Dordrecht.

7 Noble, R.D., Stern, S.A. **1995**, *Membrane Separations Technology. Principles and Applications*, Elsevier, Amsterdam.

8 Flaschel, E., Wandrey, C., Kula, M.-R. **1983**, (Ultrafiltration for the Separation of Biocatalysts), in *Advances in Biochemical Engineering/Biotechnology*, (vol. 26), ed. Fiechter A, Springer, Berlin, (pp.) 73–142.

9 Wandrey, C., Wichmann, R., Bückmann, A.F., Kula. M.-R. **1980**, (Immobilization of Biocatalysts Using Ultrafiltration Techniques), in *Enzyme Engineering 5*,

eds. Weetall, H.H., Royer, G.P., Plenum Press, New York, (pp.) 453–456.

10 Kragl, U. **1996**, *Immobilized Enzymes and Membrane Reactors. Industrial Enzymology*, Macmillan Press, London, (pp.) 275–283.

11 Thommes, J., Halfar, M., Gieren, H., Curvers, S., Takors, R., Brunschier, R., Kula, M.-R. **2001**, (Human Chymotrypsinogen B Production from *Pichia pastoris* by Integrated Development of Fermentation and Downstream Processing. Part 2. Protein Recovery), *Biotechnol. Prog.* 17, 503–512.

12 Coimbra, J.S.R., Thoemmes, J., Kula, M.-R. Silva, L.H.M., Meirelles, A.J.A. **1997**, (Separation of beta-Lactoglobulin from Cheese Whey Using an Aqueous Two-phase System), *Arquiv. Biolog. Tecnolog.* 40, 189–196.

13 Buchholz, K., Kasche, V. **1997**, *Biokatalysatoren und Enzymtechnologie*, VCH Verlagsgesellschaft, Weinheim, (p.) 343.

14 Voet, D., Voet, J. **2004**, *Biochemistry*, Wiley, New York.

15 Bisswanger, H. **1979**, *Theorie und Methoden der Enzymkinetik*, Verlag Chemie, Weinheim.

16 Segel, I.H. **1995**, *Enzyme Kinetics*, John Wiley & Sons, New York.

17 Cornish-Bowden, A. **1995**, *Fundamentals of Enzyme Kinetics*, Portland Press, London.

18 Michaelis, L., Menten, M.-L. **1913**, (Die Kinetik der Invertinwirkung), *Biochem. Z.* 49, 333–369.

19 Briggs, G.E., Haldane, J.B.S. **1925**, *Biochem. J.* 19, 338–339.

20 Biselli, M., Kragl, U., Wandrey, C. **2002**, (Reaction Engineering for Enzyme-catalyzed Biotransformations), in *Enzyme Catalysis in Organic Synthesis*, (vol. 10), eds. Drauz, K., Waldmann, H., Wiley-VCH, Weinheim, (pp.) 185–258.

21 Vasic-Racki, D., Kragl, U., Liese, A. **2003**, (Benefits of Enzyme Kinetics Modelling), *Chem. Biochem. Eng. Quart.* 17, 7–18.

22 Kragl, U., Liese, A. **1999**, (Biotransformations, Engineering Aspects), in *The Encyclopedia of Bioprocess Technology: Fermentation, Biocatalysis & Bioseparation*, eds. Flickinger, M.C., Drew, S.W., John Wiley & Sons, New York, 454–464.

23 Baerns, M., Hofmann, H., Renken, A. **1992**, *Chemische Reaktionstechnik*, Georg Thieme Verlag, Stuttgart.

24 Bailey, J.E., Ollis, D.F. **1986**, *Biochemical Engineering Fundamentals*, McGraw-Hill, New York, (pp.) 984.

25 Jakubith, M. **1998**, *Grundoperationen und chemische Reaktionstechnik*, Wiley-VCH, Weinheim.

26 Levenspiel, O. **1999**, *Chemical Reaction Engineering*, John Wiley & Sons, New York.

27 Fitzer, E., Fritz, W., Emig, G. **1995**, *Technische Chemie*, Springer-Verlag, Berlin.

28 Fogler, S. H. **1998**, *Elements of Chemical Reaction Engineering*, Prentice-Hall, Englewood Cliffs, NJ.

29 Froment, G.F., Bischoff, K.B. **1990**, *Chemical Reactor Analysis and Design*, John Wiley & Sons, New York.

30 Ertl, G., Knözinger, H., Weitkamp, J. **1997**, *Handbook of Heterogeneous Catalysis*, Wiley-VCH, Weinheim.

31 Westerterp, K.R., van Swaaij, W.P.M., Beenackers, A.A.C.M. **1984**, *Chemical Reactor Design and Operation*, Wiley, New York.

32 Richardson, J.F. Peacock, D.G. **1994**, *Chemical Engineering*, Pergamon Press, Oxford.

33 Bickerstaff, G.F. **1997**, *Immobilization of Enzymes and Cells*, Humana Press, Totowa.

34 Lütz, S., Liese, A. **2004**, *Nonimmobilized Biocatalysts in Industrial Fine Chemical Synthesis. Ullmann's Encyclopedia of Industrial Chemistry*, Electronic Release, DOI: 10.1002/14356007.h17_h01, Wiley-VCH, Weinheim.

35 Nielsen, J., Villadsen, J., Liden, G. **2003**, *Bioreaction Engineering Principles*, Kluwer Academic, New York.

36 Fang, J.-M., Lin, C.-H., Bradshaw, C.W., Wong, C.-H. **1995**, (Enzymes in Organic Synthesis: Oxidoreductions), *J. Chem. Soc., Perkin Trans. 1*, 967–978.

37 Chenault, H.K., Whitesides, G.M. **1987**, (Regeneration of Nicotinamide Cofactors for use in Organic Synthesis), *Appl. Biochem. Biotechnol.* 14, 147–197.

38 Mertens, R., Greiner, L., van den Ban, E.C.D., Haaker, H.B.C.M., Liese, A. **2003**, (Practical Applications of Hydrogenase I from *Pyrococcus furiosus* for NADPH Generation and Regeneration), *J. Mol. Catal. B: Enzymatic* 24–25, 39–52.

39 Greiner, L., Mueller, D.H., van den Ban, E.C.D., Woeltinger, J., Wandrey, C., Liese, A. **2003**, (Membrane Aerated Hydrogenation: Enzymatic and Chemical Homogeneous Catalysis. *Adv. Synth. Catal.* 345, 679–683.

40 Vuorilehto, K., Lütz, S., Wandrey, C. **2004**, (Indirect Electrochemical Reduction of Nicotinamide Coenzymes), *Bioelectrochemistry* 65, 1–7.

41 Hollmann, F., Schmid, A. **2004**, (Electrochemical Regeneration of Oxidoreductases for Cell-free Biocatalytic Redox Reactions), *Biocatal. Biotransform.* 22, 63–88.

42 Schröder, I., Steckhan, E., Liese, A. **2003**, (In situ NAD(P)+ Regeneration using 2,2′-Azinobis(3-ethylbenzothiazolin-6-sulfonate) as an Electron Transfer Mediator), *J. Electroanal. Chem.* 109–115.

43 Degenring, D., Schröder, I., Liese, A., Greiner, L. **2004**, (Resolution of 1,2-Diols by Enzyme Catalyzed Oxidation with Anodic Mediated Cofactor Regeneration in the Extractive Membrane Reactor – Gaining Insight by Adaptive Simulation), *Org. Proc. Res. Develop.* 8, 213–218.

44 Lütz, S., Steckhan, E., Liese, A. **2004**, (First Asymmetric Electroenzymatic Oxidation Catalyzed by a Peroxidase), *Electrochem. Commun.* 6, 583–587.

45 Schoevaart, R., Wolbers, M.W., Golubovic, M., Ottens, M., Kieboom, A.P.G., van Rantwijk, F., van der Wielen, L.A.M., Sheldon, R.A. **2004**, (Preparation, Optimization, and Structures of Cross-linked Enzyme Aggregates (CLEAs)), *Biotechnol. Bioeng.* 87, 754–762.

46 Laane, C., Boeren, S., Vos, K., Veeger, C. **1987**, (Rules for Optimization of Biocatalysis in Organic Solvents), *Biotechnol. Bioeng.* 30, 81–87.

47 Villela, M., Stillger, T., Muller, M., Liese, A., Wandrey, C. **2003**, (Is log P a Convenient Criterion to Guide the Choice of Solvents for Biphasic Enzymatic Reactions?) *Angew. Chem., Int. Ed. Engl.* 42, 2993–2996.

48 Dorbritz, S., Ruth, W., Kragl, U. **2003**, (Investigations on the Stability and Aggregate Formation of an Ionic Liquid), *Abstracts of Papers of the American Chemical Society* 226, U644–U644.

49 Leitner, W. **2003**, (The Better Solution? Chemical Synthesis in Supercritical Carbon Dioxide), *Chem. unserer Zeit* 37, 32–38.

50 Leitner, W. **2004**, (Recent Advances in Catalyst Immobilization using Supercritical Carbon Dioxide), *Pure Appl. Chem.* 76, 635–644.

51 Reetz, M.T., Wiesenhofer, W., Francio, G., Leitner, W. **2003**, (Continuous Flow Enzymatic Kinetic Resolution and Enantiomer Separation using Ionic Liquid/Supercritical Carbon Dioxide Media), *Adv. Synth. Catal.* 345, 1221–1228.

52 Kaftzik, N., Neumann, S., Kula, M.R., Kragl, U. **2003**, (Enzymatic Condensation Reactions in Ionic Liquids), in *Ionic Liquids as Green Solvents: Progress and Prospects*, ed. Rogers, R.D., Oxford Univ. Press (vol. 856), (pp.) 206–211.

53 Krockel, J., Kragl, U. **2003**, (Nanofiltration for the Separation of Nonvolatile Products from Solutions Containing Ionic Liquids), *Chem. Eng. Technol.* 26, 1166–1168.

54 Eckstein, M., Villela, M., Liese, A., Kragl, U. **2004**, (Use of an Ionic Liquid in a Two-phase System to Improve an Alcohol Dehydrogenase Catalysed Reduction), *Chem. Commun.* 1084–1085.

55 Pfruender, H., Amidjojo, M., Kragl, U., Weuster-Botz, D. **2004**, (Efficient Whole-cell Biotransformation in a Biphasic Ionic Liquid/Water System), *Angew. Chem., Int. Ed. Engl.* 43, 4529–4531.

56 Bornscheuer, U.T. **2002**, (Methods to Increase Enantioselectivity of Lipases and Esterases), *Curr. Opin. Biotechnol.* 13, 543–547.

57 Cirino, P.C., Arnold, F.H. **2002**, (Protein Engineering of Oxygenases for Biocatalysis), *Curr. Opin. Chem. Biol.* 6, 130–135.

6

Processes

Andreas Liese, Karsten Seelbach, Arne Buchholz, and Jürgen Haberland

In this chapter you will find industrial biotransformations sorted in the order of the enzyme classes (EC). One type of biotransformation is often carried out by several companies leading to identical or the same class of products. Here only one exemplary process is named. Only in cases where the reaction conditions differ fundamentally, resulting in a totally different process layout, are these listed separately (e.g. -aspartic acid).

It is difficult to judge which processes are applied on an industrial scale. But even if not all of the following processes are used on the ton scale, they are at least performed by industrial companies to produce compounds for research or clinical trials on kg scale.

If you know of any new biotransformation carried out on an industrial scale, or you notice that we missed any important one, we would be pleased if you could supply us with the appropriate information. For your convenience you will find a form at the end of this book.

On the next pages you will find a process example with all necessary explanations for an easy understanding of all used parameters and symbols in the flow sheets.

By reading the example you will also see the maximum number of parameters we have tried to find for each process.

Industrial Biotransformations. Andreas Liese, Karsten Seelbach, Christian Wandrey (Eds.)
Copyright © 2006 WILEY-VCH Verlag GmbH & Co. KGaA, Weinheim
ISBN: 3-527-31001-0

X.X.X.X = enzyme nomenclature number

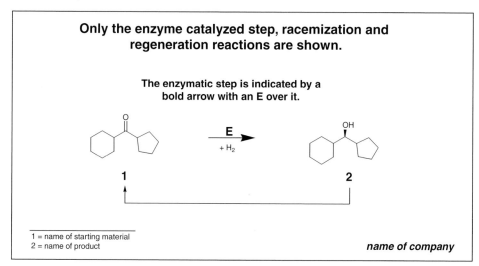

Only the enzyme catalyzed step, racemization and regeneration reactions are shown.

The enzymatic step is indicated by a bold arrow with an E over it.

E
+ H_2

1

2

1 = name of starting material
2 = name of product

name of company

Fig. X.X.X.X – 1

1) Reaction conditions

[N]:	molar concentration, mass concentration [molar mass] of component **N**
pH:	pH of reaction solution
T:	reaction temperature in °C
medium:	type of reaction medium: in most cases aqueous, but can also be several phases in combination with organic solvents
reaction type:	suggestion of enzyme nomenclature for the type of enzymatic catalyzed reaction
catalyst:	application of catalyst: solubilized / immobilized enzyme / whole cells
enzyme:	systematic name (alternative names)
strain:	name of strain
CAS (enzyme):	[CAS-number of enzyme]

2) Remarks

- Since in the chemical drawing on top of the page only the enzymatic step is shown, prior or subsequent steps, which might be part of the industrial process can be found here.

- Since it is often difficult to gain knowledge of the true industrial process conditions, those published in the past for the same reaction system are given.

- Besides the already mentioned topics you will find additional information regarding the discussed biotransformation, e.g. substrate spectrum, enzyme improvement, immobilization methods, and all other important information which does not fit to another category.

- If an established synthesis is replaced by a biotransformation, the classical, chemical synthesis can be found here as well.

3) **Flow scheme**

- The flow schemes are reduced to their fundamental steps. A list and explanation of the symbols is given in the next figures:

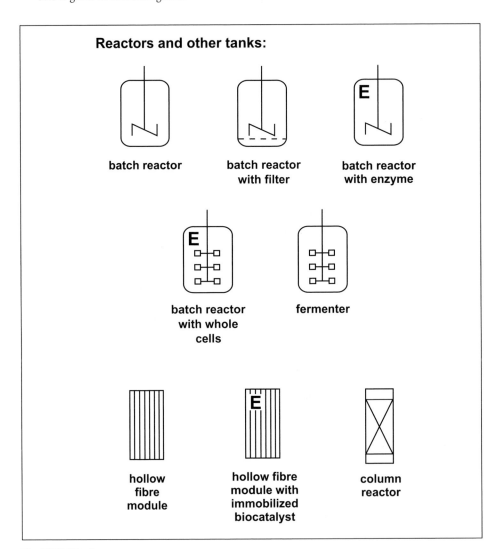

Fig. X.X.X.X – 2

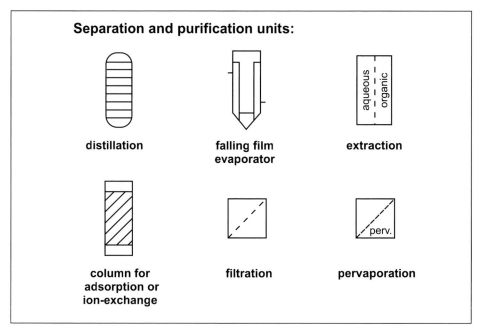

Fig. X.X.X.X – 3

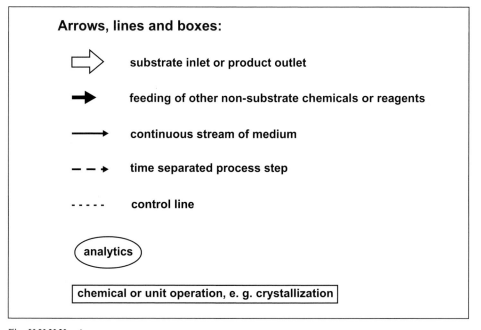

Fig. X.X.X.X – 4

4) Process parameters

conversion:	molar conversion in %
yield:	molar yield in %
selectivity	molar selectivity in %
ee:	enantiomeric excess in %
chemical purity:	purity of component in %
reactor type:	fed or repetitive batch, CSTR, plug flow reactor
reactor volume:	reactor volume in L
capacity:	mass of product per year in $t \cdot a^{-1}$
residence time:	time for one batch reaction or residence time in continuous operated reactor in hours
space-time-yield:	mass of product per time and reactor volume in $kg \cdot L^{-1} \cdot d^{-1}$
down stream processing:	purification of raw material after reaction, e. g. crystallization, filtration, distillation
enzyme activity:	in U (units = $\mu mol \cdot min^{-1}$) per mass of protein (mg) or volume of reaction solution (L)
enzyme consumption:	amount of consumed enzyme per mass of product
enzyme supplier:	company, country
start-up date:	start of production
closing date:	end of production
production site:	company, country
company:	company, country

5) Product application

- The application of the product as intermediate or the end-product are given here.

6) Literature

- You will find cited literature here. Often a personal communication or information direct from the company provided us with the necessary information.

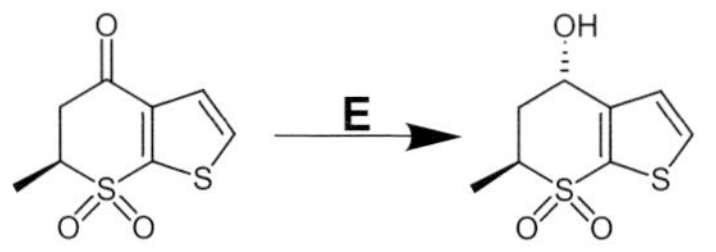

Alcohol dehydrogenase
Neurospora crassa

(6S)-**1**   (4S,6S)-**2**

1 = 5,6-dihydro-6-methyl-4H-thieno[2,3b]thiopyran-4-one-7,7-dioxide
2 = 5,6-dihydro-4-hydroxy-6-methyl-4H-thieno[2,3b]thiopyran-7,7-dioxide

AstraZeneca

Fig. 1.1.1.1 – 1

1) Reaction conditions

pH:	3.8–4.3
T:	33 °C
medium:	aqueous
reaction type:	redox reaction
catalyst:	suspended whole cells
enzyme:	alcohol NAD$^+$ oxidoreductase (alcohol dehydrogenase)
strain:	*Neurospora crassa*
CAS (enzyme):	[9031–72–5]

Alcohol dehydrogenase
Neurospora crassa

2) Remarks

- In comparison to the biotransformation the inversion of the *cis* alcohol in the chemical synthesis of Trusopt™ (see product application) is not quantitative:

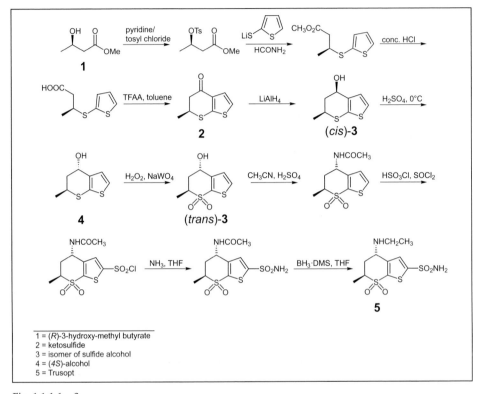

1 = (*R*)-3-hydroxy-methyl butyrate
2 = ketosulfide
3 = isomer of sulfide alcohol
4 = (*4S*)-alcohol
5 = Trusopt

Fig. 1.1.1.1 – 2

- The biological route overcomes the problem of incomplete inversion:

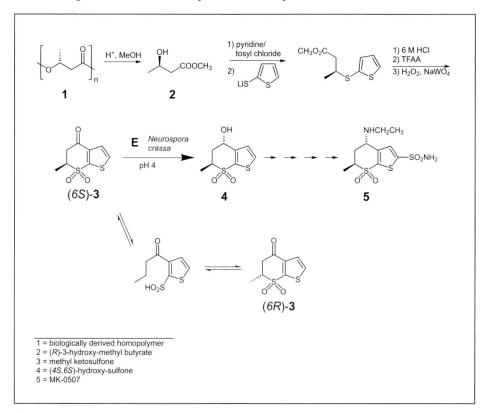

1 = biologically derived homopolymer
2 = (R)-3-hydroxy-methyl butyrate
3 = methyl ketosulfone
4 = (4S,6S)-hydroxy-sulfone
5 = MK-0507

Fig. 1.1.1.1 – 3

- The (R)-3-hydroxy-butyrate, which is responsible for the stereochemistry of the methyl group in the sulfone ring, can be produced by depolymerization of natural plastics.

- These plastics, e.g. Biopol® from Zeneca, are natural polymers produced by some microorganisms as storage compounds.

- The main problem of the chemoenzymatic synthesis is the epimerization of the (6S)-methyl ketosulfone in aqueous media above pH 5 (ring opening not possible for the reduced species).

- The reaction is carried out at pH below 5 to prevent the accumulation of the sulfone by-product.

- The NADPH-specific enzyme could be purified.

3) Flow scheme

Not published.

4) Process parameters

yield:	> 85 %
ee:	> 98 %
chemical purity:	> 99 %
reactor type:	fed batch
capacity:	multi t
down stream processing:	crystallization
start-up date:	1994
production site:	Zeneca Life Science Molecules, U.K.
company:	AstraZeneca, Sweden

5) Product application

- Intermediate in the synthesis of the carbonic anhydrase inhibitor Trusopt (see remarks).

- Trusopt (invented and marketed by Merck & Co) is a novel, topically active treatment for glaucoma.

- Glaucoma is a disease of the eye characterized by increased intraocular pressure, which results in defects in the field of vision. If left untreated, the disease can cause irreversible damage to the optic nerve, eventually leading to blindness.

6) Literature

- Blacker, A.J., Holt, R.A. (1997) Development of a multistage chemical and biological process for an optically active intermediate for an anti-glaucoma drug, in: Chirality in Industry II (Collins, A.N., Sheldrake, G.N. and Crosby, J., eds.) pp. 246–261, John Wiley & Sons, New York

- Blacklock, T.J., Sohar, P., Butcher, J.W., Lamanec, T., Grabowski, E.J.J. (1993) An enantioselective synthesis of the topically-active carbonic anhydrase inhibitor MK-0507: 5,6-dihydro-(S)-4-(ethylamino)-(S)-6-methyl-4H-thieno[2,3-b]thiopyran-2-sulfonamide 7,7-dioxide hydrochloride, J. Org. Chem. **58**, 1672–1679

- Holt, R. A. (1996) Microbial asymmetric reduction in the synthesis of a drug intermediate, Chimica Oggi **9**, 17–20

- Holt, R. A. and Rigby, S. R. (1996) Process for microbial reduction producing 4(S)-hydroxy-6(S)-methyl-thienopyran derivatives, Zeneca Limited, US 5580764

- Zaks, A., Dodds, D.R., (1997) Application of biocatalysis and biotransformations to the synthesis of pharmaceuticals, Drug Discovery Today **2**, 513–531

Alcohol dehydrogenase
Rhodococcus erythropolis

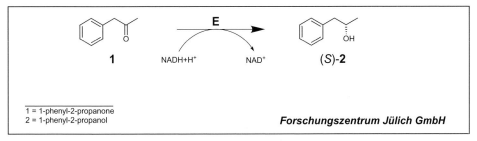

1 = 1-phenyl-2-propanone
2 = 1-phenyl-2-propanol

Forschungszentrum Jülich GmbH

Fig. 1.1.1.1 – 1

1) Reaction conditions

[**1**]:	0.015 M, 1.95 g · L^{-1} [134,18 g · mol^{-1}]
pH:	6.7
T:	25 °C
medium:	aqueous
reaction type:	redox reaction
catalyst:	solubilized enzyme
enzyme:	alcohol-NAD$^+$ oxidoreductase (alcohol dehydrogenase)
strain:	*Rhodococcus erythropolis*
CAS (enzyme):	[9031–72–5]

2) Remarks

• The cofactor regeneration is carried out with a formate dehydrogenase from *Candida boidinii* (FDH = formate dehydrogenase, EC 1.2.1.2) utilizing formate that is oxidized to CO_2:

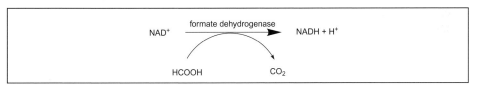

Fig. 1.1.1.1 – 2

• This reactor concept is especially attractive for starting materials of low solubility. The starting materials are directly titrated into the aqueous phase. The process consists of three loops: I: aqueous loop with a hydrophilic ultra-filtration membrane retaining the enzymes; II: permeated aqueous reaction solution products, starting materials and cofactors are passed through the tube phase of the extraction module; III: organic solvent phase, containing extracted products and starting materials.

• The charged cofactors (NAD$^+$/NADH) remain in the aqueous loops I and II. Therefore only deactivated cofactor needs to be replaced, resulting in an economically high total turnover number (= ttn).

• The extraction module consists of microporous, hydrophobic hollow-fiber membranes. The organic extraction solvent is recycled by continuous distillation. The product remains at the bottom of the distillation column.

- Using this method very good space-time yields are obtainable in spite of the low substrate solubilities:

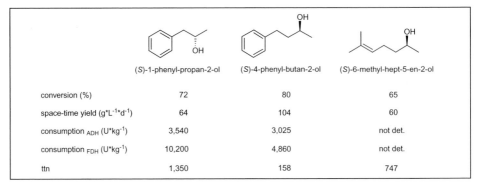

	(S)-1-phenyl-propan-2-ol	(S)-4-phenyl-butan-2-ol	(S)-6-methyl-hept-5-en-2-ol
conversion (%)	72	80	65
space-time yield (g*L^{-1}*d^{-1})	64	104	60
consumption $_{ADH}$ (U*kg^{-1})	3,540	3,025	not det.
consumption $_{FDH}$ (U*kg^{-1})	10,200	4,860	not det.
ttn	1,350	158	747

Fig. 1.1.1.1 – 3

3) **Flow scheme**

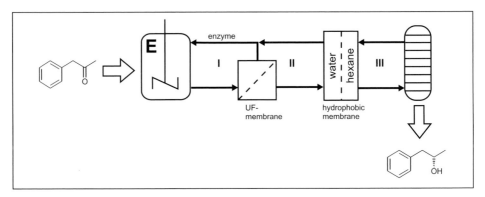

Fig. 1.1.1.1 – 4

4) **Process parameters**

conversion:	72 %
yield:	72 %
selectivity:	100 %
ee:	>99 %
reactor type:	CSTR (enzyme bimembrane reactor)
reactor volume:	0.05 L
residence time:	0.33 h
space-time-yield:	63.5 g·L^{-1}·d^{-1}
down stream processing:	distillation
enzyme activity:	0.95 U·mL^{-1}
enzyme consumption:	3.5 U·g^{-1}
enzyme supplier:	Institute of Molecular Enzyme Technology, Germany
start-up date:	1996
production site:	Forschungszentrum Jülich GmbH, Germany
company:	Forschungszentrum Jülich GmbH, Germany

5) Product application

- The trivial name for (S)-6-methyl-5-hepten-2-ol is (S)-(+)-sulcatol, a pheromone from the scolytid beetle *Gnathotrichus sulcatus / Gnathotrichus retusus*.

- (S)-1-Phenyl-2-propanol is used as an intermediate for the synthesis of amphetamines (sympathomimetics):

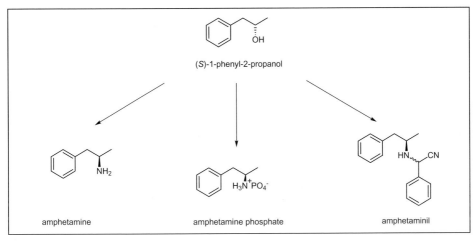

Fig. 1.1.1.1 – 5

- (S)-4-Phenyl-2-butanol is used as a precursor for anti-hypertensive agents and spasmolytics or anti-epileptics:

Fig. 1.1.1.1 – 6

6) Literature

- Bracher, F., Litz, T. (1994) Building blocks for the preparation of enantiomerically pure drugs containing a phenylalkylamine moiety, Arch. Pharm. **327**, 591–593

- Johnston, B., Slessor, K. (1979) Facile synthesis of the enantiomers of sulcatol, Can. J. Chem. **57**, 233–235

- Kragl, U., Kruse, W., Hummel, W., Wandrey, C. (1996) Enzyme engineering aspects of biocatalysis: Cofactor regeneration as example, Biotechnol. Bioeng. **52**, 309–319

- Kruse, W., Hummel, W., Kragl, U. (1996) Alcohol-dehydrogenase-catalyzed production of chiral hydrophobic alcohols. A new approach leading to a nearly waste-free process, Recl. Trav. Chim. Pays-Bas **115**, 239–243

- Kruse, W., Kragl, U., Wandrey, C. (1996) Verfahren zur kontinuierlichen enzymkatalysierten Gewinnung hydrophober Produkte, Forschungszentrum Jülich GmbH, DE 4436149 A1

- Kruse, W., Kragl, U., Wandrey, C. (1998) Process for the continuous enzymatic extraction of hydrophobic products and device suitable therefor, Forschungszentrum Jülich GmbH, Germany, US 5,795,750

- Liang, S.; Paquette, L.A. (1990) Biocatalytic-based synthesis of optically pure (C-6)-functionalized 1-(*tert*-butyldimethyl-silyloxy)-2-methyl-(*E*)-2-heptenes; Tetrahedron Asym. **1**, 445–452

- Mori, K. (1975) Synthesis of optically active forms of sulcatol – The aggregation pheromone in the scolytid beetle, *Gnathotrichus sulcatus*, Tetrahedron **31**, 3011–3012

1 = 6-benzyloxy-3,5-dioxo-hexanoic acid ethyl ester
2 = 6-benzyloxy-3,5-dihydroxy-hexanoic acid ethyl ester

Bristol-Myers Squibb

Fig. 1.1.1.1 – 1

1) Reaction conditions

[1]:	0.036 M, 10 g · L^{-1} [278.30 g · mol^{-1}]
pH:	5.9
T:	33 °C
medium:	aqueous
reaction type:	redox reaction
catalyst:	solubilized enzyme (cell extract)
enzyme:	alcohol-NAD$^+$ oxidoreductase (alcohol dehydrogenase)
strain:	*Acinetobacter calcoaceticus*
CAS (enzyme):	[9031–72–5]

2) Remarks

- The bioreduction can be carried out with whole cells as well as with cell extracts.

- To facilitate the cofactor regeneration of NADH, glucose dehydrogenase, glucose and NAD$^+$ are added to the reaction medium.

- The educt is prepared by the following method:

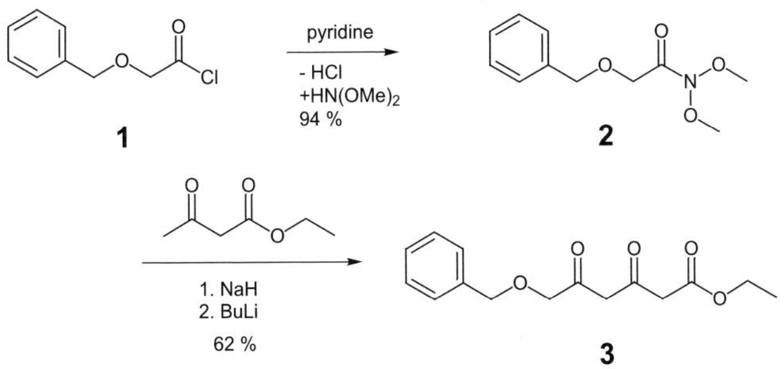

1 = benzyloxy-acetyl chloride
2 = 2-benzyloxy-*N,N*-dimethoxy-acetamide
3 = 6-benzyloxy-3,5-dioxo-hexanoic acid ethyl ester

Fig. 1.1.1.1 – 2

- This biotransformation is an alternative to the chemical synthesis via the chlorohydrin and selective hydrolysis of the acyloxy group. By chemical synthesis an overall yield of 41 % after final fractional distillation is achieved:

161

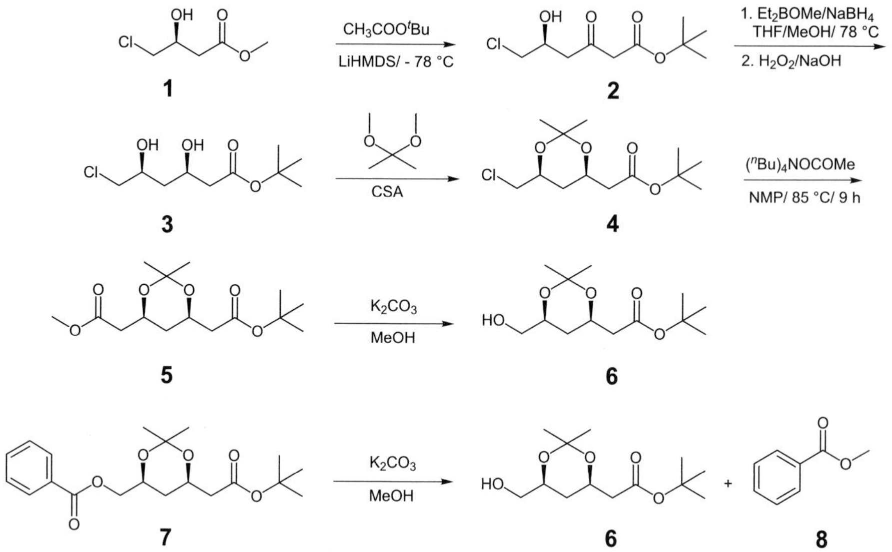

1 = 4-chloro-3-hydroxy-butyric acid methyl ester
2 = 6-chloro-5-hydroxy-3-oxo-hexanoic acid *tert*-butyl ester
3 = 6-chloro-3,5-dihydroxy-hexanoic acid *tert*-butyl ester
4 = (6-chloromethyl-2,2-dimethyl-[1,3]dioxan-4-yl)-acetic acid *tert*-butyl ester
5 = (6-*tert*-Butoxycarbonylmethyl-2,2-dimethyl-[1,3]dioxan-4-yl)-acetic acid methyl ester
6 = (6-hydroxymethyl-2,2-dimethyl-[1,3]dioxan-4-yl)-acetic acid *tert*-butyl ester
7 = (6-benzyl-2,2-dimethyl-[1,3]dioxan-4-yl)-acetic acid *tert*-butyl ester
8 = benzoic acid methyl ester

Fig. 1.1.1.1 – 3

3) Flow scheme

Not published.

4) Process parameters

yield:	92 %
ee:	99 %
reactor type:	batch
down stream processing:	centrifugation and extraction with methylene chloride
company:	Bristol-Myers Squibb, USA

5) Product application

- 6-Benzyloxy-(*3R,5S*)-dihydroxy-hexanoic acid ethyl ester is a key chiral intermediate for anti-cholesterol drugs that act by inhibition of hydroxymethylglutaryl coenzyme A (HMG-CoA) reductase. The synthesis is shown in the following scheme:

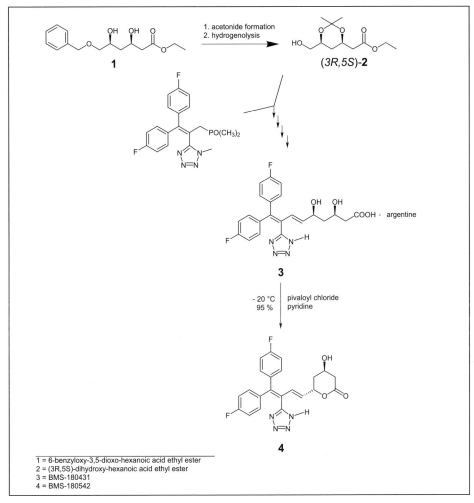

1 = 6-benzyloxy-3,5-dioxo-hexanoic acid ethyl ester
2 = (3R,5S)-dihydroxy-hexanoic acid ethyl ester
3 = BMS-180431
4 = BMS-180542

Fig. 1.1.1.1 – 4

6) Literature

- Patel, R.N., McNamee, C.G., Banerjee, A., Szarka, L.J. (1993) Stereoselective microbial or enzymatic reduction of 3,5-dioxo esters to 3-hydroxy-5-oxo, 3-oxo-5-hydroxy, and 3,5-dihydroxy esters, E.R. Squibb & Sons, Inc., EP 0569998A2
- Patel, R.N., Banerjee, A., McNamee, C.G., Brzozowski, D., Hanson, R.L., Szarka, L.J. (1993) Enantioselective microbial reduction of 3,5-dioxo-6-(benzyloxy)hexanoic acid, ethylester, Enzyme Microb. Technol. **15**, 1014–1021
- Sit, S.Y., Parker, R.A., Motoc, I., Han, W., Balasubramanian, N., Catt, J.D., Brown, P.J., Harte, W.E., Thompson, M.D., Wright J.J. (1990) Synthesis, biological profile, and quantitative structure-activity relationship of a series of novel 3-hydroxy-3-methylglutaryl coenzyme A reductase inhibitors, J.Med.Chem. **33**, 2982–2999
- Thottathil, J.K. (1998) Chiral synthesis of BMS-180542 and SQ33600; Tetrazole and phosphinic acid based HMG-CoA reductase inhibitors, Chiral USA 98, Chiral Technology – The way ahead, San Francisco, CA, USA, 18–19 May 1998

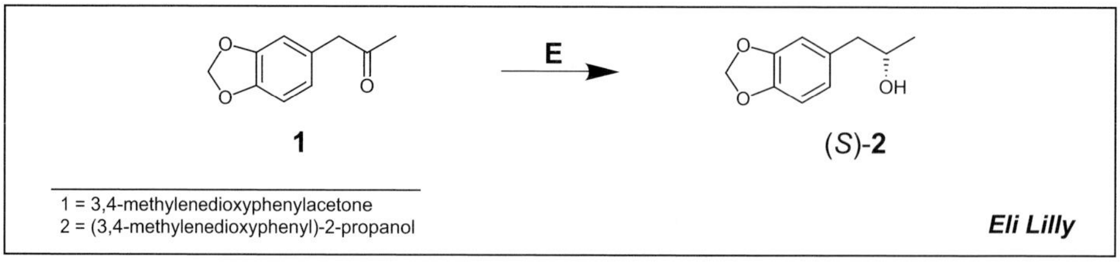

1 = 3,4-methylenedioxyphenylacetone
2 = (3,4-methylenedioxyphenyl)-2-propanol

Eli Lilly

Fig. 1.1.1.1 – 1

1) Reaction conditions

[**1**]:	< 0.011 M, 2 g · L^{-1} [178.18 g · mol^{-1}]
pH:	7.0
T:	33–35 °C
medium:	aqueous
reaction type:	reduction of keto group
catalyst:	suspended whole cells
enzyme:	alcohol-NAD$^+$ oxidoreductase (alcohol dehydrogenase)
strain:	*Zygosaccharomyces rouxii*
CAS (enzyme):	[9031–72–5]

2) Remarks

- Since the toxic limit of the substrate is 6 g · L^{-1} the substrate is adsorbed on XAD-7 resin (80 g · L^{-1} resin, allowing a reaction concentration of 40 g · L^{-1} reaction volume). Using this method the volumetric productivity can be held at a high level.

- The advantage of the XAD-7 resin in comparison to organic solvents of the same polarity (log P < 2) is its non-toxic and non-denaturating character.

- The XAD-7 resin is reused for three times without any loss in performance.

- As reactor a Rosenmund agitated filter-dryer is used. A packed bed reactor would clog and an expanded bed reactor is not applicable due to the low density of XAD-7. The resin would be blown out of the top of it. The special agitator used has a hydraulic design that enables a filtration without clogging of the filter.

- The desorption of educt is limited to the equilibrium concentration of approximately 2 g · L^{-1} in the aqueous phase.

- To separate the yeast cells from the product that is adsorbed on the resin, a 150 μm filter screen is used. In contrast to usual filtrations the resin (~ 500 μm) is retained and the yeast cells (~5 μm) stay in the filtrate. The product is liberated by washing the resin with acetone.

3) Flow scheme

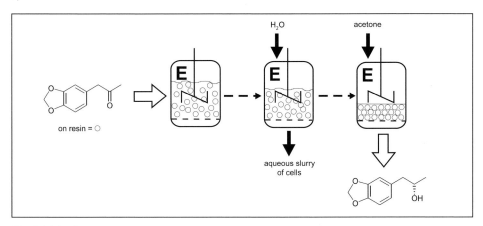

Fig. 1.1.1.1 – 2

4) Process parameters

yield:	96 %
ee:	> 99.9 %
chemical purity:	95 %
reactor type:	batch
reactor volume:	300 L
capacity:	kg scale
space-time-yield:	75 g · L^{-1} · d^{-1}
down stream processing:	filtration, resin extraction
company:	Eli Lilly and Company, USA

5) Product application

- The product is converted to LY 300164, an orally active benzodiazepine:

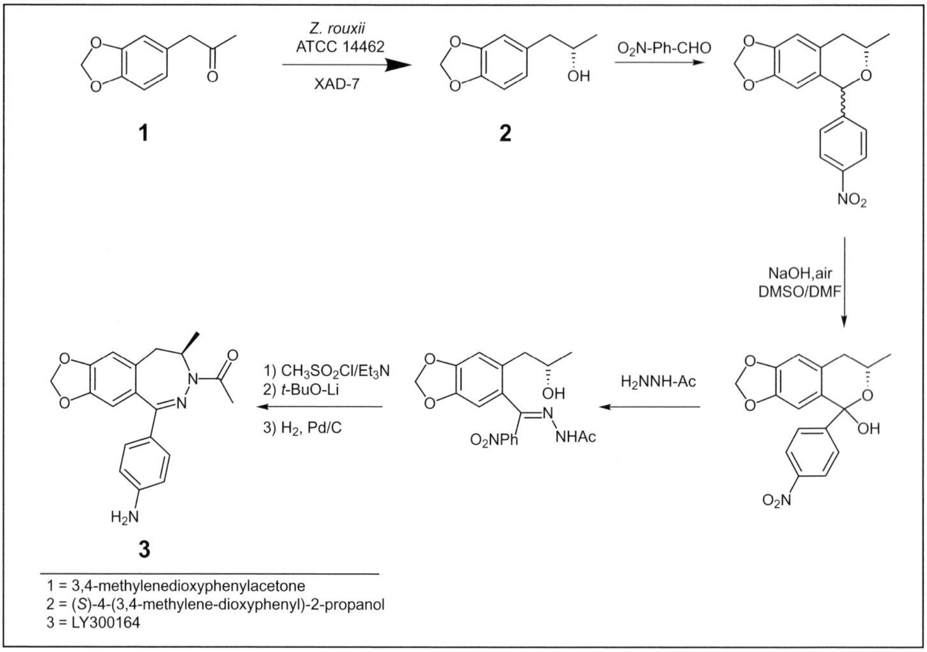

Fig. 1.1.1.1 – 3

- The product is being tested for efficacy in treating amylotropic lateral sclerosis.

6) Literature

- Anderson, B.A., Hansen, M.M., Harkness, A.R., Henry, C.L., Vicenzi, J.T., Zmijewski, M.J. (1995) Application of a practical biocatalytic reduction to an enantioselective synthesis of the 5 *H*-2,3-benzodiazepine LY300164, J. Am. Chem. Soc. **117**, 12358–12359

- Vicenzi, J.T., Zmijewski, M.J., Reinhard, M.R., Landen, B.E., Muth, W.L., Marler, P.G. (1997) Large-scale stereoselective enzymatic ketone reduction with *in-situ* product removal via polymeric adsorbent resins, Enzyme Microb. Technol. **20**, 494–499

- Zaks, A., Dodds, D.R. (1997) Application of biocatalysis and biotransformations to the synthesis of pharmaceuticals, Drug Discovery Today **2**, 513–530

- Zmijewski, M.J., Vicenzi, J., Landen, B.E., Muth, W., Marler, P.G., Anderson, B. (1997) Enantioselective reduction of 3,4-methylene-dioxyphenyl acetone using *Candida famata* and *Zygosaccharomyces rouxii*, Appl. Microbiol. Biotechnol. **47**, 162–166

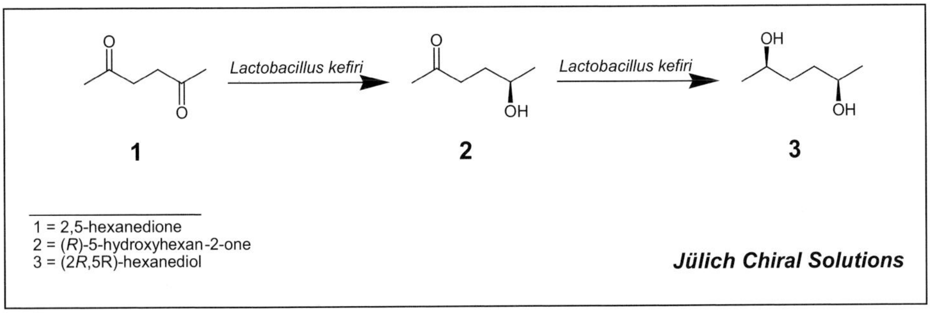

Alcohol dehydrogenase
Lactobacillus kefiri

1 = 2,5-hexanedione
2 = (R)-5-hydroxyhexan-2-one
3 = (2R,5R)-hexanediol

Jülich Chiral Solutions

Fig. 1.1.1.1 – 1

1) Reaction conditions

[1]:	0.45 M 2,5-hexanedione
pH:	6.0
T:	30 °C
medium:	aqueous
reaction type:	asymmetric reduction
catalyst:	whole cells
enzyme:	alcohol dehydrogenase
strain:	*Lactobacillus kefiri*

2) Remarks

- 2,5-Hexanedione (acetonylacetone) is routinely available from diketene chemistry.

- Process also works for other diones, such as 2,4-pentanedione (acetylacetone), 3,6-octanedione, 2,7-dimethyl-3,6-octanedione and 2,6-heptanedione.

- In contrast to the application of the isolated alcohol dehydrogenase no additional cofactor is necessary.

- Glucose is a cheap cosubstrate and hydrogen source for this process.

- The applied biocatalyst is a genetically unmodified wild-type microorganism, therefore no formalities according to the law for genetically engineered strains are necessary.

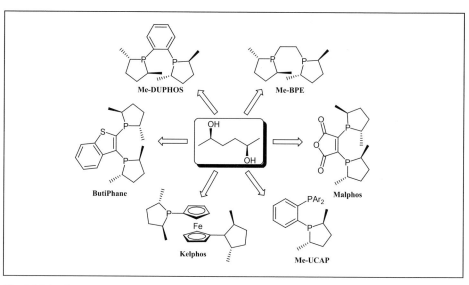

Fig. 1.1.1.1 – 2

3) Flow scheme

Not published.

4) Process parameters

yield:	90 %
ee:	99.5 % (no *S,S*-enantiomer detected)
	de 99 %
reactor type:	stirred batch reactor
capacity:	$0.25\,t \cdot a^{-1}$
space-time-yield:	$50\,g \cdot L^{-1} \cdot d^{-1}$
enzyme consumption:	$40\,g_{cdw} \cdot kg^{-1}$
production site:	Jülich
company:	Jülich Chiral Solutions, Germany

5) Product application

- The product is an important building block for chiral ligands with phospholane structure, e.g. DUPHOS.

- Chiral auxiliary for organic synthesis.

6) Literature

- Haberland, J., Hummel, W., Daußmann, T., Liese, A. (2002) New continuous production process for enantiopure (2*R*,5*R*)-hexanediol, Org. Proc. Res. Dev. **6**, 458–462

- Haberland, J., Kriegesmann, A., Wolfram, E., Hummel, W, Liese, A. (2002) Diastereoselective synthesis of optically active (2*R*,5*R*)-hexanediol, Appl. Microb. Biotechnol. **58** (5), 595–599

- Hummel, W., Liese, A., Wandrey, C. (1999) Verfahren zur Reduktion von Ketogruppen enthaltenen Verbindungen, DE 199 32 040.3

- Hummel, W., Liese, A, Wandrey, C. (2000) Verfahren zur Reduktion von Ketogruppen enthaltenen Verbindungen, EP 1067195

- Burk, M.J., Gross, M.F., Martinez, J.P. (1995) Asymmetric catalytic synthesis of β-branched amino acids via highly enantioselection hydrogenation reactions, J. Am. Chem. Soc. **117**, 9375–9376

- Braun, W., Calmuschi, B., Haberland, J., Hummel, W., Liese, A., Nickel, T., Stelzer, O. Salzer, A. (2004) Optically active pospholanes as substituents on ferrocene and chromiumarene complexes, Europ. J. Inorg. Chem. **11**, 2235–2243

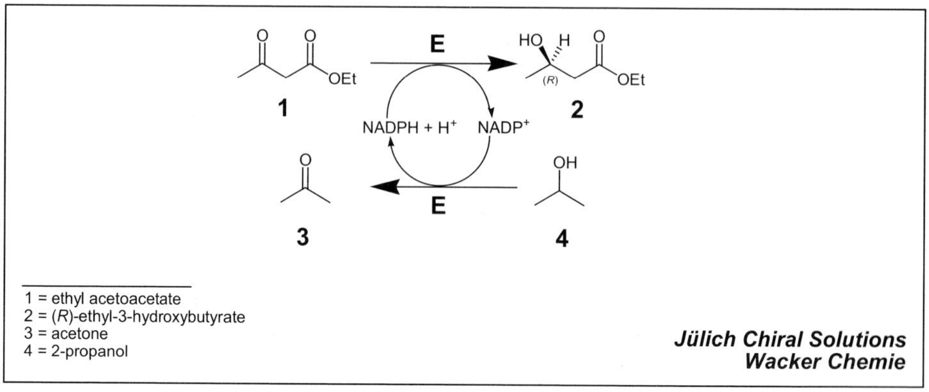

Alcohol dehydrogenase
Lactobacillus brevis

1 = ethyl acetoacetate
2 = (*R*)-ethyl-3-hydroxybutyrate
3 = acetone
4 = 2-propanol

Jülich Chiral Solutions
Wacker Chemie

Fig. 1.1.1.1 – 1

1) Reaction conditions

[1]:	0.5 M, 65 g · L⁻¹ [130.14 g · mol⁻¹]
[2]:	7.6 vol % 2-propanol
pH:	6.5
T:	28–32 °C
medium:	aqueous
reaction type:	asymmetric reduction
catalyst:	raw enzyme solution of alcohol dehydrogenase from *Lactobacillus brevis*
enzyme:	alcohol dehydrogenase
strain:	*Lactobacillus brevis*
CAS (enzyme):	[9031-72-5]

2) Remarks

- Ethyl acetoacetate is routinely available from diketene chemistry. Additional β-ketoesters are easily accessible via acylation of acetoacetates.

- Process is also applicable to the synthesis of other esters.

- Similar processes are in place for other ketones.

- Process allows for reuse of aqueous phase (drastic reduction of contaminated aqueous waste).

- Contrary to immobilization approaches even the cofactor and part of the primary reduction agent isopropanol can be reused.

- Continuous stripping of the coproduct acetone provides several benefits: it shifts the equilibrium reaction towards complete turnover, it allows for the application of low-boiling solvents for extraction and it allows a continuous reuse of the aqueous phase in standard reactor vessels.

170

3) Flow scheme

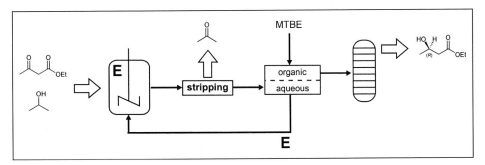

Fig. 1.1.1.1 – 2

4) Process parameters

yield:	96 %
ee:	99.8 %
reactor type:	stirred batch reactor
capacity:	$35\,t \cdot a^{-1}$
space-time-yield:	$92\,g \cdot L^{-1} \cdot d^{-1}$
down stream processing:	extraction, distillation
enzyme consumption:	$80\,kU \cdot kg^{-1}$
enzyme supplier:	Jülich Chiral Solutions, Germany
company:	Wacker Chemie, Germany

5) Product application

- Chiral ß-hydroxyesters are important chiral intermediates for organic synthesis.

- They are widely applied as building block for pharmaceuticals, agrochemicals and fragrances.

6) Literature

- Hummel, W. (1997) Biochem. Eng. **58**, 145–184

- Hummel, W., Riebel, B. (1999) US 6,225,099 and EP 0796914A2

- Rosen, T.C., Daußmann, T., Stohrer, J. (2004) Speciality Chemicals Magazine, 39–40

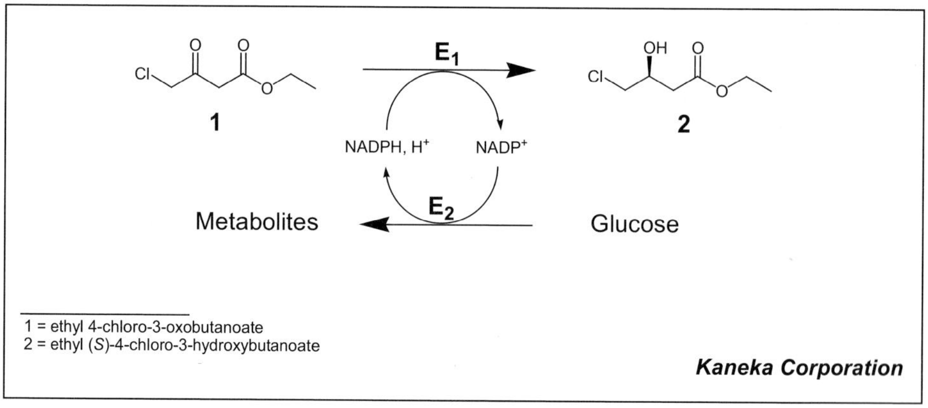

Carbonyl reductase
Escherichia coli

1 = ethyl 4-chloro-3-oxobutanoate
2 = ethyl (S)-4-chloro-3-hydroxybutanoate

Kaneka Corporation

Fig. 1.1.1.1 – 1

1) Reaction conditions

[1]:	1.82 M, 299.6 g · L⁻¹ [164.59 g · mol⁻¹]
[2]:	1.22 M glucose
pH:	6.5
T:	30 °C
medium:	two-phase-system
reaction type:	asymmetric reduction
catalyst:	whole cells
enzyme:	carbonyl reductase
enzyme:	glucose dehydrogenase
strain:	*Escherichia coli*

where the values of $[1]$ and $[2]$ are given above.

2) Remarks

- *n*-Butyl acetate is used as the organic phase.

3) Flow scheme

Not published.

4) Process parameters

yield:	85 %
ee:	99 %
company:	Kaneka Corporation, Japan

5) Product application

- (S)-4-Chloro-3-hydroxy-butanoate is an important starting material for hydroxymethylglutaryl-CoA reductase inhibitors.

172

6) Literature

- Kizaki, N., Yasohara, Y., Hasegawa, J., Wada, M., Kataoka, M., Shimizu, S. (2001) Synthesis of optically pure ethyl (*S*)-4-chloro-3-hydroxybutanoate by *Escherichia coli* transformant cells coexpressing the carbonyl reductase and glucose dehydrogenase genes, Appl. Microbiol. Biotechnol. **55**, 590–595

- Kataoka, M., Rohani, L.P.S., Yamamoto, K., Wada, M., Kawabata, H., Kita, K., Yanase, H., Shimizu, S. (1997) Enzymatic production of ethyl (*R*)-4-chloro-3-hydroxybutanoate: asymmetric reduction of ethyl 4-chloro-3-oxobutanoate by an *Escherichia coli* transformant expressing the aldehyde reductase gene from yeast, Appl. Microbiol. Biotechnol. **48**, 699–703

- Kataoka, M., Yamamoto, K., Kawabata, H., Wada, M., Kita, K., Yanase, H., Shimizu, S. (1999) Stereoselective reduction of ethyl 4-chloro-3-oxobutanoate by *Escherichia coli* transformant cells coexpressing the aldehyde reductase and glucose dehydrogenase genes, Appl. Microbiol. Biotechnol. **51**, 486–490

Aldehyde reductase
Escherichia coli

EC 1.1.1.2

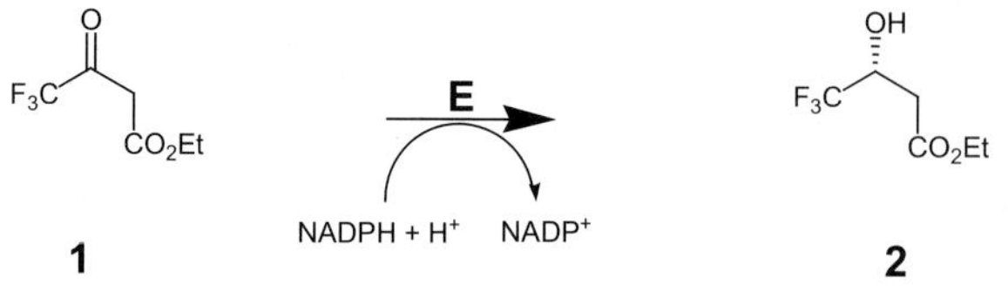

1 = ethyl-4,4,4-trifluoracetoacetate
2 = (*R*)-ethyl-4,4,4-trifluoro-3-hydroxybutanoate

Lonza

Fig. 1.1.1.2 – 1

1) Reaction conditions

[**1**]:	[184.11 g · mol^{-1}]
medium:	two-phase system: water / butyl acetate
reaction type:	redox reaction
catalyst:	whole cells
enzyme:	aldehyde reductase
enzyme:	glucose dehydrogenase
strain:	*Escherichia coli* JM109

2) Remarks

- The production strain contains two plasmids. One carries the aldehyde reductase gene from *Sporobolomyces salmonicolor*, which catalyzes the reduction step. The second one contains a glucose dehydrogenase gene from *Bacillus megaterium*, which enables cofactor regeneration of NADPH from NADP.
- The production strain is grown at 22 °C to avoid formation of inclusion bodies. Cells are harvested, washed and stored frozen until use.
- The process is carried out in a water / butyl acetate two-phase system to avoid inhibition of the reductase by the substrate and product.

3) Flow scheme

Not published.

4) Process parameters

yield:	50 %
ee:	99 %
company:	Lonza, Switzerland

5) Product application

- (*R*)-Ethyl-4,4,4-trifluoro-3-hydroxybutanoate is a building block for befloxatone, an antidepressant monoamine oxidase-A inhibitor produced by Synthelabo.

6) Literature

- Shaw, N.M., Robins, K.T., Kiener, A. (2003) Lonza: 20 Years of biotransformations, Adv. Synth. Catal. **345** (4), 425–435

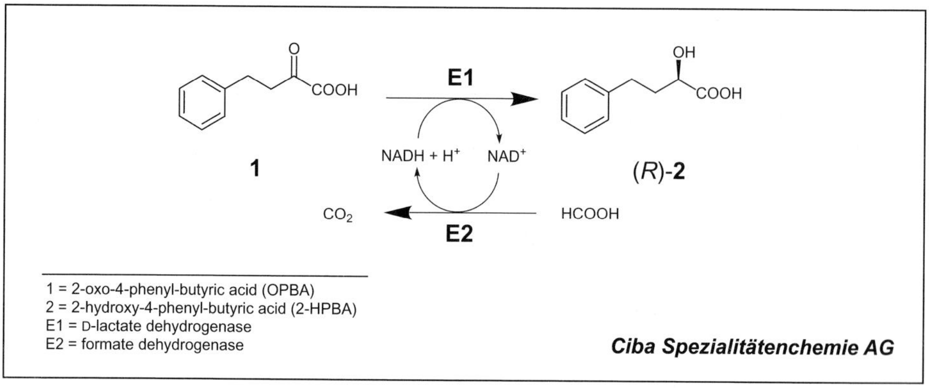

Lactate dehydrogenase
Staphylococcus epidermidis

Fig. 1.1.1.28 – 1

1) Reaction conditions

[1]:   0.2 M, 35.6 g · L^{-1} [178.18 g · mol^{-1}]
pH: 8.0
T: 30 °C
medium: aqueous
reaction type: redox reaction
catalyst: solubilized enzyme
enzyme: (*R*)-lactate-NAD oxidoreductase
strain: *Staphylococcus epidermidis*
CAS (enzyme):   [9028–36–8]

2) Remarks

- Cofactor regeneration is carried out by formate dehydrogenase from *Candida boidinii* utilizing formate and producing CO_2. During cofactor regeneration no by-product is formed that needs to be separated. In the steady state of the continuously operated stirred tank reactor a total turnover number of 900 is achieved.

- NAD$^+$ is added as cofactor at a concentration of 0.2 mM and ammonium formate at a concentration of 0.35 M.

- The optimal working pH would have been 6.5, but a pH of 8.0 is chosen due to better solubility of the hydroxy acid.

- The production is carried out in a continuously operated stirred tank reactor equipped with an ultrafiltration membrane (cut off 10,000 Da) to retain the enzymes.

- For stabilization of the enzymes 0.15 % mercaptoethanol and 1 mM EDTA are added to the substrate solution.

- At a conversion of 90 % the pH is shifted from 6.2 (substrate solution) to 8.0. The reason is the different pK_a-values of 2-oxo-4-phenylbutyric acid (pK_a = 2.3) and (*R*)-2-hydroxy-4-phenylbutyric acid (pK_a = 3.76).

- The data given relate to the above process with isolated enzymes.

Alternatively immobilized whole cells of *Proteus vulgaris* can be used in a fixed bed reactor:

- By using small units of biomass in the form of a packed bed of immobilized cells, high amounts of the hydroxy acid can be produced. The space-time-yield is high and the costs for down-stream processing are low (conversion > 99.5 %; ee = 99.8 %; space-time yield = 180 g $\cdot$ L^{-1} $\cdot$ d^{-1}; enzyme consumption = 159 U $\cdot$ kg^{-1}).

- Per liter of fermentation broth 1.45 g cell mass (80 wt.-% water) can be isolated.

- The emulsion of cells, chitosan and solid silica is dropped into a polyphosphate solution resulting in ionotropic gel-beads.

- 1,000 g cell emulsion yields 400 g (= 800 mL) beads. The beads contain 0.25 g (wet cells) $\cdot$ g^{-1} (beads).

- Here electron mediators (= V) are used instead of NADH or NADPH:

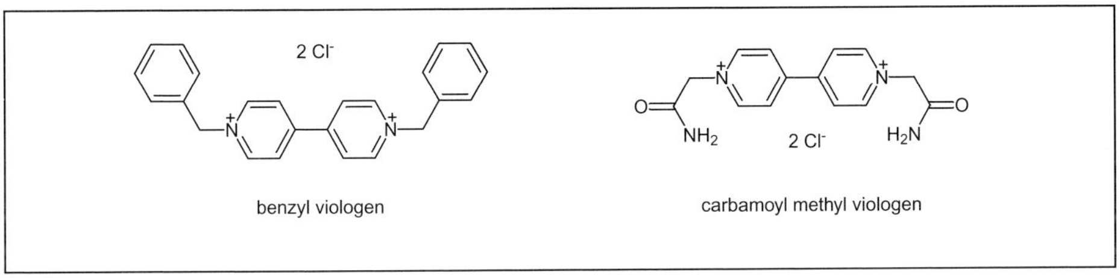

benzyl viologen   carbamoyl methyl viologen

Fig. 1.1.1.28 – 2

- The enzymes in the cells are capable of reducing and oxidizing the mediators. In this case the carbamoylmethylviologen is used. The following figure shows the regeneration cycle:

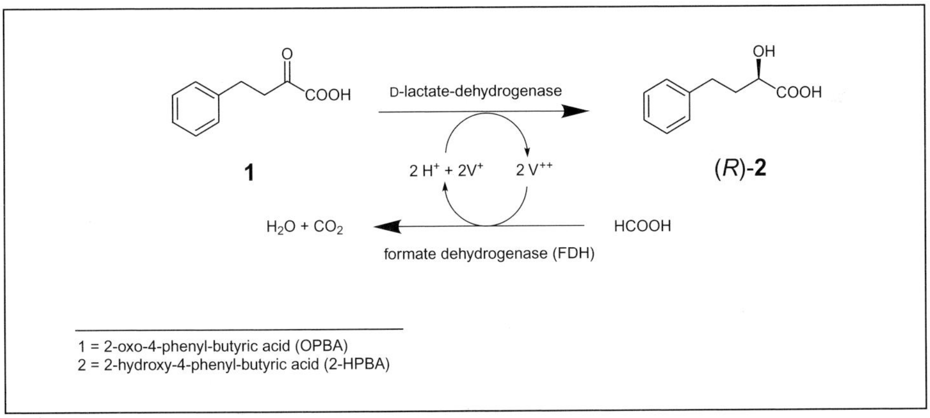

1 = 2-oxo-4-phenyl-butyric acid (OPBA)
2 = 2-hydroxy-4-phenyl-butyric acid (2-HPBA)

Fig. 1.1.1.28 – 3

- To avoid degassing of CO_2 the reaction is carried out under a pressure of 3 bar.

177

3) Flow scheme

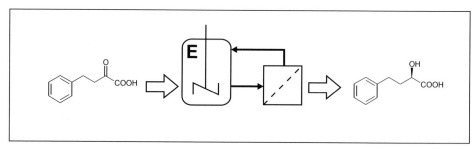

Fig. 1.1.1.28 – 4

4) Process parameters

conversion:	91 %
ee:	99.9 %
reactor type:	continuously operated stirred tank reactor (enzyme membrane reactor)
reactor volume:	0.2 L
residence time:	4.6 h (1 h)
space-time-yield:	165 g · L^{-1} · d^{-1} (410 g · L^{-1} · d^{-1})
down stream processing:	extraction, crystallization
enzyme consumption:	D-LDH: 150 U · kg^{-1} (= 1.5 % · d^{-1}); FDH = 150 U · kg^{-1}
enzyme supplier:	Sigma, Germany
company:	Ciba Spezialitätenchemie AG, Switzerland

5) Product application

- The product is a precursor for different ACE-inhibitors (ACE = angiotensin converting enzyme).

- All ACE-inhibtors have the (S)-homophenylalanine moiety in common (see page 342).

6) Literature

- Fauquex, P. F., Sedelmeier, G. (1989) Biokatalysatoren und Verfahren zu ihrer Herstellung, Ciba-Geigy AG, EP 0371408 B1

- Schmidt, E., Ghisalba, O., Gygax, D., Sedelmeier, G. (1992) Optimization of a process for the production of (R)-2-hydroxy-4-phenylbutyric acid – an intermediate for inhibitors of angiotensin converting enzyme, J. Biotechnol. **24**, 315–327

- Schmidt, E., Blaser, H.U., Fauquex, P.F., Sedelmeier, G., Spindler, F. (1992) Comparison of chemical and biochemical reduction methods for the synthesis of (R)-2-hydroxy-4-phenylbutyric acid, in: Microbial Reagents in Organic Synthesis (S. Servi, ed.), pp. 377–388, Kluwer Academic Publishers, Dordrecht, Netherlands.

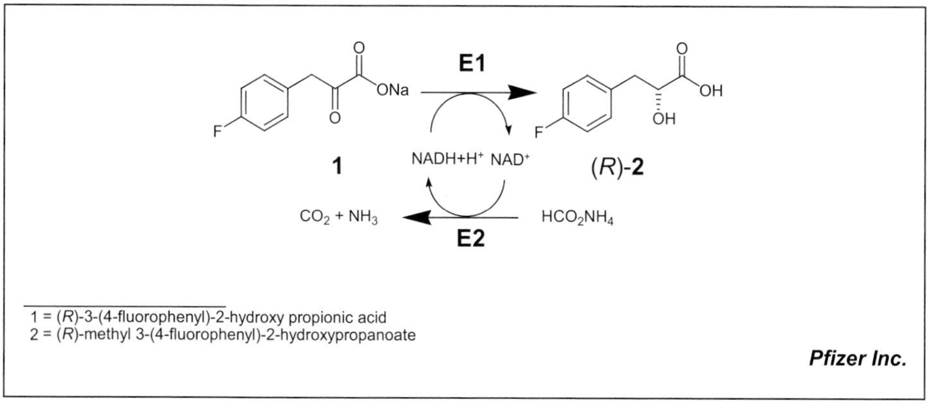

1 = (R)-3-(4-fluorophenyl)-2-hydroxy propionic acid
2 = (R)-methyl 3-(4-fluorophenyl)-2-hydroxypropanoate

Pfizer Inc.

Fig. 1.1.1.28 – 1

1) Reaction conditions

[1]:	44.4 g · L^{-1} sodium-3(-4′-fluorophenyl)-2-oxopropanoate
[2]:	0.8 M, 50.4 g · L^{-1} (ammonium formate)
[3]:	0.002 M, 1.3 g · L^{-1} (NAD$^+$)
pH:	6.3
medium:	water, EDTA (0.0005 M), mercaptoethanol (0.0001 M)
reaction type:	redox reaction
catalyst:	solubilized enzyme
enzyme:	D-lactate dehydrogenase, (R)-lactate-NAD oxidoreductase [E1]
enzyme:	formate dehydrogenase [E2]
strain:	E1: *Leuconostoc mesenteroides, Staphylococcus epidermidis*
strain:	E2: *Candida boidinii*
CAS (enzyme):	[9028-36-8]

2) Remarks

- Chemical approaches suffer from either low yields or poor stereocontrol.

- The α-keto acid salt 1 is synthesized starting from 4-fluorobenzaldehyde and a hydantoin upon condensation and saponification in 77–82 % yield on a scale of 23 kg.

- The cofactor regeneration is carried out with formate dehydrogenase from *Candida boidinii* (FDH = formate dehydrogenase, EC 1.2.1.2) utilizing formate that is oxidized to CO_2.

- Retrosynthetic approaches to (R)-methyl-3-(4-fluorophenyl)-2-hydroxypropanoate:

Fig. 1.1.1.28 – 2

- In a subsequent step the chiral hydroxy acid is esterified for integration into the chemical synthesis route of the protease inhibitor.

Fig. 1.1.1.28 – 3

3) Flow scheme

Fig. 1.1.1.28 – 4

4) Process parameters

conversion:	$\geq 90\%$
yield:	$\leq 88\%$
ee:	$>99.9\%$
reactor type:	CSTR (continuously operated enzyme membrane reactor)
reactor volume:	2.2 L
capacity:	multikilogram scale
residence time:	3 h
space-time-yield:	$560\,g \cdot L^{-1} \cdot d^{-1}$
down stream processing:	aqueous effluent solution was adjusted to pH 3.0 with 2 N HCl and extracted with MTBE, followed by evaporation of the organic layer
enzyme activity:	D-LDH: $400\,U \cdot mL^{-1}$; FDH: $20\,U \cdot mL^{-1}$
enzyme consumption:	$1\%\,d^{-1}$ loss of enzyme activity
start-up date:	2001
production site:	Pfizer Inc., USA
company:	Pfizer Inc., USA

5) Product application

- (R)-3-(4-Fluorophenyl)-2-hydroxy propionic acid is a building block for the synthesis of Rupintrivir, a rhinovirus protease inhibitor.

6) Literature

- Tao, J.H., McGee, K. (2002) Development of a continuous enzymatic process for the preparation of (R)-3-(4-fluorophenyl)-2-hydroxy propionic acid, Org. Proc. Res. Dev., **6**, 520–524

- Tao, J.H., McGee, K (2004) Development of an efficient synthesis of chiral 2-hydroxy acids, in: Asymmetric Catalysis on Industrial Scale (Blaser, H.U., Schmidt, E., eds.), pp. 323–334, Wiley-VCH, Weinheim

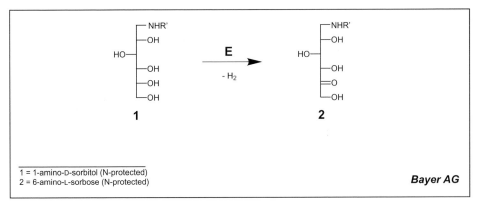

1 = 1-amino-D-sorbitol (N-protected)
2 = 6-amino-L-sorbose (N-protected)

Bayer AG

Fig. 1.1.99.21 – 1

1) Reaction conditions

[1]:	1 M (molecular weight depends on protecting group)
pH:	5.0
T:	32 °C
medium:	aqueous
reaction type:	oxidation
catalyst:	suspended whole cells
enzyme:	D-sorbitol:(acceptor)1-oxidoreductase (D-sorbitol dehydrogenase)
strain:	*Gluconobacter oxydans*
CAS (enzyme):	[9028–22–2]

2) Remarks

- The published synthesis of 1-desoxynojirimycin and its derivatives requires multiple steps and a laborious protecting group chemistry.

- To prevent undesired follow up reactions of 6-amino-D-sorbose in water the amino group has to be protected by, e.g., a benzyloxycarbonyl group. The protection of 1-amino-D-sorbitol is carried out in an aqueous medium at pH 8–10 with benzyloxycarbonyl chloride.

- The cells of *Gluconobacter oxydans* are produced by fermentation on sorbitol and used for the bioconversion step, which is carried out in water without added nutrients.

- The cells are not immobilized; the very high specific substrate conversion rate would lead to severe limitations in immobilization beads.

- 6-Amino-L-sorbose (*N*-protected) is used as an intermediate in the manufacture of miglitol via 1-desoxynojirimycin. 1-Desoxynojirimycin is produced by chemical intramolecular reductive amination of 6-amino-L-sorbose. Miglitol is used in the treatment of Type 2 (non-insulin dependent) diabetes.

D-Sorbitol dehydrogenase

Gluconobacter oxydans

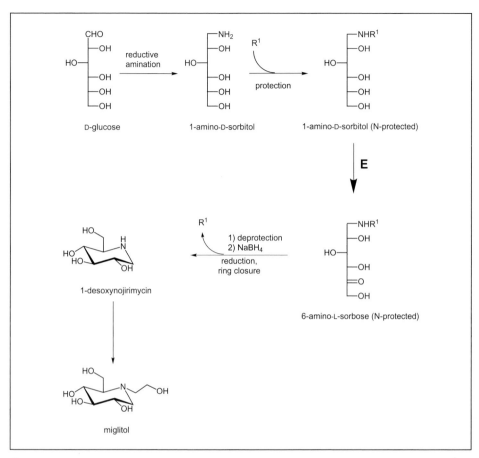

Fig. 1.1.99.21 – 2

3) **Flow scheme**

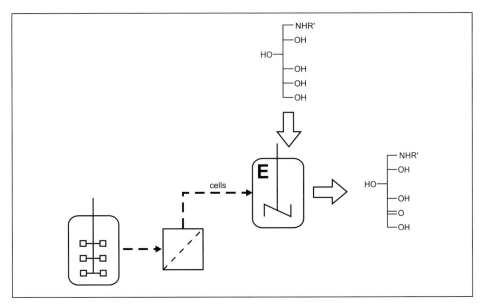

Fig. 1.1.99.21 – 3

4) **Process parameters**

yield:	90 %
reactor type:	batch
reactor volume:	> 10,000 L
company:	Bayer AG, Germany

5) **Product application**

- 6-Amino-L-sorbose is used as an intermediate for oral α-glucosidase-inhibitors.

- Derivatives of 1-desoxynojirimycin are pharmaceuticals for the treatment of carbohydrate metabolism disorders (e.g. diabetes mellitus).

6) **Literature**

- Kinast, G., Schedel, M. (1978) Verfahren zur Herstellung von 6-Amino-6-desoxy-L-sorbose, Bayer AG, DE 2834122 A1

- Kinast, G., Schedel, M. (1978) Herstellung von N-substituierten Derivaten des 1-Desoxynojiri-mycins, Bayer AG, DE 2853573 A1

- Kinast, G., Schedel, M. (1981) Vierstufige 1-Desoxynojirimycin-Synthese mit einer Biotransfor-mation als zentralem Reaktionsschritt, Angew. Chem. **93**, 799–800

- Kinast, G., Schedel, M., Koebernick, W. (1981) Verfahren zur Herstellung von N-substituierten Derivaten des 1-Desoxynojirimycins, Bayer AG, EP 0049858 A2

D-Sorbitol dehydrogenase
Gluconobacter oxydans ATCC621

EC 1.1.99.21

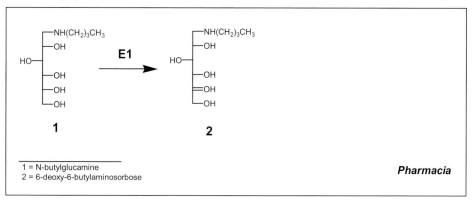

1 = N-butylglucamine
2 = 6-deoxy-6-butylaminosorbose

Pharmacia

Fig. 1.1.99.21 – 1

1) Reaction conditions

[1]:	0.927M, 220 g · L^{-1} [237.29 g · mol^{-1}]
pH:	5.5–6.0
T:	12–15 °C
reaction type:	redox reaction
catalyst:	washed cells: 14 g · L^{-1} dry cell weight
enzyme:	D-sorbitol dehydrogenase
strain:	*Gluconobacter oxydans* ATCC621
CAS (enzyme):	[9028-22-2]

2) Remarks

- Sorbitol dehydrogenase is believed to be the enzyme responsible for catalyzing the bioconversion of *N*-butylglucamine to 6-deoxy-6-butylaminosorbose.

- The scaled process utilized a 40,000 L fermenter to produce the *Gluconobacter oxydans* cells in batch fermentation.

- The cells are concentrated with a continuous discharge disc-stack centrifuge to 2000 L and washed with 40,000 L of 20 mM MgSO$_4$. The washed cells are reconcentrated and stored under refrigeration until use for the bioconversion.

3) Flow scheme

Not published.

185

4) Process parameters

conversion:	100%
yield:	about 58%
reactor type:	batch
reactor volume:	5500 L
capacity:	700 kg of product in 48 h
production site:	St. Louis, MO, USA
company:	Pfizer Inc., USA

5) Product application

- N-Butyldesoxynojirimycin (N-butyl DNJ) is an inhibitor of glycosidase activity, and as such was clinically evaluated by Pharmacia R&D as a potential therapeutic agent against retroviral infections, in particular acquired immune deficiency syndrome (AIDS).

- The activity of this compound was determined *in vitro* to prevent infection of H9 cells by the AIDS virus.

- At least four processes were given extensive consideration, three of which combined bioprocess technology with chemical synthesis.

- A novel, high-efficiency process based on the selective microbial oxidation of N-butylglucamine to an intermediate suitable for reduction to N-butyldeoxynojirimycin was selected as the best available technology for scale-up.

6) Literature

- Landis, B.H., McLaughlin, J.K., Heeren, R., Grabner, R.W., Wang, P.T. (2002) Bioconversion of N-butylglucamine to 6-deoxy-6-butylamino sorbose by *Gluconobacter oxydans*, Org. Proc. Res. Dev. **6**, 547–552

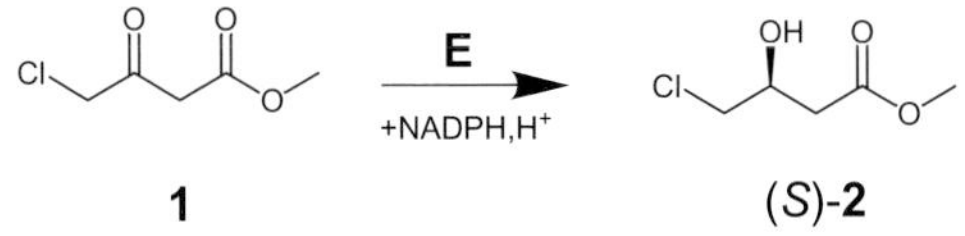

1 = 4-chloro-3-oxo-butanoic acid methyl ester
2 = 4-chloro-3-hydroxy-butanoic acid methyl ester

Bristol-Myers Squibb

Fig. 1.1.X.X – 1

1) Reaction conditions

[**1**]:	0.066 M, 10 g · L^{-1} [150.56 g · mol^{-1}]
pH:	6.8
T:	28 °C
medium:	aqueous
reaction type:	reduction of keto group
catalyst:	suspended whole cells
enzyme:	dehydrogenase
strain:	*Geotrichum candidum* SC 5469

2) Remarks

- In initial experiments an enantiomeric excess of 96.9 % was reached. After heat-treatment (at 50 °C for 30 min) of the cells prior to use the ee increased to 99 %.

- The oxidoreductase was also isolated and purified 100-fold. It is NADP-dependant and the regeneration of cofactor in case of the isolated enzyme is carried out with glucose dehydrogenase. The isolated enzyme was immobilized on Eupergit C.

3) Flow scheme

Not published.

4) Process parameters

yield:	95 %
ee:	99 %
reactor type:	batch
reactor volume:	750 L
capacity:	multi kg
down stream processing:	filtration and extraction
enzyme supplier:	Bristol-Myers Squibb, USA
company:	Bristol-Myers Squibb, USA

5) Product application

- In general, chiral β-hydroxy esters are versatile building blocks.

- (S)-4-Chloro-3-hydroxybutanoic acid methyl ester is used as chiral starting material in the synthesis of a cholesterol antagonist that inhibits the hydroxymethyl glutaryl CoA (HMG-CoA) reductase:

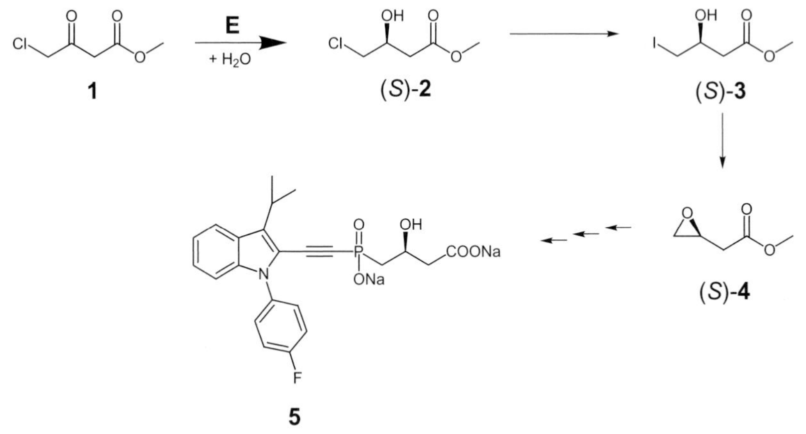

1 = 4-chloro-3-oxo-butanoic acid methyl ester
2 = 4-chloro-3-hydroxy-butanoic acid methyl ester
3 = 3-hydroxy-4-iodo-butanoic methyl ester
4 = oxiranyl-acetic acid methyl ester
5 = cholesterol antagonist

Fig. 1.1.X.X – 2

6) Literature

- Patel, R.N., McNamee, C.G., Banerjee, A., Howell, J.M., Robinson, R.S., Szarka, L.J. (1992) Stereoselective reduction of β-keto esters by *Geotrichum candidum*, Enzyme Microb. Technol. **14**, 731–738

- Patel, R.N. (1997) Stereoselective biotransformations in synthesis of some pharmaceutical intermediates, Adv. Appl. Microbiol. **43**, 91–140

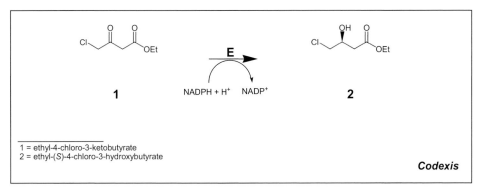

1 = ethyl-4-chloro-3-ketobutyrate
2 = ethyl-(S)-4-chloro-3-hydroxybutyrate

Codexis

Fig. 1.1.X.X – 1

1) Reaction conditions

[1]:	1.17 M,, 160 g · L^{-1} [136.53 g · mol^{-1}]
[2]:	glucose
[3]:	nicotinamide-dinucleotide phosphate, up to 80 mg · L^{-1}
pH:	7.0
T:	25 °C
reaction type:	redox reaction
catalyst:	two soluble recombinant enzymes
enzyme:	mutant alcohol dehydrogenase (ketoreductase)
enzyme:	mutant glucose dehydrogenase

2) Remarks

- The NADPH cofactor is regenerated with glucose dehydrogenase.

- This biocatalytical step is followed by the cyanation catalyzed by a mutant halohydrin dehalogenase (see page 445, EC 3.8.X.X).

3) Flow scheme

Not published.

4) Process parameters

conversion:	>99.5 %
yield:	94 %
ee:	>99.9 %
chemical purity:	>98 %
residence time:	reaction time 24 h
company:	Codexis Inc., USA

5) Literature

- Davis, C., Grate, J., Gray, D., Gruber, J., Huisman, G., Ma, S., Newman, L., Sheldon, R. (2005) Enzymatic process for the production of 4-subsitituted 3-hydroxybutyric acid derivatives, Codexis, Inc., WO04015132

- Davis, C., Jenne, S., Krebber, A., Huisman, G., Newman, L., (2005) Improved ketoreductase polypeptides and related polynucleotides, Codexis, Inc., WO05017135

- Davis, C., Fox, R., Gavrilovic, V., Huisman, G., Newman, L. (2005) Improved halohydrin dehalogenases and related polynucleotides, Codexis, Inc.,WO05017141

- Davis, C. et al (2005) Enzymatic process for the production of 4-substituted 3-hydroxybutyric acid derivatives and vicinal cyano, hydroxyl substituted carboxylic acid esters, Codexis, Inc., WO05018579

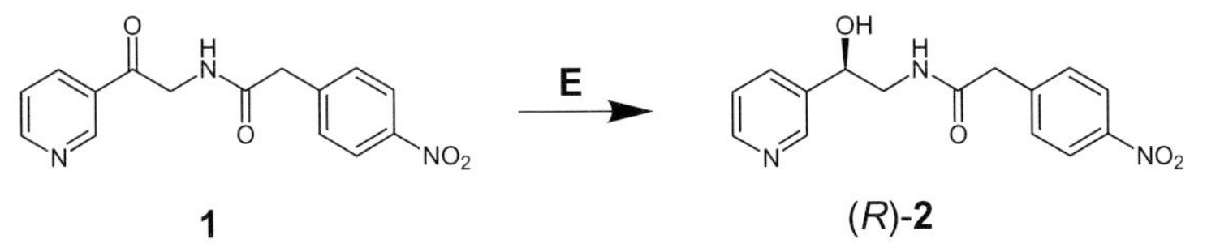

Dehydrogenase
Candida sorbophila

EC 1.1.X.X

1 = 2-(4-nitro-phenyl)-*N*-(2-oxo-2-pyridin-3-ethyl)-acetamide
2 = (*R*)-*N*-(2-hydroxy-2-pyridin-3-yl-ethyl)-2-(4-nitro-phenyl)-acetamide

Merck & Co, Inc.

Fig. 1.1.X.X – 1

1) Reaction conditions

[1]:	0.2 M, 60 g · L^{-1} [299.28 g · mol^{-1}]
pH:	5.6
T:	34 °C
medium:	two-phase: aqueous, solid
reaction type:	reduction of keto group
catalyst:	suspended whole cells
enzyme:	dehydrogenase
strain:	*Candida sorbophila*

2) Remarks

- Here the biotransformation is preferred over the chemical reduction with commercially available asymmetric catalysts (boron-based or noble-metal-based), since with the latter ones the desired elevated enantiomeric excess is not achievable.

- To prevent foaming during the biotransformation 2 mL · L^{-1} defoamer is added.

- Since the ketone has only a very low solubility in the aqueous phase, it is added as an ethanol slurry to the bioreactor. By this method no formal sterilization is possible. Heat sterilization is also not applicable because of the ketone's instability at elevated temperature. In the large scale production, 1 kg ketone is added as solution in 4 L 0.9 M H$_2$SO$_4$ to the bioreactor. This solution is sterilized prior to addition by pumping through a 0.22 μm hydrophobic Durapore microfiltration membrane (Millipore Opticap cartridge, Bedford, USA).

- After intensive investigation and optimization a glucose feed of 1.5 g · L^{-1} was found to support the highest initial bioreduction rate.

- The bioreduction is essentially carried out in a two-phase system, consisting of the aqueous phase and small beads made up of substrate and product.

- The downstream processing consists of multiple extraction steps with methyl ethyl ketone and precipitation induced by pH titration of the pyridine functional group (pK$_a$ = 4.66) with NaOH.

191

3) Flow scheme

Not published.

4) Process parameters

conversion:	> 99 %
yield:	82.5 %
ee:	> 98 %; 99.8 % after purification
chemical purity:	95 %
reactor type:	batch
reactor volume:	280 L
capacity:	multi kg
down stream processing:	extraction and precipitation
company:	Merck & Co., Inc., USA

5) Product application

- The (R)-amino alcohol is an important intermediate for the following synthesis of the β-3-agonist **13**. It can be used for obesity therapy and to decrease the level of associated type II diabetes, coronary artery disease and hypertension:

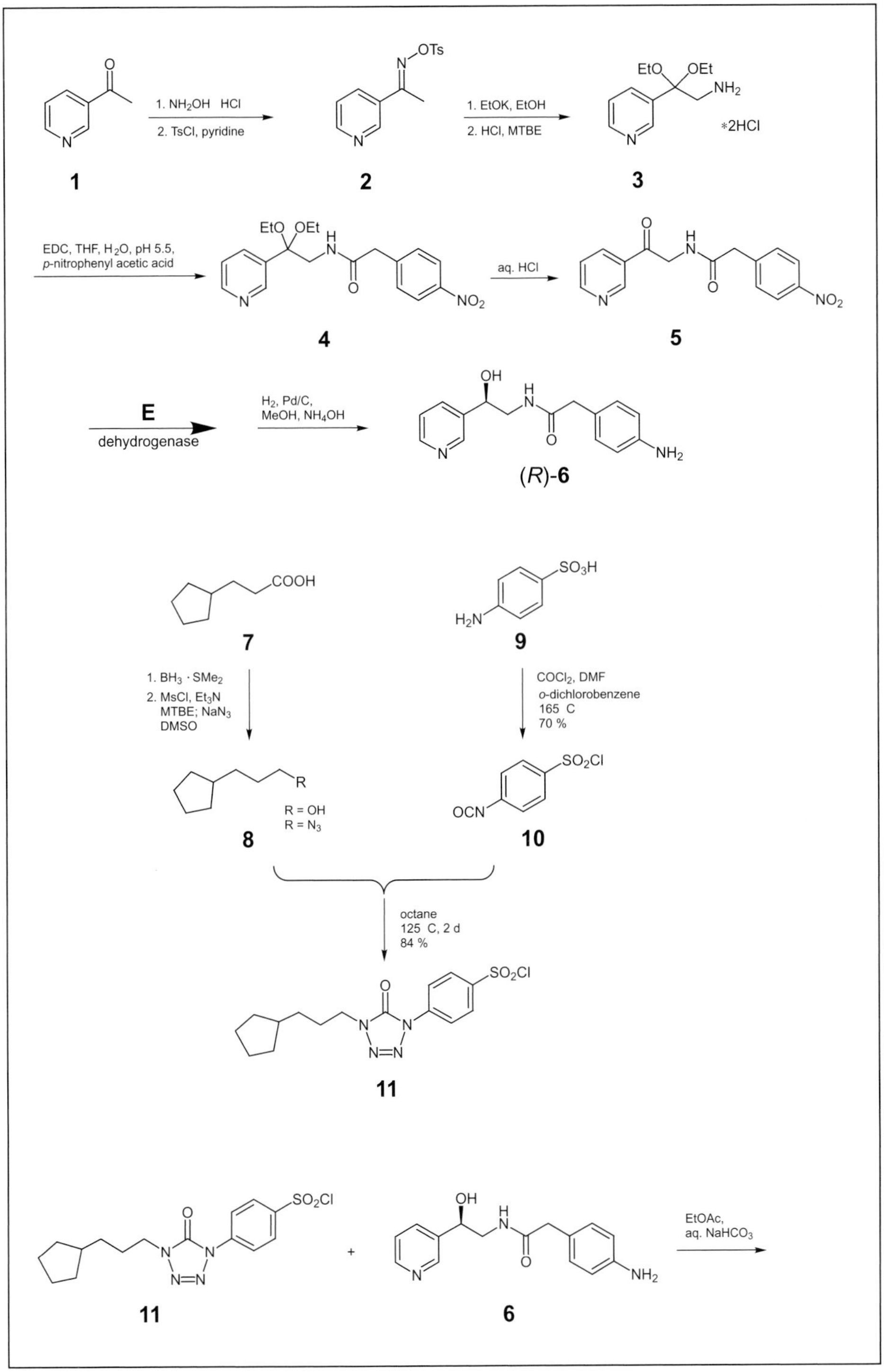
1. NH2OH HCl
2. TsCl, pyridine
1. EtOK, EtOH
2. HCl, MTBE
*2HCl
1
2
3
EDC, THF, H2O, pH 5.5,
p-nitrophenyl acetic acid
aq. HCl
4
5
E
dehydrogenase
H2, Pd/C,
MeOH, NH4OH
(R)-6
7
9
1. BH3 · SMe2
2. MsCl, Et3N
MTBE; NaN3
DMSO
COCl2, DMF
o-dichlorobenzene
165 C
70 %
R = OH
R = N3
8
10
octane
125 C, 2 d
84 %
11
EtOAc,
aq. NaHCO3
11
6

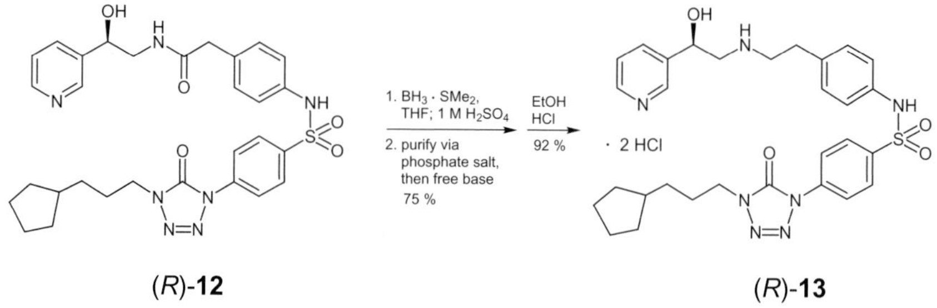

1 = 1-pyridin-3-yl-ethanone
2 = 1-pyridin-3-yl-tosyl oxime
3 = 3-pyridyl-aminomethyl ketal
4 = amide-ketal
5 = pyridyl ketone
6 = aniline-alcohol
7 = 3-cyclopentyl-propionic acid
8 = (3-azido-propyl)-cyclopentane
9 = 4-amino-benzenesulfonic acid
10 = 4-isocyanato-benzenesulfonyl chloride
11 = sulfonyl chloride
12 = sulfonylamino-acetamide
13 = di-HCl salt

Fig. 1.1.X.X – 2

6) Literature

- Chartrain, M.M., Chung, J.Y.L., Roberge, C. (1998) *N*-(*R*)-(2-hydroxy-2-pyridine-3-yl-ethyl)-2-(4-nitro-phenyl)-acetamide, Merck & Co., Inc., US 5846791

- Chung, J.Y.L., Ho, G.-J., Chartrain, M., Roberge, C., Zhao, D., Leazer, J., Farr, R., Robbins, M., Emerson, K., Mathre, D.J., McNamara, J.M., Hughes, D.L., Grabowski, E.J.J., Reider, P.J. (1999) Practical chemoenzymatic synthesis of a pyridylethanolamino beta-3 adrenergic receptor agonist, Tetrahedron Lett. **40**, 6739–6743

- Chartrain, M., Roberge, C., Chung, J., McNamara, J., Zhao, D., Olewinski, R., Hunt, G., Salmon, P., Roush, D., Yamazaki, S., Wang, T., Grabowski, E., Buckland, B., Greasham, R. (1999) Asymmetric bioreduction of (2-(4-nitro-phenyl)-*N*-(2-oxo-2-pyridin-3-yl-ethyl)-acetamide) to its corresponding (*R*)-alcohol [(*R*)-*N*-(2-hydroxy-2-pyridin-3-yl-ethyl)-2-(4-nitro-phenyl)-acetamide] by using *Candida sorbophila* MY 1833, Enzyme Microb. Technol. **25**, 489–496

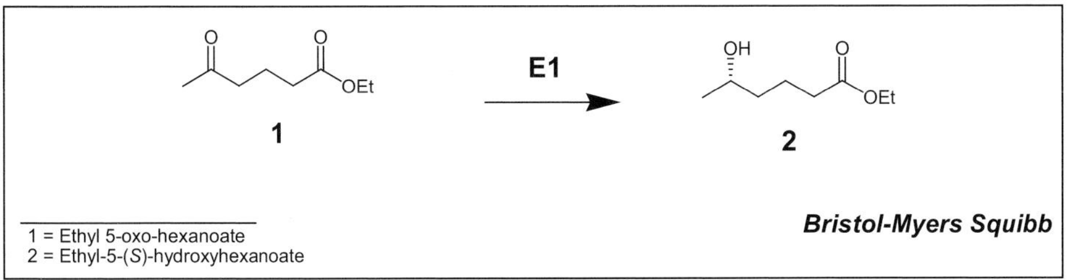

1 = Ethyl 5-oxo-hexanoate
2 = Ethyl-5-(S)-hydroxyhexanoate

Bristol-Myers Squibb

Fig. 1.1.X.X – 1

1) Reaction conditions

[1]:	0.016 M, 5 g · L^{-1} [158.19 g · mol^{-1}]
[2]:	10 % glucose
[3]:	50 mM phosphate buffer
pH:	6.5
medium:	20 % (w/v) cell suspensions (50 mM phosphate buffer)
reaction type:	reduction
catalyst:	whole cells
enzyme:	phosphoribosylaminoimidazole carboxylase
strain:	*Pichia methanolica*
CAS (enzyme):	[9032-04-6]

2) Remarks

- *Pichia methanolica* was found to be the best microbe for reduction of ethyl 5-oxo-hexanoate and 5-oxo-hexanenitrile to the corresponding S-alcohols.

- Only a few microbes have been found to possess the reductase for reduction of the ketone carbonyl group to γ- and δ-keto esters.

- Keto reduction of 5-oxo-hexanenitrile to 5-(S)-hydroxyhexanenitrile with *Pichia methanolica* was also described.

- Resolution of racemic 5-hydroxyhexanenitrile was also performed on a preparative scale with immobilized lipase PS 30 (Amano). An ee of 5-(S)-hydroxyhexanenitrile of 97 % was achieved.

3) Flow scheme

Not published.

4) Process parameters

conversion:	75 %
yield:	95 %
ee:	> 95 %
reactor type:	batch
capacity:	preparative scale
company:	Bristol-Myers Squibb, USA

5) **Literature**

- Nanduri, V.B., Hanson, R.L., Goswami, A., Wasylyk, J.M., LaPorte, T.L., Katipally, K., Chung, H.-J., Patel, R.N., (2001) Biochemical approaches to the synthesis of ethyl 5-(*S*)-hydroxyhexanoate and 5-(*S*)-hydroxyhexanenitrile, Enzyme Microb. Technol. **28**, 632–636

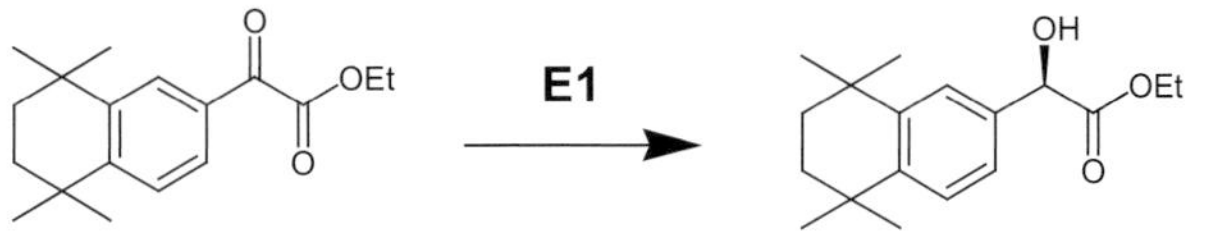

1 = 2-oxo-2-(1',2',3',4'-tetrahydro-1',1',4',4'-tetramethyl-6'-
naphthalenyl)acetate
2 = 2-(R)-hydroxy-2-(1',2',3',4'-tetrahydro-1',2',3',4'-tetrahydro-
1',1',4',4',-tetramethyl-6'-naphthalenyl)acetate

Bristol-Myers Squibb

Fig. 1.1.X.X – 1

1) Reaction conditions

[1]:	0.017 M, 5 g · L^{-1}, [288.38 g · mol^{-1}] in methanol
[1]:	25 g · L^{-1}glucose
pH:	6.8–7.0
T:	28 °C
medium:	aqueous
reaction type:	reduction
catalyst:	frozen whole cells
enzyme:	reductase
strain:	*Aureobasidium pullulans* SC 13849
CAS (enzyme):	[9025-57-4]

2) Remarks

• In a two-stage process: cells were grown in a 25-L bioreactor for 40 h and then harvested by centrifugation and stored at –60 °C until further use. Frozen cells were suspended in 50 mM potassium phosphate buffer (pH 6.8) and the resulting cell suspensions were used to carry out bioreduction.

• At the end of the reaction, hydroxy ester was adsorbed onto XAD-16 resin after filtration. It was recovered in 94 % yield from the resin by acetonitrile extraction.

• The recovered (R)-hydroxy ester was treated with Chirazyme L-2 or pig liver esterase to convert it into the corresponding (R)-hydroxy acid in quantitative yield. The enzymatic hydrolysis was used to avoid the possibility of racemization under chemical hydrolytic conditions.

• A simpler, single-stage fermentation-bioreduction process was also developed. A yield of 98 % and ee of 98 % were obtained. However, owing to low cell concentration in this single-stage process, the reaction was completed in 107 h compared with 16 h for the two-stage process.

3) Flow scheme

Not published.

4) Process parameters

yield:	99 % (overall 94 %)
ee:	97 %
reactor type:	batch
reactor volume:	5 L
company:	Bristol-Myers Squibb, USA

5) Product application

- The chiral ester and the corresponding acid are the intermediates in the synthesis of the retinoic acid receptors gamma-specific agonist, which is potentially useful as a dermatological and anti-cancer drug.

6) Literature

- Patel, R.N., Chu, L., Chidambaram, R., Zhu, J., Kant, J. (2002) Enantioselective microbial reduction of 2-oxo-2-(1′,2′,3′,4′-tetrahydro-1′,1′,4′,4′-tetramethyl-6′-naphthalenyl)acetic acid and its ester, Tetrahedron: Asymmetry **13**, 349-355

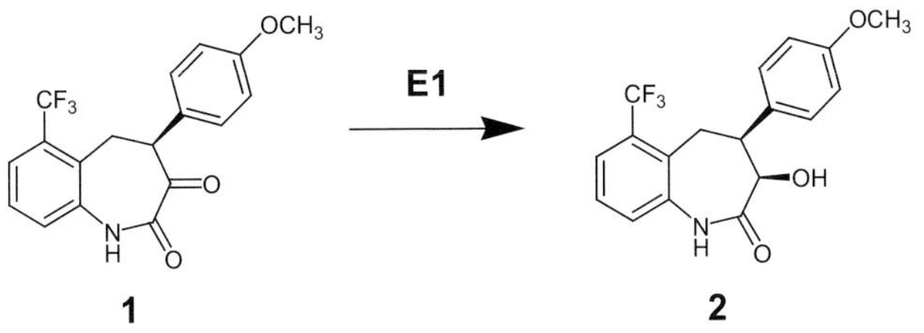

E1

1 = 4,5-dihydro-4-(4-methoxyphenyl)-6'-(trifluoromethyl)-1*H*-1-benzazepine-2,3-dione
2 = (3*R-cis*)-1,3,4,5-tetrahydro-3-hydroxy-4-(4-methoxyphenyl)-6-(trifluoromethyl)-2*H*-
1-benzazepin-2-one

Bristol-Myers Squibb

Fig. 1.1.X.X – 1

1) Reaction conditions

[1]:	0.006 M, 2 g · L^{-1} [349.3 g · mol^{-1}]
medium:	aqueous
reaction type:	reduction
catalyst:	suspended whole cells
enzyme:	reductase
strain:	*Nocardia salmonicolor* SC6310

2) Remarks

- Substrate exists predominantly in the achiral enol form in rapid equilibrium with the two enantiomeric keto forms. Reduction of the substrate could give rise to formation of four possible alcohol stereoisomers. Remarkably, conditions were found under which only a single alcohol isomer [(3 *R-cis*)-1,3,4,5-tetrahydro-3-hydroxy-4-(4-methoxyphenyl)-6'-(trifluoromethyl)-2 *H*-1-benzazepin-2-one] was obtained by microbial reduction. The most effective culture was *Nocardia salmonicolor* SC6310.

3) Flow scheme

Not published.

4) Process parameters

yield:	96 %
ee:	99.8 %
company:	Bristol-Myers Squibb, USA

5) Product application

- A key intermediates in the synthesis of longer acting and more potent antihypertensive agent: (*cis*)-3-(acetoxy)-1-[2-(dimethylamino)ethyl]-1,3,4,5-tetrahydro-4-(4-methoxyphenyl)-6-trifluoro-methyl)-2*H*-1-benzazepin-2-one is known as Diltiazem, a benzothiazepinone calcium channel blocking agent that inhibits influx of extracellular calcium through L-type voltage-operated calcium channels. This compound has been widely used clinically in the treatment of hypertension and angina.

6) Literature

- Patel, R.N. (2001) Enzymatic synthesis of chiral intermediates for drug development, Adv. Synth. Catal. **343**, (6-7) 527–546.

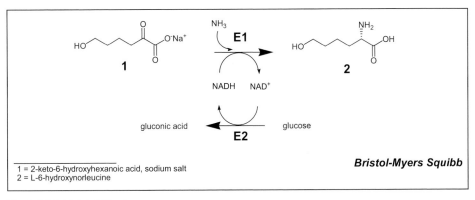

1 = 2-keto-6-hydroxyhexanoic acid, sodium salt
2 = L-6-hydroxynorleucine

Bristol-Myers Squibb

Fig. 1.4.1.3 / 1.1.1.118 – 1

1) Reaction conditions

[1]:	0.59 M, 100 g · L⁻¹ [168.12 g · mol⁻¹] 2-keto-6-hydroxy hexanoic acid, sodium salt in equilibrium with 2-hydroxytetrahydropyran-2-carboxylic acid, sodium salt
[2]:	1 M ammonium formate
[3]:	reduced nicotinamide adenine dinucleotide NADH
pH:	8.7
medium:	aqueous
reaction type:	redox reaction
catalyst:	enzymes
enzyme:	L-glutamic acid dehydrogenase
enzyme:	D-glucose dehydrogenase
strain:	Beef liver / *Bacillus megaterium*
CAS (enzyme):	[9029-12-3] / [9028-53-9]

2) Remarks

- Reductive amination of ketoacid using amino acid dehydrogenases has been shown to be a useful method for the synthesis of natural and unnatural amino acids.

- 2-Keto-6-hydroxy hexanoic acid was converted completely into L-6-hydroxynorleucine by phenylalanine dehydrogenase from *Sporosarcina* sp. or by beef liver glutamate dehydrogenase.

- Beef liver glutamate dehydrogenase was used for preparative reaction at 100 g · L⁻¹ substrate concentrations.

3) Flow scheme

Not published.

4) Process parameters

yield:	91–97 %
optical purity	> 99 %
reactor type:	batch
capacity:	preparative scale
company:	Bristol-Myers Squibb, USA

5) Product application

- L-6-hydroxynorleucine is a chiral intermediate useful for the synthesis of vasopeptidase inhibitor now in clinical trials and for the synthesis of C-7 substituted azepinones as potential intermediates for other antihypertensive metalloproteinase inhibitors.

6) Literature

- Patel, R.N. (2001) Enzymatic synthesis of chiral intermediates for Omapatrilat, an antihypertensive drug, Biomol. Eng. **17**, 167–182

- Patel, R.N. (2001) Enzymatic synthesis of chiral intermediates for drug development, Adv. Synth. Catal. **343**, (6-7) 527–546

- Hanson, L.R., Schwinden, D.M., Banerjee, A., Brzozowski, D.B., Chen, B.-C., Patel, B.P., McNamee, C.G., Kodersha, G.A., Kronenthal, D.R., Patel, R.N., Szarka, L.J. (1999) Enzymatic synthesis of L-6-hydroxynorleucine, Bioorg. Med. Chem. **7**, 2247–2252

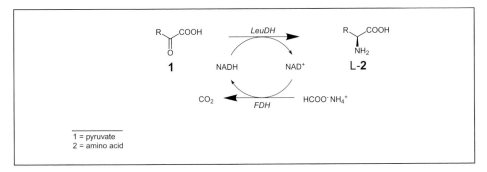

1 = trimethylpyruvic acid
2 = *tert.*-leucine

Degussa AG

Fig. 1.4.1.9 – 1

1) Reaction conditions

[1]:	0.5 M, 65.07 g · L^{-1} [130.14 g · mol^{-1}]
pH:	8.0
T:	25 °C
medium:	aqueous
reaction type:	redox reaction
catalyst:	solubilized enzyme
enzyme:	L-leucine-NAD oxidoreductase, deaminating
strain:	*Bacillus sphaericus*
CAS (enzyme):	[9028–71–7]

2) Remarks

• The expensive cofactor can be easily regenerated by formate dehydrogenase (FDH) from *Candida boidinii* utilizing ammonium formate that is oxidized to CO_2 under reduction of NAD$^+$ to NADH:

1 = pyruvate
2 = amino acid

Fig. 1.4.1.9 – 2

• The cofactor regeneration is practicably irreversible and the co-product CO_2 can be easily separated (K_{eq} = 15,000).

• Other amino acids that are synthesized by the same method:

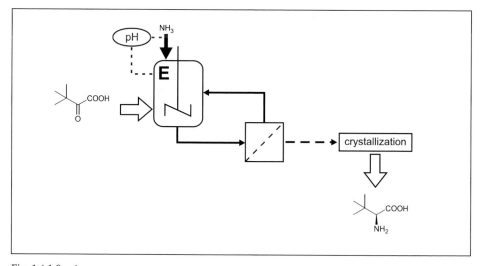

1 = neopentylglycine
2 = 3,3-dimethylpropanyl glycine
3 = 3-ethyl-3-methyl-propanyl glycine
4 = 5,5-dimethyl-butyl glycine

Fig. 1.4.1.9 – 3

- At high concentrations of the substrate trimethylpyruvate (of industrial quality), the enzymes are inhibited.

- The productivity of the process is not limited by the enzymatic reaction but by chemical reactions because at higher ammonia and keto acid concentrations the formation of by-products is increased.

- The enzymes are recycled after an ultrafiltration step.

3) Flow scheme

Fig. 1.4.1.9 – 4

4) Process parameters

yield:	74 %
reactor type:	repetitive batch with ultrafiltration (enzyme membrane reactor)
capacity:	ton scale
residence time:	2 h
space-time-yield:	$638 \text{ g} \cdot \text{L}^{-1} \cdot \text{d}^{-1}$
down stream processing:	crystallization
enzyme consumption:	leucine dehydrogenase: $0.9 \text{ U} \cdot \text{g}^{-1}$, formate dehydrogenase: $2.3 \text{ U} \cdot \text{g}^{-1}$
company:	Degussa AG, Germany

5) Product application

- The amino acids are building blocks for drug synthesis, e.g. anti-tumor agents or HIV protease inhibitors.

- They can also be used as templates in asymmetric synthesis:

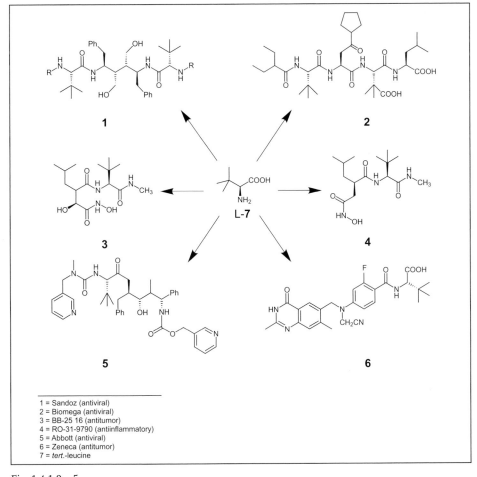

1 = Sandoz (antiviral)
2 = Biomega (antiviral)
3 = BB-25 16 (antitumor)
4 = RO-31-9790 (antiinflammatory)
5 = Abbott (antiviral)
6 = Zeneca (antitumor)
7 = *tert.*-leucine

Fig. 1.4.1.9 – 5

6) Literature

- Bommarius, A.S., Drauz, K., Hummel, W., Kula, M.-R., Wandrey, C. (1994) Some new developments in reductive amination with cofactor regeneration, Biocatalysis, **10**, 37–47

- Bommarius, A.S., Estler, M., Drauz, K. (1998) Reaction engineering of large-scale reductive amination processes; Inbio 98 Conference, Amsterdam/NL

- Bommarius, A.S., Schwarm, M., Drauz, K. (1998) Biocatalysis to amino acid-based chiral pharmaceuticals – examples and perspectives, J. Mol. Cat. B: Enzymatic **5**, 1–11

- Bradshaw, C.W., Wong, C.-H., Hummel, W., Kula, M.-R. (1991) Enzyme-catalyzed asymmetric synthesis of (S)-2-amino-4-phenylbutanoic acid and (R)-2-hydroxy-4-phenylbutanoic acid, Bioorg. Chem. **19**, 29–39

- Kragl, U., Vasic-Racki, D., Wandrey, C. (1996) Continuous production of L-*tert*-leucine in series of two enzyme membrane reactors, Bioprocess Engineering **14**, 291–297

- Krix, G., Bommarius, A.S., Drauz, K., Kottenhahn, M., Schwarm, M., Kula, M.-R. (1997) Enzymatic reduction of α-keto acids leading to L-amino acids, D- or L-hydroxy acids, J. Biotechnol. **53**, 29–39

- Wichmann, R., Wandrey, C., Bückmann, A.F., Kula, M.-R., (1981) Continuous enzymatic transformation in an enzyme membrane reactor with simultaneous NAD(H) regeneration, Biotechnol. Bioeng. **23**, 2789–2796

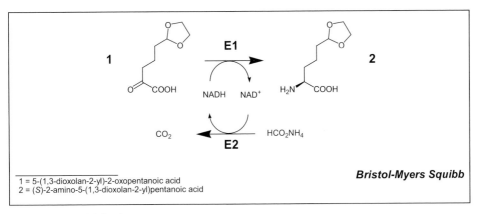

1 = 5-(1,3-dioxolan-2-yl)-2-oxopentanoic acid
2 = (S)-2-amino-5-(1,3-dioxolan-2-yl)pentanoic acid

Bristol-Myers Squibb

Fig. 1.4.1.20 / 1.2.1.2 – 1

1) Reaction conditions

[1]:	0.266 M, 50 g · L⁻¹ [188.18 g · mol⁻¹]
[2]:	19.49 g · L⁻¹ ammonium formate
[3]:	0.35 g · L⁻¹ NAD
pH:	8.0
T:	40 °C
medium:	aqueous
reaction type:	reductive amination
catalyst:	heat-dried cells of *Escherichia coli* containing cloned PheDH from *Thermoactinomyces intermedius* and heat-dried cells of *Candida boidinii*, or dried recombinant *Pichia pastoris* containing *Thermoactinomyces intermedius* PheDH
enzyme:	L-phenylalanine dehydrogenase
enzyme:	formate dehydrogenase
strain:	*Thermoactinomyces intermedius / Candida boidinii / Pichia pastoris*
CAS (enzyme):	[69403-12-9] PheDH / [9028-85-7] FDH

2) Remarks

- FDH activity in *P. pastoris* was 2.7-fold greater than for *C. boidinii*.

- Phenylalanine dehydrogenase was cloned and overexpressed in *Escherichia coli* and *Pichia pastoris*.

- Fermentation of *T. intermedius* yielded 184 units of PheDH activity per liter of whole broth in 6 h. In contrast, *E. coli* BL(DE3)pPDH155k produced over 19,000 units per liter of whole broth in about 14 h.

- Acetal amino acid had been previously prepared by an eight-step chemical synthesis from 3,4-dihydro[2*H*]pyran.

- An alternate synthesis was demonstrated by the reductive amination of ketoacid acetal using purified enzyme PheDH from *Thermoactinomyces intermedius*. The reaction required ammonia and NADH; NAD⁺ produced during the reaction was recycled to NADH by the oxidation of formate to CO_2 using purified formate dehydrogenase (FDH).

- A total of 197 kg product was produced in three 16,000-L batches using 5 % concentration of substrate with an average yield of 91 % in a procedure using heat-dried cells of *E. coli* containing cloned PheDH and heat-dried *C. boidinii*.

- *Pichia pastoris* procedure, using dried recombinant cells containing *T. intermedius* PheDH inducible with methanol and endogenous FDH, allowed both enzymes to be produced during a single fermentation. It had the following modifications over the *E. coli* /*C. boidinii* procedure: the concentration of substrate was increased to 100 g · L^{-1}; ¼ the amount of NAD$^+$ was used; dithiothreitol was omitted. The procedure was scaled up to produce 15.5 kg of product with 97 mol % yield and ee > 98 % in a 180-L batch using 10 % concentration of ketoacid.

3) Flow scheme

Not published.

4) Process parameters

conversion:	96–97.5 %
yield:	96–97.5 % (91 mol %)
ee:	> 99 %
reactor type:	batch
reactor volume:	1600 L
capacity:	197 kg of product
enzyme activity:	0.27 kU · L^{-1} PheDH / 1.66 kU · L FDH
enzyme supplier:	Bristol-Myers Squibb, USA
company:	Bristol-Myers Squibb, USA

5) Product application

- (S)-2-Amino-5-(1,3-dioxalan-2-yl)-pentanoic acid is one of three building blocks used in an alternative synthesis of Omapatrilat.

- Omapatrilat is an antihypertensive drug that acts by inhibiting angiotensin-converting enzyme (ACE) and neutral endopeptidase (NEP). Effective inhibitors of ACE have been used not only in the treatment of hypertension, but also in the clinical management of congestive heart failure.

6) Literature

- Patel, R.N. (2001) Enzymatic synthesis of chiral intermediates for Omapatrilat, an antihypertensive drug, Biomol. Eng., **17**, 167–182

- Patel, R.N. (2001) Biocatalytic synthesis of intermediates for the synthesis of chiral drug substances, Curr. Opin. Biotechnol., **12** (6), 587–604

- Patel, R.N. (2001) Enzymatic synthesis of chiral intermediates for drug development, Adv. Synth. Catal., **343**, (6-7), 527–546

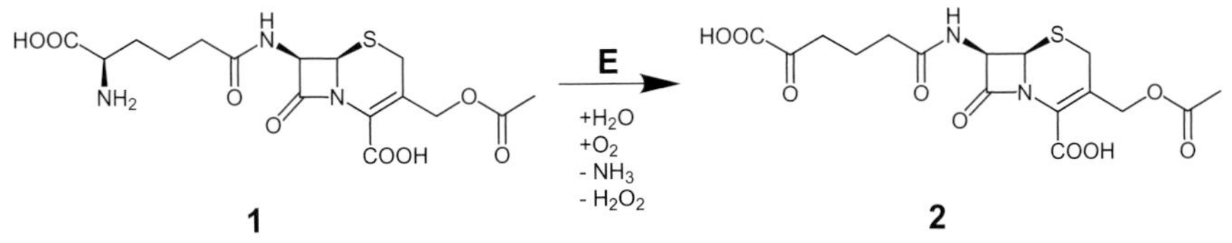

1 = Cephalosporin C
2 = N-α-ketoadipinyl-7-aminocephalosporanic acid

Sandoz AG

Fig. 1.4.3.3 – 1

1) Reaction conditions

[**1**]:	0.02 M, 7.47 g · L^{-1} [373.38 g · mol^{-1}]
pH:	7.3
T:	25 °C
medium:	aqueous
reaction type:	oxidative deamination
catalyst:	immobilized enzyme
enzyme:	D-amino-acid oxygen oxidoreductase, deaminating (D-amino-acid oxidase)
strain:	*Trijonopsis variabilis*
CAS (enzyme):	[9000–88–8]

2) Remarks

- This step is part of the 7-aminocephalosporanic acid (7-ACA) process, see page 326.
- Ketoadipinyl-7-ACA decarboxylates *in situ* in presence of H_2O_2 that is formed by the biotransformation step yielding glutaryl-7-ACA:

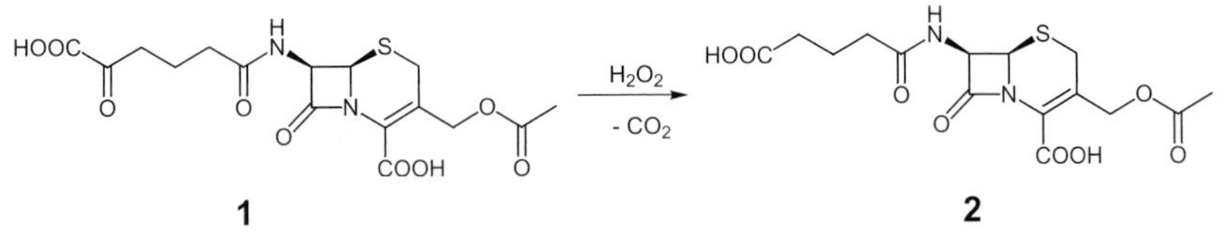

1 = α-ketoadipinyl-7-ACA
2 = glutaryl-7-ACA

Fig. 1.4.3.3 – 2

3) Flow scheme

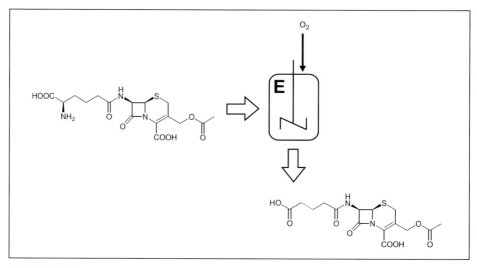

Fig. 1.4.3.3 – 3

4) Process parameters

reactor type:	batch
reactor volume:	10,000 L
capacity:	$200\ t \cdot a^{-1}$
residence time:	1.5 h
down stream processing:	reaction solution is directly transferred to the 7-ACA (7-aminocephalosporanic acid) production
enzyme consumption:	$1.1\ U \cdot g^{-1}$
enzyme supplier:	Sandoz AG, Germany
start-up date:	1996
production site:	Sandoz AG, Germany
company:	Sandoz AG, Switzerland

5) Product application

- See 7-ACA process on page 326.

6) Literature

- Matsumoto, K. (1993) Production of 6-APA, 7-ACA, and 7-ADCA by immobilized penicillin and cephalosporin amidases, in: Industrial Application of Immobilized Biocatalysts (Tanaka, A, Tosa, T., Kobayashi, T. eds.), pp. 67–88, Marcel Dekker Inc., New York

- Tanaka, T. Tosa, T., Kobayashi, T. (1993) Industrial Application of Immobilized Biocatalysts, Marcel Dekker Inc., New York

- Verweij, J., Vroom, E.D. (1993) Industrial transformations of penicillins and cephalosporins, Rec. Trav. Chim. Pays-Bas **112**, 66–81

D-Amino acid oxidase
Trigonopsis variabilis ATCC 10679

EC 1.4.3.3

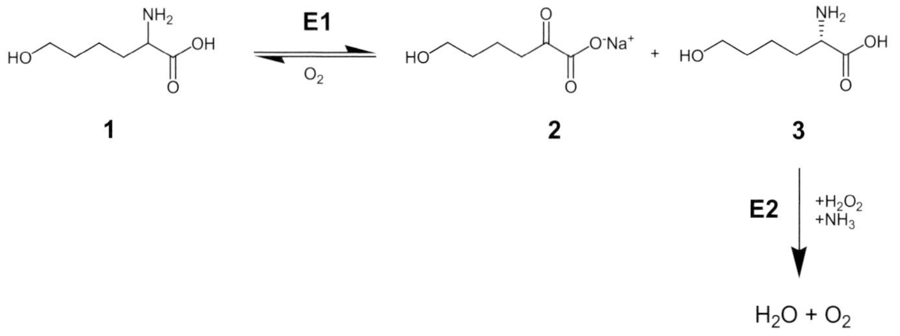

1 = (rac)-6-hydroxynorleucine
2 = 2-keto-6-hydroxyhexanoic acid, sodium salt
3 = L-6-hydroxynorleucine

Bristol-Myers Squibb

Fig. 1.4.3.3 – 1

1) Reaction conditions

[1]:	0.05 M, 7.34 g · L^{-1} [147.17 g · mol^{-1}]
pH:	7.0
T:	28 °C
medium:	50 mM potassium phosphate buffer
reaction type:	oxidative deamination
catalyst:	whole cells
enzyme:	D-amino acid oxidase
enzyme:	catalase
strain:	*Trigonopsis variabilis* ATCC 10679
CAS (enzyme):	[9000-88-8]

2) Remarks

- Racemic 6-hydroxynorleucine is produced by hydrolysis of commercially available 5-(4-hydroxybutyl)hydantoin.

- D-amino acid oxidase has been used to convert the D-amino acid into the ketoacid, leaving the L-enantiomer, which was isolated by ion exchange chromatography.

3) Flow scheme

Not published.

211

4) Process parameters

yield:	Maximum 50 %
optical purity	> 99 %
reactor type:	batch
enzyme activity:	2 g washed cells per mL
company:	Bristol-Myers Squibb, USA

5) Product application

- L-6-Hydroxynorleucine is a chiral intermediate useful for the synthesis of a vasopeptidase inhibitor now in clinical trials and for the synthesis of C-7 substituted azepinones as potential intermediates for other antihypertensive metalloproteinase inhibitors.

- 2-Keto-6-hydroxyhexanoic acid is converted further into L-6-hydroxynorleucine using amino acid dehydrogenase and beef liver glutamate dehydrogenase.

6) Literature

- Patel, R.N. (2001) Enzymatic synthesis of chiral intermediates for Omapatrilat, an antihypertensive drug, Biomol. Eng. **17**, 167-182

- Patel, R.N. (2001) Biocatalytic synthesis of intermediates for the synthesis of chiral drug substances, Curr. Opin. Biotechnol. **12**, (6) 587-604

- Patel, R.N. (2001) Enzymatic synthesis of chiral intermediates for drug development, Adv. Synth. Catal. **343**, (6-7) 527-546

- Hanson, L.R., Schwinden, D.M., Banerjee, A., Brzozowski, D.B., Chen, B.-C., Patel, B.P., McNamee, C.G., Kodersha, G.A., Kronenthal, D.R., Patel, R.N., Szarka, L.J. (1999) Enzymatic synthesis of L-6-hydroxynorleucine, Bioorg. Med. Chem. **7**, 2247–2252

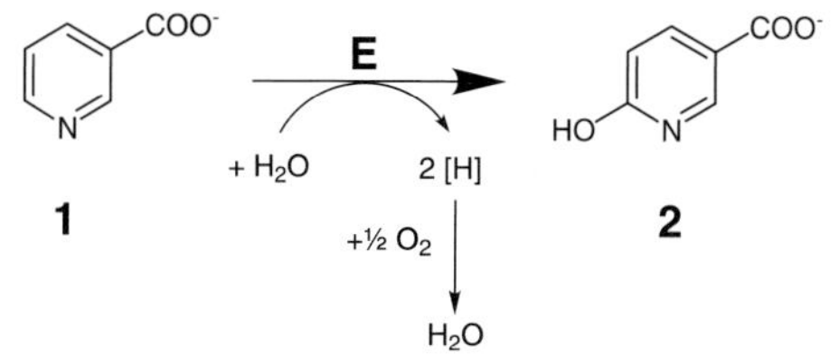

Nicotinic acid hydroxylase
Achromobacter xylosoxidans

1 = niacin = nicotinic acid = pyridine-3-carboxylate
2 = 6-hydroxynicotinate = 6-hydroxy-pyridine-3-carboxylate

Lonza AG

Fig. 1.5.1.13 – 1

1) Reaction conditions

[1]:	0.533 M, 65 g · L^{-1} [122.06 g · mol^{-1}]
pH:	7.0
T:	30 °C
medium:	aqueous
reaction type:	redox reaction (hydroxylation)
catalyst:	suspended whole cells
enzyme:	nicotinate: NADP$^+$6-oxidoreductase (nicotinic acid hydroxylase, nicotinate dehydrogenase)
strain:	*Achromobacter xylosoxidans* LK1
CAS (enzyme):	[9059-03–4]

2) Remarks

- The 6-hydroxynicotinate producing strain was found by accident, when unusually in the mother liquor of a niacin producing chemical plant white crystals of 6-hydroxynicotinate were found to precipitate.

- At niacin concentrations higher than 1 % the second enzyme of the nicotinic acid pathway, the decarboxylating 6-hydroxynicotinate hydroxylase gets strongly inhibited, whereas the niacin hydroxylase operates unaffected:

Nicotinic acid hydroxylase
Achromobacter xylosoxidans

EC 1.5.1.13

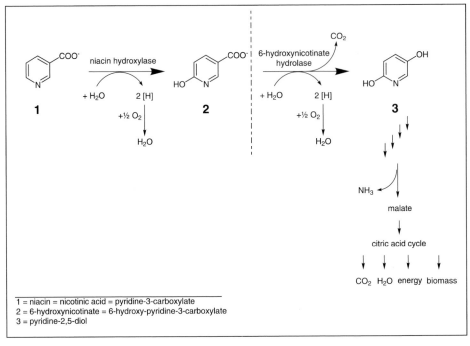

1 = niacin = nicotinic acid = pyridine-3-carboxylate
2 = 6-hydroxynicotinate = 6-hydroxy-pyridine-3-carboxylate
3 = pyridine-2,5-diol

Fig. 1.5.1.13 – 2

- The process takes place in two phases (see flow scheme):

  Phase 1: Growth of cells in a fermenter (chemostat) on niacin and subsequent storage of biomass in cooled tanks.

  Phase 2: Addition of biomass to niacin solution, incubation, separation of biomass and purification of product.

- The product is precipitated by the addition of acid.

- Alternatively, the integration of the two phases into an one reaction vessel fed-batch operation is possible (product concentration of 75 g · L^{-1} in 25 h). This procedure is not used on an industrial scale.

- Also, a continuous process was developed as a 'pseudocrystal fermentation' process. The substrate is added in its solid form and the product crystallizes out of the reaction solution. The process takes advantage of the fact that the Mg-salt of niacin is 100 times more soluble in H_2O at neutral pH than Mg-6-hydroxynicotinate. The pH is titrated to 7.0 with nicotinic acid. The concentration of Mg-nicotinate is regulated to 3 % using conductivity measurement techniques and direct addition of the salt. Mg-6-hydroxynicotinate is collected in a settler.

- Niacin hydroxylase works only in the presence of electron-transmitting systems such as cytochrome, flavine or NADP$^+$, and therefore air needs to be supplied to facilitate the cofactor regeneration. The oxygen-transfer rate limits the reaction.

- In contrast to the biotransformation the chemical synthesis of 6-substituted nicotinic acids is difficult and expensive due to difficulties in the separation of by-products.

214

3) Flow scheme

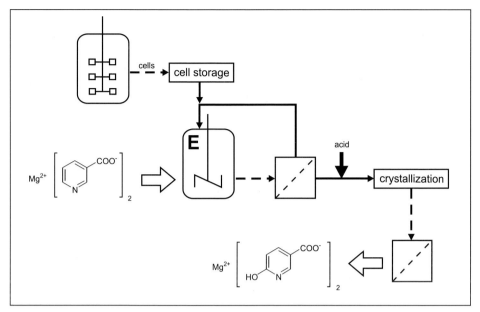

Fig. 1.5.1.13 – 3

4) Process parameters

conversion:	> 90 %
yield:	> 90 % (overall)
selectivity:	high
chemical purity:	> 99 %
reactor type:	batch
reactor volume:	12,000 L
capacity:	several tons
residence time:	12 h
down stream processing:	precipitation, centrifugation and drying
production site:	Visp, Switzerland
company:	Lonza AG, Switzerland

5) Product application

• Versatile building block chiefly in the synthesis of modern insecticides.

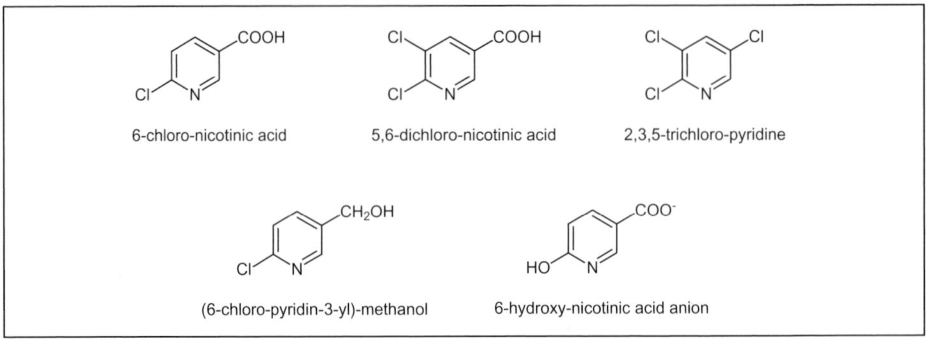

6-chloro-nicotinic acid 5,6-dichloro-nicotinic acid 2,3,5-trichloro-pyridine

(6-chloro-pyridin-3-yl)-methanol 6-hydroxy-nicotinic acid anion

Fig. 1.5.1.13 – 4

6) **Literature**

• Behrmann, E.J., Stanier, R.Y. (1957) The bacterial oxidation of nicotinic acid, J. Biol. Chem. **228**, 923–945

• Briauourt, D., Gilbert, J. (1973) Synthesis of pharmacological investigation concerning the series of 2-dialkylaminoalkoxy-5-pyridine carboxylic acids, Chim. Therap. **2**, 226

• Cheetham, P.S.J. (1994) Case studies in applied biocatalysis – from ideas to products, in: Applied Biocatalysis (Cabral, J., Best, D., Boross, L., Tramper, J., eds.) pp. 47–108. Harwood Academic Publishers, Chur

• Glöckler, R., Roduit, J.-P. (1996) Industrial bioprocesses for the production of substituted aromatic heterocycles, Chimia, **50**, 413–415

• Gsell, L. (1989) 1-Nitro-2,2-diaminoäthylenderivate, Ciba-Geigy AG, EP 0302833 A2

• Kieslich, K. (1991) Biotransformations of industrial use, 5th Leipziger Biotechnology Symposium 1990, Acta Biotechnol. **11**, 559–570

• Kulla, H.G., Lehky, A. (1985) Verfahren zur Herstellung von 6-Hydroxynikotinsäure, Lonza AG, EP 0152949 A2

• Kulla, H.G. (1991) Enzymatic hydroxylations in industrial application, Chimia **45**, 81–85

• Lehky, P., Kulla, H., Mischler, S. (1995) Verfahren zur Herstellung von 6-Hydroxynikotinsäure, Lonza AG, EP 0152948 A2

• Minamida, I., Iwanaga, K., Okauchi, T. (1989) Alpha-unsaturated amines, their production and use, Takeda Chemical Industries, Ltd.,EP 0302389 A2

• Petersen, M., Kiener, A. (1999) Biocatalysis – preparation and functionalization of N-heterocycles, Green Chem. **2**, 99–106

• Quarroz, D., (1983) Verfahren zur Herstellung von 2-Halogenpyridinderivaten, Lonza AG, EP 0084118 A1

• Sheldon, R.A. (1993) Chirotechnology, Marcel Dekker Inc., New York

• Wolf, H., Becker, B., Stendel, W., Homeyer, B., (1988) Substituierte Nitroalkene, Bayer AG, EP 0292822 A2

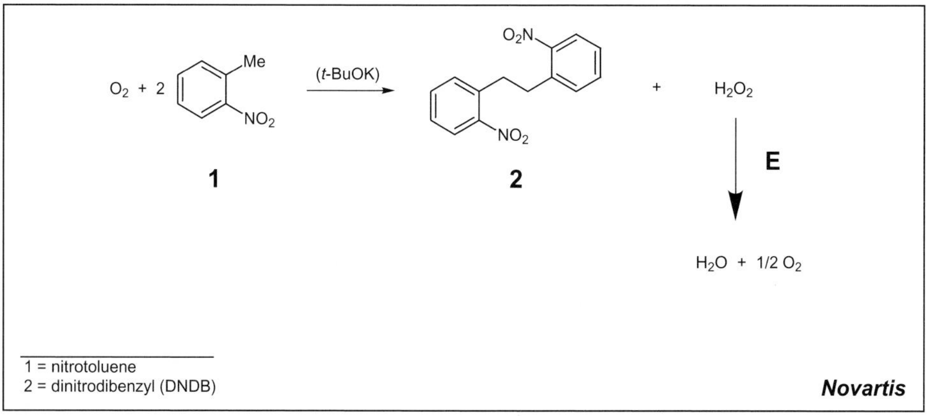

Fig. 1.11.1.6 – 1

1) Reaction conditions

pH:	6.0–9.0
medium:	3-phase-system: organic, solid, aqueous
reaction type:	redox reaction
catalyst:	solubilized
enzyme:	hydrogen-peroxide: hydrogen peroxide oxidoreductase (catalase)
strain:	*microbial source*
CAS (enzyme):	[9001-05–2]

2) Remarks

- During oxidative coupling to dinitrodibenzyl (DNDB) hydrogen peroxide is produced as a by-product. It is not possible to decompose H_2O_2 by the addition of heavy-metal catalysts because only incomplete conversion is reached. Additionally, subsequent process steps with DNDB are problematic due to contamination with heavy-metal catalyst.

- The reaction solution consists of three phases: 1) water, 2) organic phase and 3) solid phase consisting of DNDB.

- The catalase derived from microbial source has advantages compared to beef catalase since the activity remains constant over a broad pH range from 6.0 to 9.0; temperatures up to 50 °C are tolerated and salt concentrations up to 25 % do not affect the enzyme stability.

- The reaction is carried out in a cascade of CSTRs. After the chemical reaction in the first vessel, the pH is adjusted with acetic acid. Decomposition of H_2O_2 is induced by catalase solution added to the second reactor at a constant rate. Oxygen is removed by dilution with nitrogen in the third reactor.

3) Flow scheme

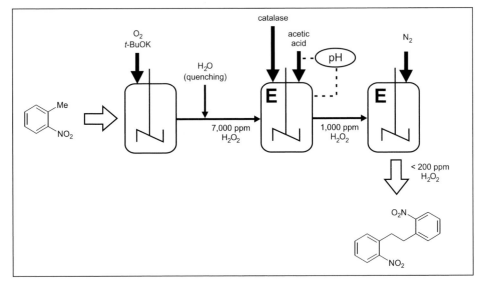

Fig. 1.11.1.6 – 2

4) Process parameters

conversion:	> 98 %
yield:	> 98 %
selectivity:	99.9 %
reactor type:	cascade of CSTR's
residence time:	> 1 h
down stream processing:	degassing with nitrogen
company:	Novartis, Switzerland

5) Product application

- The process is only relevant to remove the undesired side product H_2O_2. The dinitrodibenzyl is used as a pharmaceutical intermediate.

6) Literature

- Onken, U., Schmidt, E., Weissenrieder, T. (**1996**) Enzymatic H_2O_2 decomposition in a three-phase suspension, Ciba-Geigy; International conference on biotechnology for industrial production of fine chemicals, 93rd event of the EFB; Zermatt, Switzerland, 29.09.1996

Oxygenase
Arthrobacter sp.

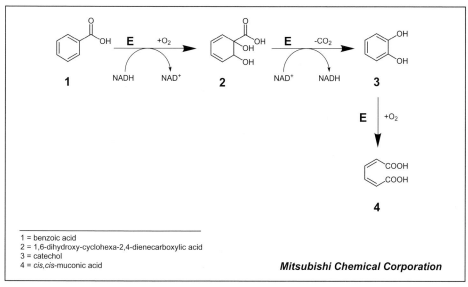

1 = benzoic acid
2 = 1,6-dihydroxy-cyclohexa-2,4-dienecarboxylic acid
3 = catechol
4 = *cis,cis*-muconic acid

Mitsubishi Chemical Corporation

Fig. 1.13.11.1 – 1

1) Reaction conditions

medium: aqueous
reaction type: redox reaction
catalyst: whole cells
enzyme: catechol-oxygen 1,2-oxidoreductase (catechase, catechol dioxygenase)
strain: *Arthrobacter* sp.
CAS (enzyme):   [9027–16–1]

2) Remarks

- Benzoic acid is continuously fed into the fermentation medium.

- The reaction solution is separated from the cells by membrane filtration (cross-flow membrane reactor).

- Since the reaction solution is colored, the first step of down stream processing is an adsorptive removal of color contaminant and the non-converted substrate benzoic acid.

- On acidification of the eluent with sulfuric acid muconic acid is precipitated and separated by an on-line filtration step.

- The residual muconic acid with a concentration of 10 % can be concentrated and precipitated by two ion-exchange steps and acidification.

3) Flow scheme

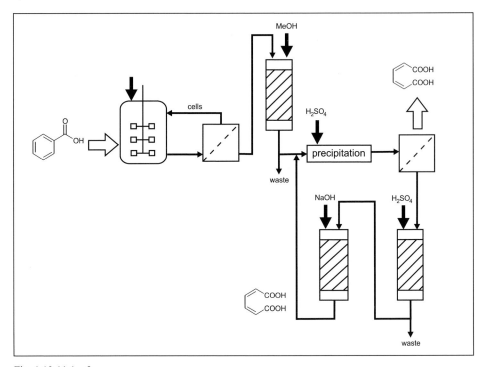

Fig. 1.13.11.1 – 2

4) Process parameters

chemical purity:	> 99 %
space-time-yield:	70 g · L^{-1} · d^{-1}
production site:	Yokohama, Japan
company:	Mitsubishi Chemical Corporation, Japan

5) Product application

- The product is a raw material for new resins, pharmaceuticals and agrochemicals. It can also be used as an intermediate for the synthesis of adipic acid (nylon production).

6) Literature

- Mizuno, S., Yoshikawa, N., Seki, M., Mikawa, T., Imada, Y., (1988) Microbial production of *cis,-cis*-muconic acid from benzoic acid, Appl. Microbiol. Biotechnol. **28**, 20–25

- Yoshikawa, N., Ohta, K., Mizuno, S., Ohkishi, H. (1993) Production of *cis,cis*-muconic acid from benzoic acid, in: Industrial Application of Immobilized Biocatalysts (Tanaka, A., Tosa, T., Kobayashi, T., eds.), pp. 131–147, Marcel Dekker Inc., New York

- Yoshikawa, N., Ohta, K., Mizuno, S., Ohkishi, H. (1993) Production of *cis,cis*-muconic acid from benzoic acid, Bioprocess Technol. **16**, 131–147

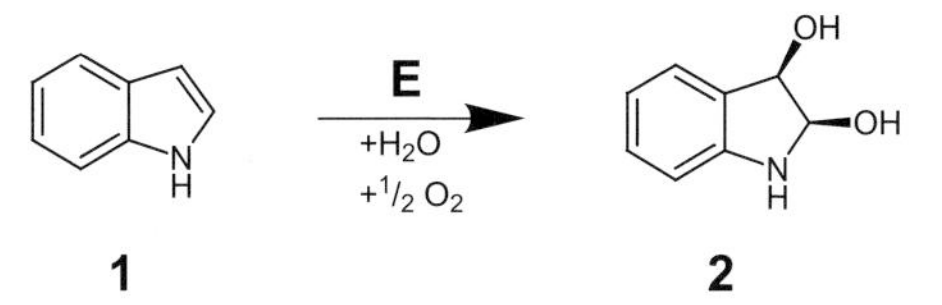

Naphthalene dioxygenase
Pseudomonas putida

1 = 1*H*-indole
2 = 2,3-dihydro-1*H*-indole-2,3-diol

Genencor International, Inc.

Fig. 1.13.11.11 – 1

1) Reaction conditions

medium:	aqueous
reaction type:	redox reaction
catalyst:	whole cells
enzyme:	naphthalene, NADH:oxygen oxidoreductase (naphthalene 1,2-dioxygenase, naphthalene oxygenase)
strain:	*Pseudomonas putida*
CAS (enzyme):	[9014–51–1]

2) Remarks

- This process demonstrates the potential of biological synthesis by constructing an adequate host cell.

- Several problems have to be solved for an effective synthesis of indigo (see figure: 1.13.11.11 – 4:

  1) The activity of naphthalene dioxygenase is very low in *Pseudomonas putida* and even lower in the first recombinant *E. coli* strains.

  2) The half-life of the indole oxidizing system is only about one to two hours. Indole concentrations above 400 mg · L^{-1} are toxic to cells due to inactivation of the ferredoxin component of the enzyme system.

  3) Indole is too expensive for commercial application. Alternatively tryptophan can be used with a possible tryptophanase step (EC 4.1.99.1; [9024-00–4]) for synthesizing the indole in the cell. But tryprophan is also too expensive.

- To overcome all these problems several improvements were necessary:

  1) The dioxygenase system is improved by using better plasmids with stronger promoters.

  2) The cloning of the ferredoxin producing genes results firstly not only in a more stable ferredoxin system but in a higher concentration of ferredoxin units. Secondly site-directed mutagenesis leads to a more stable ferredoxin system.

  3) The tryptophanase system is improved by site-directed mutagenesis.

- All improvements are combined in one single host to produce indigo from glucose. Further improvements of basic fermentation parameters lead to higher indole concentrations.

- In the following figures the chemical routes are compared to the biotransformation route.
- The classical route to indigo needs three steps starting from aniline:

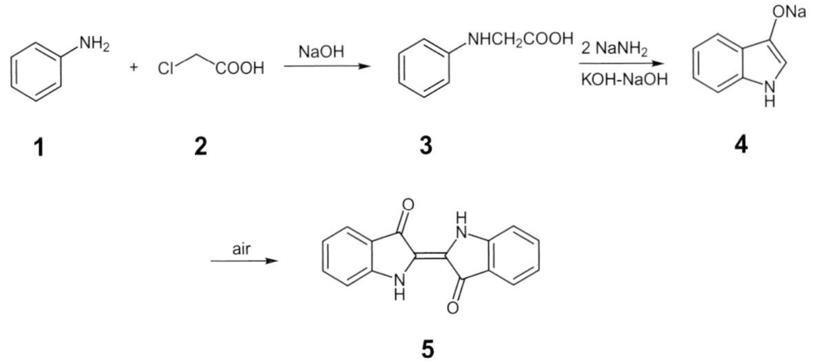

1 = phenylamine
2 = chloro-acetic acid
3 = phenylamino-acetic acid
4 = sodium salt of 1*H*-indole-3-ol
5 = indigo

Fig. 1.13.11.11 – 2

- A new catalytic route (Mitsui Chemicals, Japan) starts also from aniline but produces less inorganic salts than the classical route:

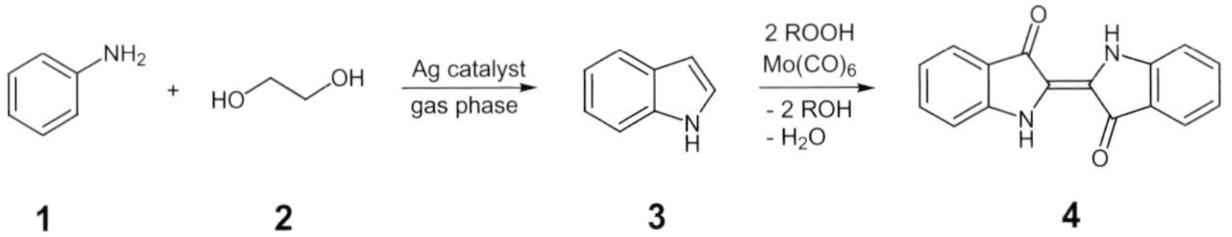

1 = phenylamine
2 = ethane-1,2-diol
3 = 1*H*-indole
4 = indigo

Fig. 1.13.11.11 – 3

- The biotransformation starts with glucose using the capability of the cells to produce tryptophan as intermediate. The gene from *Pseudomonas putida*, which codes for the enzyme naphthalene dioxygenase, is inserted into the common bacteria *E. coli*.

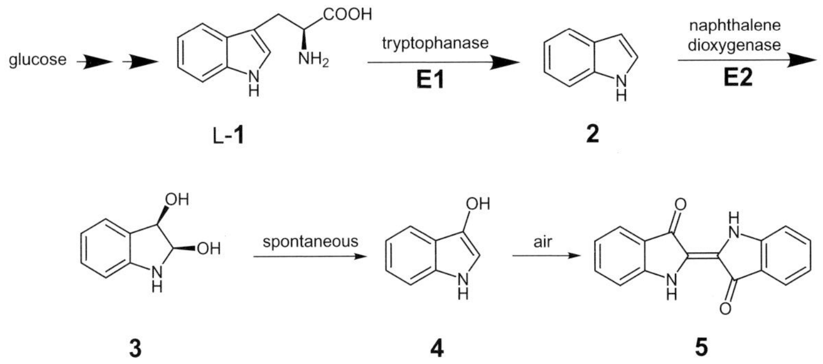

1 = L-tryptophan
2 = 1*H*-indole
3 = 2,3-dihydro-1*H*-indole-2,3-diol
4 = 1*H*-indole-3-ol
5 = indigo

Fig. 1.13.11.11 – 4

- Alternatively a toluene oxidase from *Pseudomonas mendocina* KR-I mutant strains can be used for the oxidation of indole to indole-3-ol:

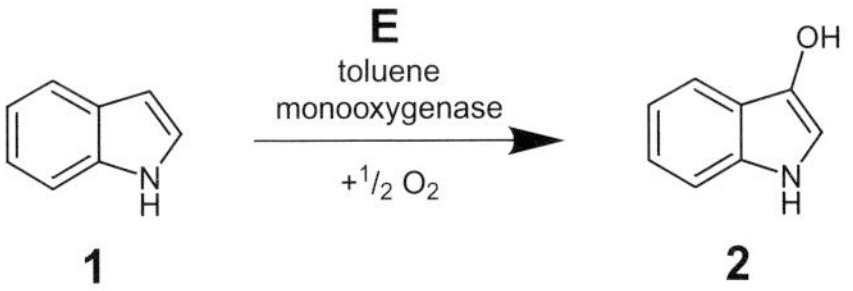

1 = 1*H*-indole
2 = 1*H*-indole-3-ol

Fig. 1.13.11.11 – 5

3) Flow scheme

Not published.

4) Process parameters

reactor type:	batch
company:	Genencor International, Inc., USA

5) Product application

- Indigo is an important brilliant blue pigment used extensively in the dyeing of cotton and woollen fabrics.

6) Literature

- Mermod, N., Harayama, S., Timmis, K.N. (1986) New route to bacterial production of indigo, Bio/Technology **4**, 321–324

- Murdock, D., Ensley, B.D., Serdar, C., Thalen, M. (1993) Construction of metabolic operons catalyzing the *de novo* biosynthesis of indigo in *Escherichia coli*, Bio/Technology, **11**, 381–385

- Serdar, C.M., Murdock, D.C., Ensley, B.D. (1992) Enhancement of naphthalene dioxygenase activity during microbial indigo production, Amgen Inc., US 5,173,425

- Sheldon, R.A. (1994) Consider the environmental quotient, Chemtech **3**, 38–47

- Yen, K.-M., Blatt, L.M., Karl, M.R. (1992) Bioconversions catalyzed by the toluene monooxygenase of *Pseudomonas mendocina* KR-1, Amgen Inc., WO 92/06208

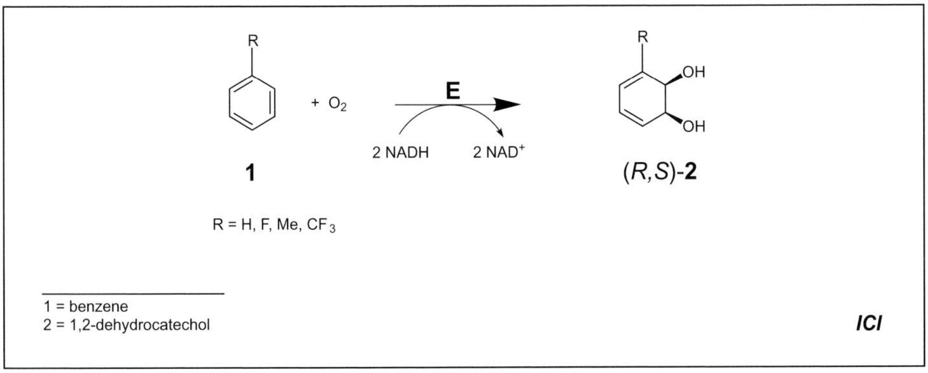

R = H, F, Me, CF$_3$

1 = benzene
2 = 1,2-dehydrocatechol

ICI

Fig. 1.14.12.10 – 1

1) Reaction conditions

reaction type: redox reaction
catalyst: suspended whole cells
enzyme: benzoate, NADH:oxygen oxidoreductase (1,2-hydroxylating)
                   (benzoate 1,2-dioxygenase, benzoate hydroxylase)
strain: *Pseudomonas putida*
CAS (enzyme):   [9059–18–1]

2) Remarks

- Benzoate dioxygenase was formerly classified as EC 1.13.99.2.

- In 1968 it was discovered that a mutant strain from *Pseudomonas putida* was lacking the activity of the dehydrogenase that normally catalyzes the dehydrogenation of *cis*-dihydrodiol to catechol. As a consequence *cis*-dihydrocatechol accumulates in the microorganism and is excreted into the medium:

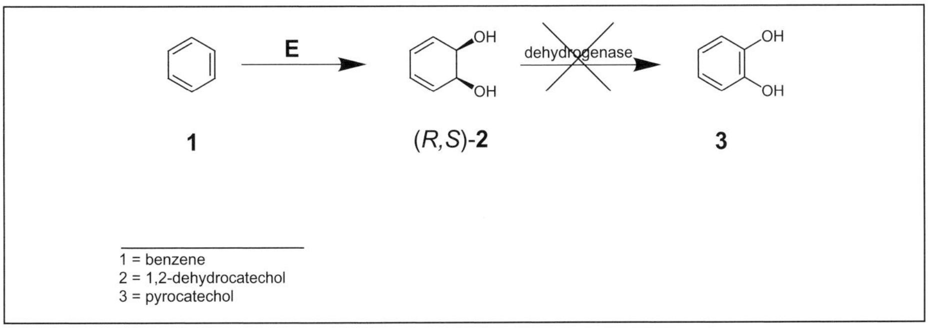

1 = benzene
2 = 1,2-dehydrocatechol
3 = pyrocatechol

Fig. 1.14.12.10 – 2

- *Pseudomas putida* exhibits a high tolerance to aromatic substrates that are normally toxic to microorganisms.

- Since *cis*-dihydrodiols show an inhibitory effect on growth of the cells, fermentation and biotransformation are physically separated in the process.

225

- The cofactor regeneration of NADH takes place cell intracellularly, driven by the addition of a carbon source.

3) Flow scheme

Not published.

4) Process parameters

selectivity:	> 99.5 %
capacity:	several $t \cdot a^{-1}$
company:	ICI, U.K., and others

5) Product application

- *cis*-Dihydrodiols are used as chiral intermediates in the synthesis of β-lactams that are effective as antiviral compounds.

- The product is also used as a polymerisation monomer.

6) Literature

- Crosby, J. (1991) Synthesis of optically active compounds: A large scale perspective, Tetrahedron **47**, 4789–4846

- Evans, C., Ribbons, D., Thomas, S., Roberts, S. (1990) Cyclohexenediols and their use, Enzymatix Ltd., WO 9012798 A2

- Sheldrake, G.N. (1992) Biologically derived arene *cis*-dihydrodiols as synthetic building blocks, in: Chirality in Industry (Collins, A.N., Sheldrake, G.N., Crosby, J., eds.) pp. 127–166, John Wiley & Sons, New York

- Taylor, S.C. (1983) Biochemical Process, ICI, EP 76606 B1

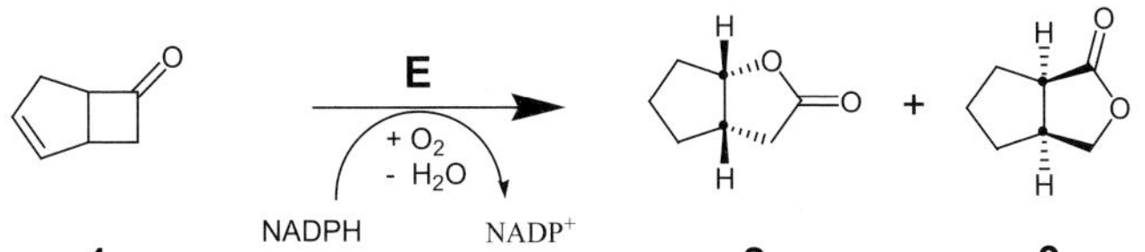

1 = bicyclo[3.2.0]hept-2-en-6-one
2 = (-)-(1*S*,5*R*)-2-oxabicyclo[3.3.0]oct-6-en-3-one
3 = (-)-(1*R*,5*S*)-3-oxabicyclo[3.3.0]oct-6-en-2-one

Sigma Aldrich

Fig. 1.14.13.22 – 1

1) Reaction conditions

[1]:	0.17 M, 18.4 g · L^{-1} (108.14 g · mol^{-1})
[2]:	10 g · L^{-1} glycerol
pH:	7.0
T:	37 °C
medium:	aqueous
reaction type:	batch
catalyst:	whole cell
enzyme:	cyclohexanone monooxygenase
strain:	cloned and overexpressed in *E. coli*

2) Remarks

- The process can be carried out with cells growing on glycerol as well as with harvested cells resolubilized in 50 mM phosphate buffer.

- The expression of the monooxygenase is induced by the addition of 0.1 % w/v of L-arabinose.

- The fermentation is carried out in a 55-L industrial fermenter with a modified sparger.

- The substrate is added loaded on Optipore L-493 resin. Per batch, 3 kg of the resin are added. By applying ISPR techniques to the reaction the performance of the reaction can be improved because at product concentrations above 3.5 g · L^{-1} the enzyme activity decreases to zero.

- The biomass is removed by centrifugation. The clear supernatant is then passed over activated charcoal. The lactones were adsorbed on the column while the substrate was eluted. After elution of the lactones with ethyl acetate the lactones were separated by liquid chromatography on silica gel. Finally both lactones could be crystallized.

3) Flow scheme

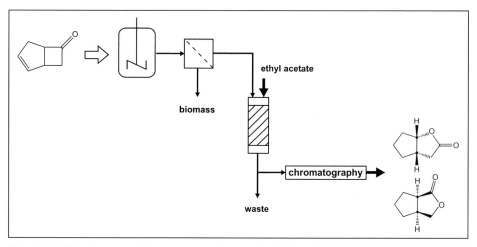

ethyl acetate

biomass

chromatography

waste

Fig. 1.14.13.22 – 2

4) Process parameters

conversion:	92 %
yield:	76 %
optical purity	98 %
reactor type:	CSTR
reactor volume:	55 L
capacity:	kg-scale
space-time-yield:	$1.1 \, g \cdot L^{-1} \cdot h^{-1}$
down stream processing:	centrifugation, adsorption on active charcoal, chromatography on silica gel, crystallization
enzyme activity:	$5 \, g_{cdw} \cdot L^{-1} / 400 \, U \cdot g^{-1} / -> 2000 \, U \cdot L^{-1}$
enzyme supplier:	Fluka, Buchs, Switzerland
production site:	Fluka AG, Switzerland
company:	Sigma Aldrich, USA

5) Product application

- The applied reaction is used as a model reaction. The feasibility of using microbiology as an alternative for Baeyer–Villiger oxidations was demonstrated.

6) Literature

- Doig, S.D., Avenell, P.J., Bird, P.A., Gallati, P., Lander, K.S., Lye, G.J., Wohlgemut, R., Woodley, J.M. (2002) Reactor operation and scale-up of whole cell Baeyer-Villiger catalyzed lactone synthesis, Biotechnol. Proc. **18**, 1039–1046

- Alphand, V., Carrea, G., Wohlgemut, R., Furstoss, R., Woodley, J.M. (2003) Towards large-scale synthetic applications of Baeyer-Villiger monooxygenases, Trends Biotechnol. **21**, (7), 318–323

Cyclohexanone monooxygenase

Acinetobacter calcoaceticus

EC 1.14.13.22

- Doig, S.D, Simpson, H., Alphand, V., Furstoss, R., Woodley, J.M. (2003) Characterization of a recombinant *Escherichia coli* TOP10 [pQR239] whole-cell biocatalyst for stereoselective Baeyer-Villiger oxidations, Enzyme Microb. Technol. **32**, 347–355

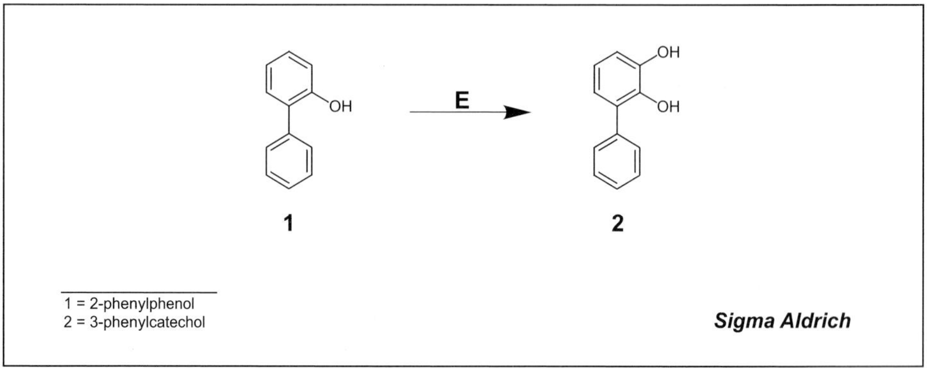

Fig. 1.14.13.44 – 1

1) Reaction conditions

[1]:	0.0012 M, 0.2 g · L^{-1} (total in fed batch 1,055 g) [170,21 g · mol^{-1}]
T:	30 °C
medium:	aqueous
reaction type:	redox reaction
catalyst:	suspended whole cells
enzyme:	2-hydroxybiphenyl,NADH:oxygen oxidoreductase (3-hydroxylating) (2-hydroxybiphenyl 3-monooxygenase)
strain:	*Escherichia coli* JM101
CAS (enzyme):	[118251–39–1]

2) Remarks

- 2-Phenylphenol and 3-phenylcatechol are highly toxic to whole cells. Therefore 2-phenylphenol is used in food preservation because of its fungicidal and bactericidal activity.

- The substrate is fed continuously to the reactor (0.45 g · L^{-1} · h^{-1}), establishing an actual concentration of the educt below the toxic level.

- Catechols are readily soluble in the aqueous phase (435 g · L^{-1}), which complicates extraction of the formed product.

- Dissolved 3-phenylcatechol readily polymerizes (product half-life at pH 7 and 37 °C is 10 h).

- The product is removed by continuous adsorption on the solid resin Amberlite™ XAD-4 (hydrophobic, polystyrene-based).

- The reaction mixture with the whole cells is continuously pumped through an external loop with a fluidized bed of the resin. Before circulating past the adsorbent all substrate is converted. Through the external loop 10 reactor volumes are circulated per hour.

- The growth rate of *Escherichia coli* is unaffected in the presence of XAD-4.

- The product is liberated from the resin by acidic methanol elution, followed by recrystallisation from *n*-hexane.

- By this process a series of other catechol derivatives are produced. This is easily possible by just changing the substrate in the feed (repetitive use of biocatalysts).

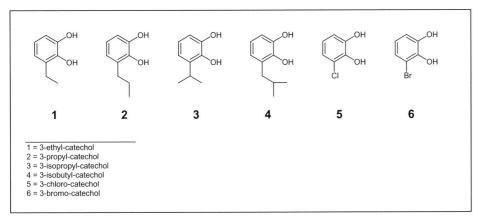

Fig. 1.14.13.44 – 2

1 = 3-ethyl-catechol
2 = 3-propyl-catechol
3 = 3-isopropyl-catechol
4 = 3-isobutyl-catechol
5 = 3-chloro-catechol
6 = 3-bromo-catechol

3) Flow scheme

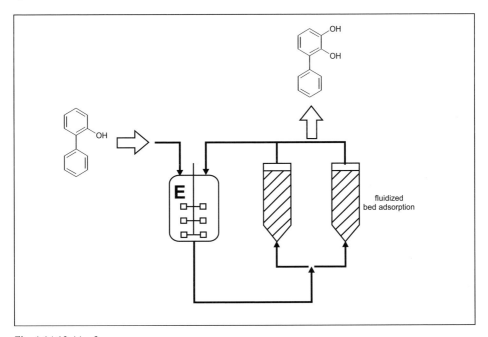

Fig. 1.14.13.44 – 3

4) Process parameters

conversion:	97 %
yield:	83 %
selectivity:	85 %
chemical purity:	77 %, 98 % after recrystallization
reactor type:	fed batch
reactor volume:	300 L

231

capacity:	multi kg
residence time:	10 h
space-time-yield:	$8 \text{ g} \cdot \text{L}^{-1} \cdot \text{d}^{-1}$
down stream processing:	adsorption, crystallization
production site:	Fluka AG, Switzerland
company:	Sigma Aldrich, USA

5) Product application

- Catechols with a substituent at the 3-position are important starting materials for the synthesis of pharmaceutical compounds (barbatusol, taxodione and L-DOPA analogues) and artificial supramolecular systems.

6) Literature

- Held, M., Panke, S., Kohler, H.-P., Feiten, H.-J., Schmid, A., Schmid, A., Wubbolts, M., Witholt, B. (1999) Solid phase extraction for biocatalytic production of toxic compounds, BioWorld **5**, 3–6

- Held, M., Schmid, A., Kohler, H.-P., Suske, W., Witholt, B., Wubbolts, M. (1999) An integrated process for the production of toxic catechols from toxic phenols based on a designer biocatalyst, Biotechnol. Bioeng. **62** (6),641–648

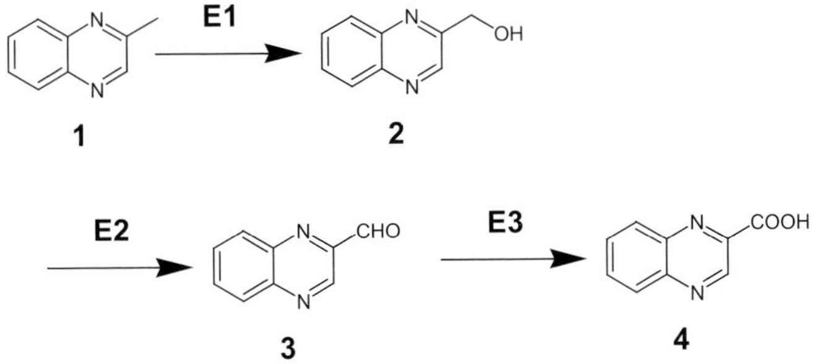

1 = 2-methylquinoxaline
2 = 2-hydroxy-methylquinoxaline
3 = 2-quinoxalinecarbaldehyde
4 = 2-quinoxalinecarboxylic acid

Pfizer Inc.

Fig. 1.14.13.62 / 1.1.1.90 / 1.2.1.28 – 1

1) Reaction conditions

[1]:	[144.17 g · mol⁻¹]
[2]:	benzyl alcohol < 1.5 g · L⁻¹ [160.17 g · mol⁻¹]
pH:	7.0–7.5
T:	29–30 °C
medium:	mineral salt medium
reaction type:	oxidation
catalyst:	whole cell
enzyme:	monooxygenase
enzyme:	benzyl alcohol dehydrogenase, aryl alcohol dehydrogenase
enzyme:	benzaldehyde dehydrogenase, benzaldehyde:NAD⁺ oxidoreductase
strain:	*Pseudomonas putida* ATCC 33015
CAS (enzyme):	[144378-37-0] / [37250-26-3] / [37250-93-4]

2) Remarks

- *P. putida* can catalyze the bioconversion of 2-methylquinoxaline using benzyl alcohol as the inducer and sole carbon source.

- The bioconversion is sensitive to the accumulation of substrate concentrations above $1.5\,g \cdot L^{-1}$ and benzyl alcohol concentration above $1\,g \cdot L^{-1}$.

- Carefully controlled additions of substrate and benzyl alcohol are very important for achieving a successful bioconversion.

- It appears that compounds with two nitrogens, such as quinoxaline and piperazine, are better substrates for *P. putida* than quinoline derivatives.

- Chemical synthesis from the di-*N*-oxide is deemed unsuitable for scale-up due to the mutagenic and thermal properties of the reactant.

- The *P. putida* process yields a product concentration greater than $10\,g \cdot L^{-1}$ and is therefore more likely to be commercially feasible.

233

3) Flow scheme

Not published.

4) Process parameters

yield:	86 %
reactor type:	fed batch, fermenter – 8 L min^{-1} airflow, 600–850 rpm agitation
reactor volume:	14 L
space-time-yield:	10.7 g · L^{-1} after 99 h
company:	Pfizer Inc.

5) Product application

- Product is used in the synthesis of a variety of biologically active compounds (WO Patent 2001034600, 2001. WO Patent 9926927, 1999. JP Patent 09328428, 1997).

6) Literature

- Wong, J.W., Watson, H.A., Bouressa Jr., J., Burns, M.P., Cawly, J.J., Doro, A.E., Guzek, D.B., Hintz, M.A., McCornik, E.L., Scully, D.A., Siderewicz, J.M., Taylor, W.J., Truesdell, S.J., Wax, R.G. (2002) Biocatalytic oxidation of 2-methylquinoxaline to 2-quinoxalinecarboxylic acid, Org. Proc. Res. Dev. **6**, 477-481

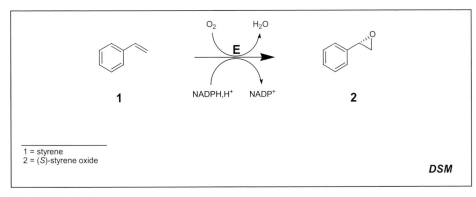

1 = styrene
2 = (S)-styrene oxide

DSM

Fig. 1.14.13.69 – 1

1) Reaction conditions

[1]:	2 % (v/v organic phase)
[2]:	4.5 g · L^{-1} · h^{-1} glucose (50 % w/v)
[3]:	1 % (v/v organic phase) n-octane
pH:	3.0
T:	30 °C
medium:	two-phase system
reaction type:	fed-batch
catalyst:	whole cell
enzyme:	styrene monooxygenase (StyA)
enzyme:	reductase (StyB)
strain:	*Escherichia coli* JM101(pSPZ10)

2) Remarks

- Glucose is added continuously with a rate of 4.5 g · L^{-1} · h^{-1} while the organic phase is added batchwise one hour after starting the glucose feed.

- The host strain *E.coli* JM101 was transformed with a plasmid containing the styrene monooxygenase from *Pseudomonas* sp. VLB120 and the *alk* regulatory system from *P. oleovorans* Gpo1.

- Bis(2-ethylhexyl)phthalate is used as an apolar carrier solvent. The organic phase also contains 2 % (v/v) n-octane, which is used as the inducer of the *alk* regulatory system.

- The system is limited by mass transfer of styrene from the organic to the aqueous phase.

- 2-Phenylethanol is produced as a by-product in approximately 10 % concentration due to the ring opening reaction of the epoxide.

Substrates	Products

Fig. 1.14.13.69 – 2

3) Flow scheme

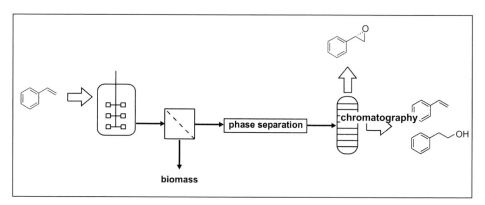

Fig. 1.14.13.69 – 3

4) Process parameters

conversion:	92.7 %
yield:	49.7 %
ee:	>99 %
selectivity:	92 %
chemical purity:	97 %
reactor type:	CSTR
down stream processing:	centrifugation, evaporation, fraction distillation
enzyme activity:	170 U · L^{-1}
enzyme supplier:	in-house
company:	DSM, The Netherlands

5) Product application

- The chiral oxiranes are versatile and important building blocks for a broad range of products.

6) Literature

- Panke, S., Held, M., Wubbolts, M.G., Witholt, B., Schmid, A. (2002) Pilot-scale production of (S)-styrene oxide from styrene by recombinant *Escherichia coli* synthesizing styrene monooxygenase, Biotechnol. Bioeng. **80** (1), 33–41

- Hollmann, F., Schmid, A., Steckhan, E. (2001) The first synthetic application of a monooxygenase employing indirect electrochemical NADH regeneration, Angew. Chem., Int. Ed. Engl. **40** (1), 169–171

- Schmid, A., Hofstetter, K., Freiten, H.-J., Hollmann, F., Witholt, B. (2001) Integrated biocatalytic synthesis on gram-scale: the highly enantioselective preparation of chiral oxiranes with styrene monooxygenase, Adv. Synth. Catal. **343** (6), 752–757

- Poechlauer, P., Skranc, W. (2002) Innovative oxidation methods in fine chemicals synthesis, Curr. Opin. Drug Discov. Develop. **5** (6), 1000–1008

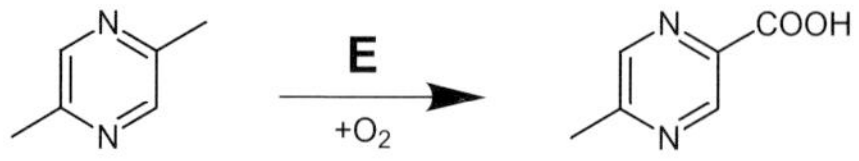

1 = 2,5-dimethylpyrazine
2 = 5-methylpyrazine-2-carboxylic acid

Lonza AG

Fig. 1.14.13.X – 1

1) Reaction conditions

[2]:	0.174 M, 24 g · L^{-1} [138.12 g · mol^{-1}]
pH:	7.0
T:	30 °C
medium:	aqueous
reaction type:	oxidation
catalyst:	suspended, growing whole cells
enzyme:	xylene oxygenase (monooxygenase)
strain:	*Pseudomonas putida* ATCC 33015

2) Remarks

- In contrast to growing cells the application of resting cells showed a strong accumulation of 2-hydroxymethyl-5-methylpyrazine that was only partially oxidized to 5-methylpyrazine-2-carboxylic acid.

- Only one of the symmetric methyl groups is selectively oxidized to a carboxylic acid.

- The desired enzymatic activity is expressed in the cells if 75 % *p*-xylene and 25 % 2,5-dimethylpyrazine is supplied as growth substrate in the fermentation.

- No bacterial metabolites from *p*-xylene were detected that could complicate the downstream processing.

- The biotransformation is terminated when bacterial growth enters the stationary phase.

- The concentration of the substrate at the end of the reaction is below 0.1 g · L^{-1}.

- The enzyme is capable of selectively oxidizing a single methyl group of heteroarenes.

- The inhibition of the *P. putida* growth begins with a product concentration of 15 g · L^{-1} so that the process is stopped at a maximum product concentration of 24 g · L^{-1}.

- *p*-Xylene and substrate concentration in the fermenter are regulated by measuring these compounds in the exhaust air of the fermenter.

- Further additional products could be synthesized by using this process (yields are given in parentheses):

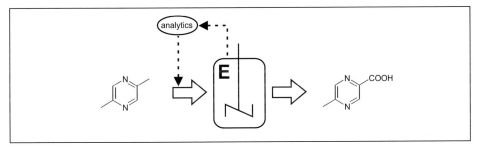

(40 %) (40 %) (70 %) (80 %) (80 %)

(50 %) (90 %) (10 %) (90 %) (90 %)

Fig. 1.14.13.X – 2

3) Flow scheme

Fig. 1.14.13.X – 3

4) Process parameters

conversion:	> 99.5 %
yield:	> 95 %
reactor type:	fed batch
reactor volume:	1,000 L
capacity:	multi t · a^{-1}
residence time:	54 h
company:	Lonza AG, Switzerland

5) Product application

- Intermediate for the drug 5-methylpyrazine-2-carboxylic acid 4-oxide (acipimox) with anti-lipolytic activity like nicotinic acid:

- Precursor in the synthesis of glipicide, a sulfonylurea compound with hypoglycemic activity. This generic drug is used in the control of diabetes:

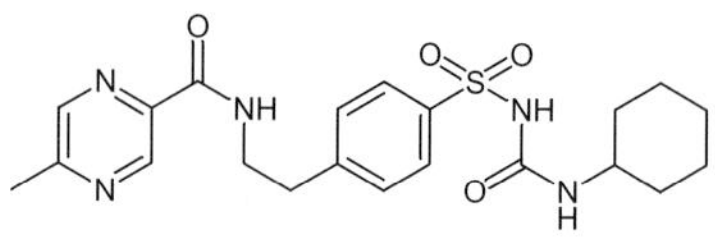

6) Literature

- Kiener, A. (1991) Mikrobiologische Oxidation von Methylgruppen in Heterocyclen, Lonza AG, EP 0442430 A2

- Kiener, A. (1992) Enzymatic oxidation of methyl groups on aromatic heterocycles: a versatile method for the preparation of heteroaromatic carboxylic acids, Angew. Chem. Int. Ed. Engl. **31**, 774–775

- Lovisolo, P. P., Briatico-Vangosa, G., Orsini, G., Ronchi, R., Angelucci, R. (1981) Pharmacological profile of a new anti-lipolytic agent: 5-methyl-pyrazine-2-carboxylic acid 4-oxide (Acipimox), Pharm. Res. Comm. **13**, 151–161

- Peterson, M., Kiener, A. (1999) Biocatalysis – Preparation and functionalization of *N*-heterocycles, Green Chem. **2**, 99–106

- Wubbolts, M.G., Panke, S., van Beilen, J. B., Witholt, B. (1996) Enantioselective oxidation by non-heme iron monooxygenases from *Pseudomonas*, Chimia, **50**, 436–437

- Zaks, A., Dodds, D. R. (1997) Application of biocatalysis and biotransformations to the synthesis of pharmaceuticals, Drug Discovery Today, **2**, 513–531

Monooxygenase
Streptomyces sp. SC 1754

EC 1.14.13.XX

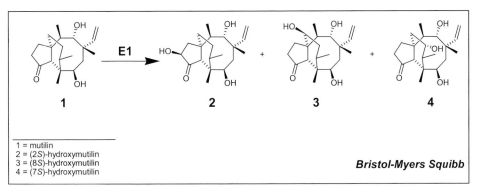

1 = mutilin
2 = (2S)-hydroxymutilin
3 = (8S)-hydroxymutilin
4 = (7S)-hydroxymutilin

Bristol-Myers Squibb

Fig. 1.14.13.XX – 1

1) Reaction conditions

[1]:	0.015 M, 0.5 g · L^{-1}, [320.47 g · mol^{-1}]
pH:	7.0
T:	28 °C
medium:	medium B: 2 % toasted nutrisoy, 0.5 % glucose, 0.5 % yeast extract, 0.5 % K$_2$HPO$_4$ and 0.1 % SAG antifoam
reaction type:	hydroxylation
catalyst:	living whole cells
enzyme:	cytochrome P-450 monooxygenase
strain:	*Streptomyces* sp. SC 1754

2) Remarks

- Sariaslani and coworkers have purified and characterized the three proteins of *S. griseus*. It is likely that the same monooxygenase enzyme system is responsible for mutilin hydroxylation by *S. griseus* strain SC 1754.

3) Flow scheme

Not published.

4) Process parameters

conversion:	77.94 % (total in 7 batches)
yield:	(8S)-hydroxymutilin (30.36 %), (7S)-hydroxymutilin (10.76 %), (2S)-hydroxymutilin (7.85 %)
reactor type:	batch
reactor volume:	100 L
capacity:	pilot-plant
down stream processing:	extraction with ethyl acetate, drying, dissolution in acetone and chloroform, silica gel column chromatography
company:	Bristol-Myers Squibb, USA

241

5) Product application

- Pleuromutilin is an antibiotic from *Pleurotus* or *Clitopilus* basidiomycete strains, which kills mainly Gram-positive bacteria and mycoplasms. A more active semi-synthetic analogue, tiamulin, has been developed for treatment of animals and poultry and has been shown to bind to a prokaryotic ribosome and inhibit protein synthesis.

- Modification of the 8-position of tiamulin is of interest as a means of preventing the metabolic hydroxylation. The target analogue would maintain the biological activity and not be susceptible to metabolic inactivation.

6) Literature

- Hanson, R.L., Matson, J.A., Brzozowski, D.B., LaPorte, T.L., Springer, D.M., Patel, R.N. (2002) Hydroxylation of mutilin by *Streptomyces griseus* and *Cunninghamella echinulata*, Org. Proc. Res. Dev. **6**, 482–487

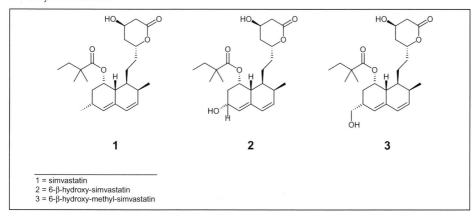

Fig. 1.14.14.1 – 1

1) Reaction conditions

[1]:	< 0.05 · 10^{-3} M, < 0.02 g · L^{-1} [416.26 g · mol^{-1}] (steady state)
[2]:	1.85 · 10^{-3} M, 0.8 g · L^{-1} [432.26 g · mol^{-1}]
pH:	6.8
T:	27 °C
medium:	aqueous
reaction type:	hydroxylation
catalyst:	suspended whole cells
enzyme:	substrate, reduced-flavoprotein: oxygen oxidoreductase (unspecific monooxygenase)
strain:	*Nocardia autotropica*
CAS (enzyme):	[62213–32–5]

2) Remarks

- Due to a strong surplus inhibition, simvastatin is continuously fed to the fermenter.

- Side products are: 6-hydroxy-simvastatin, 3-hydroxy-iso-simvastatin and 3-desmethyl-3-carboxy-simvastatin:

Fig. 1.14.14.1 – 2

- With a final product concentration of 0.8 g · L⁻¹ and a 19,000 L reactor 15.2 kg of 6-β-hydroxy-methyl-simvastatin can be produced.

- The substrate is fed at an overpressure of 20 psig through an on-line filter sterilization system (microporous stainless steel prefilter, 1–2 µm, Pall) with a concentration of 20 g · L⁻¹. The feed tank is maintained at 45 °C to prevent precipitation.

3) Flow scheme

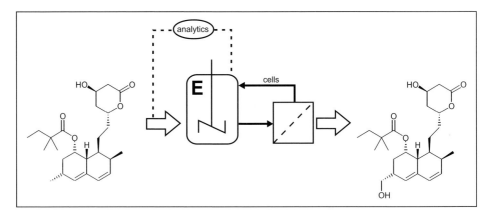

4) Process parameters

yield:	24 %
selectivity:	70 %
reactor type:	fed batch
reactor volume:	19,000 L
company:	Merck Sharp & Dohme, USA

5) Product application

- Simvastatin (and derivatives) belong to the family of HMG-CoA reductase inhibitors, which are, like lovastatin, potent cholesterol-lowering therapeutic agents.

6) Literature

- Gbewonyo, K., Buckland, B.C., Lilly, M.D. (1991) Development of a large-scale continuous substrate feed process for the biotransformation of simvastatin by *Nocardia sp.*, Biotech. Bioeng. **37**, 1101–1107

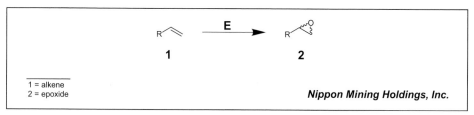

1 = alkene
2 = epoxide

Nippon Mining Holdings, Inc.

Fig. 1.14.14.1 – 1

1) Reaction conditions

T:	30 °C
medium:	two phase : aqueous / organic
reaction type:	epoxidation
catalyst:	suspended whole cells
enzyme:	substrate, reduced-flavoprotein : oxygen oxidoreductase (unspecific monooxygenase)
strain:	*Nocardia corallina* B 276
CAS (enzyme):	[62213–32–5]

2) Remarks

- The alkene monooxygenase catalyzes the epoxidation of terminal and sub-terminal alkenes:

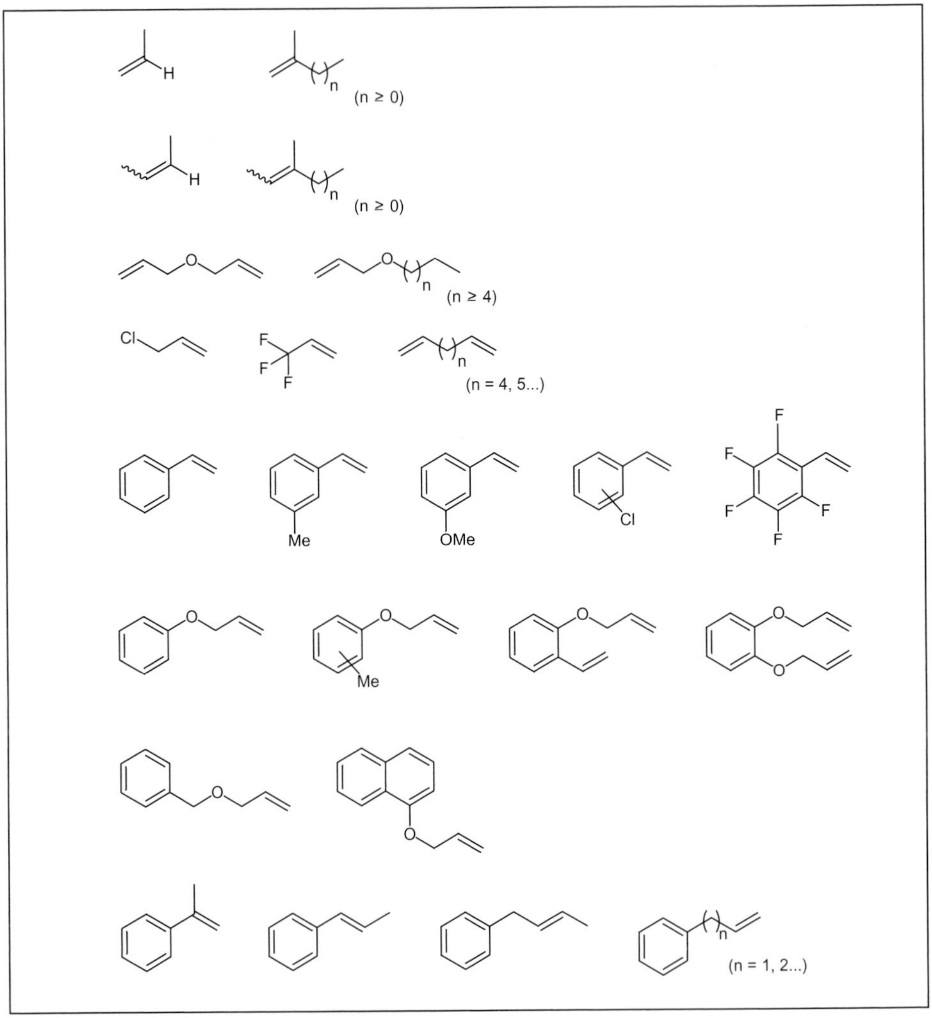

Fig. 1.14.14.1 – 2

- The stereospecific epoxidation yields predominantly the (R)-enantiomer.

- The epoxidation activity is formed constitutively since the conversion takes place independent of the type of carbon source used (alkene or glucose).

- The biotransformation is carried out in a conventional fermentation system.

- The substrates have to be divided in three main classes: short-chain gaseous (C_3-C_5), short chain – liquid (C_6-C_{12}), long-chain ($C_{>12}$) olefins.

- A two-phase system with resting cells is employed for C_6-C_{12} alk-1-enes with a non-toxic solvent (e.g. hexadecane or other alkanes). The addition of a solvent lowers the concentration of the inhibiting epoxide and allows continuous product extraction. Using alk-1-enes as solvent results in a competitive reaction leading preferentially to the shorter epoxides.

- For C_6-C_{12} alk-1-enes resting cells are used since the products are more toxic than in the case of longer chain epoxides where a growing cell reaction is applicable.

- For long chain alkenes it is advantageous to limit components of the medium during growth.

- Short chain, gaseous epoxides (C_3 to C_5) are very toxic to cells and product recovery is complicated.

- For such epoxides the cells have to be grown on glucose or similar carbon sources to guarantee the regeneration of the cofactor.

- The rate of aeration during fermentation has to be raised to extract the short chain toxic epoxide. The very low amounts of epoxides in the gas phase can be recovered by a special solvent extraction system. The process flow diagram of this special process is shown in the flow scheme.

- The microorganism was isolated from soil with propylene as the carbon source.

3) Flow scheme

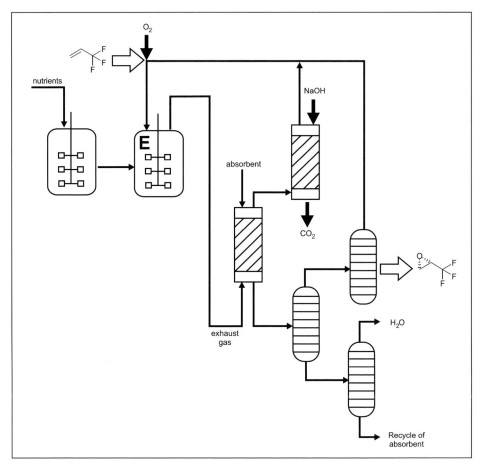

Fig. 1.14.14.1 – 3

4) Process parameters

conversion:	up to 98 %
yield:	up to 92 %
selectivity:	up to 94 %
ee:	> 80 % (up to 97 %)
reactor type:	batch with resting cells or fermenter
space-time-yield:	> 0.2 mol · L^{-1} · d^{-1}
down stream processing:	extraction and distillation
enzyme consumption:	t$_{1/2}$ > 4 d
company:	Nippon Mining Holdings, Inc., Japan

5) Product application

● From the substrates shown above the following epoxides are commercially obtained:

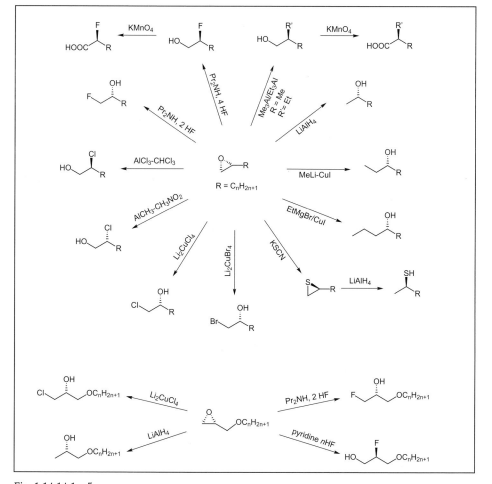

(R)-1,2-epoxyalkanes (R)-2-methyl-1,2-epoxyalkanes (R)-alkyl glycidyl ethers (R)-1,2-epoxy dec-9-ene

(R,R)-1,2,9,10-diepoxydecane pentafluorostyrene oxide 3,3,3-trifluoro-1,2-epoxypropane

Fig. 1.14.14.1 – 4

- Aliphatic epoxides can be used to synthesize various chiral intermediates for the production of ferroelectric liquid crystals:

Fig. 1.14.14.1 – 5

- 2-Methyl-1,2-epoxyalkanes are precursors for tertiary alcohols, which can be used in the synthesis of pharmaceuticals such as prostaglandins:

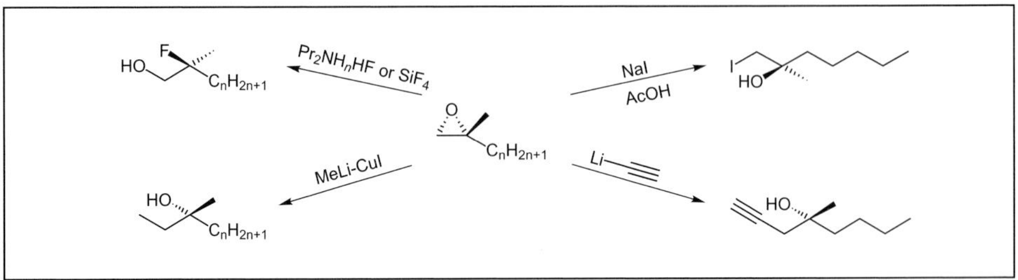

Fig. 1.14.14.1 – 6

6) Literature

- Crosby, J. (1991) Synthesis of optically active compounds: a large scale perspective, Tetrahedron **47**, 4789–4846

- Furuhashi, K., Takagi, M. (1984) Optimization of a medium for the production of 1,2-epoxytetradecane by *Nocardia corallina* B-276, Appl. Microbiol. Biotechnol. **20**, 6–9

- Furuhashi, K., Shintani, M., Takagi, M. (1986) Effects of solvents on the production of epoxides by *Nocardia corallina* B-276, Appl. Microbiol. Biotechnol. **23**, 218–223

- Furuhashi, K. (1992) Biological routes to optically active epoxides, in: Chirality in Industry (Collins, A., Sheldrake, G.N., Crosby, J., eds.), pp. 167–186, John Wiley & Sons, London

- Gallagher, S.C., Cammack, R., Dalton, H. (1997) Alkene monooxygenase from *Nocardia corallina* B-276 is a member of the class of dinuclear iron proteins capable of stereospecific epoxygenation reactions, Eur. J. Biochem. **247**, 635–641.

- Hirai, T., Fukumasa, M., Nishiyama, I., Yoshizawa, A., Shiratori, N., Yokoyama, A., Yamana, M. (1991) New ferroelectric crystal having 2-methylalkanoyl group, Ferroelectrics **114**, 251–257

- Leak, D., Aikens, P.J., Seyed-Mahmoudian, M. (1992) The microbial production of epoxides, Tibtech **10**, 256–261

- Lilly, M.D. (1994) Advances in biotransformation processes. Eighth P.V. Danckwerts memorial lecture presented at Glaziers' Hall, London, U.K. 13 May 1993, Chem. Eng. Sci. **49**, 151–159

- Shiratori, N., Yoshizawa, A., Nishiyama, I., Fukumasa, M., Yokoyama, A., Hirai, T., Yamane, M. (1991) New ferroelectric liquid crystals having 2-fluoro-2-methyl alkanoyloxy group, Mol. Cryst. Liq. Cryst. **199**, 129–140

- Takahashi, O., Umezawa, J., Furuhashi, K., Takagi, M. (1989) Stereocontrol of a tertiary hydroxyl group via microbial epoxidation. A facile synthesis of prostaglandin ω-chains, Tetrahedron Lett. **30**, 1583–1584

250

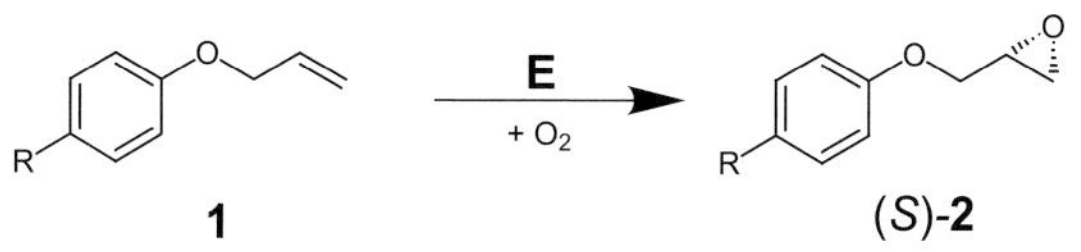

1 = 3-aryloxy-1-propene
2 = 3-aryloxy-1,2-propanoxide

DSM

Fig. 1.14.X.X – 1

1) Reaction conditions

[2]:	> 0.26 M, > 7 g · L^{-1} [267.36 g · mol^{-1}]
T:	37°C
medium:	aqueous
reaction type:	epoxidation
catalyst:	suspended whole cells
enzyme:	Oxidase
strain:	*Pseudomonas oleovarans* ATCC 29347

2) Flow scheme

Not published.

3) Process parameters

ee:	> 99.9 %
company:	DSM, The Netherlands

4) Product application

- The intermediates can be converted into β-adrenergic receptor blocking agents.

- These pharmaceuticals have the common structural feature of a 3-aryloxy- or 3-heteroaryloxy-1-alkylamino-2-propanol.

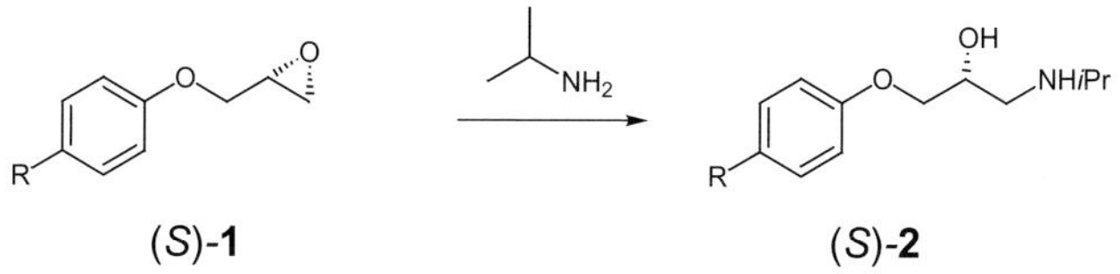

1 = 3-aryloxy-1,2-propanoxide
2 = (-)-3-aryloxy-1-alkylamino-2-propanol

Fig. 1.14.X.X – 2

- Two examples are metoprolol and atenolol:

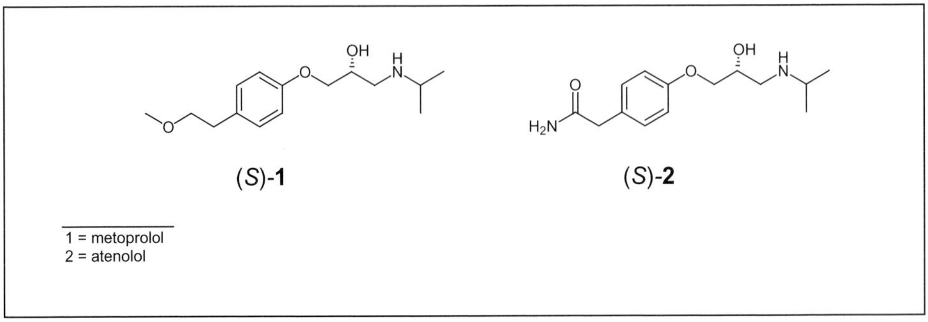

(S)-**1**   (S)-**2**

1 = metoprolol
2 = atenolol

Fig. 1.14.X.X – 3

- Since the biological activity of the enantiomers differs by more than a factor of 100, the production of the racemic mixtures of such pharmaceuticals will become outdated.

5) Literature

- Kieslich, K. (1991) Biotransformations of industrial use, Acta Biotechnol. **11**, 559–570

- Johnstone, S.L., Phillips, G.T., Robertson, B.W., Watts, P.D., Bertola, M.A., Koger, H.S., Marx, A.F. (1987) Stereoselective synthesis of S-(–)-β-blockers via microbially produced epoxide intermediates in: Laane, C., Tramper, J., Lilly, M.D. (eds.), Biocatalysis in Organic Media, Proc. Int. Symposium, Wageningen 1986, pp. 387–392, Elsevier Science Publishers, Amsterdam, Proc. Int. Symposium: Biocatalysis in organic media, 387–392, Wageningen, The Netherlands

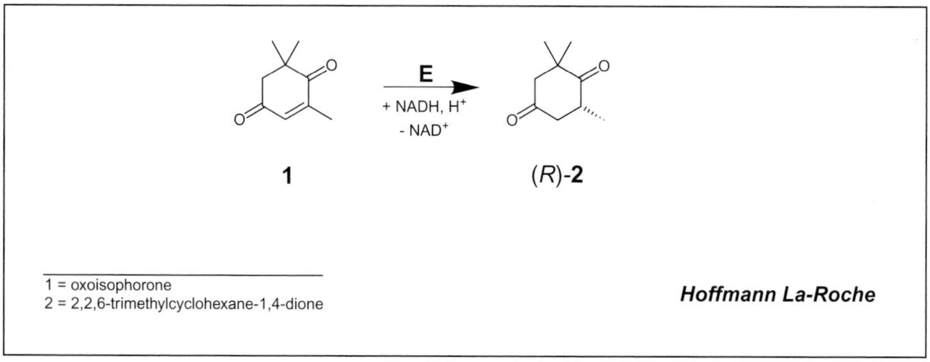

Reductase

Baker's yeast

1 = oxoisophorone
2 = 2,2,6-trimethylcyclohexane-1,4-dione

Hoffmann La-Roche

Fig. 1.X.X.X – 1

1) Reaction conditions

[1]:	0.031-0.061 M, 4.74–9.33 g · L⁻¹ [152.19 g · mol⁻¹]
[2]:	0.311 M, 47.9 g · L⁻¹ [154.21 g · mol⁻¹]

[1]: 0.031-0.061 M, 4.74–9.33 g · L^{-1} [152.19 g · mol^{-1}]
[2]: 0.311 M, 47.9 g · L^{-1} [154.21 g · mol^{-1}]
pH: 8.0–9.0
T: 20 °C
medium: aqueous
reaction type: redox reaction
catalyst: suspended whole cells
enzyme: reductase
strain: baker's yeast

2) Remarks

- On a 13 kg scale oxoisophorone is reacted by fermentative reduction with baker's yeast in an aqueous saccharose solution by periodical substrate addition and feeding with saccharose (fed-batch).

- Periodical addition of oxoisophorone is necessary to avoid toxic levels of oxoisophorone.

- At the suboptimal fermentation temperature of 20°C the growth rate is lowered but the product precipitates faster and the concentration in the aqueous phase is reduced, resulting in slower inactivation of the yeast. Additionally, the saccharose consumption is decreased relative to the production rate.

3) Flow scheme

Not published.

4) Process parameters

conversion:	87 %
yield:	80 %
selectivity:	92 %
ee:	< 97 %
chemical purity:	> 99 %
reactor type:	batch
residence time:	408 h
space-time-yield:	$2.8 \text{ g} \cdot \text{L}^{-1} \cdot \text{d}^{-1}$
down stream processing:	crystallization
enzyme supplier:	Klipfel Hefe AG, Switzerland
company:	Hoffmann La-Roche, Switzerland

5) Product application

- The dione is an intermediate for the synthesis of natural 3-hydroxycarotenoids, e.g. zeaxanthin, cryptoxanthin and structurally related compounds, and other terpenoid compounds.

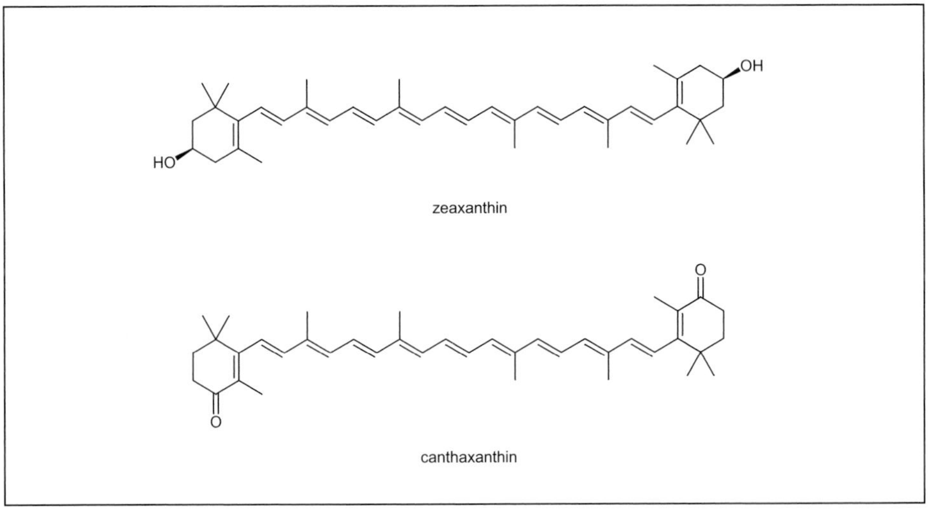

zeaxanthin

canthaxanthin

Fig. 1.X.X.X – 2

- Zeaxanthin, e.g., is the main pigment in corn and barley and can also be found in yolk.

6) Literature

- Leuenberger, H.G.W., Boguth, W., Widmer, E., Zell, R. (1976) 189. Synthese von optisch aktiven, natürlichen Carotinoiden und strukturell verwandten Naturprodukten. I. Synthese der chiralen Schlüsselverbindung (4R,6R)-4-Hydroxy-2,2,6-trimethylcyclohexanon, Helv. Chim. Acta **59**, 1832–1849

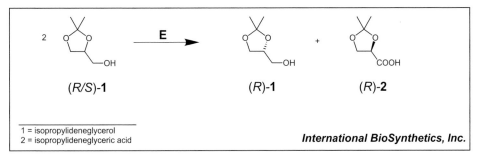

(R/S)-**1**   (R)-**1**   (R)-**2**

1 = isopropylideneglycerol
2 = isopropylideneglyceric acid

International BioSynthetics, Inc.

Fig. 1.X.X.X – 1

1) Reaction conditions

[1]:	0.076 M, 10 g · L^{-1} [132.16 g · mol^{-1}]
pH:	6.8 – 7.2
T:	30 °C
medium:	aqueous
reaction type:	oxidation
catalyst:	whole cells
enzyme:	hydroxylase
strain:	*Rhodococcus erythropolis*

2) Remarks

- The resolution is carried out by selective microbial oxidation of the (S)-enantiomer.

- The chemical synthesis of (R)-isopropylideneglycerol starting from unnatural L-mannitol is difficult and expensive:

1 = mannitol
2 = isopropylideneglyceric aldehyde
3 = isopropylideneglycerol

Fig. 1.X.X.X – 2

- Another chemical route starts from L-ascorbic acid:

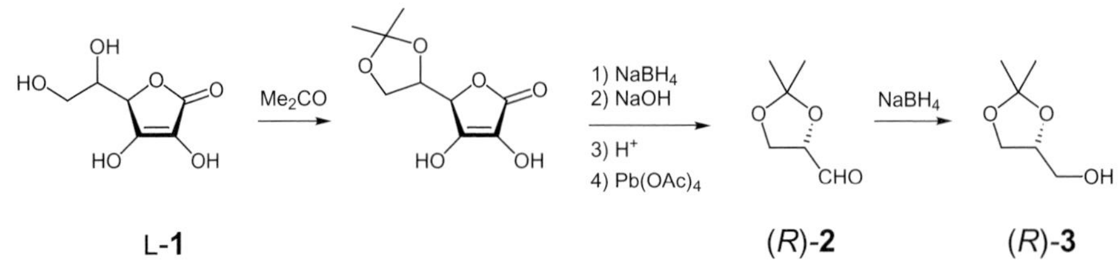

1 = ascorbic acid
2 = isopropylideneglyceric aldehyde
3 = isopropylideneglycerol

Fig. 1.X.X.X – 3

3) Flow scheme

Not published.

4) Process parameters

conversion:	50%
ee:	> 98% for (R)-**1**, > 90% for (R)-**2**
reactor type:	fed batch
company:	International BioSynthetics, Inc., The Netherlands

5) Product application

- (R)-Isopropylideneglycerol is a useful C$_3$-synthon in the synthesis of (S)-β-blockers, e.g. (S)-metoprolol.

- Also, the (R)-isopropylideneglyceric acid may be used as starting material for the synthesis of biologically active products.

6) Literature

- Crosby, J. (1991) Synthesis of optically active compounds: a large scale perspective, Tetrahedron **47**, 4789–4846

- Bertola, M. A., Koger, H. S., Phillips, G. T., Marx, A. F., Claassen, V. P. (1987) A process for the preparation of (R)- and (S)–2,2-R1,R2–1,3-dioxolane-4-methanol, Gist-Brocades NV and Shell Int. Research, EP 0244912

- Baer, E., Fischer, H. (1948) J. Am. Chem. Soc. **70**, 609

- Hirth, G., Walther, W. (1985) Synthesis of the (R)- and (S)-glycerol acetonides. Determination of the optical purity, Helv. Chim. Acta **68**, 1863

Desaturase
Rhodococcus sp.

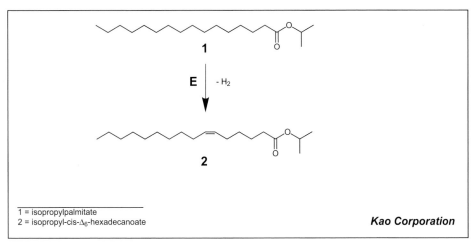

1 = isopropylpalmitate
2 = isopropyl-cis-Δ_6-hexadecanoate

Kao Corporation

Fig. 1.X.X.X – 1

1) Reaction conditions

[1]:	0.67 M, 200 g · L^{-1} [298.50 g · mol^{-1}]
[2]:	0.15 M, 45 g · L^{-1} [296.49 g · mol^{-1}]
pH:	7.0
T:	26 °C
medium:	two-phase: aqueous oil (70:30 (v:v)) emulsion
reaction type:	*cis*-desaturation (redox reaction)
catalyst:	suspended whole cells
enzyme:	desaturase
strain:	*Rhodococcus* sp. KSM-B-MT66 mutant

2) Remarks

- The dehydrogenation always takes place 9-C-atoms away from the terminal methyl group.

- With this process the production of unsaturated fatty acids is possible from low cost saturated compounds.

- The mutant KSM-B-MT66 was obtained by treatment with UV irradiation. The parent strain was originally isolated from soil samples.

- The mutant strain of *Rhodococcus* sp. *cis*-desaturates a variety of hydrocarbons and acyl fatty acids at central positions of the chains:

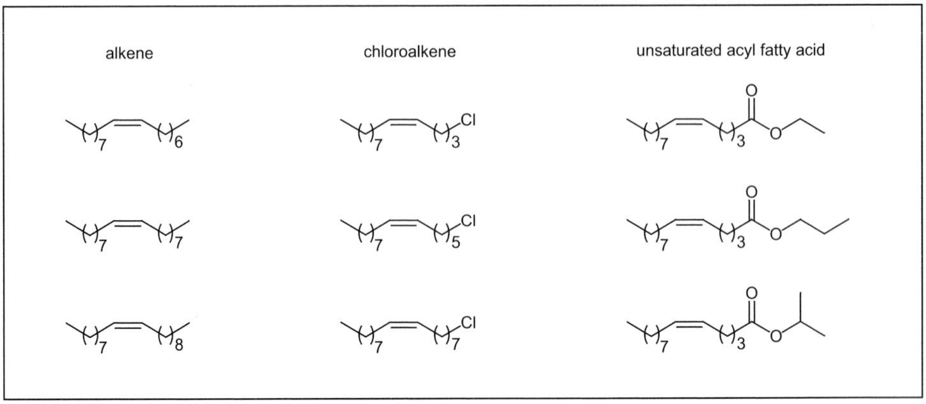

Fig. 1.X.X.X – 2

- Glutamate, thiamine and Mg^{2+} are added to enhance the *cis*-desaturation.

- The reaction is started as an O/W emulsion. Adding isopropylhexadecanoate up to 70 : 30 (v/v) leads to a phase inversion.

- The product and residual substrate in the oil phase are recovered by a hydrophobic membrane-based filtration (hydrophobic hollow-fiber module TP-113, Asahi Chemical Co., Ltd., Japan). Oil permeation flux is 5 L · m^{-2} · h^{-1}.

- Product separation is performed by the urea-adduct method and chromatography on silica gel.

3) Flow scheme

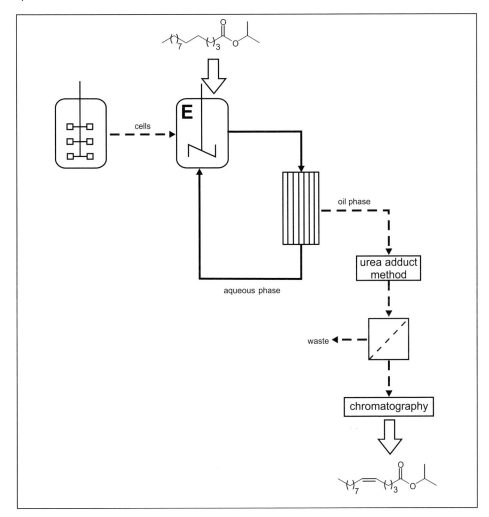

Fig. 1.X.X.X – 3

4) Process parameters

chemical purity:	97 %
reactor type:	repetitive batch
space-time-yield:	$16.8 \text{ g} \cdot \text{L}^{-1} \cdot \text{d}^{-1}$
down stream processing:	chromatography, 80 % recovery
production site:	Japan
company:	Kao Corp., Japan

5) Product application

- The products represent intermediates for the preparation of fatty acids substituted within the aliphatic chain for use in dermatological pharmaceuticals.

6) **Literature**

- Downing, D.T., Strauss, J.S. (1974) Synthesis and composition of surface lipids of human skin, J. Invest. Dermatol. **62**, 228–244

- Kieslich, K. (1991) Biotransformations of industrial use, Acta Biotechnol. **11**, 559–570

- Morello, A.M., Downing, D.T., (1976) *Trans*-unsaturated fatty acids in human skin surface lipids, J. Invest. Dermatol. **67**, 270–272

- Takeuchi, K., Koike, K., Ito, S.(1990) Production of *cis*-unsaturated hydrocarbons by a strain of *Rhodococcus* in repeated batch culture with a phase-inversion, hollow-fiber system, J. Biotechnol. **14**, 179–186

1 = 2-phenoxypropionic acid (POPS)
2 = 2-(4«-hydroxyphenoxy)propionic acid (HOPS)

BASF AG

Fig. 1.X.X.X – 1

1) Reaction conditions

[1]:	> 0.04 M, > 5 g · L^{-1} [126.17 g · mol^{-1}]
medium:	aqueous
reaction type:	hydroxylation
catalyst:	suspended whole cells
enzyme:	oxidase
strain:	*Beauveria bassiana* LU 700

2) Remarks

- The biocatalyst *Beauveria bassiana* was found by an extensive screening of microorganisms for specifically hydroxylating POPS to HPOPS regioselectively. An additional screening criteria was the tolerance to substrate concentrations > 5 g · L^{-1}.

- The production strain LU 700, having a more yeast-like morphology, was obtained by two mutations (first: UV-light, second: *N*-methyl-*N'*-nitrosoguanidine).

- The productivity was improved by 25 % through optimization of the trace elements in the media by use of a genetic algorithm (cupric acid from 0.01 to 0.75 mg · L^{-1}, manganese from 0.02 to 2.4 mg · L^{-1} and ferric ions from 0.8 to 6 mg · L^{-1}).

- The hydroxylation is not growth-associated.

- The oxidase has a very broad substrate spectrum (figure 1.X.X.X – 2):

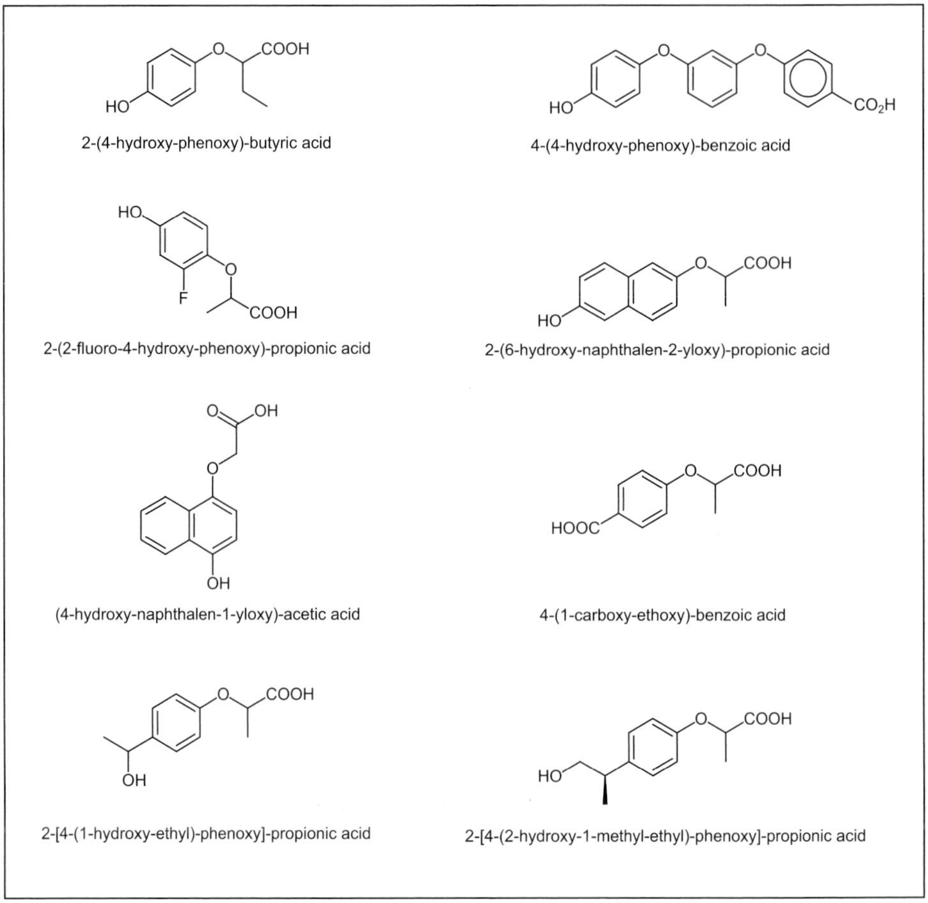

Fig. 1.X.X.X – 2

- A compound needs the structural elements of a carboxylic acid and an aromatic ring system to be a substrate for the oxidase. Hydroxylation primarily takes place in phenoxy-derivatives at the *para* position if it is free. If there is an alkyl group in the *para* position, only side-chain hydroxylation takes place. In systems with more than one ring, the most electron-rich ring is hydroxylated.

- The ee is increased during oxidation from 96 % for POPS to 98 % for HPOPS.

- The substrate (*R*)-2-phenoxypropionic acid is easily synthesized from (*S*)-2-chloropropionic acid isobutylester and phenol:

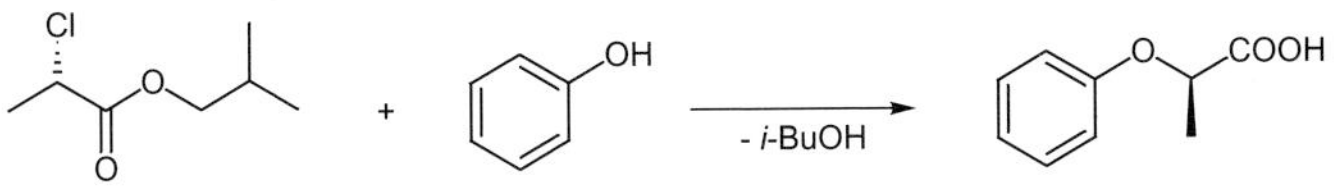

| (*S*)-**1** | **2** | (*R*)-**3** |

1 = 2-chloropropionic acid isobutyl ester
2 = phenol
3 = 2-phenoxypropionic acid (POPS)

Fig. 1.X.X.X – 3

3) Flow scheme

Not published.

4) Process parameters

ee:	> 98.0 %
chemical purity:	> 99.5 %
reactor type:	cstr
reactor volume:	120,000 L
space-time-yield:	$7 \text{ g} \cdot \text{L}^{-1} \cdot \text{d}^{-1}$
company:	BASF AG, Germany

5) Product application

- (*R*)-2-(4'-hydroxyphenoxy)propionic acid is used as an intermediate for the synthesis of enantiomerically pure aryloxypropionic acid-type herbicides.

6) Literature

- Cooper, B., Ladner, W., Hauer, B., Siegel, H. (1992) Verfahren zur fermentativen Herstellung von 2-(4-Hydroxyphenoxy)-propionsäure, EP 0465494 B1

- Dingler, C., Ladner, W., Krei, G.A., Cooper, B. Hauer, B. (1996) Preparation of (*R*)-2-(4-hydroxyphenoxy)propionic acid by biotransformation, Pestic. Sci. **46**, 33–35

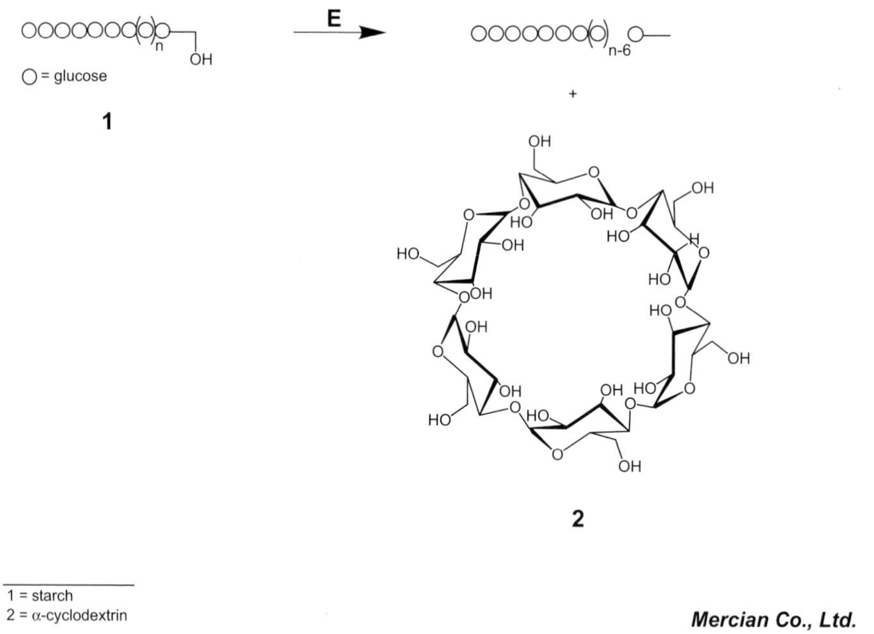

Cyclodextrin glycosyltransferase
Bacillus circulans

Fig. 2.4.1.19 – 1

1) Reaction conditions

[1]:	8.3 w/v % liquified starch
pH:	5.8 – 6.0
T:	55 °C
medium:	aqueous
reaction type:	hexosyl group transfer
catalyst:	solubilized enzyme
enzyme:	1,4-α-D-glucan 4-α-D-(1,4-α-D-glucano)-transferase (cyclizing) (cyclomaltodextrin glucanotransferase)
strain:	*Bacillus circulans*
CAS (enzyme):	[9030-09–5]

2) Remarks

- Cyclodextrins are produced by cyclodextrin glycosyltransferases as a mixture of α-, β- and γ-cyclic oligosaccharides.

- The main problem that had to be overcome to establish an economic cyclodextrin production was the separation of the cyclodextrins from the reaction media. This is important due to two points:
  1) The reaction mixtures contains many by-products.
  2) Increasing cyclodextrin concentrations inhibit the enzyme.

- The separation is established by selective adsorption of α- and β-cyclodextrins on chitosan beads with appropriate ligands. α-Cyclodextrins are selectively interacted with stearic acid and β-cyclodextrins with cyclohexanepropanamide-*n*-caproic acid. The adsorption selectivity is almost 100 %. In the case of the β-cyclodextrins a capacity of 240 g · L⁻¹ gel bed is reached.

- The process data given here are related to the production of α-cyclodextrins.

- Before entering the adsorption column the temperature is lowered to 30 °C for effective adsorption. At this temperature almost no cyclodextrins are formed during circulation. Before re-entering the main reactor the temperature of the solution is again adjusted to 55 °C by using the energy of the reaction solution leaving the reactor.

- To prevent adsorption of the cyclodextrin glycosyltransferase in the α-cyclodextrin adsorbent 3 % w/v NaCl is added. CDs adsorbed on the resin through ionic bonds are easely removed by 4 % NaCl (Okabe et al., p. 115).

3) Flow scheme

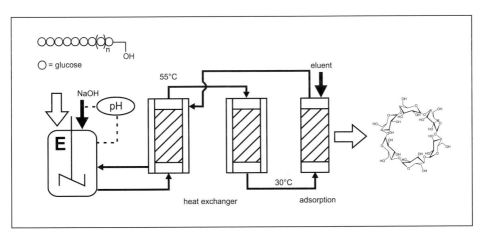

Fig. 2.4.1.19 – 2

4) Process parameters

yield:	22.3 % (α-cyclodextrin); 10.8 % (β-cyclodextrin); 5.1 % (γ-cyclodextrin)
chemical purity:	94.9 %
reactor type:	batch
down stream processing:	chromatography and crystallization
enzyme supplier:	Amano, Japan
company:	Mercian Co., Ltd., Japan

5) Product application

- Cyclodextrins serve as molecular hosts and are used in the food industry for capturing and retaining flavors. They are also used in the formulation of pharmaceuticals.

6) Literature

- Okabe, M., Tsuchiyama, Y., Okamoto, R. (1993) Development of a cyclodextrin production process using specific adsorbents, in: Industrial Application of Immobilized Biocatalysts (Tanaka, A., Tosa, T., Kobayashi, T., eds.) pp. 109–129, Marcel Dekker Inc., New York

- Tsuchiyama, Y., Yamamoto, K.-I., Asou, T., Okabe, M., Yagi, Y., Okamoto, R. (1991) A novel process of cyclodextrin production by use of specific adsorbents. Part I. Screening of specific adsorbents, J. Ferment. Bioeng. **71**, 407–412

- Yang, C.-P., Su, C.-S. (1989) Study of cyclodextrin production using cyclodextrin glycosyltransferase immobilized on chitosan, J. Chem. Technol. Biotechnol. **46**, 283–294

D-Amino acid transaminase
Bacillus sp.

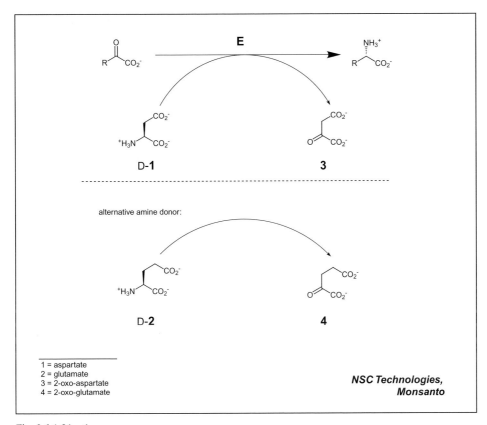

Fig. 2.6.1.21 – 1

1) Reaction conditions

pH:	8.0
T:	42 °C
medium:	aqueous
reaction type:	amino group transfer
catalyst:	whole cell
enzyme:	D-alanine; 2-oxoglutarate aminotransferase (D-aspartase transaminase)
strain:	*Bacillus* sp.
CAS (enzyme):	[37277–85–3]

2) Remarks

- The main drawback of transaminases is the equilibrium conversion of about 50 %.

- Therefore in commercial application L-aspartate is often used as the amino donor for L-amino acid transaminase. The product decarboxylates to pyruvic acid and carbon dioxide. Pyruvic acid itself is also an α-keto acid and is converted by dimerisation using acetolactate synthase to acetolactate. The latter undergoes spontaneous decarboxylation to acetoin that can be easily removed and does not participate in other reactions.

D-Amino acid transaminase
Bacillus sp.

- The production of D-amino acids proceeds in a similar way using D-aspartase or D-glutamate as amino-group donor.

- The α-keto acids are available by chemical or enzymatic methods. Amino acid deaminases generate the α-keto acids from inexpensive L-amino acids. Using amino acid racemase the amino-group donor is also accessible from cheap racemic mixtures of amino acids. The EC-number given above is chosen for D-alanine as precursor for the α-keto acid.

- The advantage of the process is the use of cloned strains implanted in *E. coli*. which are capable of conducting all steps of synthesis using cheap racemic substrates resulting in high yields of D-amino acids.

- The following scheme shows the coupling of the enzyme systems using L-aspargine as the amino-group donor for the production of D-amino acids:

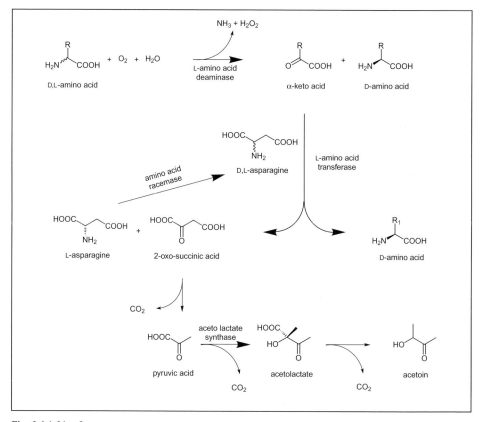

Fig. 2.6.1.21 – 2

- Possible products using this synthetic route are:

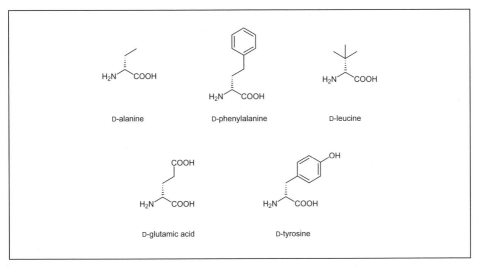

Fig. 2.6.1.21 – 3

3) Flow scheme

Not published.

4) Process parameters

ee:	100 %
reactor type:	batch
capacity:	multi t · a^{-1}
down stream processing:	crystallization
start-up date:	May 1998
production site:	NSC Technologies, USA
company:	NSC Technologies, Monsanto, USA

5) Product application

- Amino acids are used as food additives and in medicine in infusion solutions.

6) Literature

- Ager, D.J., Fotheringham, I.G., Laneman, S.A., Pentaleone D.P., Taylor, P.P. (1997) The large-scale synthesis of unnatural acids, Chim. Oggi, **15 (3/4)**, 11–14

- Fotheringham, I.G., Pentaleone, D.P., Taylor, P.P. (1997) Biocatalytic production of unnatural amino acids, mono esters, and *N*-protected derivatives, Chim. Oggi **15 (9/10)**, 33–37

- Fotheringham, I.G., Taylor, P.P., Ton, J.L. (1998) Preparation of enantiomerically pure D-amino acid by direct fermentative means, Monsanto, US 5728555

- Fotheringham, I.G., Beldig, A., Taylor, P.P. (1998) Characterization of the genes encoding D-amino acid transaminase and glutamte racemase, two D-glutamate biosynthetic enzymes of *Bacillus Sphaericus* ATCC 10208, J. Bacteriol. **180**, 4319–4323

- Taylor, P.P., Fotheringham, I.G. (1997) Nucleotide sequence of the *Bacillus licheniformis* ATCC 10716 *dat* gene and comparison of the predicted amino acid sequence with those of other bacterial species, Biochim. Biophys. Acta **1350 (1)**, 38–40

Transaminase
Bacillus megaterium

EC 2.6.1.X

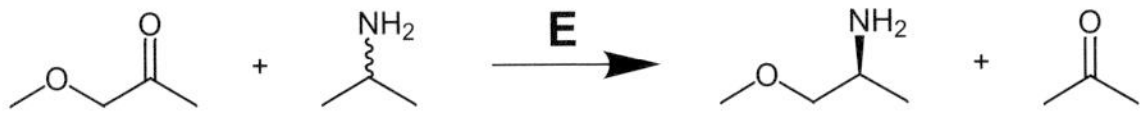

1 (*R*/*S*)-**2** (*S*)-**3** **4**

1 = 1-methoxy-propan-2-one
2 = isopropylamine
3 = methoxy isopropylamine
4 = acetone

Celgene Corporation

Fig. 2.6.1.X – 1

1) Reaction conditions

[2]:	0.7 M, 61.7 g · L^{-1} [88.11 g · mol^{-1}]
medium:	aqueous
reaction type:	amino group transfer
catalyst:	suspended whole cells
enzyme:	aminotransferase (transaminase)
strain:	*Bacillus megaterium*

2) Remarks

- Isopropylamine is the choice for the amino donor because it is a cheap and a kinetically attractive molecule.
- Main disadvantage of the wild-type transaminase is that the conversion is limited due to product inhibition.
- Starting with the wild-type transaminase, the gene encoding the enzyme could be optimized step by step by mutation.
- A single mutation in the gene coding the transaminase increases the possible product concentration from 0.16 M to 0.45 M.
- The screening criteria for better genes included, beside higher inhibitor concentrations, better reaction rates, higher stability and lower K_M-values.
- Since no recycling of the catalyst is integrated, the residual activity after one batch run is of no interest.

3) Flow scheme

Not published.

4) Process parameters

selectivity:	94 %
ee:	> 99 %
reactor type:	batch
enzyme consumption:	0.15 $kg_{catalyst\ powder} \cdot kg^{-1}_{product}$ (t > 6 h)
enzyme supplier:	Celgene Corporation, USA
company:	Celgene Corporation, USA

5) Product application

- Possible products derived from (S)-methoxyisopropylamine are several agrochemicals, e. g.:

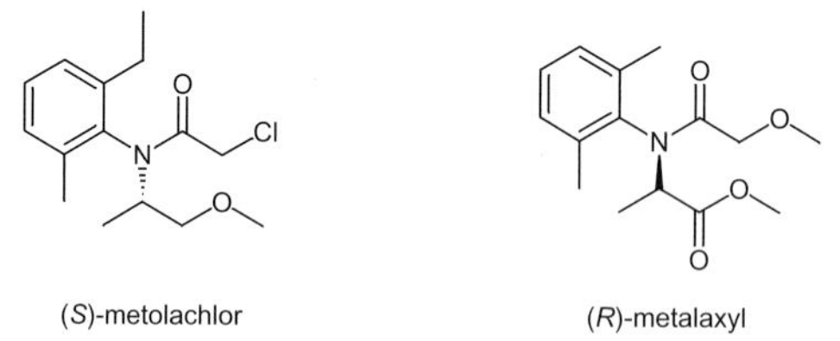

(S)-metolachlor   (R)-metalaxyl

Fig. 2.6.1.X – 2

- Ciba spent about 10 years to find a chemical catalyst for the production of an enantiomeren-riched (S)-metolachlor. But only an ee of 79 % is possible resulting in a reduced application rate of only 38 %. Other companies developed different approaches leading to higher product costs than the biotransformation route.

6) Literature

- Matcham, G.W. (1997) Chirality and biocatalysis in agrochemical production, INBIO Europe 97 Conference, Spring Innovations Ltd, Stockport, UK

- Matcham, G.W., Lee, S. (1994) Process for the preparation of chiral 1-aryl-2-aminopropanes, Celgene Corporation, US 5,360,724

- Stirling, D.I., Matcham, G.W., Zeitlin, A.L. (1994) Enantiomeric enrichment and stereoselective synthesis of chiral amines, Celgene Corporation, US 5,300,437

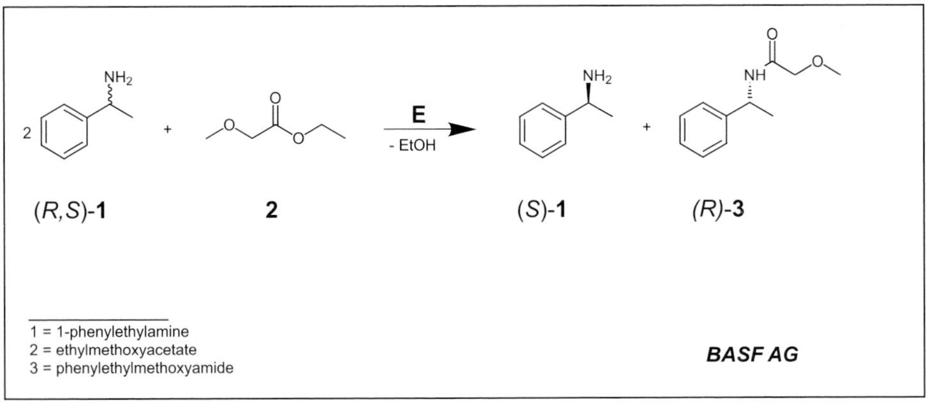

Lipase
Burkholderia plantarii

1 = 1-phenylethylamine
2 = ethylmethoxyacetate
3 = phenylethylmethoxyamide

BASF AG

Fig. 3.1.1.3 – 1

1) Reaction conditions

[(R/S)-1]:	1.65 M, 200 g · L^{-1} [121.18 g · mol^{-1}] in MTBE (= methyl-*tert*-butylether)
[(S)-1]:	1.4 M, 170 g · L^{-1} [121.18 g · mol^{-1}] in MTBE
pH:	8.0–9.0
T:	25 °C
medium:	MTBE-ethylmethoxyacetate
reaction type:	carboxylic ester hydrolysis
catalyst:	immobilized enzyme
enzyme:	triacylglycerol acylhydrolase (triacylglycerol lipase)
strain:	*Burkholderia plantarii*
CAS (enzyme):	[9001–62–1]

2) Remarks

- The lipase is immobilized on polyacrylate.

- The lowering in activity caused by the use of organic solvent can be offset (about 1,000 times and more) by freeze-drying a solution of the lipase together with fatty acids (e.g. oleic acid).

- The E-value of the reaction is greater than 500.

- The (*R*)-phenylethylmethoxy amide can be easily hydrolyzed to get the (*R*)-phenylethylamine:
- The (*S*)-enantiomer can be racemized using a palladium catalyst.

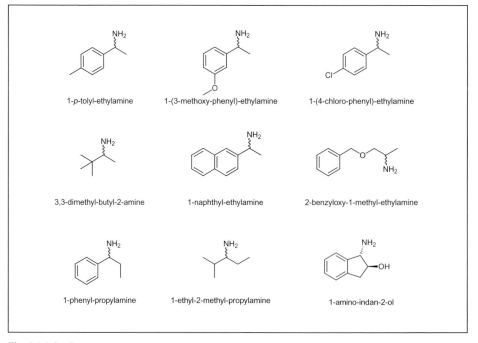

1 = phenylethylmethoxyamide
2 = 1-phenylethylamine

Fig. 3.1.1.3 – 2

- The following amines can also be used in this process:

1-*p*-tolyl-ethylamine

1-(3-methoxy-phenyl)-ethylamine

1-(4-chloro-phenyl)-ethylamine

3,3-dimethyl-butyl-2-amine

1-naphthyl-ethylamine

2-benzyloxy-1-methyl-ethylamine

1-phenyl-propylamine

1-ethyl-2-methyl-propylamine

1-amino-indan-2-ol

Fig. 3.1.1.3 – 3

3) Flow scheme

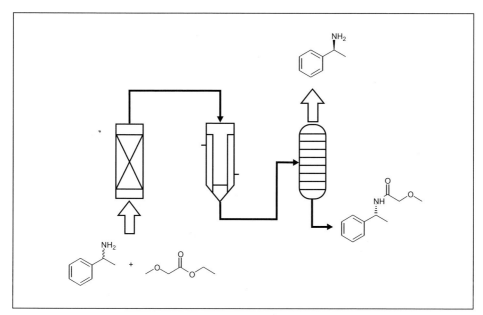

Fig. 3.1.1.3 – 4

4) Process parameters

conversion:	50 %
yield:	> 90 %
ee:	> 99 % (S); 93 % (R)
reactor type:	plug-flow reactor or batch
capacity:	> 100 t · a⁻¹
residence time:	5–7 h
down stream processing:	distillation or extraction
production site:	BASF AG, Germany
company:	BASF AG, Germany
start-up date:	1993

5) Product application

- Products are intermediates for pharmaceuticals and pesticides.
- They can also be used as chiral synthons in asymmetric synthesis.

6) Literature

- Balkenhohl, F., Hauer, B., Ladner, W., Schnell, U., Pressler, U., Staudenmaier, H.R. (1995) Lipase katalysierte Acylierung von Alkoholen mit Diketenen, BASF AG, DE 4329293 A1

- Balkenhohl, F., Ditrich, K., Hauer, B., Ladner, W. (1997) Optisch aktive Amine durch Lipase-katalysierte Methoxyacetylierung, J. prakt. Chem. **339**, 381–384

- Reetz, M.T., Schimossek, K. (1996) Lipase-catalyzed dynamic kinetic resolution of chiral amines: use of palladium as the racemization catalyst, Chimia **50**, 668–669

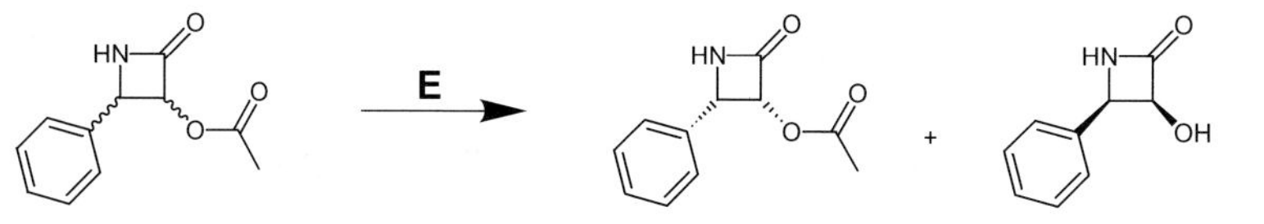

Lipase
Pseudomonas cepacia

1 = cis-azetidinone acetate
2 = cis-azetidinone

Bristol-Myers Squibb

Fig. 3.1.1.3 – 1

1) Reaction conditions

[1]:	0.049 M, 10 g · L^{-1} [205.21 g · mol^{-1}]
pH:	7.0
T:	29 °C
medium:	aqueous
reaction type:	carboxylic ester hydrolysis
catalyst:	immobilized enzyme
enzyme:	triacylglycerol acylhydrolase (lipase, triacylglycerol lipase)
strain:	*Pseudomonas cepacia*
CAS (enzyme):	[9001–62–1]

2) Remarks

- The enzyme is immobilized by adsorption onto polypropylene beads. It can be recycled several times.

- The immobilized enzyme is reused for ten cycles without any loss in activity, productivity or optical purity of the product.

- The rate of hydrolysis is determined to be 0.12 g · L^{-1} · h^{-1}. It remains constant over ten cycles.

- At the end of the reaction the temperature is lowered to 5 °C and the agitation from 200 rpm to 50 rpm. The product (3R,4S)-azetidinone acetate precipitates from the reaction mixture.

- The immobilized enzyme floats on top of the reactor due to its hydrophobicity and is separated by draining.

3) Flow scheme

Not published.

4) Process parameters

yield:	> 96 %
ee:	> 99.5 %
reactor type:	repetitive batch
reactor volume:	150 L
capacity:	1.2 kg · batch^{-1}
down stream processing:	precipitation
enzyme supplier:	Amano Enzyme Inc., Japan
company:	Bristol-Myers Squibb, USA

5) Product application

- (3R,4S)-Azetidinone acetate is an intermediate for the synthesis of paclitaxel (tradename of the product is taxol):

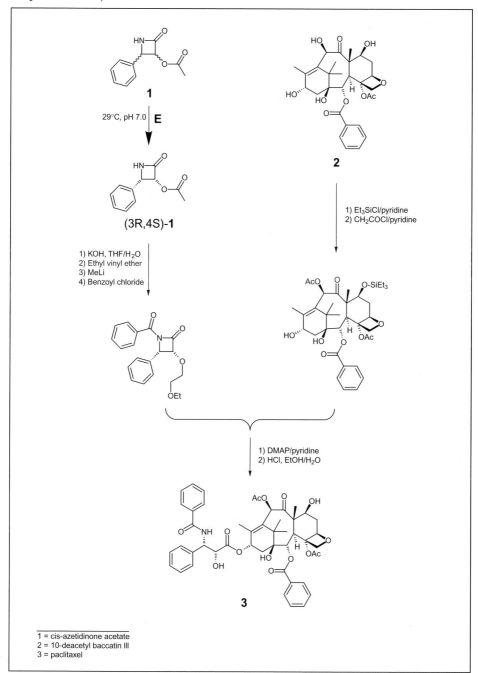

1 = cis-azetidinone acetate
2 = 10-deacetyl baccatin III
3 = paclitaxel

Fig. 3.1.1.3 – 2

- It is indicated for the treatment of various cancers.

- The main problem until now was that the source of taxol – the bark of the Pacific yew *Taxus brevifolia* – contains taxol at a very low concentration (< 0.02 wt.-%).

- The alternate semisynthetic pathway, in which the tetracyclic diterpene 10-deacetyl baccatin can be isolated from the leaves of the European yew *Taxus baccata* (0.1 wt.-%), leads directly to paclitaxel.

6) Literature

- Holton, R.A. (1992) Method for preparation of taxol using ß-lactam, Florida State University, US 5,175,315

- Patel, R.N., Szarka, L.J., Partyka, R.A. (1993) Enzymatic process for resolution of enantiomeric mixtures of compounds useful as intermediates in the preparation of taxanes, E.R. Squibb Sons & Inc., EP 552041 A2

- Patel, R.N., Banerjee, A., Ko, R.Y., Howell, J.M., Li, W.-S., Comezoglu, F.T., Partyka, R.A., Szarka, L.J. (1994) Enzymic preparation of (3*R-cis*)-3-(acetyloxy)-4-phenyl-2-azetidinone: a taxol side-chain synthon, Biotechnol. Appl. Biochem. **20**, 23–33

- Zaks, A., Dodds, D.R. (1997) Application of biocatalysis and biotransformations to the synthesis of pharmaceuticals, Drug Discovery Today **2**, 513–531

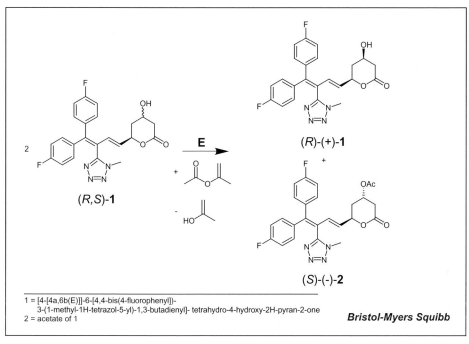

2

(R,S)-1

E

(R)-(+)-1

+

(S)-(-)-2

1 = [4-[4a,6b(E)]]-6-[4,4-bis(4-fluorophenyl])-
   3-(1-methyl-1H-tetrazol-5-yl)-1,3-butadienyl]- tetrahydro-4-hydroxy-2H-pyran-2-one
2 = acetate of 1

Bristol-Myers Squibb

Fig. 3.1.1.3 – 1

1) Reaction conditions

[1]:	0.009 M, 4 g · L^{-1} [438.43 g · mol^{-1}]
T:	37 °C
medium:	toluene
reaction type:	acetylation
catalyst:	immobilized enzyme
enzyme:	triacylglycerol acylhydrolase (triacylglycerol lipase)
strain:	*Pseudomonas cepacia*
CAS (enzyme):	[9001–62–1]

2) Remarks

- The lipase is immobilized on a polypropylene support (Accurel PP).

- The reaction suspension additionally contains 2 % (v/v) isopropenyl acetate, 0.05 % distilled water and 1 % (w/v) lipase (Amano PS-30).

- Isopropenyl acetate is used as acyl donor, since here acetone is produced. If vinyl acetate is applied as acyl donor then acetaldehyde is formed, which is much more incompatible with lipases than acetone.

- Immobilized lipase is reused for 5 times in a repetitive batch mode.

- The (S)-acetate can be hydrolyzed in a subsequent reaction by the same lipase in an aqueous/organic two-phase system. The cleaved lactone is resynthesized using pivaloyl chloride.

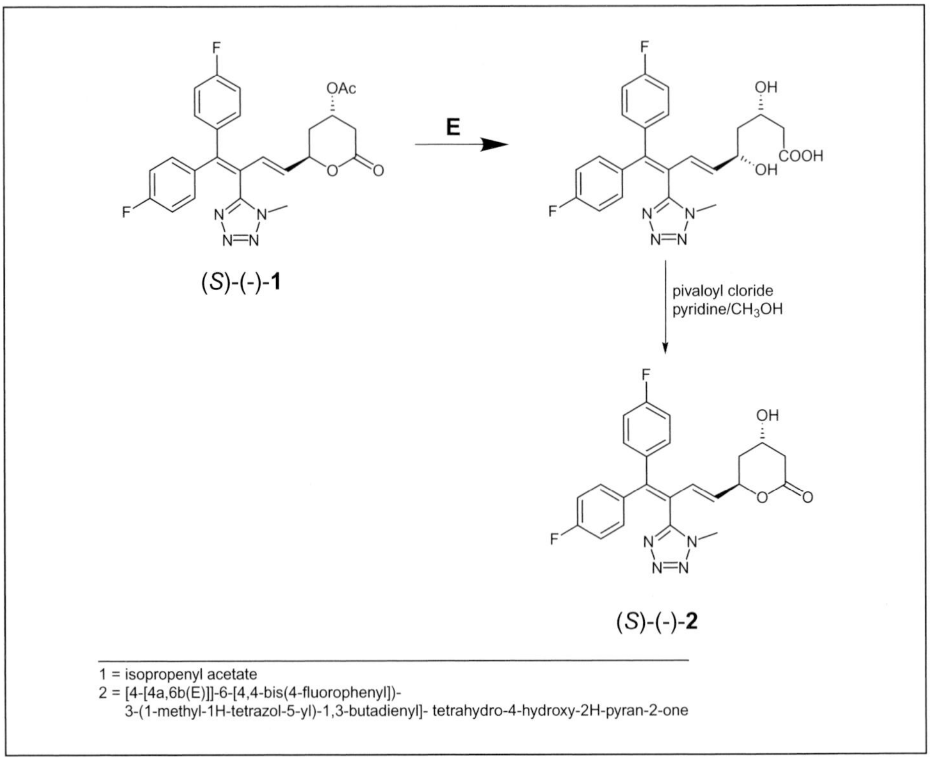

(S)-(-)-**1**

pivaloyl cloride
pyridine/CH_3OH

(S)-(-)-**2**

1 = isopropenyl acetate
2 = [4-[4a,6b(E)]]-6-[4,4-bis(4-fluorophenyl])-
   3-(1-methyl-1H-tetrazol-5-yl)-1,3-butadienyl]- tetrahydro-4-hydroxy-2H-pyran-2-one

Fig. 3.1.1.3 – 2

3) Flow scheme

Not published.

4) Process parameters

yield:	48 %
ee:	98.5 %
chemical purity:	99.5 %
reactor type:	repetitive batch
reactor volume:	640 L
capacity:	multi kg
residence time:	20 h
down stream processing:	filtration, distillation, crystallization
enzyme supplier:	Amano, Japan
company:	Bristol-Myers Squibb, USA

5) Product application

- The product is a hydroxymethyl glutaryl coenzyme A (HMG-CoA) reductase inhibitor and a potential anticholesterol drug candidate.

6) Literature

- Patel, R.N. (1992) Enantioselective enzymatic acetylation of racemic [4-[4α,6β(*E*)]]-6-[4,4-bis(4-fluorophenyl])-3-(1-methyl-1H-tetrazol-5-yl)-1,3-butadienyl]-tetrahydro-4-hydroxy-2H-pyran-2-one, Appl. Microbiol. Biotechnol. **38**, 56–60

- Patel, R.N. (1997) Stereoselective biotransformations in synthesis of some pharmaceutical intermediates, Adv. Appl. Microbiol. **43**, 91–140

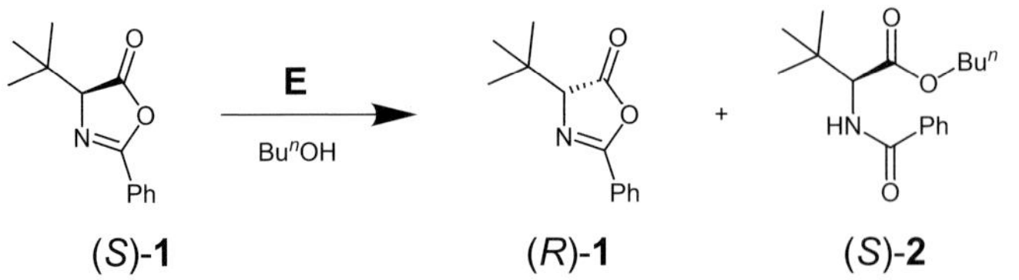

(S)-1 → (R)-1 + (S)-2

1 = azlactone of *tert*-leucine
2 = ester amide

Celltech Group plc

Fig. 3.1.1.3 – 1

1) Reaction conditions

medium:	organic solvent
reaction type:	carboxylic ester hydrolysis
catalyst:	immobilized enzyme
enzyme:	triacylglycerol acylhydrolase (triacylglycerol lipase)
strain:	*Mucor miehei*
CAS (enzyme):	[9001–62–1]

2) Remarks

- At Degussa, the synthesis of (S)-*tert*-leucine is carried out as an asymmetric reductive amination of the prochiral keto acid (dehydrogenase and cofactor recycling); see page 300. Chiroscience uses the hydrolase-catalyzed resolution of azlactones instead.

- Azlactones easily racemize and can be prepared by cyclodehydration of an *N*-acylated amino acid with acetic anhydride.

- Although the opening of the ring with water can also be achieved non-enzymatically, the reaction in presence of water is very unselective. An alcohol in a water-free system has to be used instead.

- Even if the process is formally described as a kinetic resolution it is ultimately a deracemization. The azlactone is racemized *in situ* resulting in an overall yield of 90 % for the chiral amino acid. The following figure shows the reaction steps:

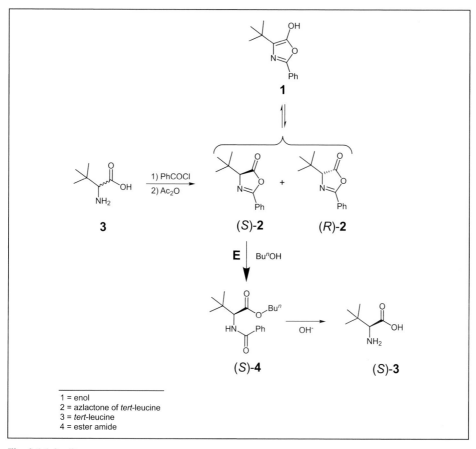

Fig. 3.1.1.3 – 2

- A high substrate concentration promotes racemization of the (R)-lactone and improves yield and ee.

- Since the substrate can also be synthesized via the glycine azlactone this methodology can be applied to other *tert*-alkyl glycine derivatives. The following figure shows the route to the azlactone derivative of *tert*-leucine:

1 = *N*-benzoylglycine (hippuric acid)
2 = glycine azlactone
3 = azlactone of *tert*-leucine

Fig. 3.1.1.3 – 3

- The concentration of the substrate could be increased to 20 %, and conditions were established whereby the reaction was completed within 24 h.

- The deprotection of the ester amide to the amino acid could be carried out with potassium hydroxide without racemization (the cleavage by acid hydrolysis leads to partial racemization caused by transient recyclization to the stereochemically labile azlactone).

3) Flow scheme

Not published.

4) Process parameters

yield:	>90 %
ee:	>97 %
residence time:	24 h
company:	Celltech Group, U.K.

5) Product application

- The product is useful as a lipophilic, hindered component of peptides. Cleavage by peptidases is disfavoured because of its bulky nature, resulting in peptides of improved metabolic stability.

- The amino acids are also useful building blocks for a number of chiral auxiliaries and ligands where the presence of the bulky *tert*-butyl group makes these compounds particularly effective for asymmetric synthesis:

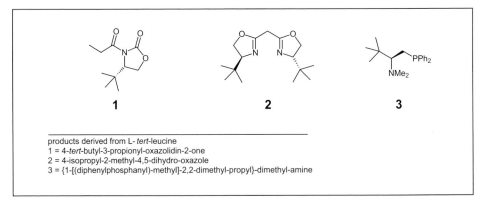

1   **2**   **3**

products derived from L-*tert*-leucine
1 = 4-*tert*-butyl-3-propionyl-oxazolidin-2-one
2 = 4-isopropyl-2-methyl-4,5-dihydro-oxazole
3 = {1-[(diphenylphosphanyl)-methyl]-2,2-dimethyl-propyl}-dimethyl-amine

Fig. 3.1.1.3 – 4

6) Literature

- McCague, R., Taylor, S.J.C. (1997) Dynamic resolution of an oxazolinone by lipase biocatalysis: Synthesis of *(S)-tert*-leucine, in: Chirality In Industry II (Collins, A.N., Sheldrake, G.N., Crosby, J., eds.), pp. 201–203, John Wiley & Sons, New York

- Turner, N.J., Winterman, J.R., McCague, R., Parratt, J.S., Taylor, S.J.C. (1995) Synthesis of homochiral L-*(S)-tert*-leucine via a lipase catalysed dynamic resolution process, Tetrahedron Lett. **36**, 1113–1116

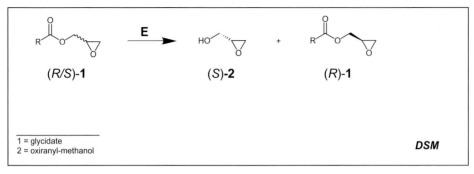

(R/S)-**1**   (S)-**2**   (R)-**1**

1 = glycidate
2 = oxiranyl-methanol

DSM

Fig. 3.1.1.3 – 1

1) Reaction conditions

reaction type: carboxylic ester hydrolysis
catalyst: immobilized enzyme
enzyme: triacylglycerol acylhydrolase (triacylglycerol lipase, porcine pancreas lipase = PPL)
strain: porcine pancreas (organism)
CAS (enzyme):   [9001–62–1]

2) Remarks

- This is the oldest biocatalytic process at DSM, still operated campaign-wise.

3) Flow scheme

Not published.

4) Process parameters

yield: > 85 %
ee: > 99.9 %
reactor type: batch
company: DSM, The Netherlands

5) Product application

- The propane oxirane can be converted into (*S*)-beta blockers:

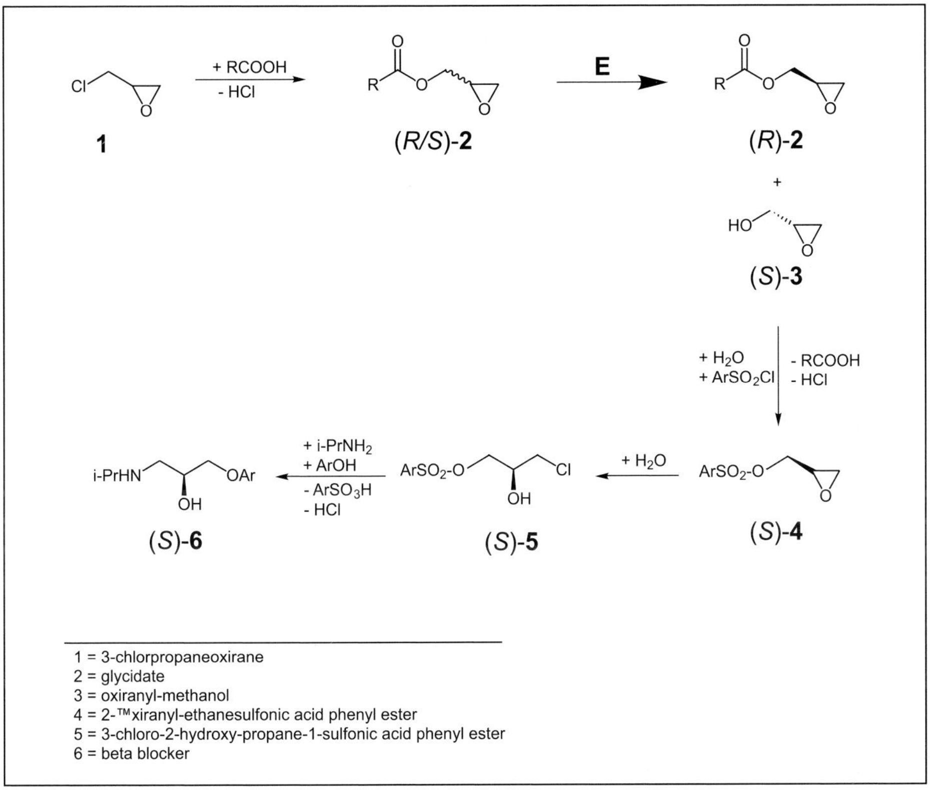
+ RCOOH
- HCl
E
1
(R/S)-2
(R)-2
(S)-3
+ H2O
+ ArSO2Cl
- RCOOH
- HCl
+ i-PrNH2
+ ArOH
- ArSO3H
- HCl
+ H2O
i-PrHN
OAr
OH
ArSO2-O
Cl
(S)-6
(S)-5
(S)-4
1 = 3-chlorpropaneoxirane
2 = glycidate
3 = oxiranyl-methanol
4 = 2-™xiranyl-ethanesulfonic acid phenyl ester
5 = 3-chloro-2-hydroxy-propane-1-sulfonic acid phenyl ester
6 = beta blocker

Fig. 3.1.1.3 – 2

- Typical examples of beta blockers are shown in the following figure:

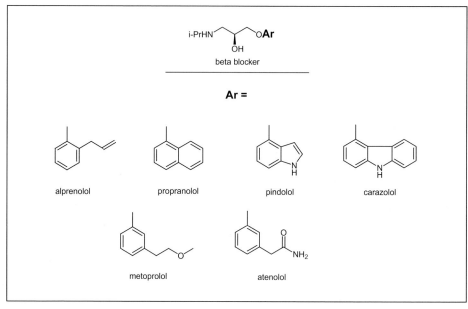

Fig. 3.1.1.3 – 3

6) Literature

- Sheldon, R.A. (1993) Chirotechnology, Marcel Dekker Inc., New York

- Elferink, V.H.M. (1995) Progress in the application of biocatalysis in the industrial scale manufacture of chiral molecules, Chiral USA 96, 11th International Spring Innovations Chirality Symposium, pp. 79–80, Boston

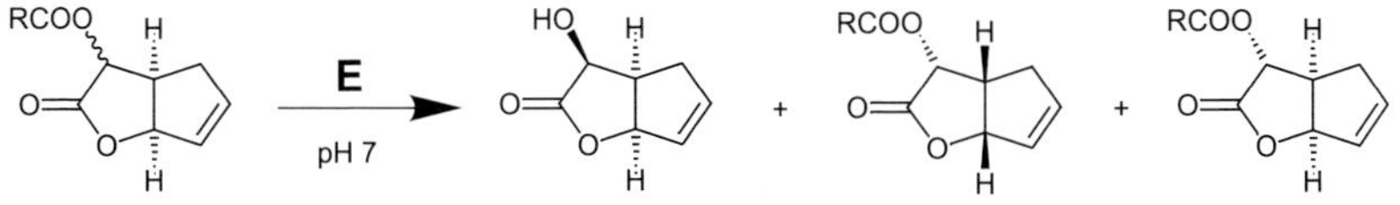

(±)-1 → (-)-2 + (+)-1 + (±)-3

1 = 4-hydroxy-2-oxabicyclo[3.3.0]oct-7-en-3-one butyrate ester (R = propyl)
   (crude mixture of *endo*- and *exo*-butyrate esters)
2 = 4-*endo*-hydroxy-2-oxabicyclo[3.3.0]oct-7-en-3-one
3 = 4-*exo*-hydroxy-2-oxabicyclo[3.3.0]oct-7-en-3-one butyrate ester

Celltech Group plc

Fig. 3.1.1.3 – 1

1) Reaction conditions

pH:	7.0
T:	25 °C
medium:	aqueous
reaction type:	carboxylic ester hydrolysis
catalyst:	solubilized enzyme
enzyme:	triacylglycerol acylhydrolase (lipase, triacylglycerol lipase)
strain:	*Pseudomonas fluorescens*
CAS (enzyme):	[9001–62–1]

2) Remarks

- The hydrolysis is preferred over the transesterification in this case since the reaction rate and enantioselectivity of the acylation are drastically reduced.

- Continuous addition of NaOH to the reaction mixture during hydrolysis is necessary to maintain neutral pH.

- If a hydrophobic ester (e.g. butyrate) is used, the ester can be extracted into the organic phase (heptane), while the alcohol remains in the aqueous phase.

- The butyrate ester is insoluble in water, so that after centrifugation and separation of the aqueous phase the alcohol can be easily extracted and purified by crystallization.

- About ten percent of the ester are lost because of pH-dependent ring-opening to an unstable carboxylic acid salt.

3) Flow scheme

Not published.

4) Process parameters

yield: 22 %
ee: > 92 % (after recrystallization >99 %)
reactor type: batch
capacity: multi-kg
residence time: 75 h
down stream processing: extraction, crystallization
production site: Celltech Group plc, U.K.
company: Celltech Group plc, U.K.

5) Product application

- The product can be used as intermediate for the anti-HIV agent carbovir.

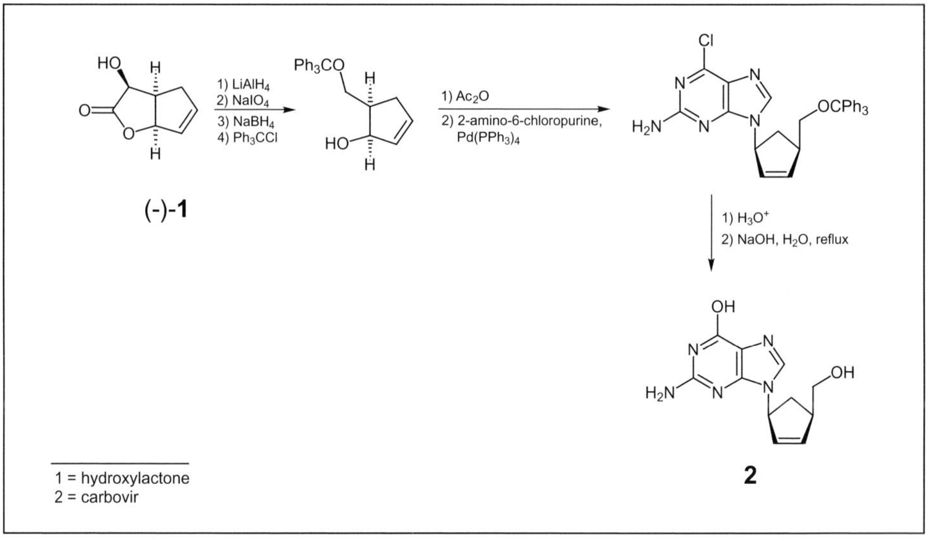

1 = hydroxylactone
2 = carbovir

Fig. 3.1.1.3 – 2

- As (–)-hydroxylactone it can be used as intermediate for the synthesis of hypocholesteremic reagents and the antifungal agent brefeldin A:

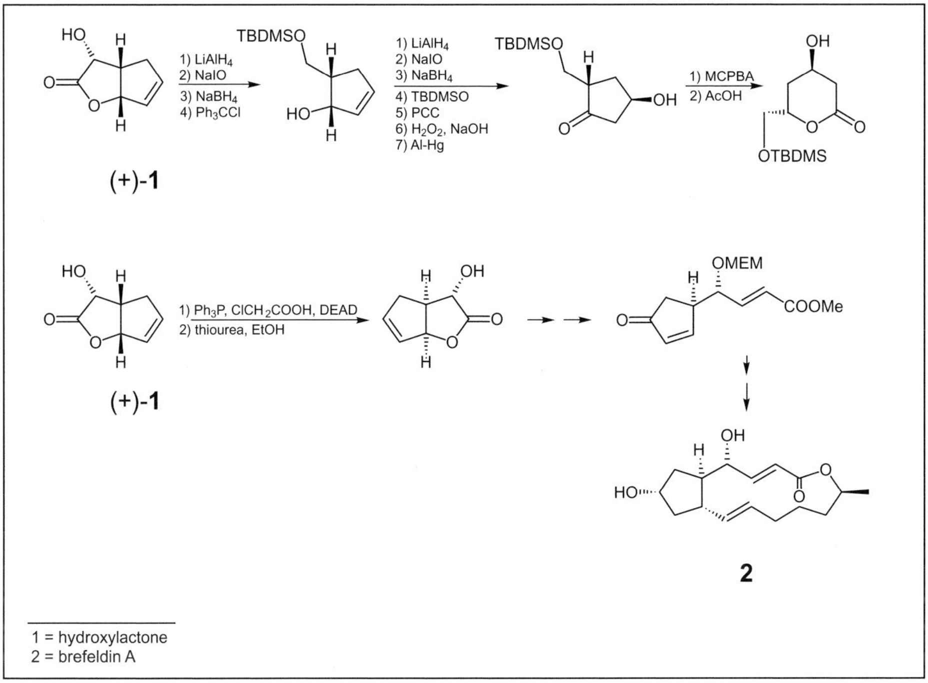

Fig. 3.1.1.3 – 3

6) Literature

- Evans, C.T., Roberts, S.M., Shoberu, K.A., Sutherland, A.G. (1992) Potential use of carbocylic nucleosides for the treatment of AIDS: Chemo-enzymatic syntheses of the enantiomers of carbovir; J. Chem. Soc. Perkin Trans. 1, 589–592

- MacKeith, R.A., McCague, R., Olivo, H.F., Palmer, C.F., Roberts, S.M. (1993) Conversion of (–)-4-hydroxy-2-oxabicyclo[3.3.0]oct-7-en-3-one into the anti-HIV agent Carbovir, J. Chem. Soc. Perkin Trans. 1 1, 313–314

- MacKeith, R.A., McCague, R., Olivo, H.F., Roberts, S.M., Taylor, S.J.C., Xiong, H. (1994) Enzyme-catalysed kinetic resolution of 4-endo-hydroxy-2-oxabicyclo[3.3.0]oct-7-en-3-one and employment of the pure enantiomers for the synthesis of anti-viral and hypocholesteremic agents, Bioorg. Med. Chem. 2, 387–394

- Taylor, S.J.C., McCague, R. (1997) Resolution of a versatile hydroxylactone synthon 4-endo-hydroxy-2-oxabicyclo[3.3.0]oct-7-en-3-one by lipase deesterification, in: Chirality In Industry II (Collins, A. N., Sheldrake, G. N. and Crosby, J., eds.), pp. 190–193, John Wiley & Sons, New York

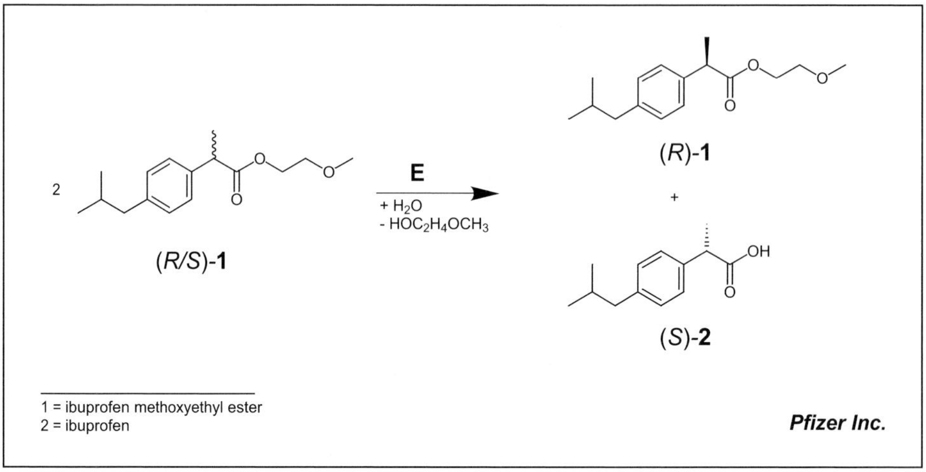

1 = ibuprofen methoxyethyl ester
2 = ibuprofen

Pfizer Inc.

Fig. 3.1.1.3 – 1

1) Reaction conditions

pH:	5.0
T:	20 °C
medium:	multiphase: aqueous / organic / solid
reaction type:	carboxylic ester hydrolyis
catalyst:	immobilized enzyme
enzyme:	triacylglycerol acylhydrolase (triacylglycerol lipase, lipase)
strain:	*Candida cylindraceae*
CAS (enzyme):	[9001–62–1]

2) Remarks

- Although the lipase shows good activity over a broad pH range, a low pH value has to be employed because the enzyme is deactivated by ibuprofen. At low pH-values the low solubility of ibuprofen ester can be used to prevent deactivation.

- The main problem of the enzymatic synthesis is the low solubility of substrates in water. In this case the ester solubility is below 1 mM. To circumvent problems of handling big volumes of water, a membrane based concept is realized.

- A hollow fibre membrane is used, where the lipase is immobilized (non-covalently, entrapped) in the pores of the membrane. The hydrophobic ibuprofen methoxyethylester is delivered solubilized in the organic phase to the outside of the asymmetric membrane. After conversion, the ibuprofen is extracted by the aqueous phase into the lumen of the hollow fibres.

- The advantage of this reactor setup is that the membrane stabilizes the aqueous/organic interface providing a high surface area for contact between the organic and aqueous phases without dispersing one phase into the other.

- In combination with another membrane module adjusted to a higher pH, the product can be easily separated from the unconverted ester, which can be easily recycled to the first membrane system. These techniques allows a low ibuprofen concentration at low pH leading to high catalyst stability.

- An alternative process starting from racemic 2-arylpropionitrile by the action of nitrilase was investigated by Asahi Chemical Ind. Co., Japan.

3) Flow scheme

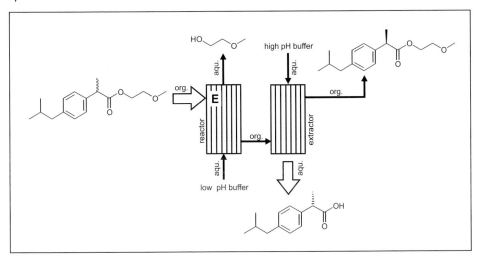

Fig. 3.1.1.3 – 2

4) Process parameters

ee:	96 %
reactor type:	multiphase membrane reactor in batch mode
capacity:	multi kg
space-time-yield:	$3.6 \ g \cdot d^{-1} \cdot g_{Lipase}^{-1}$ (related to membrane area: $18 \ g \cdot d^{-1} \cdot m^{-2}$)
down stream processing:	extraction and distillation
enzyme activity:	$5 \ g \cdot m^{-2}$
enzyme consumption:	$t_{1/2} = 30 \ d$
enzyme supplier:	Genzyme Corp., USA
company:	Pfizer Inc., USA

5) Product application

- Ibuprofen is an important nonsteroidal antiinflammatory drug.

- The *in vitro* activity of the (S)-enantiomer is 100 times that of the (R)-enantiomer.

6) Literature

- Adams, S.S., Bresloff, P., Mason, C.G. (1976) Pharmacological differences between the optical isomers of ibuprofen: evidence for metabolic inversion of the (–)-isomer, J. Pharm. Pharmacol. **28**, 256–257.

- Cesti, P., Piccardi, P. (1986) Process for the biotechnological preparation of optically active α-arylalkanoic acids, Montedison S.p.A., Italy, EP 195717 A2

- Lopez, J.L., Wald, S.A., Matson, S.L., Quinn, J.A. (1990) Multiphase membrane reactors for separating stereoisomers, Ann. N. Y. Acad. Sci., **613**, 155–166

- McConville, F.X., Lopez, J.L., Wald, S.A. (1990) Enzymatic resolution of ibuprofen in a multiphase membrane reactor, in: Biocatalysis (Abramowicz, D.A., ed.) pp.167–177, van Nostrand Reinhold, New York

- Sheldon, R.A. (1993) Chirotechnology, Marcel Dekker Inc., New York

- Sih, C. J. (1987) Process for preparing (S)-α-methylarylacetic acids, Wisconsin Alumni Research Foundation, USA, EP 227078 A1

- Yamamoto, K., Ueno, Y., Otsubo, K., Kawakami, K., Komatsu, K.I. (1990) Production of (S)-(+)-ibuprofen from a nitrile compound by *Acinetobacter sp*. Strain AK226, Appl. Environ. Microbiol. **56**, 3125–3129

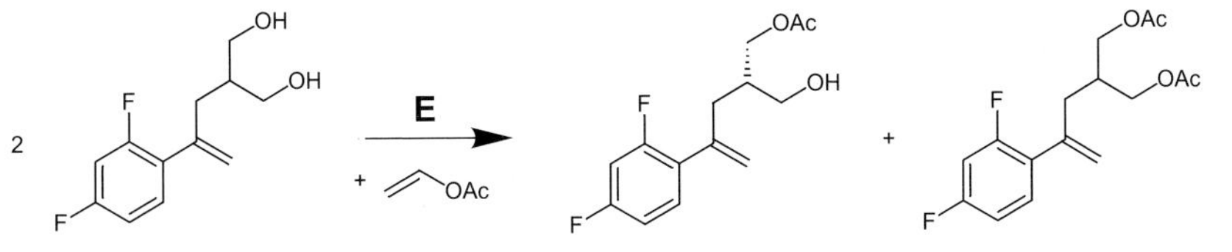

1 = 2-[2-(2,4-difluoro-phenyl)-allyl]-propane-1,3-diol
2 = acetic acid 4-(2,4-difluoro-phenyl)-2-hydroxymethyl-pent-4-enyl ester
3 = acetic acid 2-acetoxymethyl-4-(2,4-difluoro-phenyl)-pent-4-enyl ester

Schering Plough

Fig. 3.1.1.3 – 1

1) Reaction conditions

[1]:	0.876 M, 200 g · L^{-1} [228.24 g · mol^{-1}]
T:	0 °C
medium:	vinyl acetate in acetonitrile
reaction type:	carboxylic ester hydrolysis
catalyst:	immobilized enzyme (Novozyme 435)
enzyme:	triacylglycerol acylhydrolase
strain:	*Candida antarctica*
CAS (enzyme):	[9001–62–1]

2) Remarks

- The lipase from *Candida antarctica* catalyzes the pro-*S* acetylation of the diol.

- In addition to 74 % of the (*S*)-monoacetate, 26 % of the diacetate is formed.

- Acetonitrile is selected as solvent since the subsequent iodocyclization is also carried out in acetonitrile (see product application). Therefore, the reaction solution can be directly transferred to the chemical step after separating the lipase by filtration. To do so, it is important to reach a very high conversion, since the racemic diol also reacts in the iodocyclization. With the diacetate no reaction takes place.

- As acylating agent 1.25 equivalents of vinyl acetate are used.

3) Flow scheme

Not published.

4) Process parameters

conversion:	99 %
yield:	74 %
selectivity:	73 %
ee:	> 99 %
enzyme supplier:	Novo Nordisk, Denmark
company:	Schering-Plough, USA

5) Product application

- The product is used as an improved azole antifungal. The complete synthesis is shown in the following figure:

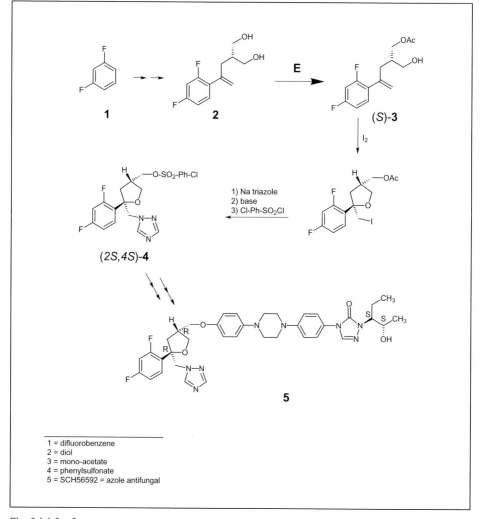

1 = difluorobenzene
2 = diol
3 = mono-acetate
4 = phenylsulfonate
5 = SCH56592 = azole antifungal

Fig. 3.1.1.3 – 2

- It shows activity against systemic *Candida* and pulmonary *Aspergillus* infections (phase II, clinical).

- The increased activity in comparison to other azole antifungals results from the tetrahydrofuran ring that replaces the 1,3-dioxolane ring present in other azole drugs.

6) Literature

- Morgan, B., Dodds, D.R., Zaks, A., Andrews, D.R., Klesse, R. (1997) Enzymatic desymmetrization of prochiral 2-substituted-1,3-propanediols: A practical chemoenzymatic synthesis of a key precursor of SCH51048, a broad-spectrum orally active antifungal agent, J. Org. Chem. **62**, 7736–7743

- Morgan, B., Stockwell, B.R., Dodds, D.R., Andrews, D.R., Sudhakar, A.R., Nielsen, C.M., Mergelsberg, I., Zumbach, A. (1997) Chemoenzymatic approaches to SCH 56592, a new azole antifungal, J. Am. Oil Chem. Soc. **74**, 1361–1370

- Pantaleone, D.P. (1999) Biotransformations: "Green" processes for the synthesis of chiral fine chemicals, in: Handbook of Chiral Chemicals (Ager, D.J., ed.) pp. 245–286, Marcel Dekker Inc., New York

- Saksena, A.K., Girijavallabhan, V.M., Pike, R.E., Wang, H., Lovey, R.G., Liu, Y.-T., Ganguly, A.K., Morgan, W.B., Zaks, A. (1995) Process for preparing intermediates for the synthesis of antifungal agents, Schering Corporation, US 5,403,937

- Saksena, A.K., Girijavallabhan, V.M., Wang, H., Liu, Y.-I., Pike, R.R., Ganguly, A.K. (1996) Concise asymmetric routes to 2,2,4-trisubstituted tetrahydrofurans via chiral titanium imide enolates: Key intermediates towards synthesis of highly active azole antifungals SCH 51048 and SCH 56592, Tetrahedron Lett. **37**, 5657–5660

- Zaks, A., Dodds, D.R. (1997) Application of biocatalysis and biotransformations to the synthesis of pharmaceuticals, Drug Discovery Today **2** (6), 513–531

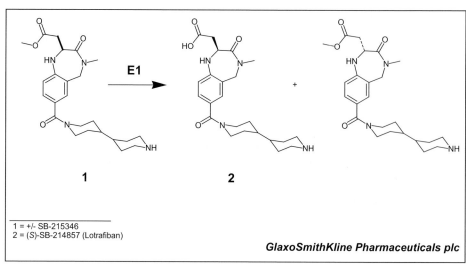

1 = +/- SB-215346
2 = (S)-SB-214857 (Lotrafiban)

GlaxoSmithKline Pharmaceuticals plc

Fig. 3.1.1.3 – 1

1) Reaction conditions

[1]:	0.07 M, 31 g · mol^{-1} [442.55 g · mol^{-1}]
[2]:	1.0 N HCl
[3]:	1.5 M ammonia solution
pH:	6.0–6.8
T:	29–30 °C
medium:	water
reaction type:	carboxylic ester hydrolysis
catalyst:	immobilized enzyme
enzyme:	triacylglycerol lipase, Chirazyme L-2, lipase B
strain:	*Candida antarctica*
CAS (enzyme):	[9001-62-1]

2) Remarks

- The biocatalyst supplied by Boehringer–Mannheim uses a macroporous cross-linked resin to support the enzyme, but the protein is covalently bound to the polymer, making it ideal for use in aqueous environments.

- A few kilograms of resin are sufficient to produce hundreds of kilograms of product.

- After use, some biocatalyst was stored for over a year in 3.5 M ammonium sulfate and showed no loss of activity or chiral selectivity on reuse.

- Allowing for multiple reuses, the calculated cost of enzyme resin per kg of drug substance is between €7 and €14.

- The product is easily separated. On treating the 1:1 aqueous solution of (*S*)-acid (main product) and (*R*)-ester with a dichloromethane solution of CBZ chloride with the pH maintained at 7.0, the (*R*)-ester partitions into organic solvent as the carbamate, whilst the required (*S*)-acid remains unreacted in the aqueous phase. The desired product is then isolated simply by concentration and cooling.

- The (*R*)-ester can be recovered in quantitative yield from the dichloromethane layer, racemized and deprotected to regenerate racemic methyl ester to be recycled back into the resolution process.

- In contrast the chemical synthesis was derived using (*S*)-aspartic acid. The reaction was low-yielding (~40 %). The chiral purity of product fell to ~96 %. This translated into drug substance of an unacceptable *ee* of 91 %.

- A biocatalytic resolution seems an ideal solution.

3) Flow scheme

Not published.

4) Process parameters

conversion:	90 %
yield:	84 % (overall 42 %)
selectivity:	100 %
optical purity:	99 %
chemical purity:	99 %
reactor type:	batch
reactor volume:	2400 L
capacity:	several hundred kilograms of product
down stream processing:	extraction, distillation, cooling, centrifugation, drying in vacuum
enzyme consumption:	few kg per hundred kg product (at least 100 reuses of the resin possible)
enzyme supplier:	Roche Diagnostics, Switzerland
production site:	GlaxoSmithKline plc, U.K.
company:	GlaxoSmithKline Pharmaceuticals, U.K.

5) Product application

- Lotrafiban is an orally active GPIIb/IIIa fibrinogen receptor antagonist designed for the prevention of thrombotic events.

6) Literature

- Walsgrove, T.C., Powell, L., Wells, A. (2002) A practical and robust process to produce SB-2124857, Lotrafiban, [(2*S*)-7-(4,4′-bipiperidinylcarbonyl)-2,3,4,5,-tetrahydro-4-methyl-4-oxo-1*H*-1,4-benzodiazepine-2-acetic acid] utilising an enzymatic resolution as the final step, Org. Proc. Res. Dev. **6**, 488–491

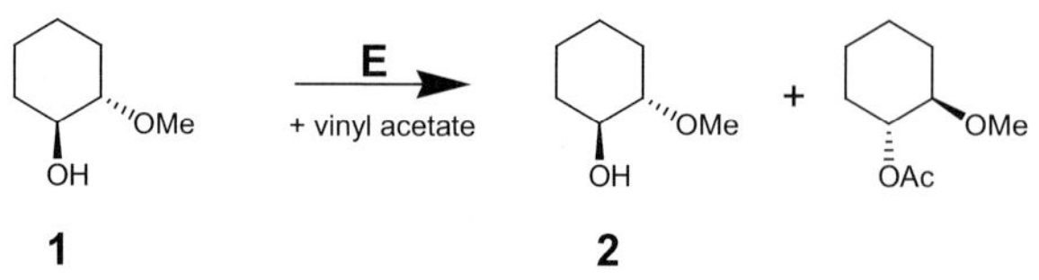

Fig. 3.1.1.3 – 1

1 = (+/-)-*trans*-2-methoxycyclohexanol
2 = (+)-(1S,2S)-2-methoxycyclohexanol

GlaxoSmithKline plc

1) Reaction conditions

[1]:	1.4 M, 182 g · L^{-1} (130.18 g · mol^{-1})
[2]:	1.7 M vinyl acetate
[3]:	0.16 M triethylamine
T:	room temperature
medium:	cyclohexane
reaction type:	lipase mediated transesterification
catalyst:	immobilized enzyme (novozyme 435)
enzyme:	triacylglycerol acylhydrolase
strain:	*Candida antarctica*
CAS (enzyme):	[9001-62-1]

2) Remarks

- After nine cycles the immobilized enzyme retained more than half of its initial activity. Free *Pseudomonas fluorescens* lipase lost 75 % of activity after 4 cycles. However, by immobilizing this lipase on celite powder the same stability as for the commercially-available one could be obtained.

- After DSP the recovered enzyme is stored at 4 °C for future use.

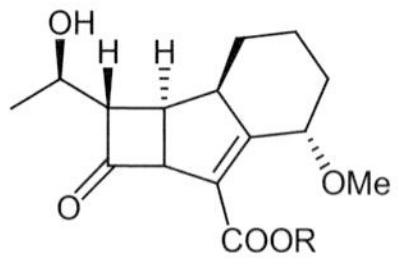

Fig. 3.1.1.3 – 2

1 R = Na
2 R = CH(CH$_3$)OC(O)OC$_6$H$_{11}$

3) Flow scheme

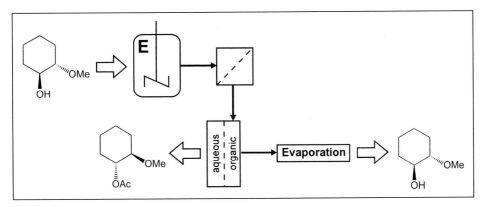

Fig. 3.1.1.3 – 3

4) Process parameters

conversion:	55 %
yield:	36 %
ee:	>98 %
reactor type:	batch
residence time:	6–8 h
down stream processing:	filtration, extraction (water), extraction (ethylacetate), evaporation
enzyme supplier:	Novo Nordisk
company:	GlaxoSmithKline plc, U.K.

5) Product application

- The product is used as building block for Trinems, a new class of totally synthetic β-lactam antibiotics containing a tricyclic skeleton (see Scheme 2). Both compounds are in full clinical development.

6) Literature

- Stead, P., Marley, H., Mahmoudian, M., Webb, G., Noble, D., Ip, Y.T., Piga, E., Rossi, T., Roberts, S., Dawson, M.J. (1996) Efficient procedures for the large-scale preparation of (1S,2S)-*trans*-2-methoxycyclohexanol, a key chiral intermediate in the synthesis of tricyclic β-lactam antibiotics, Tetrahedron: Asymmetry **7** (8), 2247–2250

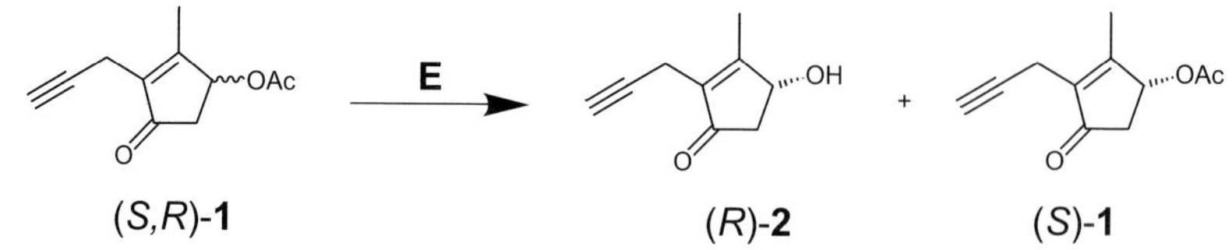

1 = acetic acid 2-methyl-4-oxo-3-prop-2-ynyl-cyclopent-2-enyl ester
2 = 4-hydroxy-3-methyl-2-prop-2-ynyl-cyclopent-2-enone

Sumitomo Chemical Co.

Fig. 3.1.1.3 – 1

1) Reaction conditions

[**1**]:	4.16 M, 800 g · L^{-1} [192.21 g · mol^{-1}]
pH:	7.0
T:	40 °C
medium:	two-phase: aqueous / organic
reaction type:	carboxylic ester hydrolysis
catalyst:	suspended enzyme
enzyme:	triacylglycerol acylhydrolase (triacylglycerol lipase)
strain:	*Arthrobacter* sp.
CAS (enzyme):	[9001–62–1]

2) Remarks

- Subsequent to the lipase-catalyzed hydrolysis, the cleaved alcohol is sulfonated in the presence of the acylated compound with methanesulfonyl chloride. The hydrolysis of the sulfonated enantiomer in the presence of small amounts of calcium carbonate takes place under inversion of the chiral center as opposed to the hydrolysis of the acylated enantiomer, which is carried out under retention of the chiral center. By this means, an enantiomeric excess of 99.2 % and a very high yield is achieved for the (*R*)-alcohol:

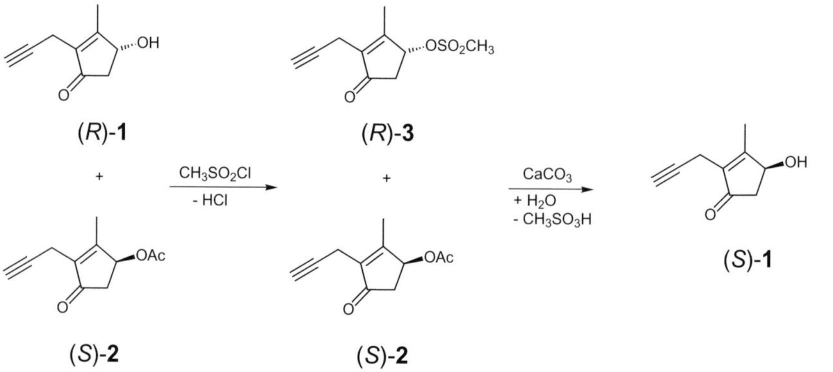

Fig. 3.1.1.3 – 2

- For this resolution an E-value of 1,300 was determined.

3) Flow scheme

Not published.

4) Process parameters

conversion:	49.9 %
ee:	99.2 % (alcohol)
reactor type:	batch
company:	Sumitomo Chemical Co., Japan

5) Product application

- The (S)-alcohol is used as an intermediate in the synthesis of pyrethroids, which are used as insecticides. They show excellent insecticidal activity and a low toxicity in mammals.

6) Literature

- Hirohara, H., Nishizawa, M. (1998) Biochemical synthesis of several chemical insecticide intermediates and mechanism of action of relevant enzymes, Biosci. Biotechnol. Biochem. **62**, 1–9

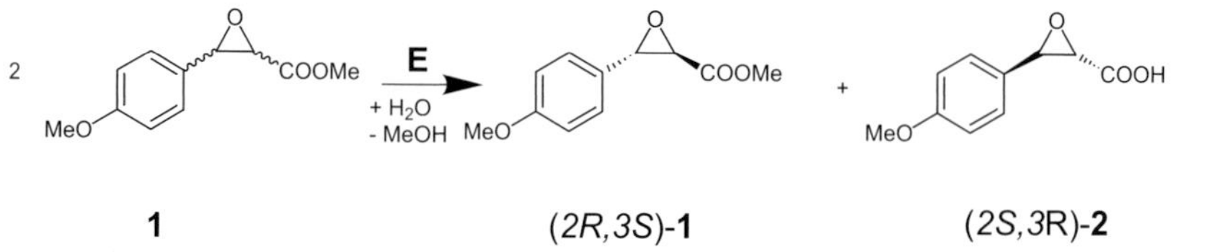

1 = 3-(4-methoxyphenyl) glycidic acid methyl ester = MPGM
2 = 3-(4-methoxyphenyl) glycidic acid

Tanabe Seiyaku Co., Ltd.

Fig. 3.1.1.3 – 1

1) Reaction conditions

[**1**]:	< 0.6 M, < 125 g · L^{-1} [208.21 g · mol^{-1}]
pH:	8.5 (aqueous solution in lumen loop)
T:	22 °C
medium:	two-phase-system, aqueous / toluene
reaction type:	carboxylic ester hydrolysis
catalyst:	immobilized enzyme
enzyme:	triacylglycerol acylhydrolase (triacylglycerol lipase)
strain:	*Serratia marescens* Sr41 8000
CAS (enzyme):	[9001–62–1]

2) Remarks

- A hydrophilic hollow fibre membrane is used (polyacrylonitrile) as reactor unit (see flow scheme).
- The lipase is immobilized onto a spongy layer by pressurized adsorption.
- The lipase does not attack *(2R,3S)*-(4-methoxyphenyl)glycidic acid methyl ester which acts as a competitive inhibitor.
- The formed acid (hydrolyzed (+)-methoxyphenylglycidate) is unstable and decarboxylates to give 4-methoxyphenylacetaldehyde; this aldehyde strongly inhibits and deactivates the enzyme. It can be removed continuously by filtration as bisulfite adduct. The bisulfite acts also as buffer to maintain constant pH during synthesis.
- The lipase is also inhibited by Co^{2+}, Ni^{2+}, Fe^{2+}, Fe^{3+} and EDTA, but can be activated by Ca^{2+} and Li^{+}.
- Apparent V_{max} for hydrolysis of (+)MPGM is 1.7 U · $mg_{Protein}^{-1}$.
- The substrate specificity of the hydrolase from *S. marcescens* is shown in the following table:

Substrate	R	Specific activity (U*mg^{-1})[a]	Substrate	R	Specific activity (U*mg^{-1})[b]
MeO–phenyl-epoxide–COOR	—CH$_3$	570	R—COOMe	—CH$_3$	61
	—C$_2$H$_5$	320		—C$_3$H$_7$	450
	—C$_3$H$_7$	510		—C$_5$H$_{11}$	91
	—C$_4$H$_9$	640		—C$_7$H$_{15}$	620
	—C$_5$H$_{11}$	650		—C$_9$H$_{19}$	570
				—C$_{11}$H$_{23}$	380
				—C$_{13}$H$_{27}$	210
				—C$_{15}$H$_{31}$	91
R–phenyl-epoxide–COOMe	—CH$_3$	640			
	—H	560			
MeO–phenyl–CH(OH)–C(OH)–COOMe		350	CH$_2$–O–COR / CH–O–COR / CH$_2$–O–COR	—CH$_3$	270
				—C$_3$H$_7$	3300
				—C$_5$H$_{11}$	2500
				—C$_7$H$_{15}$	3500
				—C$_9$H$_{19}$	1500
R–phenyl–CH=CH–COOMe	—OCH$_3$	0		—C$_{11}$H$_{23}$	140
	—CH$_3$	0		—C$_{13}$H$_{27}$	120
	—H	0		—C$_{15}$H$_{31}$	56

a)/ b) for conditions of assays see lit. H. Matsumane and T. Shibatani

Fig. 3.1.1.3 – 2

3) Flow scheme

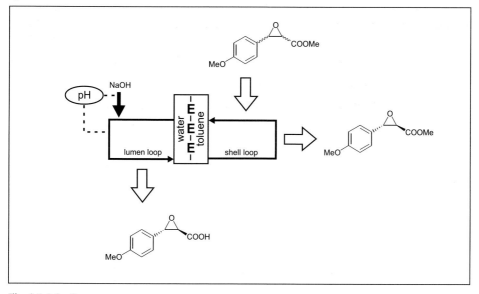

Fig. 3.1.1.3 – 3

4) Process parameters

yield:	40–45 %
ee:	99.9 %
reactor type:	batch, hollow fibre reactor (Sepracor Inc. Massachusetts, USA)
capacity:	40 kg (–)-MPGM $\cdot$ m^{-2} $\cdot$ a^{-1}
down stream processing:	crystallization
enzyme activity:	$1.6 \cdot 10^5$ U $\cdot$ m^{-2}
enzyme stability:	$t_{1/2} = 127$ h
start-up date:	1993
production site:	Tanabe Seiyaku Co., Japan
company:	Tanabe Seiyaku Co. Ltd., Japan

5) Product application

- The product is an intermediate in the synthesis of diltiazem.

- Diltiazem hydrochloride is a coronary vasodilator and a calcium channel blocker (for anti-anginal and anti-hypertensive actions) and is produced worldwide in excess of 100 t $\cdot$ a^{-1}.

- In comparison to the chemical route only 5 steps (instead of 9 steps) are necessary. The kinetic resolution is carried out in an earlier step during the synthesis resulting in reduction of waste:

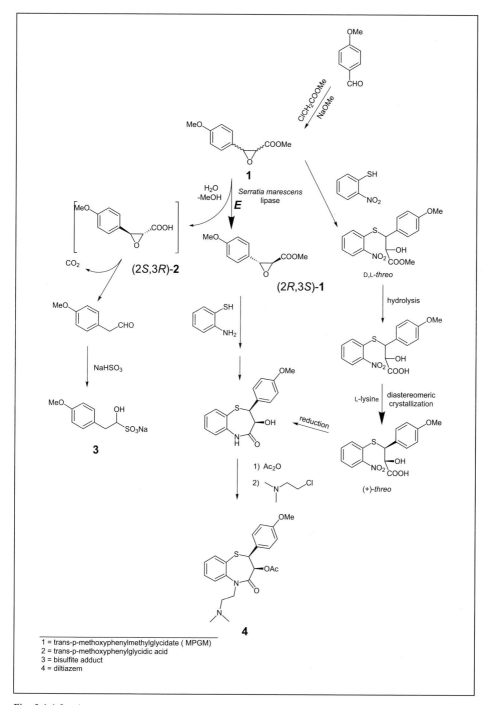

1 = trans-p-methoxyphenylmethylglycidate (MPGM)
2 = trans-p-methoxyphenylglycidic acid
3 = bisulfite adduct
4 = diltiazem

Fig. 3.1.1.3 – 4

6) Literature

- Elferink, V.H.M. (1996) Progress in the application of biocatalysts in the industrial scale manufacture of chiral molecules, Proc. 11th Int. Spring Innovations Chirality Symposium, USA

- Kierkels, J.G.T., Peeters, W.P.H. (1994) Process for the enzymatic preparation of optically active transglycidic acid esters, DSM NV, EP 602740 A1

- López, J.L., Matson, S.L., Stanley, T.J., Quinn, J.A. (1991) Liquid-liquid extractive membrane reactors, Bioproc. Technol. **11**, 27–66

- Matson, S.L. (1987) Method and apparatus for catalyst containment in multiphase membrane reactor systems, PCT WO 87/02381, PCT US 86/02089

- Matsumane, H., Furui, M., Shibatani, T., Tosa, T. (1994) Production of optically active 3-phenylglycidic acid ester by the lipase from *Serratia marcescens* on a hollow-fiber membrane reactor, J. Ferment. Bioeng. **78**, 59–63

- Matsumane, H., Shibatani, T. (1994) Purification and characterization of the lipase from *Serratia marcescens* Sr41 8000 responsible for asymmetric hydrolysis of 3-phenylglycidic acid esters, J. Ferment. Bioeng. **77**, 152–158

- Tosa, T., Shibatani, T. (1995) Industrial application of immobilized biocatalysts in Japan, Ann. N. Y. Acad. Sci. **750**, 364–375

- Zaks, A., Dodds, D.R. (1997) Application of biocatalysis and biotransformations to the synthesis of pharmaceuticals, Drug Discovery Today, **2**, 513–531

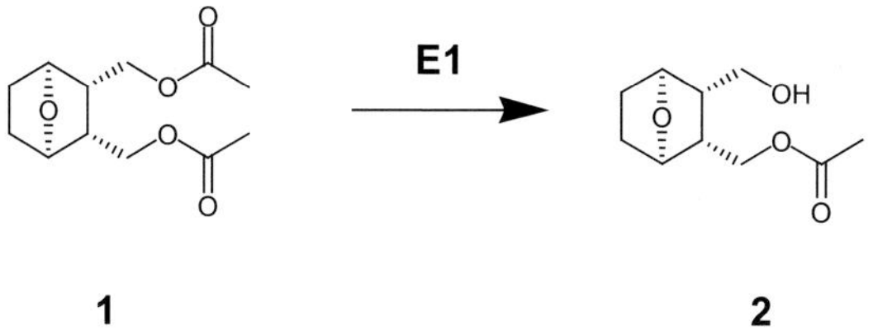

E1

1 → **2**

Bristol-Myers Squibb

1 = (exo,exo)-7-oxabicyclo[2.2.1]heptane-2,3-dimethanol diacetate ester
2 = ((1R,2R,3S,4S)-3-(hydroxymethyl)-7-oxa-bicyclo[2.2.1]heptan-2-yl)methyl acetate

Fig. 3.1.1.3 – 1

1) Reaction conditions

[1]:	0.02 M, 5 g · L^{-1} (242.27 g · mol^{-1}]
pH:	7.0
T:	5 °C
medium:	biphasic system with 10% toluene in 50 mM potassium phosphate buffer
reaction type:	carboxylic ester hydrolysis
catalyst:	lipase PS-30 immobilized on Accurel® polypropylene
enzyme:	triacylglycerol lipase (lipase PS-30)
strain:	*Pseudomonas cepacia*
CAS (enzyme):	[9001-62-1]

2) Remarks

- The immobilized enzyme was reused (5 cycles) without loss of enzyme activity, productivity or ee of product.

3) Flow scheme

Not published.

4) Process parameters

conversion:	90 %
yield:	90 mol %
ee:	> 98 %
reactor volume:	80 L
enzyme supplier:	Amano Enzyme, Inc., Japan
company:	Bristol-Myers Squibb, USA

5) Product application

- The (S)-monoacetate ester was oxidized to its corresponding aldehyde, which was hydrolyzed to give the (S)-lactol. The (S)-lactol was used in the chemo-enzymatic synthesis of thromboxane A$_2$ antagonist. Thromboxane A$_2$ (TxA2) is an exceptionally potent vasoconstrictor substance.

Together with its potent anti-aggregatory and vasodilator activities, TxA2 plays an important role in the maintenance of vascular homeostasis and contributes to the pathogenesis of a variety of vascular disorders.

6) Literature

- Patel, R.N. (2001) Enzymatic synthesis of chiral intermediates for drug development, Adv. Synth. Catal. **343**, (6-7) 527–546

- Patel, R.N., (2001) Enzymatic preparation of chiral pharmaceutical intermediates by lipases, J. Liposome Res. **11**, (4) 355–393

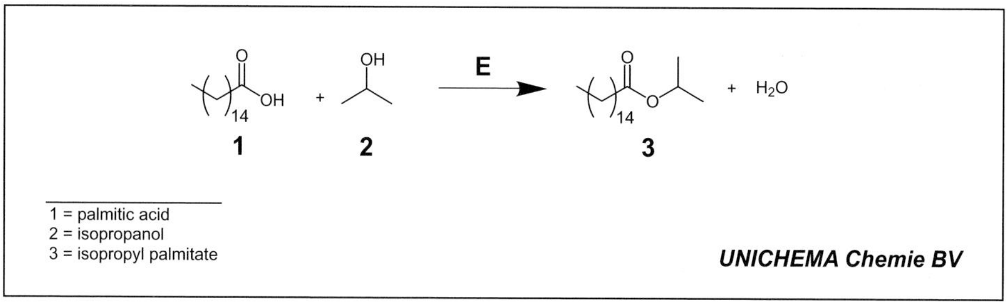

Fig. 3.1.1.3 – 1

1 = palmitic acid
2 = isopropanol
3 = isopropyl palmitate

UNICHEMA Chemie BV

1) Reaction conditions

[1]:	3.1 M, 800 g · L^{-1} [256.43 g · mol^{-1}]
pH:	7.0
T:	60 °C
medium:	2-propanol
reaction type:	carboxylic ester hydrolysis
catalyst:	immobilized enzyme
enzyme:	triacylglycerol acylhydrolase (triacylglycerol lipase, lipase)
strain:	*Candida antarctica*
CAS (enzyme):	[9001–62–1]

2) Remarks

- The problem during ester synthesis is the produced water, which leads to equilibrium conditions meaning forward and backward reaction have the same rates.

- Two possible process layouts are published:

  1) The reaction water is removed by azeotropic distillation (alcohol/water) at 0.26 bar. 2-Propanol is continuously fed to the reactor (58 g · h^{-1}) to replace the distilled one. The catalyst can be easily removed by filtration.

  2) Alternatively the reaction water is removed during esterification by pervaporation at only 80 °C. The reaction solution with the lower water level is cooled to 65 °C and passed to the second reactor unit. After a second pervaporation step the water content is lowered to 0.2 wt.-%.

- The process can be adapted to other alcohols and acids. By the same process isopropyl myristate is produced from myristic acid (H_3C-$(CH_2)_{12}$-COOH).

3) Flow scheme

1) Process with azeotropic distillation:

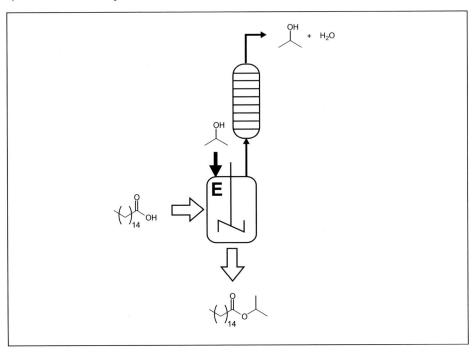

Fig. 3.1.1.3 – 2

2) Process with pervaporation:

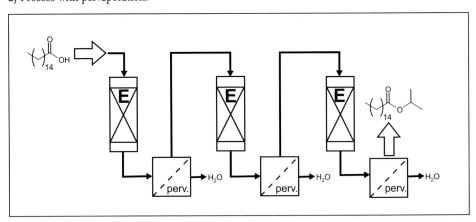

Fig. 3.1.1.3 – 3

4) Process parameters

yield:	99 %
chemical purity:	> 99 %
reactor type:	batch
residence time:	14 h
enzyme supplier:	Novo Nordisk, Denmark
company:	UNICHEMA Chemie BV, The Netherlands

5) Product application

- Isopropyl palmitate and isopropyl myristate are used in the preparation of soaps, skin creams, lubricants and greases.

6) Literature

- Hills, G.A., McCrae, A.R., Poulina, R.R. (1990) Ester preparation, Unichema Chemie BV, EP 0383405 B1

- Kemp, R.A., Macrae, A.R. (1992) Esterification process, Unichema Chemie BV, EP 0506159 A1

- McCrae, A.R., Roehl, E.-L., Brand, H.M. (1990) Bio-Ester – Bio-Esters, Seifen-Öle-Fette-Wachse **116** (6), 201–205

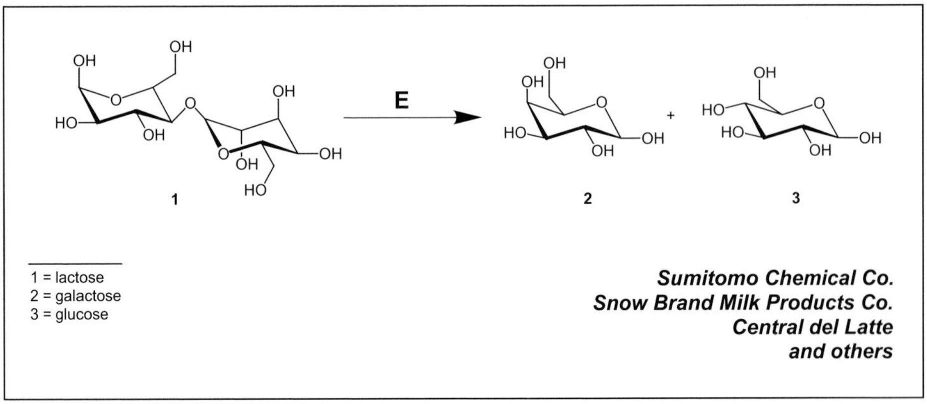

1 = lactose
2 = galactose
3 = glucose

Sumitomo Chemical Co.
Snow Brand Milk Products Co.
Central del Latte
and others

Fig. 3.2.1.23 – 1

1) Reaction conditions

[1]:	~ 5 % in milk [342.30 g · mol^{-1}]
T:	35 °C
medium:	milk
reaction type:	O-glycosyl bond hydrolysis
catalyst:	immobilized enzyme
enzyme:	β-D-galactoside galactohydrolase (lactase, hydrolact, β-galactosidase)
strain:	*Aspergillus oryzae, Saccharomyces lactis*, and others
CAS (enzyme):	[9031–11–2]

2) Remarks

- Milk with hydrolyzed lactose is much sweeter since the sweetening powers of lactose, glucose and galactose are 20, 70, and 58 % respectively.

- Different processes for hydrolyzing lactose in milk have been established by different companies.

- Snamprogretti's enzyme is immobilized in the microcavities of fibers made from cellulose triacetate. The advantage of fibers as support is the enormous surface area. The fibers are stretched between two metallic bars in a column.

- Snow Brand Milk Products, Japan, uses a horizontal rotary column reactor. Here the fibrous immobilized β-galactosidase is placed on a wire mesh cylinder.

3) Flow scheme

Not published.

4) Process parameters

conversion:	70 – 81 %
reactor type:	plug-flow reactor
reactor volume:	typical size: 1,500 to 250,000 L
capacity:	e.g. 8,000 L · d^{-1} (Central del Latte, Italy)
enzyme supplier:	Snamprogretti, Italy, and others
company:	Sumitomo Chemical Co., Japan; Snow Brand Milk Products Co., Ltd., Japan; Central del Latte, Italy; and others

5) Product application

- Lactose needs to be removed before consumption by babies and people (e.g. Asians and Italians) who are not able to produce or do not have enough β-galactosidase activity. These people are called lactose-intolerant.

6) Literature

- Cheetham, P.S.J. (1994) Case studies in applied biocatalysis – from ideas to products, in: Applied Biocatalysis (Cabral, J.M.S., Best, D., Boross, L., Tramper, H., eds.) pp. 47–108, Harwood Academic Publishers, Chur

- Honda, Y., Kako, M., Abiko, K., Sogo, Y., (1993) Hydrolysis of lactose in milk, in: Industrial Application of Immobilized Biocatalysts (Tanaka, A., Tosa, T., Kobayashi, T., eds.) pp. 109–129, Marcel Dekker Inc., New York

- Marconi, W., Morisi, F., (1979) Industrial application of fiber-entrapped enzymes, in: Appl. Biochem. Bioeng. 2, Enzyme Technology, (Wingard, L.B., Katchalski-Katzir, E., Goldstein, L., eds.) pp. 219–258, Academic Press, New York

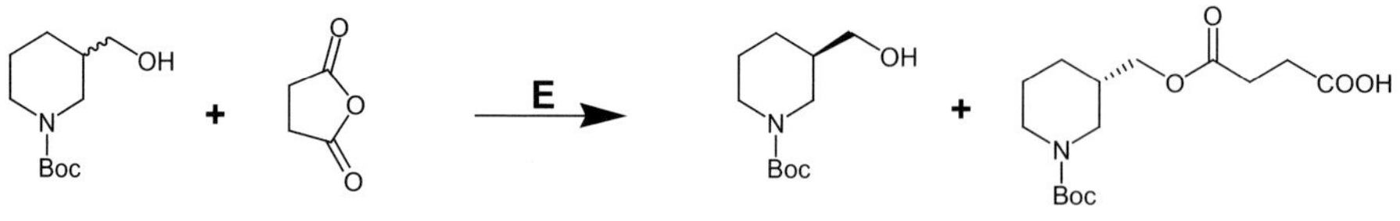

Lipase
Pseudomonas cepacia

1 = (R,S)-N-(*tert*-butoxycarbonyl)-3-hydroxymethylpiperidine
2 = succinic anhydride
3 = (R)-N-(*tert*-butoxycarbonyl)-3-hydroxymethylpiperidine
4 = (S)-hemisuccinate

Bristol-Myers Squibb

Fig. 3.1.1.3 – 1

1) Reaction conditions

[1]:	0.46 M, 99 g · L^{-1} [215.29 g · mol^{-1}]
[2]:	0.39 M, 39 g · L^{-1} [100.07 g · mol^{-1}]
T:	20 °C
medium:	biphasic: water–toluene
reaction type:	hydrolysis
catalyst:	immobilized enzyme (Amano lipase PS)
enzyme:	triacylglycerol lipase
strain:	*Pseudomonas cepacia*
CAS (enzyme):	[9001-62-1]

2) Remarks

- The lipase of *Pseudomonas cepacia* from various commercial sources, especially the lipase PS from Amano, was found to be the best enzyme for the stereospecific hydrolysis of the (R,S) esters.

- Enzymatic resolution of the (R,S) esters by lipase P from *Pseudomonas fluorescens* was reported for the large-scale production (Roche) as a process with low substrate concentration and chromatographic separation. Such processes will have limited practical use in large-scale industrial application.

- The process for enzymatic resolution followed by easy separation involved lipase PS/PP enzyme-catalyzed esterification of (R,S)-alcohol with succinic anhydride followed by separation of the unreacted R-alcohol from the S-hemisuccinate by extraction with base (5 % NaHCO$_3$). Subsequent hydrolysis of the S-hemisuccinate with NaOH provided the desired S-alcohol.

- By using toluene as solvent, the S-alcohol was isolated in 23 % yield (maximum theoretical yield is 50 %) and having ee >95 % The E-value was found to be 65–70.

- The reaction rate was also increased by increasing the amount of succinic anhydride.

- The reaction can be controlled by choice of solvent, amount of succinic anhydride and amount of enzyme.

- The enzyme-catalyzed succinylation process provided *S*-alcohol with 95 % ee at about 30 % conversion. To improve the ee and also the yield, repeated esterification of the enriched alcohol was pursued.

- The objective was to carry out the first esterification of the *(R,S)*-alcohol to a higher conversion and even a slightly lower ee, isolate the product ester, hydrolyze to the *S*-alcohol and then re-esterify the enriched *S*-alcohol a second time to obtain finally the *S*-alcohol with high (>98 %) ee and in high yield.

- Also, this process involves repetition, the unit operations are simple and do not involve any chromatographic separations.

3) Flow scheme

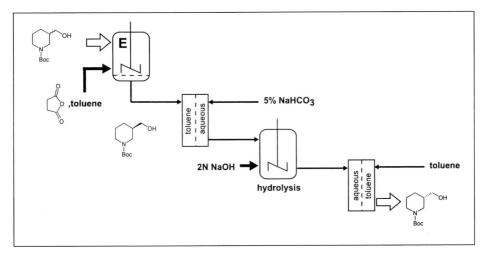

Fig. 3.1.1.3 – 2

4) Process parameters

yield:	overall 32 % (theoretical 50 %)
ee:	98.9 %
reactor type:	batch
reactor volume:	~3 L
down stream processing:	filtration, extraction, hydrolysis, extraction, evaporation, crystallization
enzyme supplier:	Amano Enzyme, Co., Japan
company:	Bristol-Myers Squibb

5) Product application

- (*S*)-*N*-(*tert*-butoxycarbonyl)-3-hydroxymethylpiperidine is a key intermediate in the synthesis of a potent tryptase inhibitor.

6) Literature

• Goswami, A., Howell, J.M., Hua, E.Y., Mirfakhrae K.D., Soumeillant, M.C., Swaminathan, S., Qian, X., Quiroz, F.A., Vu, T.C., Wang, X., Zheng, B., Kronenthal, D.R., Patel, R.N. (2001) Chemical and enzymatic resolution of (R,S)-N-(tert-butoxycarbonyl)-3-hydroxymethylpiperidine, Org. Proc. Res. Dev. **5**, 415–420

Lactonase

Fusarium oxysporum

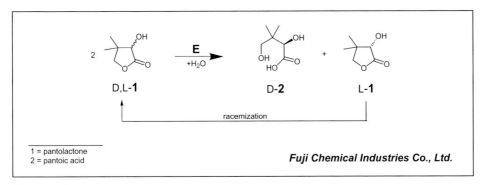

D,L-**1**   D-**2**   L-**1**

1 = pantolactone
2 = pantoic acid

Fuji Chemical Industries Co., Ltd.

Fig. 3.1.1.25 – 1

1) Reaction conditions

[1]:	2.69 M, 350 g · L^{-1} [130.14 g · mol^{-1}]
pH:	6.8 – 7.2
T:	30 °C
medium:	aqueous
reaction type:	carboxylic ester hydrolysis
catalyst:	immobilized whole cells
enzyme:	1,4-lactone hydroxyacylhydrolase (γ-lactonase)
strain:	*Fusarium oxysporum*
CAS (enzyme):	[37278–38–9]

2) Remarks

- The reverse reaction, the lactonization of aldonic acid, is catalyzed under acidic conditions. The reverse reaction does not take place with aromatic substrates.

- The lactonase from *Fusarium oxysporum* has a very broad substrate spectrum:

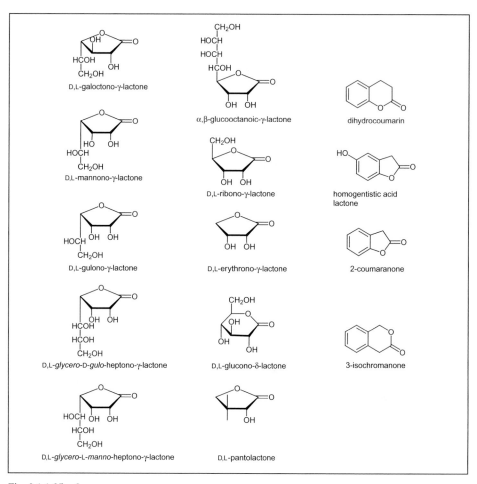

D,L-galoctono-γ-lactone

α,β-glucooctanoic-γ-lactone

dihydrocoumarin

D,L-mannono-γ-lactone

D,L-ribono-γ-lactone

homogentistic acid lactone

D,L-gulono-γ-lactone

D,L-erythrono-γ-lactone

2-coumaranone

D,L-*glycero*-D-*gulo*-heptono-γ-lactone

D,L-glucono-δ-lactone

3-isochromanone

D,L-*glycero*-L-*manno*-heptono-γ-lactone

D,L-pantolactone

Fig. 3.1.1.25 – 2

- For the synthesis whole cells are immobilized in calcium alginate beads and used in a fixed bed reactor.

- The immobilized cells retain more than 90 % of their initial activity even after 180 days of continuous use.

- At the end of the reaction L-pantolactone is extracted and reracemized to D,L-pantolactone that is recycled into the reactor. The D-pantoic acid is chemically lactonized to D-pantolactone and extracted:

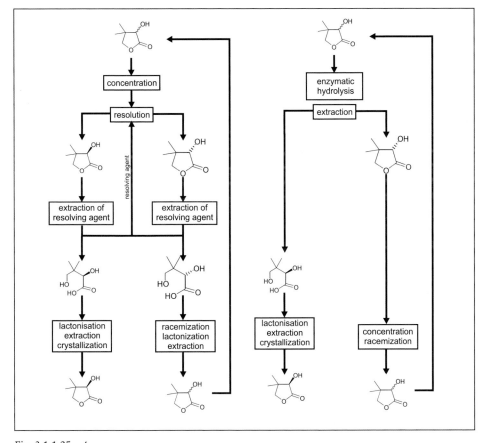

Fig. 3.1.1.25 – 3

- The biotransformation skips several steps that are necessary in the chemical resolution process:

Fig. 3.1.1.25 – 4

- By using the lactonase from *Brevibacterium protophormia* L-lactones can be obtained:

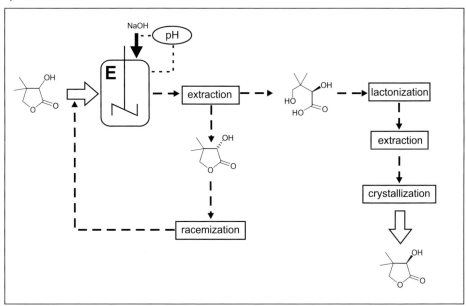

Fig. 3.1.1.25 – 5

3) Flow scheme

Fig. 3.1.1.25 – 6

4) Process parameters

conversion:	90–95 %
ee:	90–97 %
reactor type:	plug-flow reactor
residence time:	21 h
down stream processing:	lactonization, extraction and crystallization
production site:	Fuji Chemical Industries, Co. Ltd., Japan
company:	Fuji Chemical Industries, Co. Ltd., Japan

5) Product application

- The pantoic acid is used as a vitamin B_2-complex.

- D- and L-pantolactones are used as chiral intermediates in chemical synthesis.

6) Literature

- Shimizu, H., Ogawa, J., Kataoka, M., Kobayashi, M. (1997) Screening of novel microbial enzymes for the production of biologically and chemically useful compounds, in: Advances in biochemical engineering biotechnology, Vol. 58: New Enzymes for Organic Synthesis (Scheper, T., ed.) pp. 45–88, Springer, New York

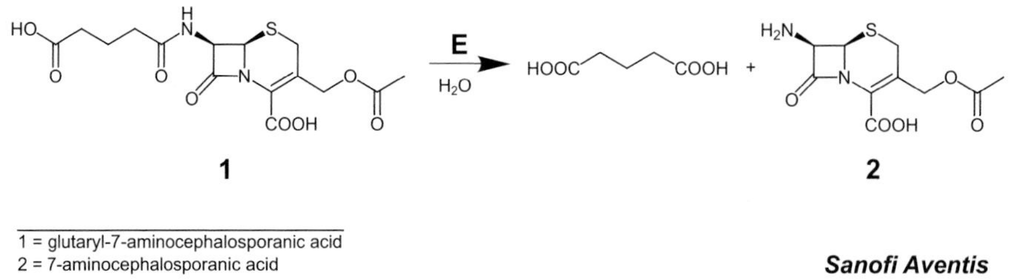

1 = glutaryl-7-aminocephalosporanic acid
2 = 7-aminocephalosporanic acid

Sanofi Aventis

Fig. 3.1.1.41 – 1

1) Reaction conditions

pH:	8.3
T:	30 °C
medium:	aqueous
reaction type:	carboxylic ester hydrolysis
catalyst:	immobilized enzyme
enzyme:	ω-amidodicarboxylate amidohydrolase (ω-amidase, α-keto acid-ω-amidase)
strain:	*Escherichia coli*
CAS (enzyme):	[9025–19–8]

2) Remarks

- Second step of the 7-aminocephalosporanic acid process; for first step see page 209.

- The glutaryl amidase is immobilized on a spherical carrier and can be reused many times.

- Here an existing chemical process has been replaced by an enzymatic process due to environmental considerations:

Glutaryl amidase
Escherichia coli

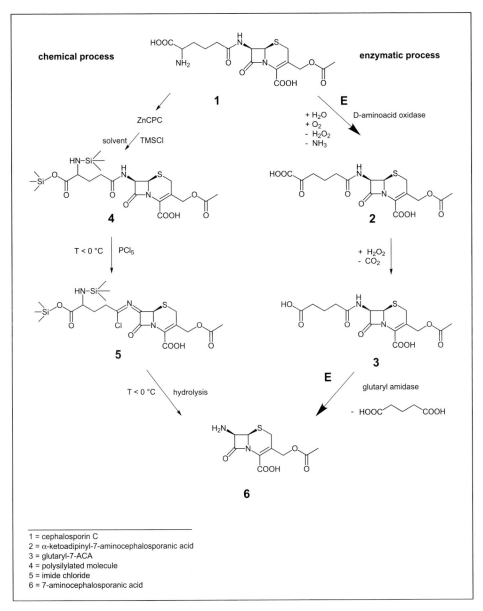

1 = cephalosporin C
2 = α-ketoadipinyl-7-aminocephalosporanic acid
3 = glutaryl-7-ACA
4 = polysilylated molecule
5 = imide chloride
6 = 7-aminocephalosporanic acid

Fig. 3.1.1.41 – 2

- In the first step the zinc salt of cephalosporin C (ZnCPC) is produced, followed by the protection of the functional groups (NH_2 and COOH) with trimethylchlorosilane. The imide chloride is synthesized in the subsequent step at 0 °C with phosphorous pentachloride. Hydrolysis of the imide chloride yields 7-ACA. By replacement of this synthesis with the biotransformation the usage of heavy-metal salts ($ZnCl_2$), chlorinated hydrocarbons and precautions for highly flammable compounds can be circumvented. For every tonne of 7-ACA produced, the waste-gas emission is reduced from 7.5 to 1.0 kg. Mother liquors requiring incineration are reduced from 29 to 0.3 t. Residual zinc that is recovered as $Zn(NH_4)PO_4$ is completely reused, consumption is reduced from 1.8 to 0 t.

- The absolute costs of environmental protection are reduced by 90 % per ton of 7-ACA.

3) Flow scheme

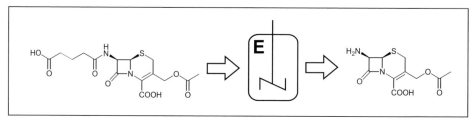

Fig. 3.1.1.41 – 3

4) Process parameters

reactor type:	batch
reactor volume:	10,000 L
capacity:	200 t · a⁻¹
residence time:	1.5 h
down stream processing:	crystallization
enzyme supplier:	Sanofi Aventis, France
start-up date:	1996
production site:	Sanofi Aventis, Germany
company:	Sanofi Aventis, France

5) Product application

- 7-ACA is an intermediate for semi-synthetic cephalosporins.

6) Literature

- Aretz, W. (1998) Hoechst Marion Roussel, personal communication

- Christ, C. (1995) Biochemical production of 7-aminocephalosporanic acid, in: Ullmann's Encyclopedia of Industrial Chemistry, Vol. **B8** (Arpe, H.-J. ed.) pp. 240–241, VCH Verlagsgesellschaft, Weinheim

- Verweij, J., Vroom, E.D. (1993) Industrial transformations of penicillins and cephalosporins, Rec. Trav. Chim. Pays-Bas **112** (2), 66–81

- Matsumoto, K. (1993) Production of 6-APA, 7-ACA, and 7-ADCA by immobilized penicillin and cephalosporin amidases, in: Industrial Application of Immobilized Biocatalysts (Tanaka, A., Tosa, T., Kobayashi, T. eds.) pp. 67–88, Marcel Dekker Inc., New York

Glutaryl amidase
Pseudomonas sp.

Fig. 3.1.1.41 – 1

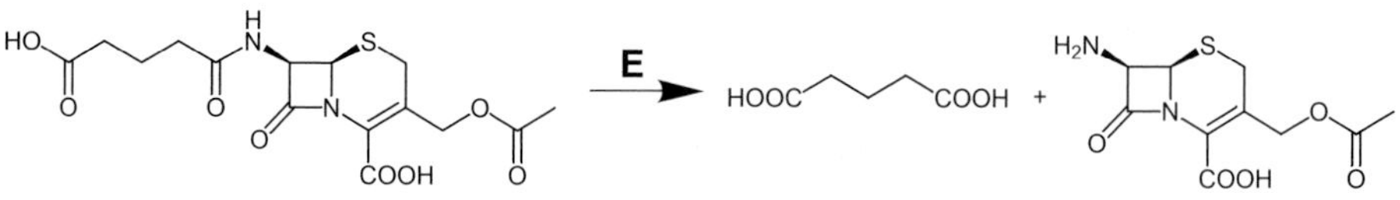

1 = glutaryl-7-ACA
2 = 7-aminocephalosporanic acid

Toyo Jozo Co.
Asahi Kasei Chemical Co. Ltd.

1) Reaction conditions

[1]:	0.026 M, 10 g · L^{-1} [386.38 g · mol^{-1}]
pH:	7.5 – 8.5
T:	30 °C
medium:	aqueous
reaction type:	carboxylic ester hydrolysis
catalyst:	immobilized enzyme
enzyme:	ω-amidodicarboxylate amidohydrolase (ω-amidase, α-keto acid-ω-amidase)
strain:	*Pseudomonas* GK-16
CAS (enzyme):	[9025–19–8]

2) Remarks

- Second step of the 7-aminocephalosporanic acid process; for the first step see page 209.

- The glutaryl amidase is immobilized by adsorption onto a porous styrene anion-exchange resin and subsequent crosslinking with 1 % glutaraldehyde.

- The liberated glutaric acid is an inhibitor of the glutaryl amidase and additionally lowers the pH. Therefore the pH is controlled by an autotitrator.

- The process is started at 15 °C. To compensate enzyme deactivation during the production, it is gradually increased to 25 °C. After 70 cycles the enzyme is replaced.

- The reaction solution circulates at the rate of 10,000 L · h^{-1}.

- Here an existing chemical process has been replaced by an enzymatic process due to environmental reasons.

3) Flow scheme

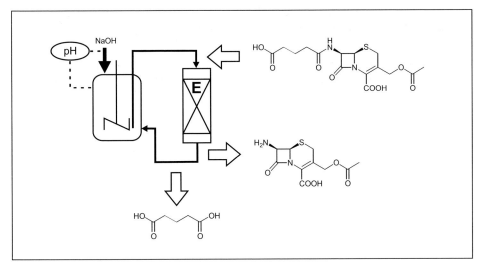

Fig. 3.1.1.41 – 2

4) Process parameters

yield:	95 %
reactor type:	repetitive batch, fixed bed reactor
reactor volume:	1,000 L
capacity:	$90 \, t \cdot a^{-1}$
residence time:	4 h
start-up date:	1973
company:	Asahi Kasei Chemical Industry Co., Ltd., Japan, and Toyo Jozo, Japan

5) Product application

- 7-ACA is an intermediate for semisynthetic cephalosporins.

6) Literature

- Christ, C. (1995) Biochemical production of 7-aminocephalosporanic acid, in: Ullmann's Encyclopedia of Industrial Chemistry, Vol. **B8** (Arpe, H.-J. ed.) pp. 240–241, VCH Verlagsgesellschaft, Weinheim

- Matsumoto, K. (1993) Production of 6-APA, 7-ACA, and 7-ADCA by immobilized penicillin and cephalosporin amidases, in: Industrial Application of Immobilized Biocatalysts (Tanaka, A, Tosa, T., Kobayashi, T. eds.) pp. 67–88, Marcel Dekker Inc., New York

- Tsuzuki, K., Komatsu, K., Ichikawa, S., Shibuya, Y. (1989) Enzymatic synthesis of 7-aminocephalosporanic acid (7-ACA), Nippon Nogei Kagaku Kaishi **63** (12), 1847

- Verweij, J., de Vroom, E. (1993) Industrial transformations of penicillins and cephalosporins, Rec. Trav. Chim. Pays-Bas **112** (2), 66–81

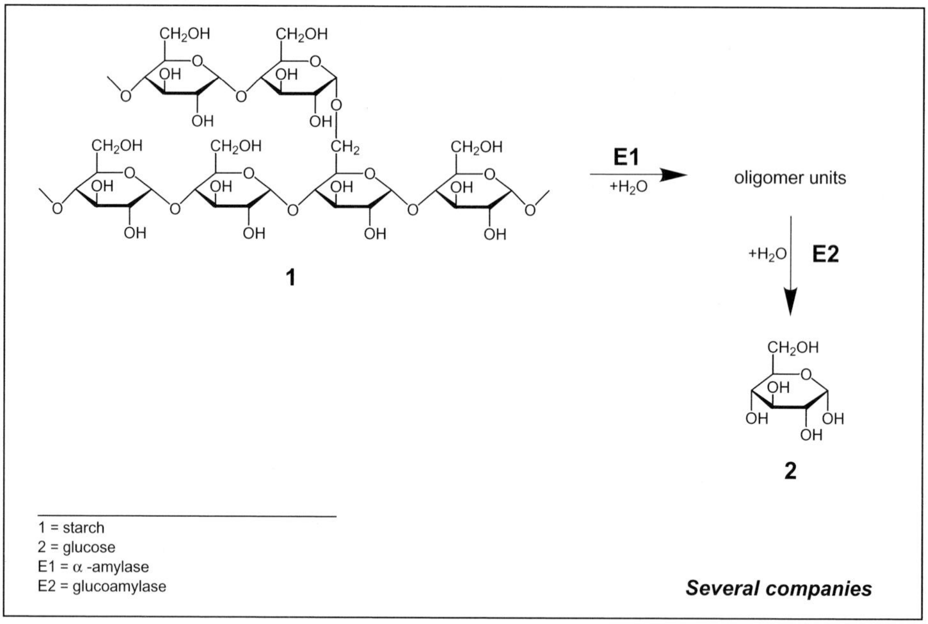

Fig. 3.2.1.1/3.2.1.3 – 1

1) Reaction conditions

pH: liquefying: 6.0 – 6.5 / glucosidation: 4.2
T: 115°C – 95 °C / 60 °C
medium: aqueous
reaction type: O-glucosyl-bond hydrolysis (endo / exo)
catalyst: solubilized enzyme
enzyme: 1,4-α-D-glucan glucanohydrolyse (α-amylase, glycogenase) / glucan 1,4-a-gluco-
                 sidase (glucoamylase, amyloglucosidase)
strain: *Bacillus licheniformis / Aspergillus niger*
CAS (enzyme):   [9000–90–2] / [9032-08-0]

2) Remarks

- The process is a part of the production process for high fructose corn syrup (see page 508).

- After several improvements this process provides an effective way for an important, low-cost sugar substitute derived from grain.

- At various stages enzymes are employed in this process.

- The corn kernels are softened by treatment with sulfur dioxide and lactic acid bacteria to separate oil, fiber and proteins. The enzymatic steps are cascaded to yield the source product for the invertase process after liquefaction in continuous cookers, debranching and filtration.

- Since starches from different natural sources have different compositions the procedure is not unique. The process ends if all starch is completely broken down to limit the amount of oligomers of glucose and dextrins. Additionally, recombination of molecules has to be prevented.

α-Amylase / Amyloglucosidase
Bacillus licheniformis / Aspergillus niger

- The thermostable enzymes can be used up to 115 °C. The enzymes need Ca^{2+} ions for stabilization and activation. Since several substances in corn can complex cations, the cation concentration is increased requiring a further product purification causing the necessity to refine the product.

3) Flow scheme

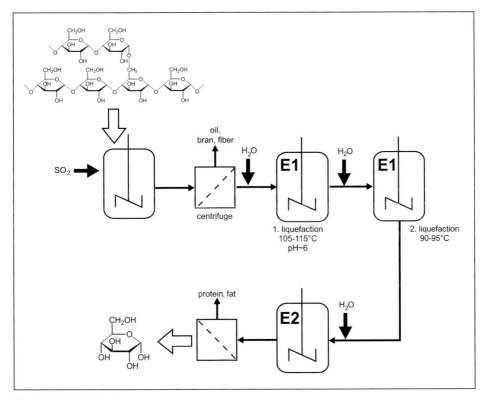

Fig. 3.2.1.1/3.2.1.3 – 2

4) Process parameters

conversion:	> 95 %
yield:	depending on natural source
selectivity:	> 99 %
reactor type:	continuously operated stirred tank reactor
capacity:	> 10,000,000 t · a^{-1} (worldwide)
residence time:	2–3 h / 48–72 h
down stream processing:	filtration
production site:	world wide
company:	several companies

5) Product application

• The product is a feed stock for high fructose syrup (see page 508).

6) Literature

• Gerhartz, W.(1990) Enzymes in Industry: Production and Applications, VCH, Weinheim

• Holm, J., Bjoerck, I., Ostrowska, S., Eliasson, A.-C., Asp, N.-G., Larsson, K., Lundquist, I. (1983) Digestibility of amylose-lipid complexes *in vitro* and *in vivo*, Stärke, **35**, 294–297.

• Holm, J., Bjoerck, I., Eliasson, A.-C. (1985) Digestibility of amylose-lipid complexes *in vitro* and *in vivo*, Prog. Biotechnol., **1**, 89–92

• Kainuma, K.(1998) Applied glycoscience-past, present and future, Foods Food Ingredients J. Jpn., **178**, 4–10

• Labout, J.J.M. (1985) Conversion of liquefied starch into glucose using a novel glucoamylase system, Starch/Stärke, **37**, 157–161

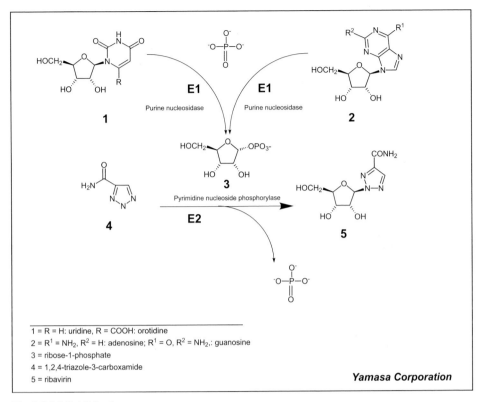

1 = R = H: uridine, R = COOH: orotidine
2 = R^1 = NH$_2$, R^2 = H: adenosine; R^1 = O, R^2 = NH$_2$,: guanosine
3 = ribose-1-phosphate
4 = 1,2,4-triazole-3-carboxamide
5 = ribavirin

Yamasa Corporation

Fig. 3.2.2.1/2.4.2.2 – 1

1) Reaction conditions

T:	60 °C
medium:	aqueous
reaction type:	*N*-glycosyl bond hydrolysis / pentosyl group transfer
catalyst:	solubilized enzyme
enzyme:	purine nucleosidase / purine – nucleoside phosphorylase
strain:	*Erwinia carotovora*
CAS (enzyme):	[9025–44–9] / [9055-35-0]

2) Remarks

- The enzymes were isolated and purified from *Erwinia carotovora*.

- Alternatively biocatalysts from *Brevibacterium acetylicum* and *Bacillus megaterium* can be used.

3) Flow scheme

Not published.

4) Process parameters

company: Yamasa Corporation, Japan

5) Product application

- Ribavirin is an antiviral drug.

6) Literature

- Shirae, H., Yokozeki, K., Uchiyama, M., Kubota, K. (1988) Enzymatic production of ribavirin from purine nucleosides by *Brevibacterium acetylicum* ATCC 954, Agric. Biol. Chem. **52**, 1777–1784

- Shirae, H., Yokozeki, K. (1991) Purification and properties of purine nucleoside phosphorylase from *Brevibacterium acetylicum* ATCC 954, Agric. Biol. Chem. **55**, 493–499

- Shirae, H., Yokozeki, K. (1991) Purifications and properties of orotidine phosphorolyzing enzyme and purine nucleoside phosphorylase from *Erwinia carotovora* AJ 2992, Agric. Biol. Chem. **55** (7), 1849–1857

- Zaks, A., Dodds, D.R. (1997) Application of biocatalysis and biotransformations to the synthesis of pharmaceuticals, Drug Discovery Today **2**, 513–531

Aminopeptidase
Pseudomonas putida

EC 3.4.11.1

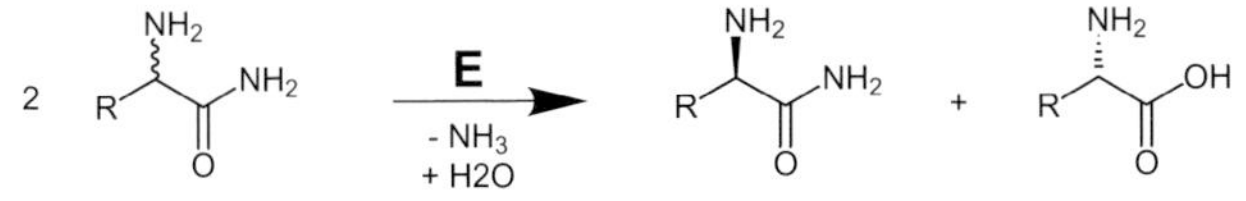

D,L-**1** D-**1** L-**2**

1 = α-amino acid amide
2 = α-amino acid

DSM

Fig. 3.4.11.1 – 1

1) Reaction conditions

[**1**]:	up to 20 g · L^{-1}
pH:	8.0 – 10.0
medium:	aqueous
reaction type:	carboxylic acid amide hydrolysis
catalyst:	suspended whole cells
enzyme:	α-aminoacyl-peptide hydrolase (cytosol aminopeptidase, aminopeptidase, leucyl peptidase)
strain:	*Pseudomonas putida* ATCC 12633
CAS (enzyme):	[9001–61-0]

2) Remarks

- The substrates for this biotransformation can be readily obtained from the appropriate aldehyde via the Strecker synthesis.
- Conversion to the racemic amide in one step is possible under alkaline conditions in the presence of a catalytic amount of ketone with yields above 90 %.
- Subsequent to the biotransformation with soluble permeabilized cells benzaldehyde is added so that the Schiff base of the D-amide is precipitated and can be easily isolated by filtration. An acidification step leads to the D-amino acid.
- The L-amino acid can be recycled by racemization so that a theoretical yield of 100 % is possible.
- The same process can be used for the synthesis of L-amino acids by racemizing the Schiff base of the D-amide in a short time using small amounts of base in organic solvents.
- The following figure shows all the synthetic routes:

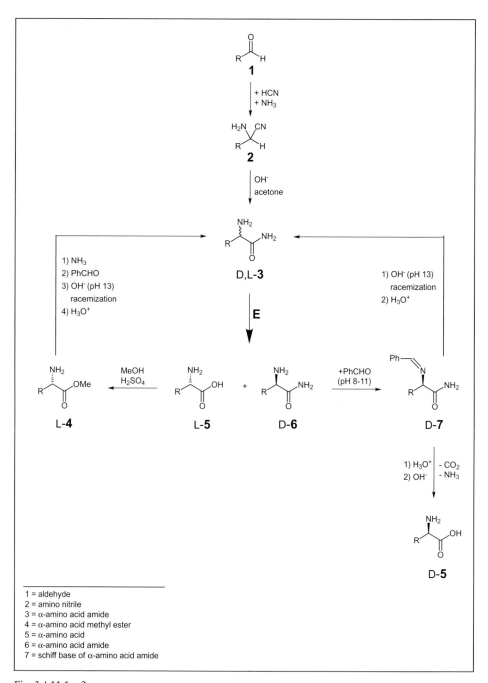

1 = aldehyde
2 = amino nitrile
3 = α-amino acid amide
4 = α-amino acid methyl ester
5 = α-amino acid
6 = α-amino acid amide
7 = schiff base of α-amino acid amide

Fig. 3.4.11.1 – 2

338

Aminopeptidase
Pseudomonas putida

- The whole cell catalyst of *Pseudomonas putida* accepts a wide range of substrates:

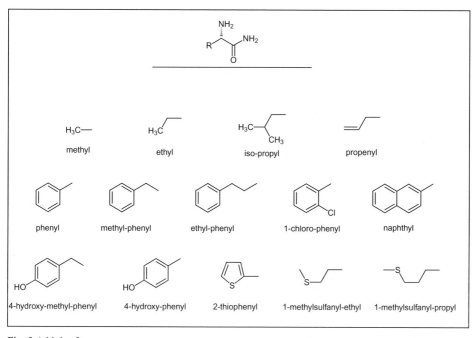

Fig. 3.4.11.1 – 3

- Addition of Mn^{2+} to the purified enzyme (up to 20 mM) in a preincubation step result in a 12-fold increase in activity, whereas Cu^{2+} and Ca^{2+} inhibit the enzyme at a concentration of 1 mM.

- The next table shows the K_M and V_{max} values for different substrates:

substrate	K_m (mM)	V_{max} (U·mg^{-1})
	65	1,565
	15	80
	130	110

Fig. 3.4.11.1 – 4

- Using *in vivo* protein engineering not only mutant strains of *Pseudomonas putida* exhibiting L-amidase and also D-amidase but also amino acid amide racemase activities were obtained. Using these mutants a convenient synthesis of α-H-amino acids with 100 % yield would be possible with a single cell system.

- It is noteworthy that only α-H-substrates can be used. By screening techniques a new bio-catalyst of the strain *Mycobacterium neoaurum* was found, which is capable of converting α-substituted amino acid amides. The next figure shows possible substrates for *Mycobacterium neoaurum*:

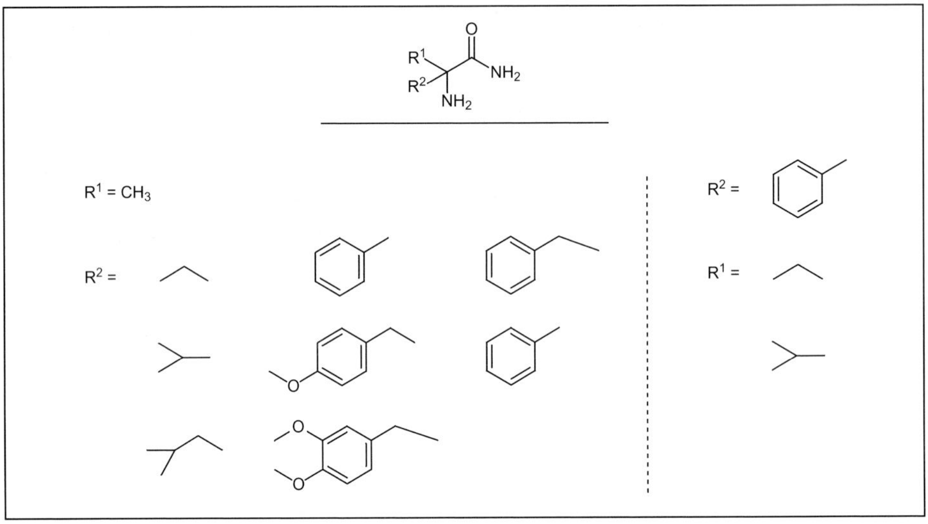

Fig. 3.4.11.1 – 5

3) Flow scheme

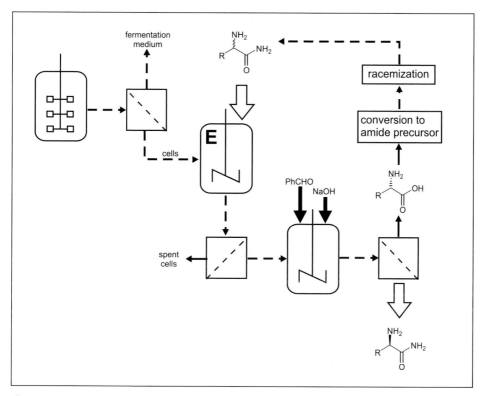

Fig. 3.4.11.1 – 6

4) Process parameters

ee:	> 99 %
reactor type:	batch
capacity:	ton scale
enzyme activity:	338,000 U · $g_{protein}^{-1}$
enzyme supplier:	Novo Nordisk, Denmark
start-up date:	1988
company:	DSM, The Netherlands

5) Product application

- The α-H-amino acids are intermediates in the synthesis of antibiotics, injectables, food and feed additives.

- Examples are the antibiotics ampicillin, amoxicillin (see page 290), the sweetener aspartame (see page 373), several ACE-inhibitors and the pyrethroid insecticide fluvalinate (Fig. 3.4.11.1 – 7).

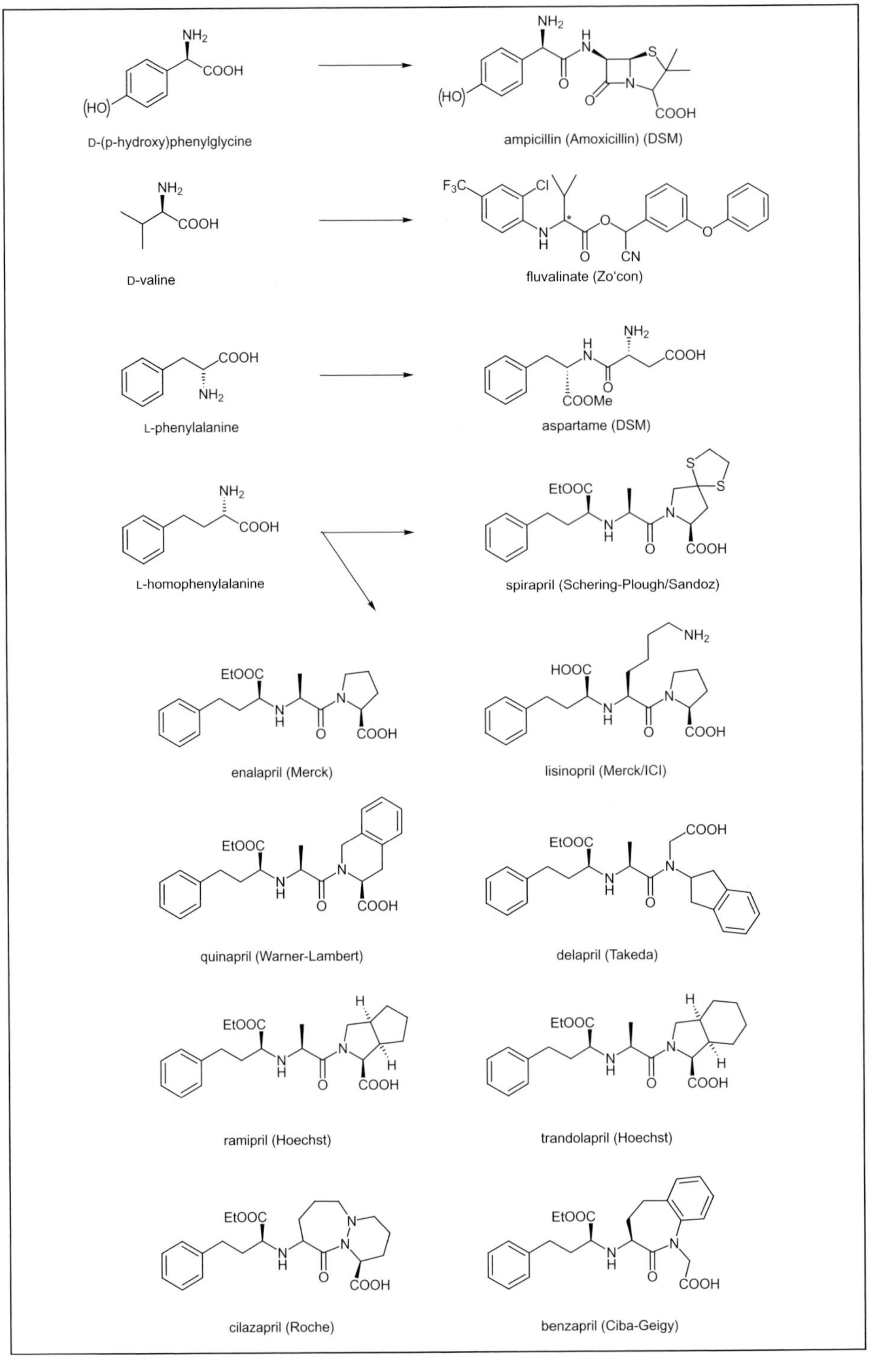

D-(p-hydroxy)phenylglycine → ampicillin (Amoxicillin) (DSM)

D-valine → fluvalinate (Zo'con)

L-phenylalanine → aspartame (DSM)

L-homophenylalanine → spirapril (Schering-Plough/Sandoz)

enalapril (Merck)

lisinopril (Merck/ICI)

quinapril (Warner-Lambert)

delapril (Takeda)

ramipril (Hoechst)

trandolapril (Hoechst)

cilazapril (Roche)

benzapril (Ciba-Geigy)

Fig. 3.4.11.1 – 7

- The antihypertensive drug L-methyl-DOPA (see page 461) and the herbicide arsenal can be synthesized using α-alkyl amino acids:

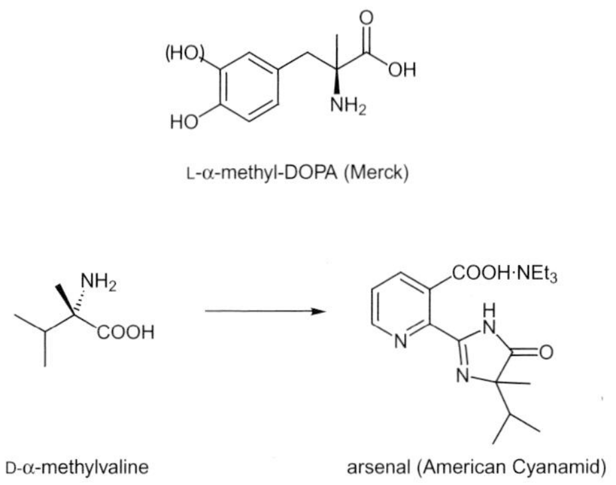

L-α-methyl-DOPA (Merck)

D-α-methylvaline   arsenal (American Cyanamid)

Fig. 3.4.11.1 – 8

6) Literature

- Crosby, J. (1991) Synthesis of optically active compounds: A large scale perspective, Tetrahedron **47**, 4789–4846

- Kamphuis, J., Hermes, H.F.M., van Balken, J.A.M., Schoemaker, H.E., Boesten, W.H.J., Meijer, E.M., (1990) Chemo-enzymatic synthesis of enantiomerically pure alpha-H and alpha-alkyl alpha-amino acids and derivates, in: Amino acids: Chemistry, Biology, Medicine; (Lubec, G., Rosenthal, G.A., eds.) pp. 119–125, ESCOM Science Pupl., Leiden

- Kamphuis, J., Meijer, E.M., Boesten, W.H.J., Sonke, T., van den Tweel, W.J.J., Schoemaker, H.E. (1992) New developments in the synthesis of natural and unnatural amino acids, in: Enzyme Engineering XI, Vol. 672 (Clark, D.S., Estell, D.A., eds.), pp. 510–527, Ann. N. Y. Acad. Sci.

- Meijer, E.M., Boesten, W.H.J., Schoemaker, H.E., van Balken, J.A.M. (1985) Use of biocatalysts in the industrial production of speciality chemicals, in: Biocatalysis in Organic Synthesis, (Tramper, J., Van der Plas, H.C., Linko, P.), eds.) pp. 135–156 Elsevier, Amsterdam, The Netherlands

- Schoemaker, H.E., Boesten, W.H.J., Kaptein, B., Hermes, H.F.M., Sonke, T., Broxterman, Q.B., van den Tweel, W.J.J., Kamphuis, J. (1992) Chemo-enzymatic synthesis of amino acids and derivatives, Pure & Appl. Chem. **64**, 1171–1175

- Sheldon, R.A. (1993) Chirotechnology, Marcel Dekker, New York

- van den Tweel, W.J.J., van Dooren, T.J.G.M., de Jonge, P.H., Kaptein, B., Duchateau, A.L.L., Kamphuis, J. (1993) *Ochrobactrum anthropi* NCIMB 40321: a new biocatalyst with broad-spectrum L-specific amidase activity, Appl. Microbiol. Biotechnol. **39**, 296–300

Carboxypeptidase B
Pig Pancreas

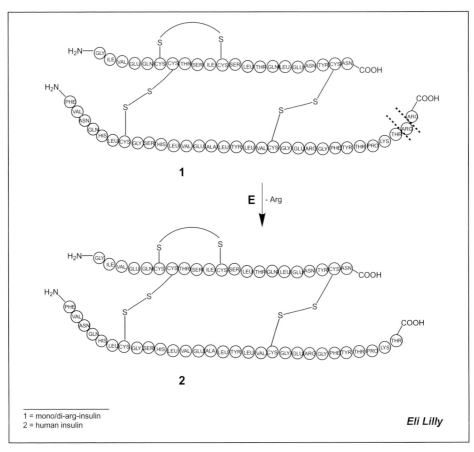

1 = mono/di-arg-insulin
2 = human insulin

Eli Lilly

Fig. 3.4.17.2 – 1

1) Reaction conditions

T:	30–35 °C
medium:	aqueous
reaction type:	carboxylic acid amide hydrolysis
catalyst:	solubilized enzyme
enzyme:	peptidyl-L-arginine hydrolyase (carboxypeptidase)
strain:	pig pancreas
CAS (enzyme):	[9025–24–5]

2) Remarks

- See preparation of educt on page 354.

- For an overview of alternative synthetic path see page 346.

3) Flow scheme

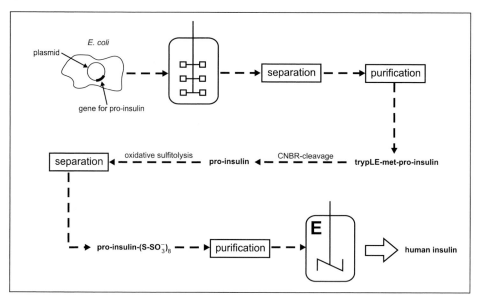

Fig. 3.4.17.2 – 2

4) Process parameters

conversion:	> 99 %
yield:	> 95 %
selectivity:	> 99 %
down stream processing:	chromatography
company:	Eli Lilly, USA

5) Product application

- For details about insulin see process of Hoechst Marion Roussel on page 346.

6) Literature

- Frank, B.H., Chance, R.E. (1983) Two routes for producing human insulin utilizing recombinant DNA technology, Münch. Med. Wschr. **125**, 14–20

- Jørgensen, L.N., Rasmussen, E., Thomsen, B. (1989) HM(ge), Novo's biosynthetic insulin; Med. View. **III**, No. 4, 1–7

- Ladisch, M.R., Kohlmann, K.L. (1992) Recombinant human insulin, Biotechnol. Prog. **8**, 469–478

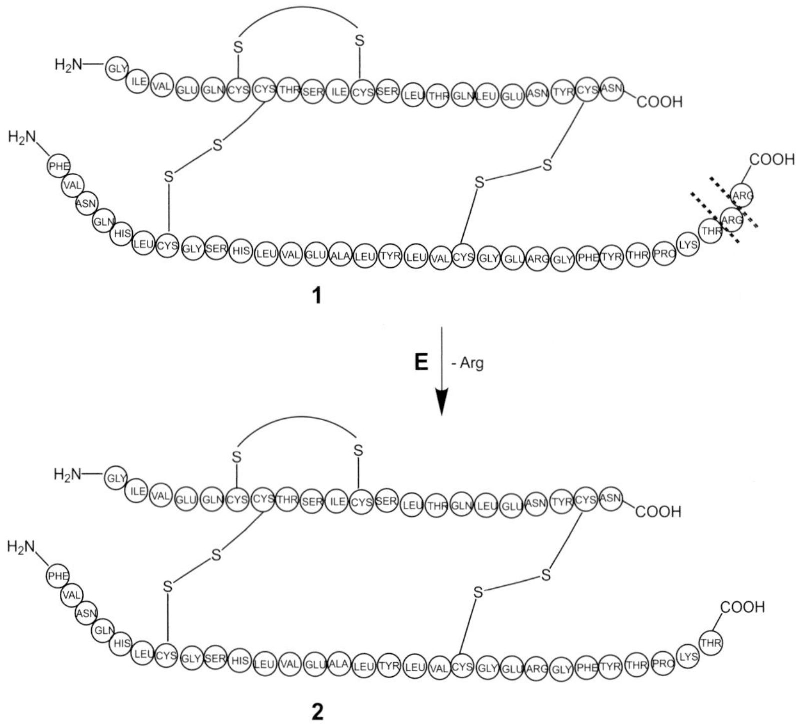

1 = mono/di-arg-insulin
2 = human insulin

Sanofi Aventis

Fig. 3.4.17.2 – 1

1) Reaction conditions

[**1**]:	$0.4\ g \cdot L^{-1}$
[**2**]:	$0.36\ g \cdot L^{-1}$
pH:	8.0
T:	30–35 °C
medium:	aqueous
reaction type:	carboxylic acid amide hydrolysis
catalyst:	solubilized enzyme
enzyme:	peptidyl-L-arginine hydrolyase (carboxypeptidase)
strain:	Pig pancreas
CAS (enzyme):	[9025–24–5]

2) Remarks

- See preparation of educt on page 352.

- Historically, insulin has been purified from animal tissues. Frozen bovine or porcine pancreas are diced and in a multi-step procedure of acidification, neutralization and concentration and after several chromatography steps insulin can be isolated.

- Nowadays there are four main routes to produce insulin:
  1) Extraction from human pancreas,
  2) chemical synthesis from individual amino acids,
  3) conversion of porcine insulin to human insulin and
  4) fermentation of genetically engineered microorganisms.

- The last fermentation procedure can be divided into four main syntheses:

  1) Synthesis of chain A and chain B in separated genetically modified *E. coli* strains. The chains are combined by several chemical steps and the crude insulin is purified (Genentech):

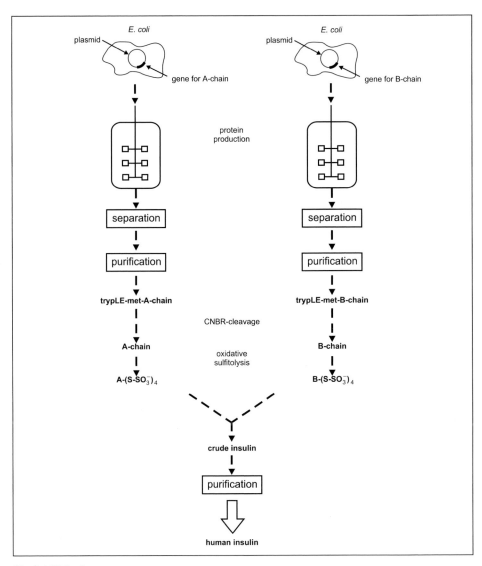

Fig. 3.4.17.2 – 2

2) Production of TrpLE-met-pro-insulin by genetically modified *E. coli* and further synthetic and purification steps (Eli Lilly, see process on page 346.)

3) Production of pre-pro-insulin by genetically modified *E. coli* (HMR, process shown above).

4) Production of a pro-insulin which is converted to insulin by transpeptidation (Novo Nordisk, see page 354).

- Insulin belongs to the first mammal proteins that were synthesized with identical amino acid sequence using recombinant DNA-technology. Today the world capacity is about $2 \text{ t} \cdot \text{a}^{-1}$.

3) Flow scheme

Not published.

4) Process parameters

conversion:	>99.9 %
yield:	> 90 %
selectivity:	> 90 %
reactor type:	batch
reactor volume:	7,500 L
capacity:	$> 0.5 \text{ t} \cdot \text{a}^{-1}$
residence time:	4–6 h
space-time-yield:	$1.7 \text{ g} \cdot \text{L}^{-1} \cdot \text{d}^{-1}$
down stream processing:	chromatography
enzyme activity:	$20 \text{ U} \cdot \text{g}^{-1}_{enzyme}$
enzyme supplier:	Calbiochem, USA
start-up date:	1998
production site:	Sanofi Aventis, Germany
company:	Sanofi Aventis, France

5) Product application

- In 1921 insulin could be isolated from dog pancreas.

- The sequence was identified by Sanger in 1955.

- Insulin regulates the blood sugar level. It is used for the treatment of diabetes mellitus. Up to 5 % of the population of the western world suffers from diabetes.

- A glucose level below 0.25 mM leads to problems in energy supply to the brain. Consequences can include coma, irreversible brain damage and death.

- Since insulin is decomposed by proteases it cannot be administered orally. A new application method is inhalation of modified insulin with faster effects due to exchange of single hydrophobic amino acid with several hydrophilic amino acids. The continuous application after measurement of the blood sugar level is the most comfortable and efficient way.

6) Literature

- Frank, B.H., Chance, R.E. (1983) Two routes for producing human insulin utilizing recombinant DNA technology, Münch. Med. Wschr. **125**, 14–20

- Jørgensen, L.N., Rasmussen, E., Thomsen, B. (1989) HM(ge), Novo's biosynthetic insulin; Med. View. **III**, No. 4, 1–7

- Ladisch, M.R., Kohlmann, K.L. (1992) Recombinant human insulin, Biotechnol. Prog. **8**, 469–478

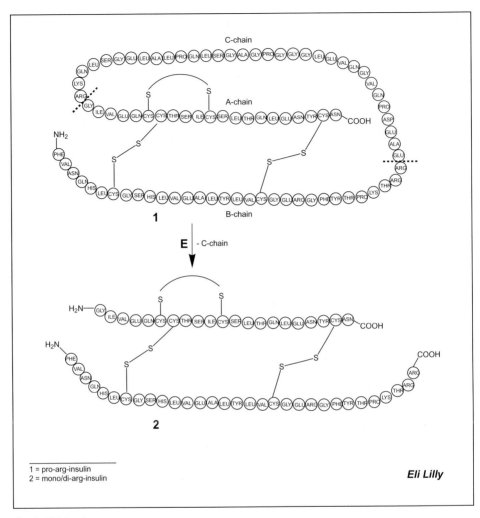

Fig. 3.4.21.4 – 1

1) Reaction conditions

pH:	< 7.0
T:	6 °C
medium:	aqueous
reaction type:	carboxylic acid amide hydrolysis
catalyst:	solubilized enzyme
enzyme:	tryptase (trypsin)
strain:	pig pancreas
CAS (enzyme):	[9002-07–7]

2) Remarks

- The precursor pro-insulin is directly produced by fermentation of recombinant *E. coli* (see second step of process on page 344).

- See page 346 for an overview of possible syntheses of insulin.

3) Flow scheme

- See page 243.

4) Process parameters

yield:	> 70 %
reactor type:	batch
company:	Eli Lilly, USA

5) Product application

- See process of Hoechst Martion Roussel on page 346.

6) Literature

- Frank, B.H., Chance, R.E. (1983) Two routes for producing human insulin utilizing recombinant DNA technology, Münch. Med. Wschr. **125**, 14–20

- Jørgensen, L.N., Rasmussen, E., Thomsen, B. (1989) HM(ge), Novo's biosynthetic insulin; Med. View. **III**, No. 4, 1–7

- Ladisch, M.R., Kohlmann, K.L. (1992) Recombinant human insulin, Biotechnol. Prog. **8**, 469–478

Trypsin
Pig Pancreas

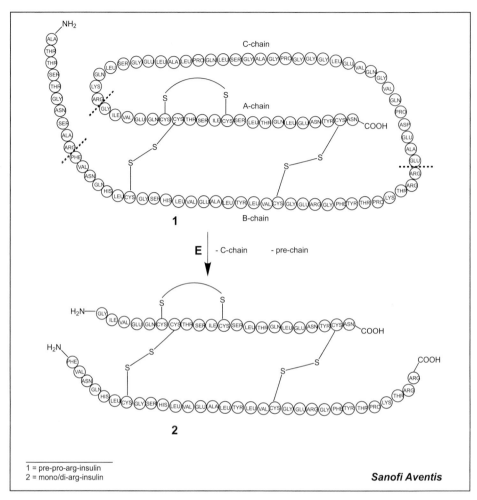

Fig. 3.4.21.4 – 1

1) Reaction conditions

[1]:	$0.8 \text{ g} \cdot \text{L}^{-1}$
pH:	8.3
T:	6 °C
medium:	aqueous
reaction type:	carboxylic acid amide hydrolysis
catalyst:	solubilized enzyme
enzyme:	tryptase (trypsin)
strain:	Pig pancreas
CAS (enzyme):	[9002-07–7]

2) Remarks

- The precursor pre-pro insulin (PPI) is directly produced by fermentation of *E. coli* using the recombinant DNA-technology.

3) Flow scheme

Not published.

4) Process parameters

conversion:	> 99.9 %
yield:	> 65 %
selectivity:	> 65 %
reactor type:	batch
reactor volume:	10,000 L
capacity:	$> 0.5\ t \cdot a^{-1}$
residence time:	6 h
space-time-yield:	$2.1\ g \cdot L^{-1} \cdot d^{-1}$
enzyme activity:	$80\ U \cdot L_{reaction\ solution}^{-1}$
enzyme supplier:	Calbiochem, USA
start-up date:	1998
production site:	Sanofi Aventis, Germany
company:	Sanofi Aventis, France

5) Product application

- See next step of process on page 346.

6) Literature

- Frank, B.H., Chance, R.E. (1983) Two routes for producing human insulin utilizing recombinant DNA technology, Münch. Med. Wschr. **125**, 14–20

- Jørgensen, L.N., Rasmussen, E., Thomsen, B. (1989) HM(ge), Novo's biosynthetic insulin; Med. View. **III**, No. 4, 1–7

- Ladisch, M.R., Kohlmann, K.L. (1992) Recombinant human insulin, Biotechnol. Prog. **8**, 469–478

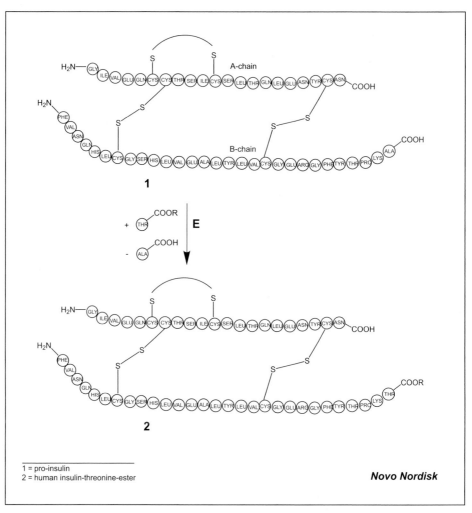

1 = pro-insulin
2 = human insulin-threonine-ester

Novo Nordisk

Fig. 3.4.21.4 – 1

1) Reaction conditions

pH:	< 7.0
T:	6 °C
medium:	acetic acid with low water content
reaction type:	carboxylic acid amide hydrolysis
catalyst:	solubilized enzyme
enzyme:	tryptase (trypsin)
strain:	pig pancreas
CAS (enzyme):	[9002-07-7]

2) Remarks

- The precursor pro-insulin is directly produced by fermentation of *Saccharomyces cerevisiae* using recombinant DNA-technology. The fermenters with a volume of 80 m³ run continuously for 3–4 weeks with addition substrate at the same speed as the broth is drawn.

- The difference between this process and the synthetic routes of Hoechst Marion Roussel and Eli Lilly is that a threonine ester is synthesized by a transpeptidation that can be easily hydrolyzed to insulin.

- To prevent the possible cleavage beside trypsin at position 22 of the B-chain, the following conditions are applied: low water concentration by adding organic solvents, surplus of threonine ester, low temperature and low pH.

- See page 346 for an overview of possible synthesis of insulin.

3) Flow scheme

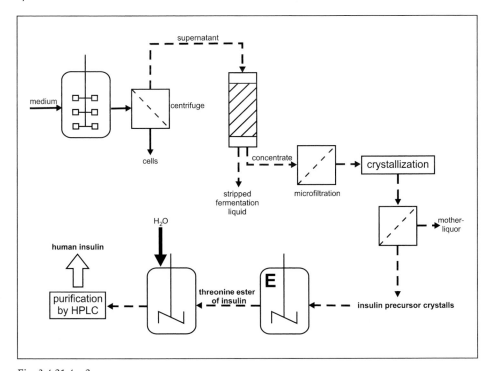

Fig. 3.4.21.4 – 2

4) Process parameters

conversion:	> 99.9 %
yield:	> 97 %
selectivity:	> 97 %
reactor type:	batch
down stream processing:	chromatography
company:	Novo Nordisk, Denmark

5) Product application

- The threonine ester of human insulin can be converted into human insulin by simple hydrolysis with subsequent purification steps (see also flow scheme):

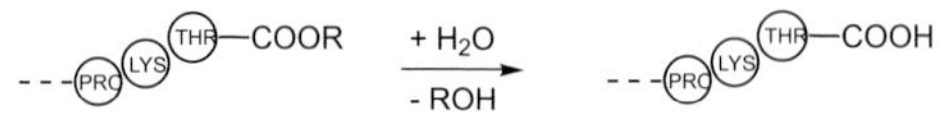

1 = human insulin-threonine-ester
2 = human insulin

Fig. 3.4.21.4 – 3

- For details about insulin see the process by Hoechst Marion Russel on page 346.

6) Literature

- Frank, B.H., Chance, R.E. (1983) Two routes for producing human insulin utilizing recombinant DNA technology, Münch. Med. Wschr. **125**, 14–20

- Jørgensen, L.N., Rasmussen, E., Thomsen, B. (1989) HM(ge), Novo's biosynthetic insulin; Med. View. **III**, No. 4, 1–7

- Ladisch, M.R., Kohlmann, K.L. (1992) Recombinant human insulin, Biotechnol. Prog. **8**, 469–478

- Markussen, J. (1981) Process for preparing insulin esters, Novo Industri, GB 2069502

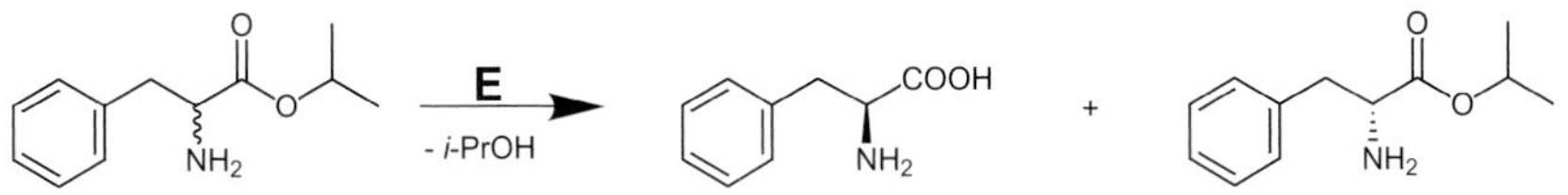

1 = phenylalanine-isopropylester
2 = phenylalanine

Coca-Cola

Fig. 3.4.21.62 – 1

1) Reaction conditions

[1]:	1.2 M, 248.76 g $\cdot$ L^{-1} [207.3 g $\cdot$ mol^{-1}]
[pH]:	7.5
T:	25 °C
medium:	aqueous two-phase
reaction type:	peptide bond hydrolysis
catalyst:	solubilized enzyme
enzyme:	subtilisin Carlsberg (hydrolase)
strain:	*Bacillus licheniformis*
CAS (enzyme):	[9014-01–1]

2) Remarks

- The EC number was formerly 3.4.4.16 and 3.4.21.14.

- The availability of (S)-phenylalanine at low cost is critical to the manufacture of the sweet dipeptide (S)-aspartyl-(S)-phenylalanine methyl ester (aspartame).

- The non-converted enantiomer is continuously extracted via a supported liquid membrane (=SLM) that is immobilized in a microporous membrane. The liquid membrane consists of 33 % N,N-diethyldodecanamide in dodecane.

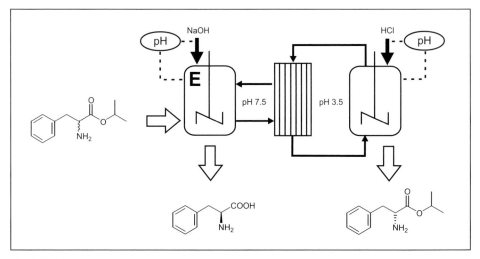

Fig. 3.4.21.62 – 2

- The uniqueness of this process is the continuous extraction with a second aqueous phase of pH 3.5. The hydrolysis product, the amino acid, is charged at pH 7.5 and is not extracted into the acidic aqueous phase.

- The non-converted (R)-amino ester can be racemized by refluxing in anhydrous toluene in the presence of an immobilized salicylaldehyde catalyst.

3) **Flow scheme**

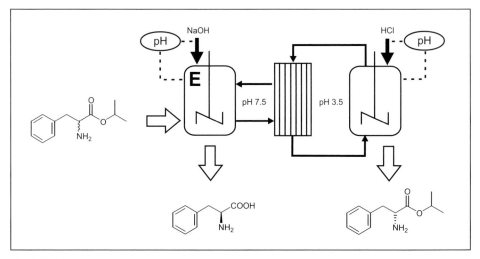

Fig. 3.4.21.62 – 3

4) Process parameters

yield:	73 % (= 36,6 % with reference to racemic mixture)
ee:	95 %
reactor type:	cstr with continuous extraction of non-converted enantiomer
residence time:	0.2 h
space-time-yield:	$14 \text{ g} \cdot \text{L}^{-1} \cdot \text{d}^{-1}$
enzyme supplier:	Sigma Aldrich, USA
company:	Coca-Cola, USA

5) Product application

- (S)-Phenylalanine is used as an intermediate for the synthesis of aspartame.

6) Literature

- Mirviss, S.B. (1987) Racemization of amino acids, US 4713470

- Ricks, E.E., Estrada-Valdes, M.C., McLean, T.L., Iacobucci, G.A. (1992) Highly enantioselective hydrolysis of (R,S)-phenylalanine isopropyl ester by subtilisin Carlsberg. Continuous synthesis of (S)-phenylalanine in a hollow fibre/liquid membrane reactor, Biotechnol. Prog. **8**, 197–203

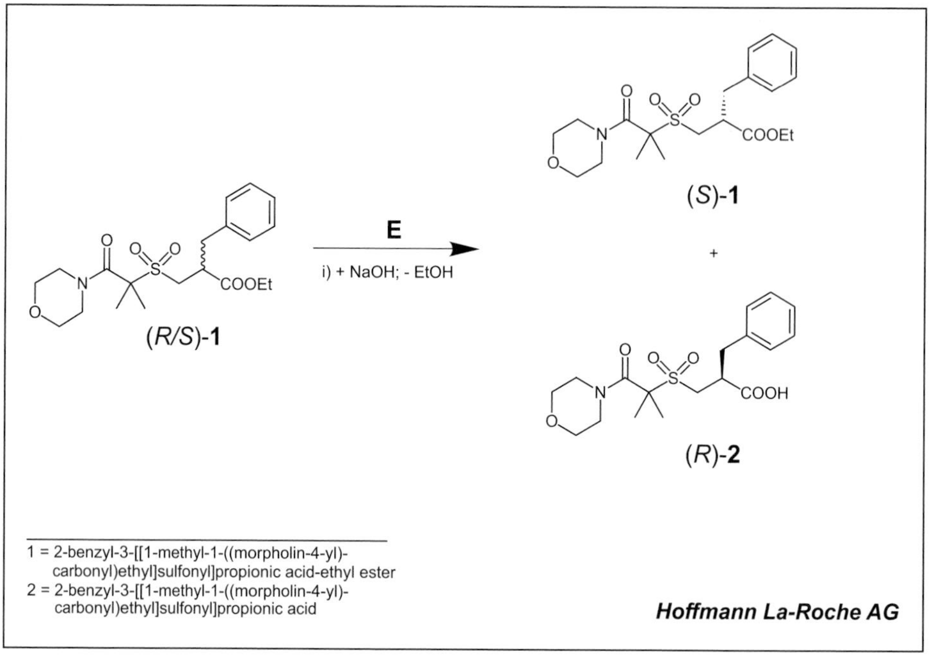

Fig. 3.4.21.62 – 1

1) Reaction conditions

[**1**]:	0.198 M, 80.68 g · L⁻¹ [407.45 g · mol⁻¹]
pH:	7.5
T:	40 °C
medium:	two-phase: aqueous/organic
reaction type:	ester bond hydrolysis
catalyst:	solubilized enzyme
enzyme:	subtilisin Carlsberg (hydrolase)
strain:	*Bacillus licheniformis*
CAS (enzyme):	[9014-01–1]

2) Remarks

- The EC number was formerly 3.4.4.16 and 3.4.21.14.

- On a 17 kg scale (R,S)-**1** was reacted as an 8 % aqueous suspension using solid Optimase® M 440 (6.5 % with respect to (R,S)-**1**, pH 7.5, 40 °C, 69 h) to yield 43 % of (S)-**2**.

- Alternatively, (R,S)-**1** is reacted as a 10 % solution in an aqueous buffer / *tert*-butyl methyl ether mixture using Protease® L 660 (10 % with respect to (R,S)-**1**, pH 7.5, 40°C) to give 46 % of (S)-**2**.

- Optimase® M 440 is a granular subtilisin preparation used in detergent formulations.

- Protease® L 660 is a stabilized, water-miscible, liquid food-grade subtilisin preparation used in detergent formulations.

- The major component of both the preparations is subtilisin Carlsberg (subtilisin A).

- Even under completely non-physiological conditions (substrate to buffer ratio of 1:1), the enzyme shows excellent chemical and enantiomeric purity (> 99 %).

- Optimase® M 440 is cheap; has high tolerance to high substrate concentrations, salt concentrations and organic solvents; is stable at high temperatures; shows high enantioselectivity (even at high temperatures) and has an acceptable reaction rate.

3) Flow scheme

Not published.

4) Process parameters

conversion:	50 %
yield:	43 %
selectivity:	high
ee:	> 99 %
chemical purity:	> 99 %
reactor type:	batch
reactor volume:	200 L
residence time:	69 h
space-time-yield:	$11 \ g \cdot L^{-1} \cdot d^{-1}$
down stream processing:	disk separator, crystallization
enzyme supplier:	Novo Nordisk, Denmark
start-up date:	1992
closing date:	1992
production site:	Switzerland
company:	Hoffmann La-Roche AG, Switzerland

5) Product application

- (*S*)-**2** was prepared as an intermediate for the renin inhibitors ciprokiren and remikiren:

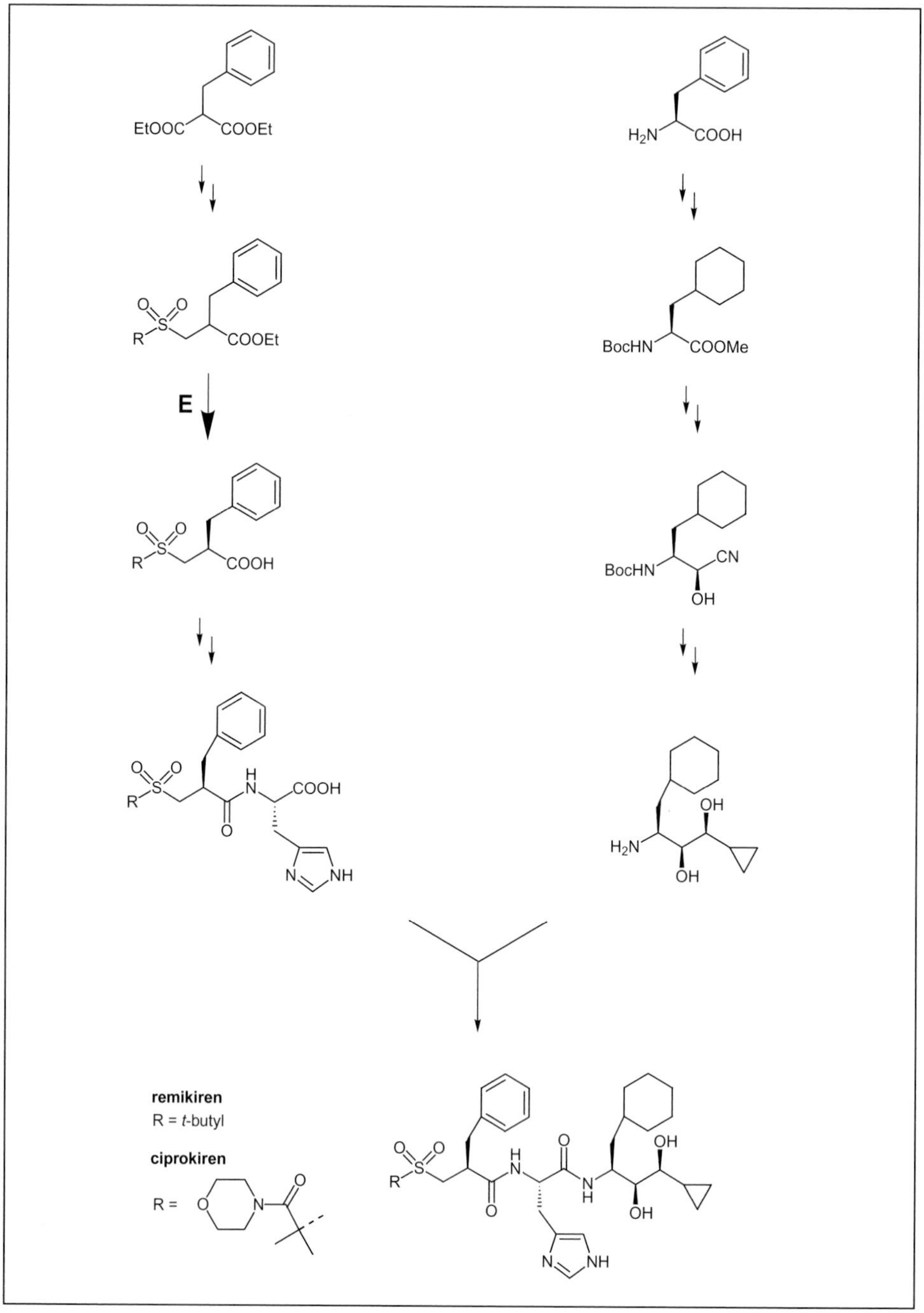

Fig. 3.4.21.62 – 2

6) Literature

- Doswald, S., Estermann, H., Kupfer, E., Stadler, H., Walther, W., Weisbrod, T., Wirz, B., Wostl, W. (1994) Large scale preparation of chiral building blocks for the P_3 site of renin inhibitors, Bioorg. Med. Chem. **2**, 403–410

- Wirz, B., Weisbrod, T., Estermann, H. (1995) Enzymatic reactions in process research – The importance of parameter optimization and workup, Chimica Oggi. **14**, 37–41

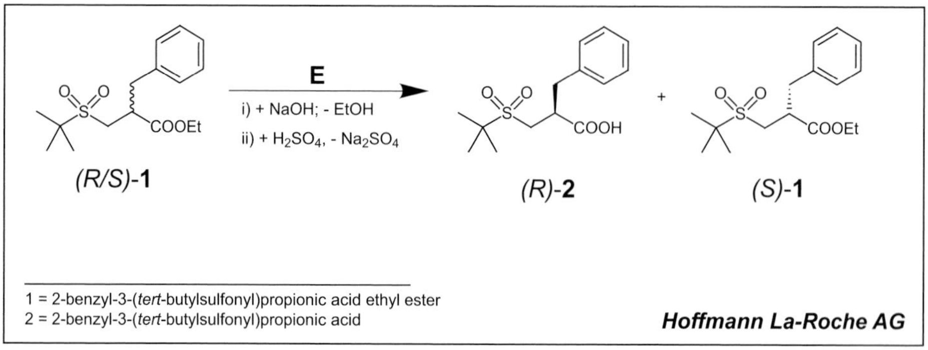
E
i) + NaOH; - EtOH
ii) + H2SO4, - Na2SO4
COOEt
COOH
COOEt
(R/S)-1
(R)-2
(S)-1
1 = 2-benzyl-3-(tert-butylsulfonyl)propionic acid ethyl ester
2 = 2-benzyl-3-(tert-butylsulfonyl)propionic acid
Hoffmann La-Roche AG

Fig. 3.4.21.62 – 1

1) Reaction conditions

[**1**]:	0.733 M, 228.25 g · L^{-1} [311.39 g · mol^{-1}]
pH:	7.5–8.5
T:	39–40 °C
medium:	two-phase: aqueous / organic
reaction type:	ester bond hydrolysis
catalyst:	solubilized enzyme
enzyme:	subtilisin Carlsberg (hydrolase)
strain:	*Bacillus licheniformis*
CAS (enzyme):	[9014-01-1]

2) Remarks

- The EC number was formerly 3.4.4.16 and 3.4.21.14.
- On a 150 kg scale, (*R,S*)-**1** is reacted under non-optimized conditions as a 22 % aqueous emulsion using solid Optimase® M 440 (5 % with respect to (*R,S*)-**1**, pH 7.5, 40 °C, 70 h) to yield 41 % of (*S*)-**2**.
- Under optimized conditions (*R,S*)-**1** is reacted as a 50 % emulsion in a 130 mM borax buffer using liquid Protease® L 660 (10 % with respect to (*R,S*)-**1**, pH 8.5, 40 °C, 42.3 h) to give 42 % of (*S*)-**2** (space-time yield of 550 g · L^{-1} · d^{-1}).
- Optimase® M 440 is a granular subtilisin preparation used in detergent formulations.
- Protease® L 660 is a stabilized, food-grade, water-miscible, liquid subtilisin preparation used in detergent formulations. It has a much higher activity than Optimase® M 440 based on weight.
- Substilin Carlsberg is extremely tolerant to high concentrations of substrate, salt and organic solvent, to elevated temperature and to a combination of all these factors.
- Reduction of the reaction volume is crucial for lowering production costs. This can be achieved either by increasing the initial concentration of substrate (to about 20 %; higher concentrations lead to emulsification problems) or by using more concentrated sodium hydroxide solutions (2 M NaOH; higher concentrations affect hydrolysis rate) for pH control during reaction.

- Due to substrate precipitation at lower temperatures, it is necessary to keep the temperature above the melting point of the substrate (36 – 39 °C). The reaction temperature is maintained at 38 – 43 °C. The reaction rate decreases slightly above 45 °C.

- The poor substrate solubility can also be overcome by dissolving the substrate in a water-miscible cosolvent.

- Immobilization and reuse of the cheap enzyme are not commercially attractive in this case.

- A continuously operated disk separator reduces the number of extraction and filtration steps at neutral and acidic pH.

3) Flow scheme

Not published.

4) Process parameters

conversion:	50 %
yield:	41 – 46 %
selectivity:	high
ee:	> 99 %
chemical purity:	> 99 %
reactor type:	batch
capacity:	> 100 kg
residence time:	70 h
space-time-yield:	$20 \ g \cdot L^{-1} \cdot d^{-1}$
down stream processing:	extraction, filtration, crystallization
enzyme supplier:	Solvay Enzymes, germany
start-up date:	1991
closing date:	1992
company:	Hoffmann La-Roche AG, Switzerland

5) Product application

- (*S*)-**1** was prepared as an intermediate for the renin inhibitor remikiren:

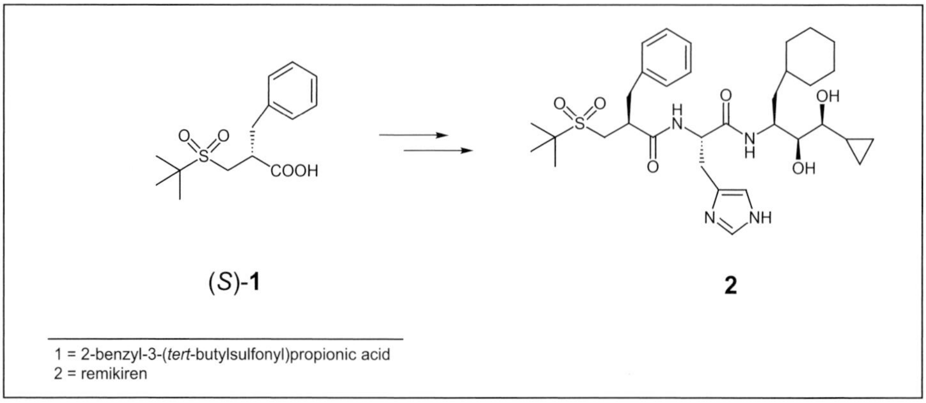

(*S*)-**1**   **2**

1 = 2-benzyl-3-(*tert*-butylsulfonyl)propionic acid
2 = remikiren

Fig. 3.4.21.62 – 2

6) Literature

- Doswald, S., Estermann, H., Kupfer, E., Stadler, H., Walther, W., Weisbrod, T., Wirz, B., Wostl, W. (1994) Large scale preparation of chiral building blocks for the P_3 site of renin inhibitors, Bioorg. Med. Chem. **2**, 403–410

- Wirz, B., Weisbrod, T., Estermann, H. (1995) Enzymatic reactions in process research – The importance of parameter optimization and workup, Chimica Oggi. **14**, 37–41

Subtilisin
Bacillus sp.

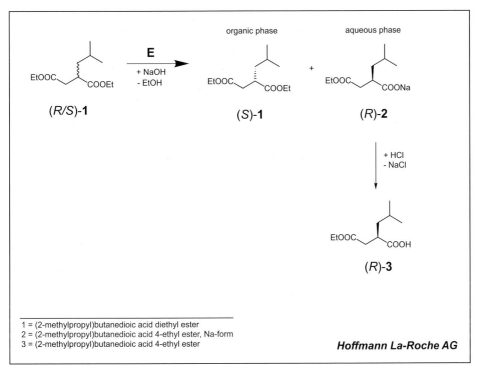

Fig. 3.4.21.62 – 1

1 = (2-methylpropyl)butanedioic acid diethyl ester
2 = (2-methylpropyl)butanedioic acid 4-ethyl ester, Na-form
3 = (2-methylpropyl)butanedioic acid 4-ethyl ester

Hoffmann La-Roche AG

1) Reaction conditions

[**1**]: 0.84 M, 197.73 g · L⁻¹ [235.39 g · mol⁻¹]
pH: 8.5
T: 22–25 °C
medium: two-phase: aqueous / organic
reaction type: ester bond hydrolysis
catalyst: solubilized enzyme
enzyme: subtilisin Carlsberg (hydrolase)
strain: *Bacillus* sp.
CAS (enzyme): [9014-01–1]

2) Remarks

- The EC number was formerly 3.4.4.16 and 3.4.21.14.

- The diester **1** is utilized as a 20 % emulsion in 30 mM aqueous NaHCO₃ using Protease® L 660 or Alcalase® 2.5 L (9 % each, with respect to **1**).

- The unconverted (*S*)-**1** can be extracted in a solvent such as toluene and racemized by heating the anhydrous extract with a catalytic amount of sodium ethoxide. The resulting racemic precursor **1** can be recycled, thus improving the overall yield from 45 % up to 87 %.

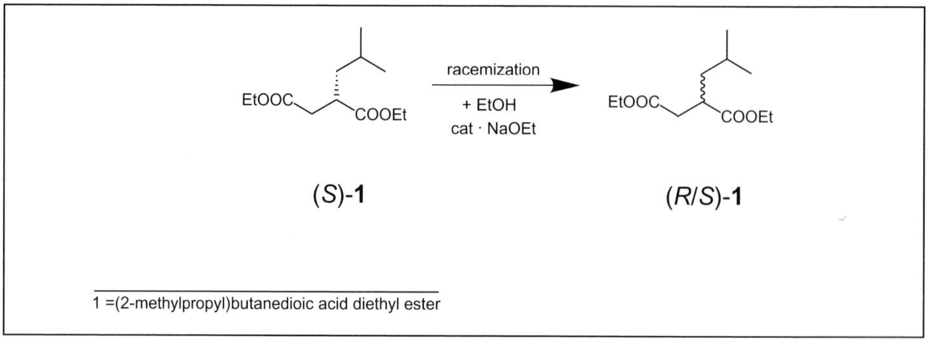

Fig. 3.4.21.62 – 2

- The reaction was repeatedly carried out on a 200 kg scale with respect to **1**.

- The enzyme is highly stereoselective even at high substrate concentrations (20 %).

- The existing chemical research synthesis for bulk amounts was replaced by the chemo-enzymatic route starting from the cheap bulk agents maleic anhydride and isobutylene:

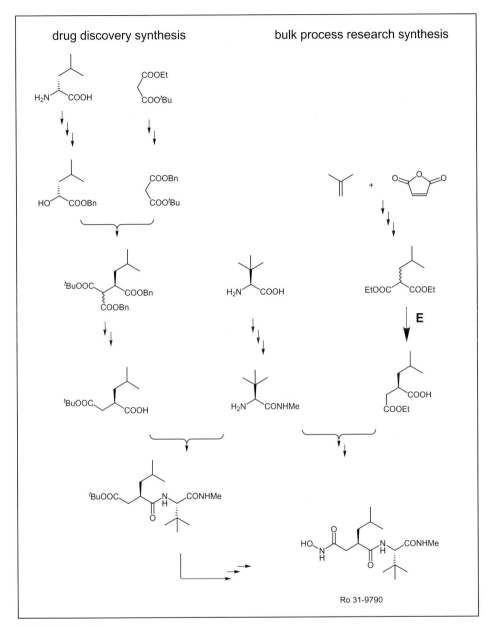

drug discovery synthesis   bulk process research synthesis

Ro 31-9790

Fig. 3.4.21.62 – 3

- The key step of this new synthetic route is the enantio- and regio-selective monohydrolysis of diethyl 2-isobutylsuccinate to generate the (*R*)-monoester.

3) Flow scheme

Not published.

4) Process parameters

conversion:	45 %
ee:	> 99 %
reactor type:	batch
residence time:	48 h
space-time-yield:	$32 \text{ g} \cdot \text{L}^{-1} \cdot \text{d}^{-1}$
enzyme supplier:	Novo Nordisk, Denmark
start-up date:	1994
closing date:	1994
production site:	Switzerland
company:	Hoffmann La-Roche AG, Switzerland

5) Product application

- (R)-**3** is used as a building block for potential collagenase inhibitors (e.g. Ro 31–9790) in the treatment of osteoarthritis.

6) Literature

- Doswald, S., Estermann, H., Kupfer, E., Stadler, H., Walther, W., Weisbrod, T., Wirz, B., Wostl, W. (1994) Large scale preparation of chiral building blocks for the P3 site of renin inhibitors, Bioorg. Med. Chem. **2**, 403–410.

- Wirz, B., Weisbrod, T., Estermann, H. (1995) Enzymic reactions in process research – The importance of parameter optimization and work-up, Chimica Oggi. **14**, pp. 37–41

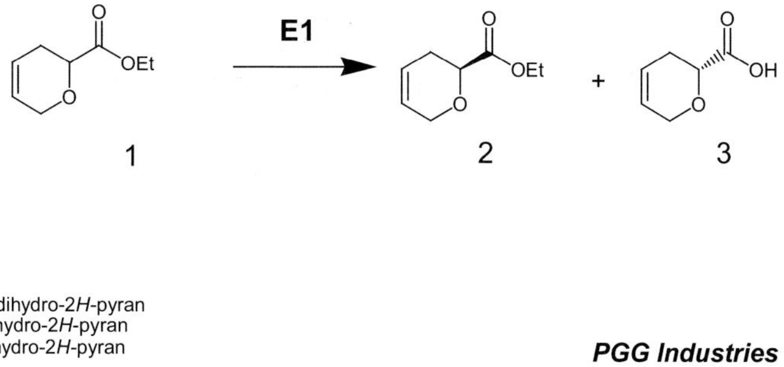

1 = (*R*,*S*)-2-carboethoxy-3,6-dihydro-2*H*-pyran
2 = (*R*)-2-carboethoxy-3,6-dihydro-2*H*-pyran
3 = (*S*)-2-carboethoxy-3,6-dihydro-2*H*-pyran

PGG Industries

Fig. 3.4.21.62 – 1

1) Reaction conditions

[**1**]:   [156.18 g · mol^{-1}]
pH: 7.0–8.0
T: 20–30 °C
medium: aqueous, phosphate buffer
reaction type: hydrolysis
catalyst: solubilized enzyme
enzyme: subtilisin protease ChiroCLEC^{TM} –BL
                 (*Bacillus lentus* subtilisin protease formulation, ca. 5 % solution of the enzyme)
strain: engineered *Bacillus lentus*
CAS (enzyme):   [9014-01-1]

2) Remarks

- The engineered *Bacillus lentus* subtilisin is a serine protease.

- Work-up of the product is simple: after the reaction, toluene is added and product is extracted and recovered.

- Low cost of the enzyme also eliminates the need for its recovery and reuse, thus making this resolution an easy process to scale-up.

3) Flow scheme

Not published.

4) Process parameters

conversion: 42.5–52 %
yield: 35–40 % (recovery)
ee: ≥ 99.5 %
chemical purity: > 97 %
reactor type: batch
reactor volume: 1 L

371

down stream processing: extraction of the product with toluene
enzyme supplier: Altus Biologics Inc., USA
company: PGG Industries, USA

5) Product application

- For the synthesis of a key intermediate used in the synthesis of a protein kinase C inhibitor drug.

- Synthesis is a combination of hetero Diels–Alder and biocatalytic reactions.

6) Literature

- Caille, J.C., Govindan, C.K., Junga, H., Lalonde, J., Yao, Y. (2002) Hetero Diels-Alder-biocatalysis approach for the synthesis of (S)-3-[2-{(methylsulfonyl) oxy}ethoxy]-4-(triphenylmethoxy)-1-butanol methanesulfonate, a key intermediate for the synthesis of the PKC inhibitor LY333531, Org. Proc. Res. Dev. **6** , 471–476

- Kuhn, P., Knapp, M., Soltis, M., Granshaw, G., Thoene, M., Bott, R. (1998) The 0.78 Å structure of a serine protease: *Bacillus lentus* Subtilisin, Biochemistry **37**, 13446–13452

- Graycar, T., Knapp, M., Granshaw, G., Dauberman, J., Bott, R. (1999) Engineered *Bacillus lentus* Subtilisins having altered flexibility, J. Mol. Biol. **292**, 97–109

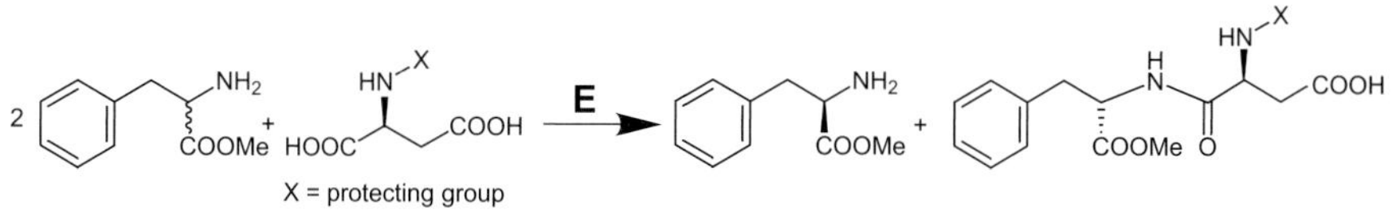

D,L-**1** L-**2** D-**1** L-**3**

1 = phenylalanine methylester
2 = aspartic acid (protected)
3 = α-aspartame (protected)

DSM

Fig. 3.4.24.27 – 1

1) Reaction conditions

pH:	7.0–7.5
T:	50 °C
medium:	aqueous
reaction type:	homogeneous
catalyst:	solubilized enzyme
enzyme:	thermolysin (thermoase)
strain:	*Bacillus proteolicus / thermoproteolyticus*
CAS (enzyme):	[9073–78–3]

2) Remarks

- The main problem in chemical synthesis is the formation of by-product β-aspartame. This isomer is of bitter taste and has to be completely removed from the α-isomer.
- The advantages of the enzymatic route are:
  1) no β-isomer is produced,
  2) the enzyme is completely stereoselective, so that racemic mixtures of the substrate or the appropriate enantiomer of the amino acid can be used,
  3) no racemization occurs during synthesis, and
  4) the reaction takes place in aqueous media under mild conditions.
- The bacterial strain was found in the Rokko Hot Spring in central Japan. Consequently it is very stable up to temperatures of 60 °C.
- The enzyme contains a zinc ion which is responsible for its activity and 4 calcium ions which play an important role in its stability.
- Since the reaction is limited by the equilibrium the products have to be removed from the reaction mixture to achieve high yields.

- If an excess of phenylalanine methylester (which is inert to the reaction) is added, the carboxylic anion of the protected aspartame forms a poorly soluble adduct which precipitates from the reaction mixture. The precipitate can be removed easily by filtration.

- Final steps of the process are the removal of protecting groups and racemization of the formed L-amino acid:

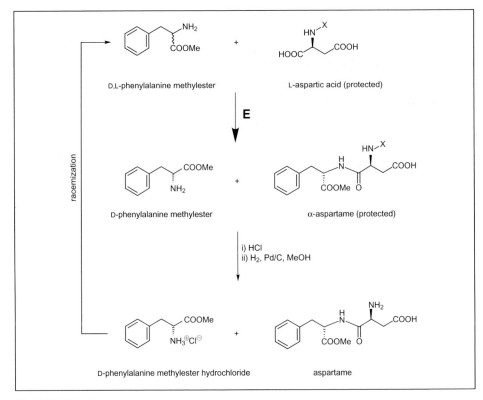

Fig. 3.4.24.27 – 2

Thermolysin
Bacillus thermoproteolyticus

3) Flow scheme

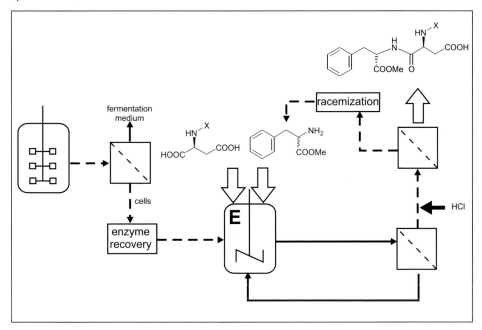

Fig. 3.4.24.27 – 3

4) Process parameters

selectivity:	> 99.9 %
ee:	> 99.9 %
reactor type:	batch
capacity:	~ 10,000 t a^{-1} worldwide; ~ 2,500 t a^{-1} by enzymatic coupling
down stream processing:	filtration
start-up date:	1988
company:	DSM, The Netherlands

5) Product application

- Aspartame is used as a low-calorie sweeteener in food, including soft drinks, table-top sweet-eners, dairy products, instant mixes, dressings, jams, confectionery, toppings and in pharma-ceuticals.

- Aspartame is a sweetener which is 200 times sweeter than sucrose.

- It was first approved in France in 1979.

- The first trademark was Nutrasweet™ (by Monsanto).

375

6) Literature

- Harada, T., Irino, S., Kunisawa, Y., Oyama, K. (1996) Improved enzymatic coupling reaction of N-protected-L-aspartic acid and phenylalanine methyl ester, Holland Sweetener Company, The Netherlands, EP 0768384 A1

- Kirchner, G., Salzbrenner, E., Werenka, C., Boesten, W. (1997) Verfahren zur Herstellung von Z-L-Asparaginsäure-Dinatriumsalz aus Fumarsäure, DSM Chemie Linz GmbH, Austria, Holland Sweetener Company, The Netherlands, EP 0832982 A2

- Oyama, K. (1992) The industrial production of aspartame. In: Chirality in Industry (Collins, A.N., Sheldrake, G.N., Crosby, J., eds.), pp. 237–247, John Wiley & Sons Ltd., New York

- Sheldon, R.A. (1993) Chirotechnology, Marcel Dekker, New York, 235

- Wiseman, A. (1995) Handbook of Enzyme Biotechnology, Ellis Horwood, London

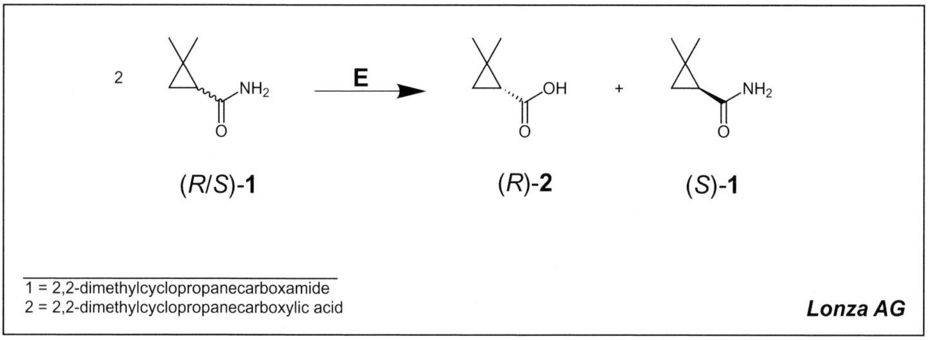

Amidase
Comamonas acidovorans

1 = 2,2-dimethylcyclopropanecarboxamide
2 = 2,2-dimethylcyclopropanecarboxylic acid

Lonza AG

Fig. 3.5.1.4 – 1

1) Reaction conditions

[1]:	0.247 M, 28 g · L^{-1} [113.17 g · mol^{-1}]
pH:	7.0
T:	37 °C
medium:	aqueous (10 mM sodium phosphate buffer)
reaction type:	amide hydrolysis
catalyst:	suspended whole cells
enzyme:	acylamide amidohydrolase (amidase)
strain:	*Comamonas acidovorans* A18 DSM: 6351, expressed in *Escherichia coli* XL1Blue/pCAR6
CAS (enzyme):	[9012–56-0]

2) Remarks

- The reaction rate and selectivity are increased by the addition of 5 % (v/v) ethanol.

- The product is purified by extraction, electrodialysis or drying.

- The isolation of the hydrolase enzyme is also possible. Prior to the application of the isolated enzyme in the biotransformation the hydrolase is immobilized on Eupergit C (1.5 U · g^{-1}$_{\text{wet weight Eupergit C}}$), thereby increasing its stability.

3) Flow scheme

Not published.

4) Process parameters

yield:	44 %
ee:	98.6 %
reactor type:	batch
capacity:	> 1 t · a^{-1}
residence time:	7 h
company:	Lonza AG, Switzerland

5) Product application

- Intermediate for the synthesis of the dehydropeptidase-inhibitor cilastatin. This drug is administered in combination with penem-carbapenem antibiotics to prevent deactivation by renal dehydropeptidase in the kidney.

6) Literature

- Robins, K., Gilligan, T. (1992) Biotechnologisches Verfahren zur Herstellung von (*S*)-(+)-2,2-Dimethylcyclopropancarboxamid und (*R*)-(−)-2,2-Dimethylcyclopropancarbonsäure, Lonza AG, EP 0502525 A1

- Zimmermann, T., Robins, K., Birch, O.M., Böhlen, E. (1992) Gentechnologisches Verfahren zur Herstellung von (*S*)-(+)-2,2-Dimethylcyclopropancarboxamid mittels Mikroorganismen, Lonza AG, EP 0524604 A2

Amidase
Klebsiella terrigena

EC 3.5.1.4

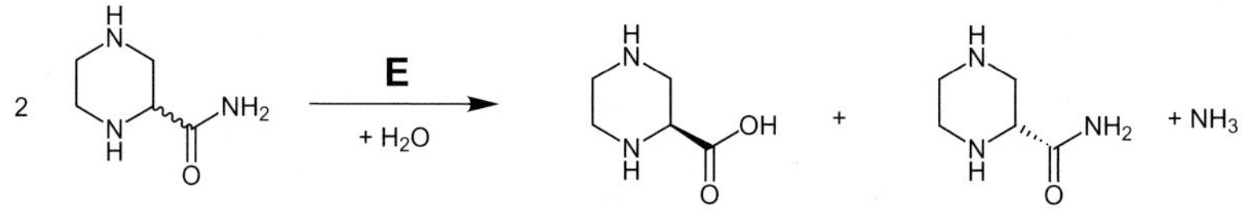

(*R*/*S*)-**1** (*S*)-**2** (*R*)-**1**

1 = piperazine-2-carboxamide
2 = piperazine-2-carboxylic acid

Lonza AG

Fig. 3.5.1.4 – 1

1) Reaction conditions

[**1**]:	0.15 M, 20 g · L^{-1} [129.16 g · mol^{-1}]
pH:	8.0
T:	40 – 47 °C
medium:	aqueous
reaction type:	carboxylic acid amide hydrolysis (kinetic resolution)
catalyst:	suspended whole cells
enzyme:	acylamide amidohydrolase (amidase)
strain:	*Klebsiella terrigena* DSM 9174
CAS (enzyme):	[9012–56-0]

2) Remarks

- Primary screening was done starting from soil samples using growth media that contained racemic carboxamides. The cell-free media of the individual clones were then analysed on TLC plates, only strains with approximately 50 % of the racemic carboxamides were chosen.
- This kinetic resolution is attractive because the starting material can be easily prepared:

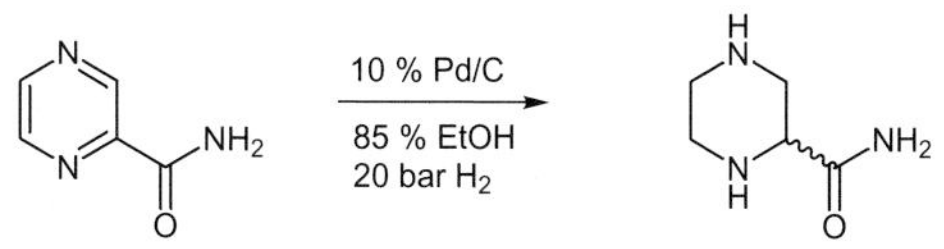

1 (*R*/*S*)-**2**

1 = pyrazine-2-carboxamide
2 = piperazine-2-carboxamide

Fig. 3.5.1.4 – 2

- The microorganism can be grown in fermenters on the racemic carboxamides at the same time as the biotransformations are taking place.

- After completition of conversion the cells are removed by centrifugation and the supernatant is concentrated 10-fold at 60 °C under reduced pressure. The product precipitates by acidifying with conc. HCl (pH 1).

- Using the strain *Burkholderia* DSM 9925 instead of *Klebsiella* DSM 9174 leads to (R)-piperazine-2-carboxylic acid (99 % ee, 22 % yield).

- (S)-Piperazine-2-carboxylic acid has also been prepared by kinetic resolution of racemic 4-(*tert*-butoxycarbonyl)piperazine-2-carboxamide with leucine aminopeptidase and of (R,S)-*n*-octyl-pipecolate with *Aspergillus* lipase. Both processes show practical disadvantages in starting material preparation and biocatalyst availability.

3) Flow scheme

Not published.

4) Process parameters

yield:	41.0 %
ee:	99.4 %
reactor type:	batch
capacity:	multi kg
residence time:	32 h – 76 h
down stream processing:	precipitation
company:	Lonza AG, Switzerland

5) Product application

- The product is used as an intermediate for pharmaceuticals, e.g. the orally active HIV protease inhibitor crixivan from Merck:

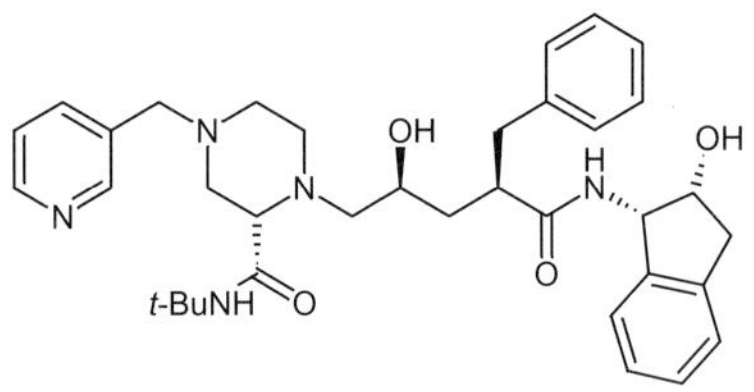

Fig. 3.5.1.4 – 3

- It is also a precursor of numerous bioactive compounds.

6) Literature

- Aebischer, B., Frey, P., Haerter, H.-P., Herrling, P. L., Mueller, W., Olverman, H. J., Watkins, J. C. (1989) Synthesis and NMDA antagonistic properties of the enantiomers of 4-(3-phosphono-propyl)piperazine-2-carboxylic acid (CPP) and of the unsaturated analogue (E)-4-(3-phosphono-prop-2-enyl)piperazine-2-carboxylic acid (CPP-ene), Helv. Chim. Acta **72**, 1043–1051

- Askin, D., Eng, K. K., Rossen, K., Purick, R. M., Wells, K. M., Volante, R. P., Reider, P. J. (1994) Highly diastereoselective reaction of a chiral, non-racemic amide enolate with (S)-glyci-dyl tosylate. Synthesis of the orally active HIV-1 protease inhibitor L-735,524, Tetrahedron Lett. **35**, 673–676

- Bigge, C. F., Johnson, G., Ortwine, D. F., Drummond, J. T., Retz, D. M., Brahce, L. J., Coughenour, L. L., Marcoux, F. W., Probert Jr., A. W. (1992) Exploration of N-phosphonoalkyl-, N-phosphonoalkenyl-, and N-(phosphonoalkyl)phenyl-spaced α-amino acids as competitive N-methyl-D-aspartic acid antagonists, J. Med. Chem. **35**, 1371–1384

- Bruce, M. A., St Laurent, D. R., Poindexter, G. S., Monkovic, I., Huang, S., Balasubramanian, N. (1995) Kinetic resolution of piperazine-2-carboxamide by leucine aminopeptidase. An application in the synthesis of the nucleoside transport blocker (–)Draflazine, Synth. Commun. **25**, 2673–2684

- Eichhorn, E., Roduit, J.-P., Shaw, N., Heinzmann, K., Kiener, A. (1997) Preparation of (S)-piperazine-2-carboxylicacid, (R)-piperazine-2-carboxylic acid, and (S)-piperidine-2-carboxylic acid by kinetic resolution of the corresponding racemic carboxamides with stereoselective amidase in whole bacterial cells, Tetrahedron Asymm. **8**, 2533–2536

- Kiener, A., Roduit, J.-P., Kohr, J., Shaw, N. (1995) Biotechnologisches Verfahren zur Herstel-lung von cyclischen (S)-α-Aminocarbonsäuren und (R)-α-Aminocarbonsäureamiden, Lonza AG, EP 0686698 A2

- Kiener, A., Roduit, J. P., Heinzmann, K. (1996) Biotechnical production process of piperazine (R)-α-carboxylic acids and piperazine (S)-α-carboxylic acid amide, Lonza AG, WO 96/35775

- Petersen, M., Kiener, A.(1999) Biocatalysis: Preparation and functionalization of N-hetero-cycles, Green Chem. **2**, 99–106

- Shiraiwa, T., Shinjo, K., Kurokawa, H. (1991) Asymmetric transformations of proline and 2-piperidinecarboxylic acid via formation of salts with optically active tartaric acid, Bull. Chem. Soc. Jpn. **64**, 3251–3255

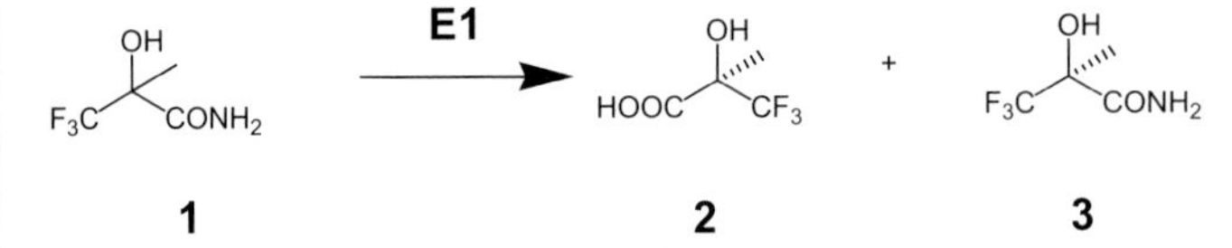

1 = 2-(trifluoromethyl)-2-hydroxypropanamide
2 = (R)-3,3,3-trifluoro-2-hydroxy-2-methylpropanoic acid
3 = (S)-2-(trifluoromethyl)-2-hydroxypropanamide

Lonza AG

Fig. 3.5.1.4 – 1

1) Reaction conditions

[1]:	0.636 M, 100 g · L⁻¹ [157.09 g · mol⁻¹]
pH:	8.0
T:	37 °C
reaction type:	carboxylic acid amide hydrolysis
catalyst:	suspended whole cells
enzyme:	acylamide amidohydrolase (amidase)
strain:	*E. coli* strain containing recombinant *Klebsiella oxytoca* amidase
CAS (enzyme):	[9012 ± 56-0]

Values above with scientific notation: 0.636 M, $100\,\text{g}\cdot\text{L}^{-1}$ $[157.09\,\text{g}\cdot\text{mol}^{-1}]$

2) Remarks

- The amidase was purified and characterized from wild-type strain *Klebsiella oxytoca* PRS1.

- The amidase gene was cloned into *E. coli* to improve safety and the productivity of the biotrans-formation. In addition, this avoids the slime-capsule problem and gives the possibility to use other microorganisms as hosts for the cloned gene.

3) Flow scheme

Not published.

4) Process parameters

yield:	almost the theoretical maximum of 50 %
ee:	100 %
optical purity	> 98 % of (R)-acid
reactor volume:	1500 L
capacity:	(R)-acid: 100 kg; (S)-acid: 100 g
down stream processing:	Microfiltration to remove cells. Ultrafiltration through a 70-kDa membrane to remove proteins. Concentration by thin-film evaporation. Extraction of (S)-amide with ethyl acetate. Extraction of (R)-acid with methyl *tert*-butyl ether. Vacuum distillation of ether. The precipitated product was collected by centrifugation and dried at 30 °C and *p* 50 mbar. The drying conditions must be closely controlled to avoid sublimation of the product.

Amidase
Klebsiella oxytoca

company: Lonza AG, Switzerland

5) Product application

- The (R)- and (S)-acids, respectively, are intermediates for the synthesis of a number of potential pharmaceuticals, which include ATP-sensitive potassium channel openers for the treatment of incontinence and inhibitors of pyruvate dehydrogenase kinase for treatment of diabetes.

6) Literature

- Shaw, N.M., Naughton, A., Robins, K., Tinschert, A., Schmid, E., Hischier, M.-L., Venetz, V., Werlen, J., Zimmermann, T., Brieden, W., de Riedmatten, P., Roduit, J.-P., Zimmermann, B., Neumüller, R. (2002) Selection, purification, characterization, and cloning of a novel heat-stable stereo-specific amidase from *Klebsiella oxytoca*, and its application in the synthesis of enantiomerically pure (R)- and (S)-3,3,3-trifluoro-2-hydroxy-2-methylpropionic acids and (S)-3,3,3-trifluoro-2-hydroxy-2-methylpropionamide, Org. Proc. Res. Dev. **6**, 497–504

Urease

Lactobacillus fermentum

1 = urea
2 = ammonia

Asahi Kasei Chemicals Corporation
Toyo Jozo

Fig. 3.5.1.5. – 1

1) Reaction conditions

[1]:	12–37 ppm
pH:	4.0–4.5 (sake or other alcoholic beverages)
T:	10–15 °C
medium:	aqueous
reaction type:	carboxylic acid amide hydrolysis
catalyst:	immobilized enzyme
enzyme:	urea amidohydrolase (urease, acid urease)
strain:	*Lactobacillus fermentum*
CAS (enzyme):	[9002–13–5]

2) Remarks

- Ethyl carbamate is a natural component of sake or other fermented beverages. It is known to be carcinogenic, teratogenic, and mutagenic.

- It is also formed in alcoholic beverages under heat:

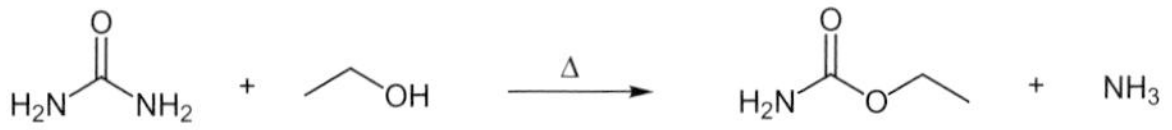

1 = urea
2 = ethanol
3 = ethyl carbamate
4 = ammonia

Fig. 3.5.1.5. – 2

- To prevent formation of ethyl carbamate the urea concentration has to be decreased to 3 ppm.

- The continuous process works in a stable manner over more than 150 days.

- The isolated enzyme is immobilized on a polyacrylonitrile support. For high-speed treatment of urea the urease is immobilized on Chitopearl® (porous chitosan) beads.

Urease
Lactobacillus fermentum

EC 3.5.1.5.

- This method has been and is used for urea removal from sake by many companies in Japan.

- For table wine the limiting level of ethyl carbamate is 30 ppm. The removal of urea from red and white wine cannot be realized if tannin, an urease inhibitor, is present in wine.

3) Flow scheme

Not published.

4) Process parameters

reactor type:	plug-flow reactor
reactor volume:	100 L
enzyme activity:	$140 \ U \cdot g^{-1}_{wet \ carrier}$
enzyme supplier:	Takeda Chemical Ind.
start-up date:	1988
production site:	Asahi Kasei Chemicals, Japan
company:	Asahi Kasei Chemicals Corporation, Japan

5) Literature

- Matsumoto, K. (1993) Removal of urea from alcoholic beverages by immobilized acid urease, in: Industrial Application of Immobilized Biocatalysts (Tanaka, A., Tosa, T., Kobayashi, T., eds.), pp. 255–273, Marcel Dekker Inc., New York

- Yoshizawa, K., Takahashi, K. (1988) Utilization of urease for decomposition of urea in sake, J. Brew. Soc. Jpn. **83**, 142–144

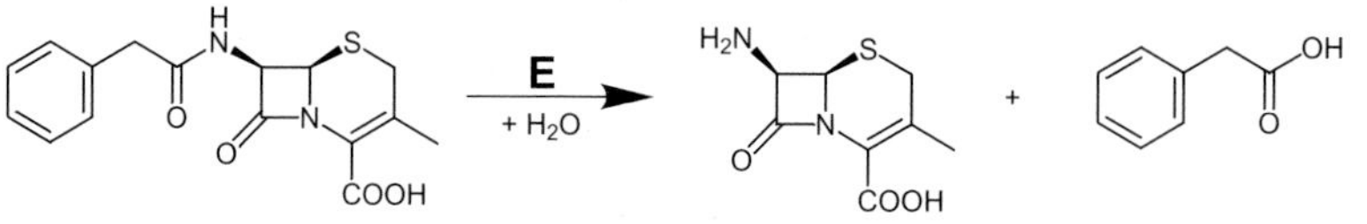

1 = cephalosporin G
2 = 7-amino deacetoxy cephalosporinic acid (7-ADCA)
3 = phenylacetic acid

Dr. Vig Medicaments

Fig. 3.5.1.11 – 1

1) Reaction conditions

[1]:	0.3 M, 100 g · L^{-1} [332.38 g · mol^{-1}]
pH:	8.0
T:	37 °C
medium:	aqueous
reaction type:	carboxylic acid amide hydrolysis
catalyst:	immobilized enzyme
enzyme:	penicillin amidohydrolase (penicillin acylase)
strain:	*E. coli*, optimized
CAS (enzyme):	[9014-06–6]

2) Remarks

- The reaction vessel is equipped with a filter sieve at the bottom to retain the immobilized penicillin acylase.

- By the same process 6-amino penicillanic acid (6-APA) is produced starting from penicillin G. In this case a yield of 87 % is achieved.

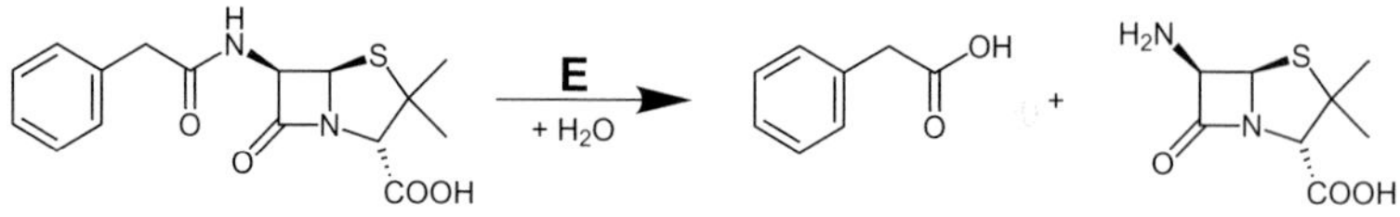

1 = penicillin G
2 = phenylacetic acid
3 = 6-amino penicillianic acid (6-APA)

Fig. 3.5.1.11 – 2

- In the case of 6-APA production the enzyme consumption is reduced by a factor of about 2 to 250 U · kg^{-1} in comparison to 7-ADCA production.

3) Flow scheme

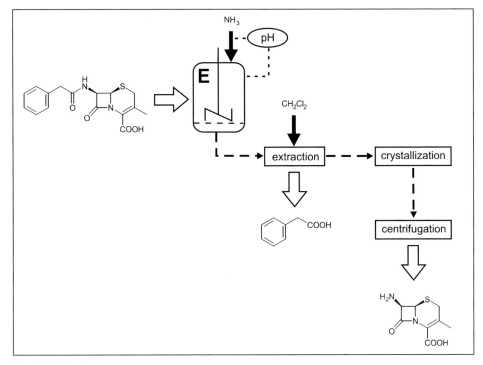

Fig. 3.5.1.11 – 3

4) Process parameters

conversion:	99 %
yield:	93 %
selectivity:	94 %
reactor type:	repetitive batch
reactor volume:	2,000 L
capacity:	$300 \ t \cdot a^{-1}$
residence time:	1.5 – 2 h
space-time-yield:	$30 \ g \cdot L^{-1}$
down stream processing:	crystallization
enzyme activity:	$1,000 \ U \cdot g^{-1}$
enzyme consumption:	$450 \ U \cdot kg^{-1}$
production site:	Dr. Vig Medicaments, India
company:	Dr. Vig Medicaments, India and others

5) Product application

- In the manufacture of semi-synthetic β-lactam antibiotics. The worldwide capacity is more than $20,000 \ t \cdot a^{-1}$.

6) Literature

- Vig, C.B. (1999) personal communication

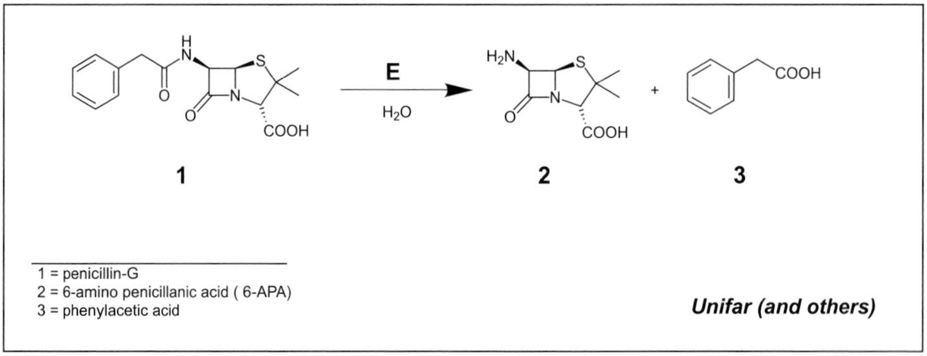

1 = penicillin-G
2 = 6-amino penicillanic acid (6-APA)
3 = phenylacetic acid

Unifar (and others)

Fig. 3.5.1.11 – 1

1) Reaction conditions

[1]:	0.24 M, 80 g · L^{-1} [334.39 g · mol^{-1}]
pH:	8.0
T:	30–35 °C
medium:	aqueous
reaction type:	carboxylic acid amide hydrolysis
catalyst:	immobilized enzyme
enzyme:	penicillin amidohydrolase (penicillin acylase, penicillin amidase)
strain:	*Escherichia coli* and others (e.g. *Bacillus megaterium*)
CAS (enzyme):	[9014-06–6]

2) Remarks

- The enzyme is isolated and immobilized on Eupergit-C (Röhm, Germany).

- The production is carried out in a repetitive batch mode. The immobilized enzyme is retained by a sieve with a mesh size of 400.

- The time for filling and emptying the reactor is approximately 30 min.

- The residual activity of biocatalyst after 800 batch cycles, which is one production campaign, is about 50 % of the initial activity.

- The hydrolysis time after 800 batch cycles increases from the initial 60 min to 120 min.

- Phenylacetic acid is removed by extraction and 6-APA can be crystallized.

- The yield can be increased by concentrating the split-solution and/or the mother liquor of crystallization via vacuum evaporation or reverse osmosis.

- The production operates for 300 days per year with an average production of 12.8 batch cycles per day (production campaigns of 800 cycles per campaign).

- Several chemical steps are replaced by a single enzyme reaction. Organic solvents, the use of low temperature (–40 °C) and the need for absolutely anhydrous conditions, which made the process difficult and expensive, are no longer necessary in the enzymatic process:

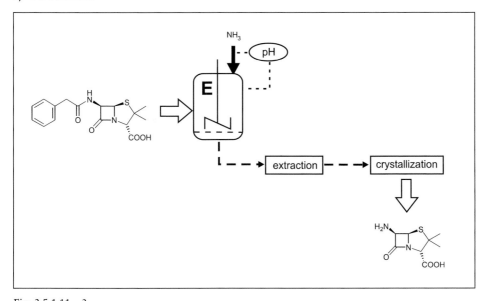

Fig. 3.5.1.11 – 2

3) Flow scheme

Fig. 3.5.1.11 – 3

4) Process parameters

conversion:	98 %
yield:	86 % (in reaction: 97 %)
selectivity:	> 99 %
chemical purity:	99 %
reactor type:	repetitive batch
reactor volume:	3,000 L
capacity:	300 t $\cdot$ a^{-1}
residence time:	1.5 h (average over 800 cycles; initial: 1 h)
space-time-yield:	445 g $\cdot$ L^{-1} $\cdot$ d^{-1} (which is the average for a production campaign of 800 batch cycles)
down stream processing:	extraction, crystallization (see remarks)
enzyme activity:	22 M U, corresponding to approx. 100 kg wet biocatalyst (27.5 kg of dry Eupergit-C)
enzyme consumption:	345 U $\cdot$ kg^{-1} (6-APA)
start-up date:	1973
production site:	Unifar, Turkey (and elsewhere)
company:	Unifar, Turkey (and others)

5) Product application

- 6-APA is used as an intermediate for the manufacture of semi-synthetic penicillins.

6) Literature

- Cheetham, P. (1995) The application of enzymes in industry, in: Handbook of Enzyme Biotechnology (Wiseman, A. ed.), pp. 493–498, Ellis Horwood, London

- Krämer, D., Boller, C. (1998) personal communication.

- Matsumoto, K. (1993) Production of 6-APA, 7-ACA, and 7-ADCA by immobilized penicillin and cephalosporin amidases, in: Industrial Application of Immobilized Biocatalysts (Tanaka, A., Tosa, T., Kobayashi, T. eds.) pp. 67–88, Marcel Dekker Inc., New York

- Tramper, J. (1996) Chemical versus biochemical conversion: when and how to use biocatalysts, Biotechnol. Bioeng. **52**, 290–295

- Verweij, J., de Vroom, E., (1993) Industrial transformations of penicillins and cephalosporins, Rec. Trav. Chim. Pays-Bas, **112**, 66–81

Penicillin amidase
Bacillus megaterium

EC 3.5.1.11

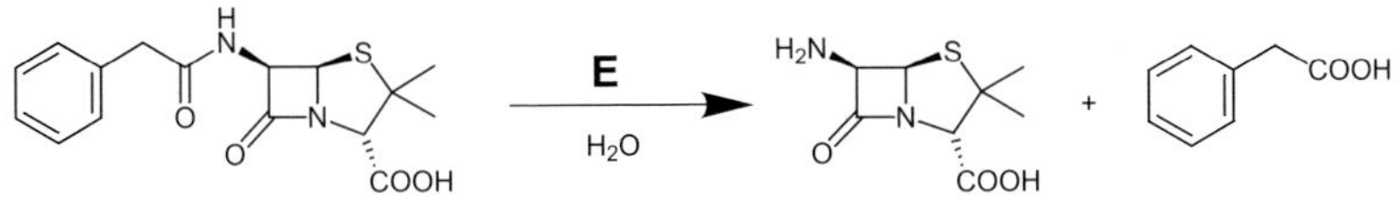

1 = penicillin-G
2 = 6-amino penicillanic acid (6-APA)
3 = phenylacetic acid

Asahi Kasei Chemicals Corporation

Fig. 3.5.1.11 – 1

1) Reaction conditions

[1]:	0.3 M, 100 g · L^{-1} [334.39 g · mol^{-1}]
pH:	8.4
T:	30–36 °C
medium:	aqueous
reaction type:	carboxylic acid amide hydrolysis
catalyst:	immobilized enzyme
enzyme:	penicillin amidohydrolase (penicillin acylase, penicillin amidase)
strain:	*Bacillus megaterium*
CAS (enzyme):	[9014-06-6]

2) Remarks

- The enzyme from *Bacillus megaterium* is an exoenzyme and immobilized using an aminated porous polyacrylonitrile (PAN) fiber as solid support.

- The production is carried out in a recirculation reactor consisting of 18 parallel columns with immobilized enzyme. Each column has a volume of 30 L. The circulation of the reaction solution is established with a flow rate off 6,000 L · h^{-1}. One cycle time takes 3 hours.

- The lifetime of each column is 360 cycles.

- Purification of 6-APA is done by isoelectric precipitation at pH 4.2 with subsequent filtration and washing with methanol.

391

3) Flow scheme

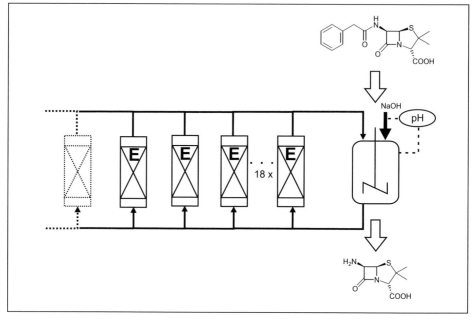

Fig. 3.5.1.11 – 2

4) Process parameters

yield:	86 %
chemical purity:	> 98 %
reactor type:	repetitive batch
reactor volume:	> 500 L
residence time:	3 h
space-time-yield:	800 g · L⁻¹ · d⁻¹
down stream processing:	isoelectric precipitation
enzyme activity:	1,800 U · g⁻¹(dry)
company:	Asahi Kasei Chemicals Corporation, Japan

space-time-yield: $800 \text{ g} \cdot \text{L}^{-1} \cdot \text{d}^{-1}$

enzyme activity: $1,800 \text{ U} \cdot \text{g}^{-1}(\text{dry})$

5) Product application

- 6-APA is used as an intermediate for manufacturing semi-synthetic penicillins.

6) Literature

- Matsumoto, K. (1993) Production of 6-APA, 7-ACA, and 7-ADCA by immobilized penicillin and cephalosporin amidases, in: Industrial Application of Immobilized Biocatalysts (Tanaka, A., Tosa, T., Kobayashi, T. eds.) pp. 67–88, Marcel Dekker Inc., New York

- Tramper, J. (1996) Chemical versus biochemical conversion: when and how to use biocatalysts, Biotechnol. Bioeng. **52**, 290–295

- Verweij, J., de Vroom, E. (1993) Industrial transformations of penicillins and cephalosporins, Rec. Trav. Chim. Pays-Bas, **112**, 66–81

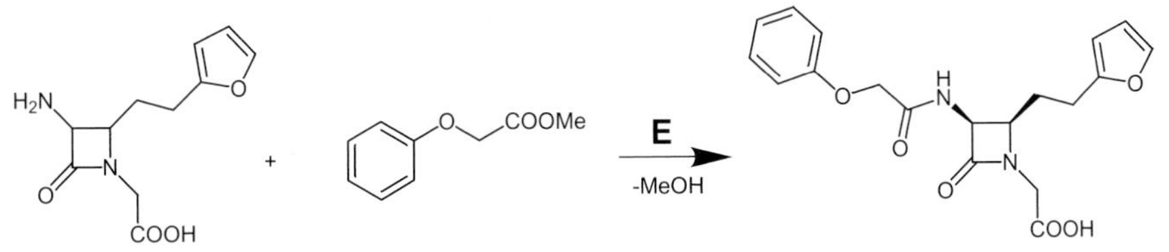

[(2R,3S),(2S,3R)]-**1**   **2**   (2R,3S)-**3**

1 = *cis*-3-amino-azetidinone
2 = phenoxy-acetic acid methyl ester
3 = β-lactam intermediate

Eli Lilly

Fig. 3.5.1.11 – 1

1) Reaction conditions

[1]:	0.042 M, 10 g · L^{-1} [238.24 g · mol^{-1}]
[2]:	0.063 M, 10.5 g · L^{-1} [166.17 g · mol^{-1}]
pH:	6.0
T:	28 °C
medium:	aqueous
reaction type:	carboxylic acid amide hydrolysis
catalyst:	immobilized enzyme
enzyme:	penicillin amidohydrolase (penicillin acylase, gen G amidase)
strain:	*Escherichia coli*
CAS (enzyme):	[9014-06-06]

2) Remarks

- The chemical resolution of the racemic azetidinone gives low yields.

- It was thought that the pen G amidase would exhibit only a limited substrate spectrum, since it does not hydrolyze the phenoxyacetyl side chain of penicillin V. Nevertheless the Lilly process shows that the pen G amidase acylates the 3-amino function with the methyl ester of phenoxyacetic acid.

- The acylation occurs using methyl phenylacetate (MPA) or methyl phenoxyacetate (MPOA) as the acylating agents.

- The enzyme displays similiar enantioselectivity with MPA or MPOA. It is immobilized on Eupergit (Röhm GmbH, Germany).

3) Flow scheme

Not published.

4) Process parameters

yield: 45 %
ee: > 99.9 %
reactor type: batch
down stream processing: filtration
company: Eli Lilly, USA

5) Product application

- The (2R,3S)-azetidinone is a key intermediate in the synthesis of loracarbef, a carbacephalosporin antibiotic:

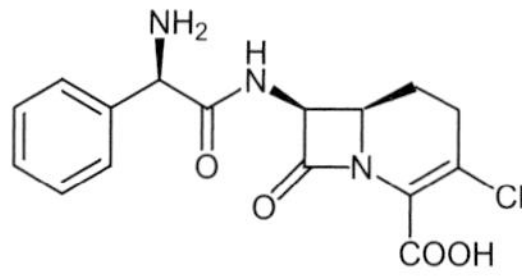

Loracarbef

Fig. 3.5.1.11 – 2

- Loracarbef is a stable analog of the antibiotic cefaclor.

6) Literature

- Zaks, A., Dodds, D.R. (1997) Application of biocatalysis and biotransformations to the synthesis of pharmaceuticals, Drug Discovery Today **2**, 513–530

- Zmijewski, M.J., Briggs, B.S., Thompson, A.R., Wright, I.G. (1991) Enantioselective acylation of a beta-lactam intermediate in the synthesis of Loracarbef using penicillin G amidase, Tetrahedron Lett. **32**, 1621–1622

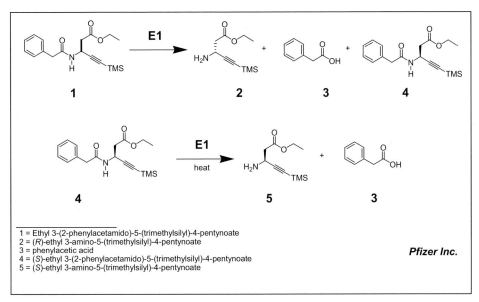

1 = Ethyl 3-(2-phenylacetamido)-5-(trimethylsilyl)-4-pentynoate
2 = (*R*)-ethyl 3-amino-5-(trimethylsilyl)-4-pentynoate
3 = phenylacetic acid
4 = (*S*)-ethyl 3-(2-phenylacetamido)-5-(trimethylsilyl)-4-pentynoate
5 = (*S*)-ethyl 3-amino-5-(trimethylsilyl)-4-pentynoate

Pfizer Inc.

Fig. 3.5.1.11 – 1

1) Reaction conditions

[1]:	0.47 M, 100 g · L^{-1} ethyl 3-amino-5-(trimethylsilyl)-4-pentynoate
[2]:	0.37 M, 50 g · L^{-1} phenylacetic acid
pH:	5.4–5.6
T:	25 °C
medium:	water
reaction type:	carboxylic acid amide hydrolysis
catalyst:	immobilized enzyme PGA-450
enzyme:	Penicillin amidohydrolase (penicillin amidase)
strain:	*Penicillin G amidohydrolase* expressed in *E. coli* ATCC 9637
CAS (enzyme):	[9014-06-6]

2) Remarks

- Although penicillin G acylase (PGA) has traditionally been used to hydrolyze penicillin G, it has also been used to resolve other amino compounds.

- The racemic ethyl 3-amino-5-(trimethylsilyl)-4-pentynoate needs to be acylated prior to the deacylation by the enzyme PGA.

- Acylation and deacylation are performed by the same enzyme under different reaction conditions.

- On a scale up to 70 L, 25 cycles were shown to be plausible.

3) Flow scheme

Not published.

4) Process parameters

conversion:	98 %
yield:	43–46 % (S)-amine
ee:	96–98 %
selectivity:	99.5 %

reactor type:	batch
reactor volume:	70 L
down stream processing:	MTBE extraction followed by separation of the enzyme through a filter screen; evaporation of solvent
enzyme consumption:	Approximately 25 reaction cycles possible before enzyme loses 50 % of its initial activity
company:	Pfizer Inc., USA

5) Product application

- The β-amino acid, (S)-ethyl-3-amino-4-pentynoate, is a chiral synthon used in the synthesis of xemilofiban hydrochloride, an anti-platelet agent.

6) Literature

- Topgi, R.S., Ng, J.S., Landis, B., Wang, P., Behling, J.R. (1999) Use of enzyme penicillin acylase in selective amidation/amide hydrolysis to resolve ethyl 3-amino-4-pentynoate isomers, Bioorg. Med. Chem. **7**, 2221–2229

- Landis, B.H., Mullins, P.B., Mullins, K.E., Wang, T. (2002) Kinetic resolution of ß-amino esters by acylation using immobilized penicillin amidohydrolase, Org. Proc. Res. Dev. **6**, 539–546

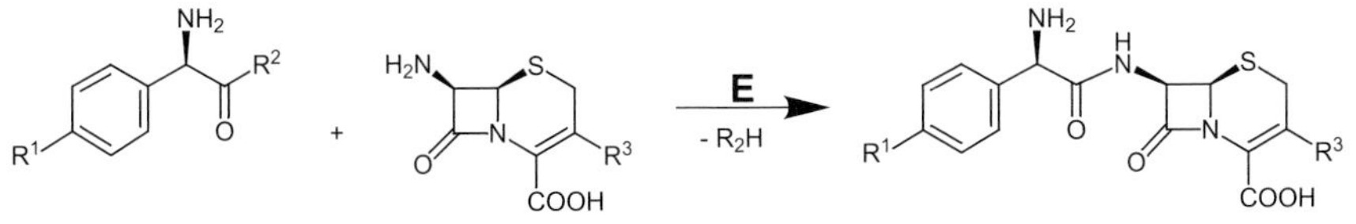

Fig. 3.5.1.11 – 1

1a = phenylglycineamide (R^1=H, R^2=NH$_2$) = PGA
1b = phenylglycinemethylester (R^1=H, R^2=OMe) = PGM
1c = hydroxyphenylglycineamide (R^1=OH, R^2=NH$_2$) = HPGA
1d = hydroxyphenylglycinemethylester (R^1=OH, R^2=OMe) = HPGM
2a = 7-aminodeacetoxycephalosporanic acid (R^3=Me) = 7-ADCA
2b = 7-aminodeacetoxymethyl-3-chlorocephalosporanic acid (R^3=Cl) = 7-ACCA
3a = cefaclor (R^1=H, R^3=Cl)
3b = cephalexin (R^1=H, R^3=Me)
3c = cefadroxil (R^1=OH, R^3=Me)

DSM

1) Reaction conditions

medium:	aqueous
reaction type:	carboxylic acid amide hydrolysis
catalyst:	immobilized whole cells or enzyme
enzyme:	penicillin amidase (penicillin amidase, α-acylamino-β-lactam acylhydrolase)
strain:	*Escherichia coli* and others
CAS (enzyme):	[9014-06-06]

2) Remarks

- The established chemical synthesis started from benzaldehyde and included fermentation of penicillin G. The process consisted of ten steps with a waste stream of 30–40 kg waste per kg product. The waste contained methylene chloride, other solvents, silylating agents and many by-products from side chain protection and acylating promoters.

- In comparison, the chemoenzymatic route needs only six steps including three biocatalytic ones.

- The following figure compares the chemical and chemo-enzymatic routes (Bruggink, 1996):

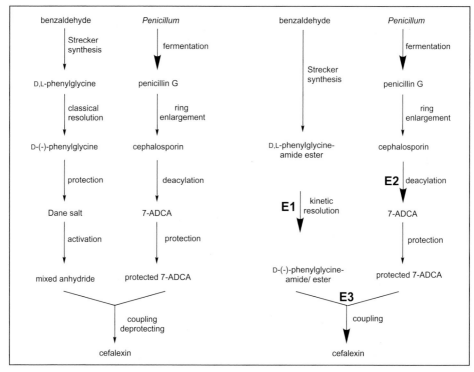

Fig. 3.5.1.11 – 2, E1 = penicillin amidase, E2 = aminopeptidase, E3 = penicillin acylase

- Beside *E. coli*, e.g., the strains *Klyveromyces citrophila* and *Bacillus megaterium* produce penicillin acylase.

- The yields of the kinetically controlled synthesis catalyzed by penicillin G acylase can be greatly increased by continuous extraction of water-soluble products such as cefalexin.

- The penicillin acylases do not accept charged amino groups. Therefore phenylglycine itself cannot be used since at a pH value at which the carboxyl function is necessarily uncharged the amino group will be charged.

- To reach non-equilibrium concentrations of the product, the substrate must be activated as an ester or amide. By this means the amino group can be partly uncharged at the optimal pH value of the enzyme. In biological systems the activation energy is delivered by ATP.

- The enzyme can be covalently attached on a gelatin-based carrier. Consequently the catalyst becomes water insoluble and can be easily separated from the reaction solution. Additionally the selectivity can be improved by choice of the right carrier composition. By-products resulting from hydrolysis of the educt can be avoided.

- To reach high conversions and high yields D-(–)-phenylglycine has to be added in a high molar excess.

- Since the characteristics of the shown substances are different for each antibiotic, a special synthetic method had to be established for each of them.

- The production of **cefalexin** was the first successful application.

- If an excess of D-(–)PGA is used, surplus or non-converted D-(–)PGA has to be separated and recycled.

- The separation of D-(–)PGA can be done by addition of benzaldehyde and formation of the poorly soluble Schiff base which can be filtered off subsequent to separation of enzyme and solid products by filtration.

- Also the D-(–)-PGA is almost insoluble in aqueous solution, so that at the end of the reaction three solids (d-(–)-PGA, cefalexin, D-(–)-HPGA) have to be separated.

- One solution of this problem was the use of a special immobilized enzyme, which floats after stopping of the stirrer (Novo Nordisk, Denmark). The reaction solution containing the solid products can be removed from the bottom of the reactor.

- A better technique uses enzyme immobilizates with a defined diameter. At the end the reaction solution and solid substances can be removed from the reactor using a special sieve that is not permeable to the immobilized enzyme. This technique is shown in the flow scheme (Fig. 3.5.1.11–4).

- 7-ACCA (= 7-aminodeacetoxymethyl-3-chlorocephalosporanic acid) can be obtained by ozonolysis and chlorination of 3-methylene cephams. It is the precursor for the synthesis of cefaclor. Cefaclor is unstable at pH values above 6.5 while the solubility of 7-ACCA is very low at pH values under 6.5.

- One strategy, also established for cefalexin, is to add a complexing agent (β-naphthol). The complex crystallizes out and yields above 90 % are possible.

- Using this technique the concentration of product in the reaction mixture is very low, so that succeeding reactions can be suppressed and the mother liquor can be discarded. The disadvantage is the necessity of an organic solvent to yield a two phase system in which the decomplexation at low pH is possible.

- Using the same synthetic pathway alternatively to 7-ADCA and 7-ACCA 6-APA derivatives can also be synthesized:

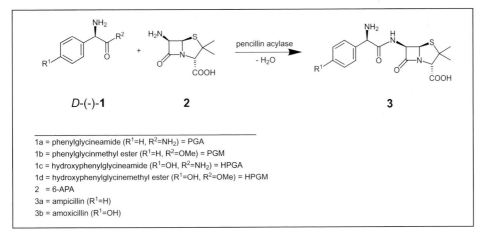

D-(-)-**1**   **2**   **3**

1a = phenylglycineamide (R^1=H, R^2=NH$_2$) = PGA
1b = phenylglycinmethyl ester (R^1=H, R^2=OMe) = PGM
1c = hydroxyphenylglycineamide (R^1=OH, R^2=NH$_2$) = HPGA
1d = hydroxyphenylglycinemethyl ester (R^1=OH, R^2=OMe) = HPGM
2 = 6-APA
3a = ampicillin (R^1=H)
3b = amoxicillin (R^1=OH)

Fig. 3.5.1.11 – 3

- In contrast to cefalexin, **ampicillin** has a higher solubility, so that by using the recovery strategy of cefalexin too much product would be lost.

- Since the penicillanic acid derivatives are more sensitive towards degradation than cephalos-poranic acid derivatives at almost all pH values, the conversion of 6-APA has to be complete and the product has to be recovered rapidly by crystallization (Fig. 3.5.1.11–5).

- The biocatalyst can be retained in the reactor by the sieve method analogous to the cephalexin procedure. Solubilized and precipitated product and D-(–)-PGA crystals are dissolved at acidic pH. Pure ampicillin is precipitated by adjusting the pH to its isoelectric point.

- In a similiar manner **amoxicillin** can be recovered. The advantage in this case is the low solubility of the product under reaction conditions so that hydrolysis of the product is sup-pressed since it precipitates at first. A semi-continuous reactor system with high substrate con-centrations can be used. (Fig. 3.5.1.11–6).

3) Flow scheme

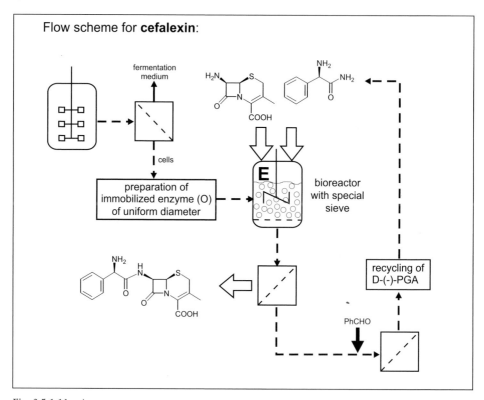

Fig. 3.5.1.11 – 4

Penicillin acylase
Escherichia coli

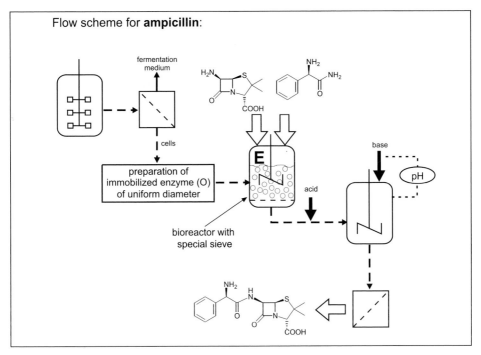

Fig. 3.5.1.11 – 5

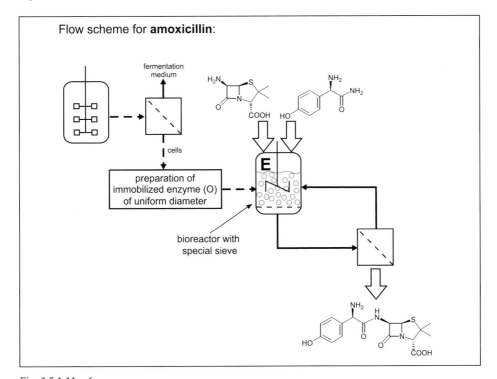

Fig. 3.5.1.11 – 6

4) Process parameters

yield:	> 90 %
selectivity:	> 95 %
ee:	> 99 %
reactor type:	repetitive batch (see flow schemes)
capacity:	2000 t · a^{-1} (worldwide)
down stream processing:	filtration
company:	DSM, The Netherlands

5) Product application

- The products are β-lactam antibiotics.

6) Literature

- Bruggink, A. (1996) Biocatalysis and process integration in the synthesis of semi-synthetic antibiotics, CHIMIA **50**, 431–432

- Bruggink, A., Roos, E.C., de Vroom, E. (1998) Penicillin acylase in the industrial production of β-lactam antibiotics, Org. Proc. Res. Dev. **2**, 128–133

- Clausen, K. (1995) Method for the preparation of certain β-lactam antibiotics, Gist-Brocades N. V., US 5,470,717

- Hernandez-Justiz, O., Fernandez-Lafuente, R., Terrini, M., Guisan, J.M. (1998) Use of aqueous two-phase systems for *in situ* extraction of water soluble antibiotics during their synthesis by enzymes immobilized on porous supports, Biotech. Bioeng. **59**, 1, 73–79

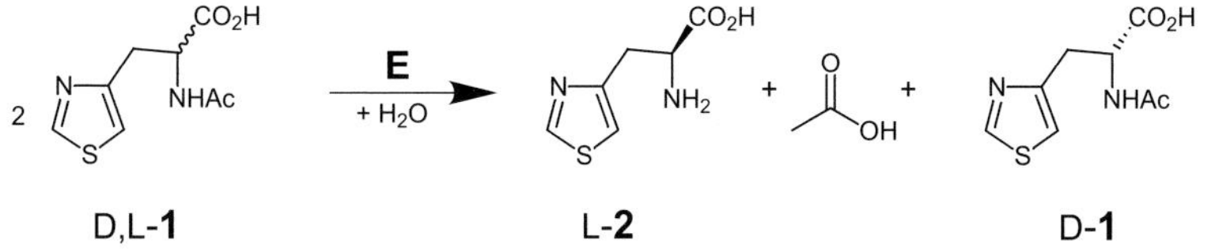

1 = *N*-acetyl-D,L-3-(4-thiazolyl)alanine
2 = 3-(4-thiazolyl)alanine
3 = *N*-acetyl-D-3-(4-thiazolyl)alanine

Celltech Group plc

Fig. 3.5.1.14 – 1

1) Reaction conditions

pH:	7.0
medium:	aqueous
reaction type:	carboxylic acid amide hydrolysis
catalyst:	immobilized enzyme
enzyme:	*N*-acetyl-L-amino-acid amidohydrolase (aminoacylase)
strain:	*Aspergillus niger*
CAS (enzyme):	[9012–37–7]

2) Remarks

- The whole process consisting of the following reaction steps is performed in one vessel:

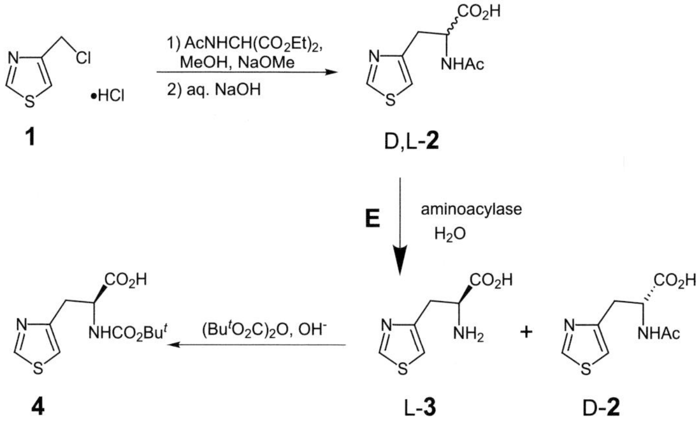

1 = 4-chloromethylthiazole hydrochloride
2 = *N*-acetyl-3-(4-thiazolyl)alanine
3 = 3-(4-thiazolyl)alanine
4 = *t*-butoxycarbonyl-L-thiazoylalanine (BOC-protected thiazoylalanine)

Fig. 3.5.1.14 – 2

- The hydrolysis is performed by slow addition of hydroxide solution, maintaining the mixture at pH 9–10; at this pH the initial product undergoes decarboxylation; this causes a reduction in pH and, by careful control of the base addition, the resulting mixture can be kept approximately neutral.

- The neutral solution can be directly used for the biotransformation.

- The product is extracted directly from the aqueous reaction mixture with methyl-*tert*-butyl ether (MTBE) and is of high enantiomeric purity.

- The D-isomer can be recycled via an oxazolinone that tautomerizes to the enol:

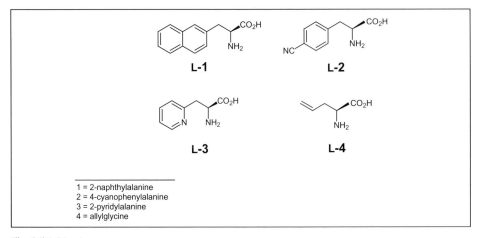

1 = *N*-acetyl-3-(4-thiazolyl)alanine
2 = oxazolinone (azlactone)
3 = enol

Fig. 3.5.1.14 – 3

- Since the enzyme is immobilized and employed in a packed bed reactor it can be used several times resulting in a protein-free solution after down-stream processing.

- The process can be applied to other unnatural alanine derivatives, e.g.:

1 = 2-naphthylalanine
2 = 4-cyanophenylalanine
3 = 2-pyridylalanine
4 = allylglycine

Fig. 3.5.1.14 – 4

3) Flow scheme

Not published.

4) Process parameters

ee:	>99 %
reactor type:	plug-flow reactor
capacity:	several kg
down stream processing:	solvent extraction
company:	Celltech Group plc, U.K.

5) Product application

- The product is used as a component of antihypertensive inhibitors of the enzyme renin, where it acts as a mimic of histidine.

- Two examples of L-thiazolylalanine-containing inhibitors are:

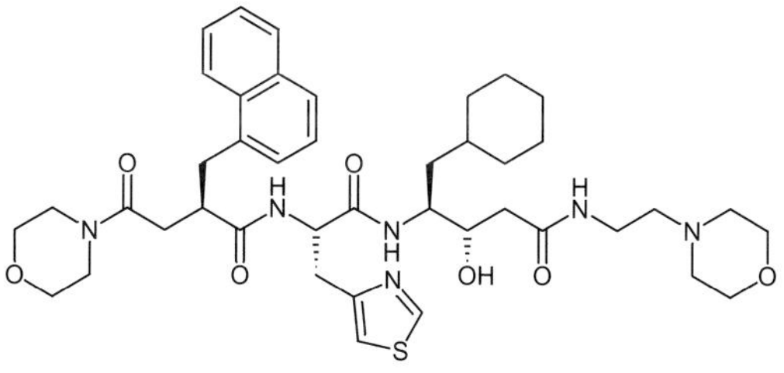

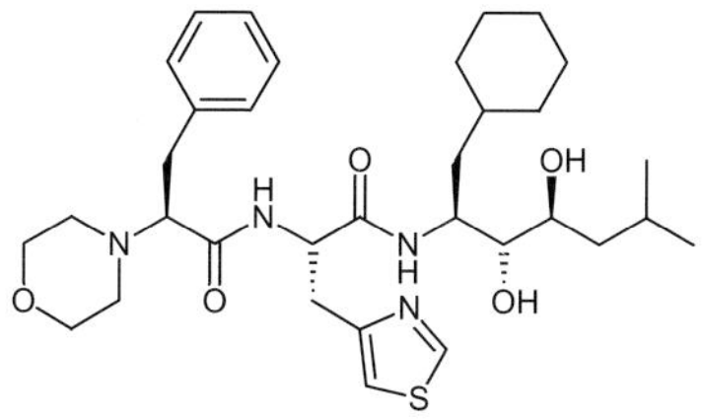

1 = thiazolalanine containing inhibitor (Sankyo)
2 = thiazolalanine containing inhibitor (Abbott)

Fig. 3.5.1.14 – 5

6) Literature

- Hsiao, C.-N., Leanna, M.R., Bhagavatula, L., Lara, E., Zydowsky, T.M. (1990) Synthesis of N-(*tert*-butoxycarbonyl)-3-(4-thiazolyl)-L-alanine, Synth. Commun. **20**, 3507–3517

- McCague, R., Taylor, S.J.C. (1997) Integration of an acylase biotransformation with process chemistry: a one-pot synthesis of NtBoc-L-3-(4-thiazolyl)alanine and related amino acids, in: Chirality In Industry II (Collin, S.A.N., Sheldrake, G.N., Crosby, J., eds.), pp. 194–200, John Wiley & Sons, New York

- Miyazawa, T., Iwanaga, H., Ueji, S., Yamada, T., Kuwata, S. (1989) Porcine pancreatic lipase catalyzed enantioselective hydrolysis of esters of N-protected unusual amino acids, Chem. Lett. 2219–2222

- Nishi, T., Saito, F., Nagahori, H., Kataoka, M., Morisawa, Y. (1990) Syntheses and biological activities of renin inhibitors containing statine analogues, Chem. Pharm. Bull. **38**, 103–109

- Rosenberg, S.H., Spina, K.P., Woods, K.W., Polakowski, J., Martin, D.L., Yao, Z., Stein, H.M., Cohen, J., Barlow, J.L., Egan, D.A., Tricarico, K.A., Baker, W.R., Kleinert, H.D. (1993) Studies directed toward the design of orally active renin inhibitors. 1. Some factors influencing the absorption of small peptides, J. Med. Chem. **36**, 449–459

- Taylor, S.J.C., McCague, R. (1997) Dynamic resolution of an oxazolinone by lipase biocatalysis: Synthesis of (S)-*tert*-leucine, in: Chirality In Industry II (Collins, A.N., Sheldrake, G.N., Crosby, J., eds.), pp. 194–200, John Wiley & Sons, New York

Aminoacylase

Aspergillus oryzae

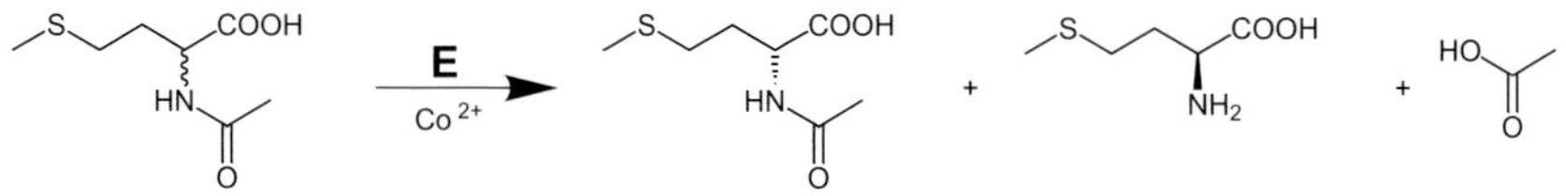

D,L-**1**   D-**2**   L-**3**   **4**

1 = *N*-acetyl-methionine
2 = *N*-acetyl-methionine
3 = methionine
4 = acetic acid

Degussa AG

Fig. 3.5.1.14 – 1

1) Reaction conditions

[**1**]:	0.6 M, 97.96 g · L^{-1} [163.27 g · mol^{-1}]
[Co^{2+}]:	0.5 · 10^{-3} M, 0.029 g · L^{-1} [58.93 g · mol^{-1}] (activator)
pH:	7.0
T:	37 °C
medium:	aqueous
reaction type:	hydrolysis
catalyst:	solubilized enzyme
enzyme:	*N*-acyl-L-amino-acid amidohydrolase (aminoacylase, acylase 1)
strain:	*Aspergillus oryzae*
CAS (enzyme):	[9012–37–7]

2) Remarks

- The *N*-acetyl-D,L-amino acid precursors are conveniently accessible through acetylation of D,L-amino acids with acetyl chloride or acetic anhydride under alkaline conditions in a Schotten-Baumann reaction.

- As effector Co^{2+} is added to increase the operational stability of the acylase.

- The unconverted acetyl-D-methionine is racemized by acetic anhydride under alkaline conditions and the racemic acetyl-D,L-methionine is recycled.

- The racemization can also be carried out in a molten bath or by racemase.

- Product recovery of L-methionine is achieved by crystallization, because L-methionine is much less soluble than the substrate.

- A polyamide ultrafiltration membrane with a cutoff of 10,000 dalton is used.

- Several proteinogenic and non-proteinogenic amino acids are produced in the same way by Degussa-Hüls:

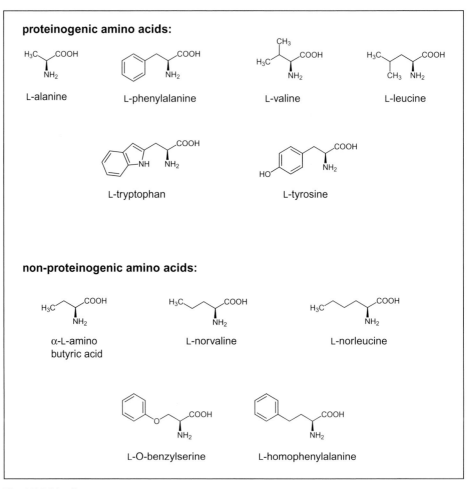

Fig. 3.5.1.14 – 2

3) Flow scheme

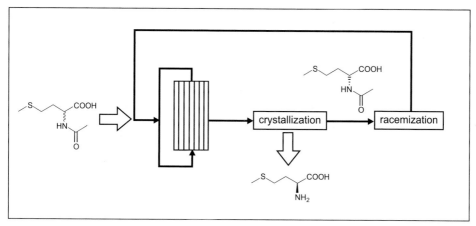

Fig. 3.5.1.14 – 3

4) Process parameters

yield:	80 %
ee:	99.5 %
reactor type:	cstr, UF-membrane reactor
capacity:	> 300 t · a^{-1} Drauz (Artikel-Korrektur)
residence time:	2.9 h
space-time-yield:	592 g · L^{-1} · d^{-1}
down stream processing:	crystallization
enzyme activity:	1,067 U · g$^{-1}_{\text{protein}}$
enzyme consumption:	2,067 U · kg^{-1}
enzyme supplier:	Amano Enzymes, Inc., Japan
production site:	Degussa-Rexime, China
company:	Degussa AG, Germany

5) Product application

- L-Amino acids are used for parenteral nutrition (infusion solutions), feed and food additives, cosmetics, pesticides and as intermediates for pharmaceuticals as well as chiral synthons for organic synthesis.

6) Literature

- Bommarius, A.S., Drauz, K., Klenk, H., Wandrey, C. (1992) Operational stability of enzymes – acylase-catalyzed resolution of *N*-acetyl amino acids to enantiomerically pure L-amino acids, Ann. N. Y. Acad. Sci. **672**, 126–136

- Chenault, H.K., Dahmer, J., Whitesides, G.M. (1989) Kinetic resolution of unnatural and rarely occuring amino acids: enantioselective hydrolysis of *N*-acyl amino acids catalyzed by acylase I, J. Am. Chem. Soc. **111**, 6354–6364

- Leuchtenberger, W., Karrenbauer, M., Plöcker, U. (1984) Scale-up of an enzyme membrane reactor process for the manufacture of L-enantiomeric compounds, Enzyme Engineering 7, Ann. N. Y. Acad. Sci. **434**, 78–86

- Takahashi, T., Izumi, O., Hatano, K. (1989) Acetylamino acid racemase, production and use thereof, Takeda Chemical Industries, Ltd., EP 0 304 021 A2

- Wandrey, C., Flaschel, E. (1979) Process development and economic aspects in enzyme engineering. Acylase L-methionine system. In: Advances in Biochemical Engineering 12 (Ghose, T.K., Fiechter, A., Blakebrough, N., eds.),pp. 147–218, Springer-Verlag, Berlin

- Wandrey, C., Wichmann, R., Leuchtenberger, W., Kula, M.R. (1981) Process for the continuous enzymatic change of water soluble α-ketocarboxylic acids into the corresponding amino acids, Degussa AG, US 4,304,858

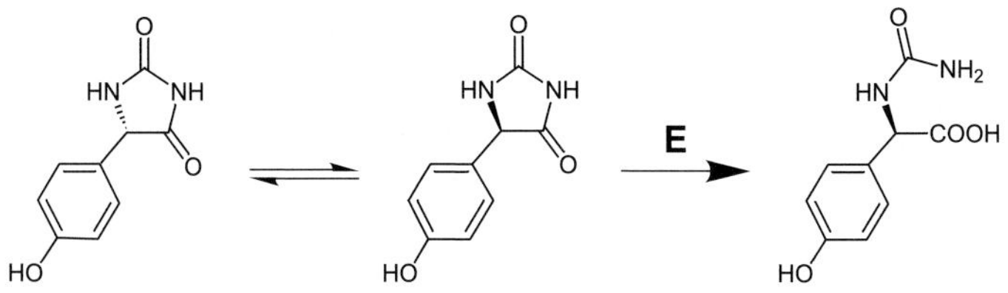

L-1   D-1   D-2

1 = 5-(*p*-hydroxyphenyl)-hydantoin
2 = D-*N*-carbamoyl *p*-hydroxyphenyl glycine

Kanegafuchi Chemical Industries Co., Ltd.

Fig. 3.5.2.2 – 1

1) Reaction conditions

pH:	8.0
medium:	aqueous
reaction type:	carboxylic acid amide hydrolysis
catalyst:	immobilized whole cells
enzyme:	5,6-dihydropyridine amidohydrolase (dihydropyrimidinase, hydantoin peptidase, hydantoinase)
strain:	*Bacillus brevis*
CAS (enzyme):	[9030–74–4]

2) Remarks

- The hydantoinase is D-specific.

- The unreacted L-hydantoins are readily racemized under the conditions of enzymatic hydrolysis.

- Quantitative conversion is achieved because of the *in situ* racemization.

- L-Specific hydantoinases are also known.

- This process enables the stereospecific preparation of various amino acids, such as D-tryptophan, D-phenylalanine, D-valine, D-alanine and D-methionine.

- The carbamoyl group can be removed by use of a carbamoylase (EC 3.5.1.77, see page 415) or alternatively by chemical treatment with sodium nitrite:

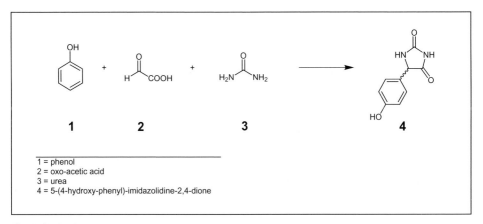

2 = D-*N*-carbamoyl *p*-hydroxyphenyl glycine
3 = D-4-hydroxyphenyl glycine

Fig. 3.5.2.2 – 2

- Racemic hydantoins are synthesized starting from phenol derivatives, glyoxylic acid and urea via the Mannich condensation:

1 = phenol
2 = oxo-acetic acid
3 = urea
4 = 5-(4-hydroxy-phenyl)-imidazolidine-2,4-dione

Fig. 3.5.2.2 – 3

- Re-use of cells through as immobilized cell catalyst is possible.

- Several other companies have developed patented processes leading to D-hydroxyphenyl gly-cine (Ajinomoto, DSM, Snamprogetti, Recordati and others, for example see page 415).

3) Flow scheme

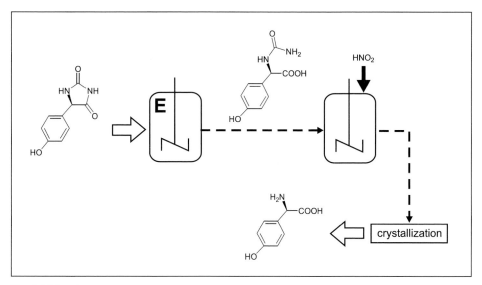

Fig. 3.5.2.2 – 4

4) Process parameters

conversion:	100 %
capacity:	300–700 t · a^{-1}
enzyme activity:	17.14 U · g^{-1}
start-up date:	1983
company:	Kanegafuchi Chemical Industries Co., Ltd., Japan

5) Product application

- D-*p*-Hydroxyphenyl glycine is a key raw material for the semisynthetic penicillins, ampicillin and amoxycillin.

- It is also used in photographic developers.

6) Literature

- Cheetham, P.S.J. (1994) Case studies in applied biocatalysis, in: Applied Biocatalysis (Cabral, J.M.S., Best, D., Boross, L., Tramper, J.; eds.) pp. 68–70, Harwood Academic Publishers, Chur, Switzerland

- Crosby, J. (1991) Synthesis of optically active compounds: a large scale perspective, Tetrahedron **47**, 4789–4846

- Ikemi, M. (1994) Industrial chemicals: enzymatic transformation by recombinant microbes, Bioprocess Technology **19**, pp. 797–813

- Schmidt-Kastner, G., Egerer, P. (1984) Amino acids and peptides. In: Biotechnology, Vol. 6a (Kieslich, K., ed.), pp. 387–419, Verlag Chemie, Weinheim

- Sheldon, R.A. (1993) Chirotechnology, Marcel Dekker Inc., New York

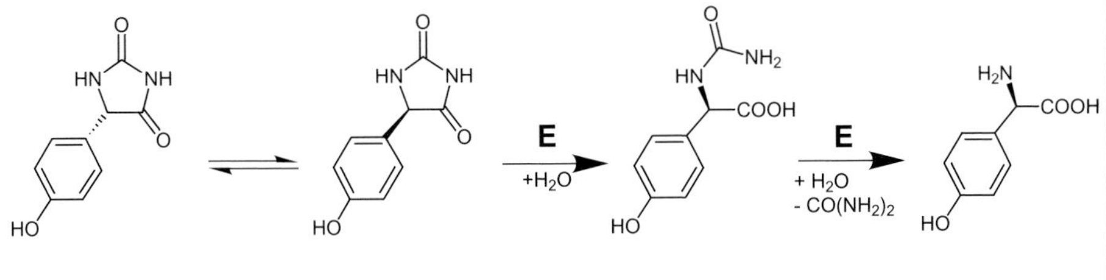

1 = *p*-hydroxyphenylhydantoin
2 = *N*-carbamoyl amino acid
3 = α–(-) phenyl glycine

Dr. Vig Medicaments

Fig. 3.5.2.4 / 3.5.1.77 – 1

1) Reaction conditions

[1]:	>0.21 M, 40 g · L⁻¹ [192.17 g · mol⁻¹]
pH:	8.0
T:	38 °C
medium:	aqueous
reaction type:	carboxylic acid amide hydrolysis
catalyst:	whole cells
enzyme:	L-5-carboxymethylhydantoin amidohydrolase / *N*-carbamoyl-D-amino acid hydrolase
strain:	*Pseudomonas* sp.
CAS (enzyme):	[9025–14–3] / –

The reaction conditions rendered above in LaTeX:

[1]: >0.21 M, 40 g $\cdot$ L^{-1} [192.17 g $\cdot$ mol^{-1}]

2) Remarks

- The strain contains both enzymes (hydantoinase and carbamoylase).

- The cell biomass is used directly in the process. Alternatively the enzymes may be extracted and immobilized.

- Instead of hydroxyphenylglycine the non-hydroxylated phenylglycine can be produced in the same way.

- Kanegafuchi also uses the hydantoinase to hydrolyze D,L-5-(*p*-hydroxyphenyl)hydantoin (see page 411). The second step, the hydrolysis of the D-*N*-carbamoyl-D-hydroxyphenyl glycine is performed either chemically with HNO₂ or enzymatically with carbamoylase.

3) Flow scheme

Not published.

415

4) Process parameters

conversion:	95 %
yield:	80 %
selectivity:	84 %
chemical purity:	98.5 %
reactor type:	batch
reactor volume:	15,000 L
capacity:	pilot scale
residence time:	15 – 20 h
space-time-yield:	57.6 g $\cdot$ L^{-1} $\cdot$ h^{-1}
down stream processing:	concentration and crystallization
enzyme supplier:	Captive Production, India
production site:	Dr. Vig Medicaments, India
company:	Dr. Vig Medicaments, India

5) Product application

- The product is used as a chemical intermediate in the synthesis of amoxycillin/cefadroxil through 'Dane'-salt formation and of ampicillin/cefalexin through acid chloride formation with 6-APA/7-ADCA respectively.

6) Literature

- Vig, C. (1997) personal communication

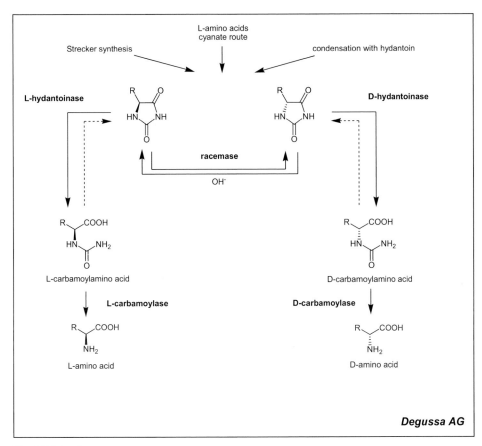

Fig. 3.5.2.4 – 1

1) Reaction conditions

medium: aqueous
reaction type: carboxylic acid amide hydrolysis
catalyst: whole cells
enzyme: ʟ-hydantoinase (hydantoin hydrolase, carboxymethylhydantoinase, ʟ-5-carboxy-
               methylhydantoin amidohydrolase)
enzyme: ʟ-carbamoylase
enzyme: hydantoin racemase
strain: recombinant *E. coli* overexpressing hydantoinase, carbamoylase, and hydantoin
               racemase from *Arthrobacter* sp. DSM 9771
CAS (enzyme): [9025-14-3]

2) Remarks

• Process under development.

417

- Processes for the production of L-amino acids are still limited. Therefore Degussa developed a tailor-made recombinant whole cell biocatalyst. Further reduction of biocatalyst cost by use of recombinant *Escherichia coli* cells overexpressing a hydantoinase, carbamoylase and hydantoin racemase from *Arthrobacter sp.* DSM 9771 were achieved.

- This highly active recombinant whole-cell biocatalyst can be produced in high-cell density fermentation on a m³-scale at concentrations above $50\,g \cdot L^{-1}$ dry cell weight.

- Although the L-selectivity of the designed enzyme is not impressive and leaves room for further improvements, the productivity of the process can be dramatically improved. These improvements have been confirmed on a m³-scale using a simple batch reactor concept coupled to a continuous centrifuge for cell separation.

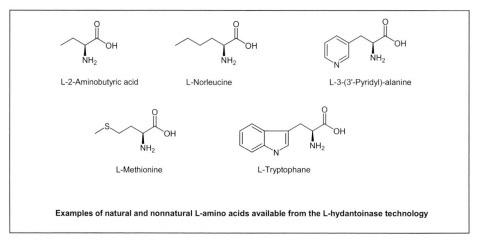

L-2-Aminobutyric acid   L-Norleucine   L-3-(3'-Pyridyl)-alanine

L-Methionine   L-Tryptophane

Examples of natural and nonnatural L-amino acids available from the L-hydantoinase technology

Fig. 3.5.2.4 – 2

3) Flow scheme

Not published.

4) Process parameters

conversion:	40 %
yield:	100 %
ee:	20 %
reactor type:	batch
reactor volume:	m³
down stream processing:	continuously operated centrifuge for cell separation
company:	Degussa AG, Germany

5) Product application

- Optically pure natural and non-natural L-amino acids.

6) Literature

- May, O., Verseck, S., Bommarius, A., Drauz, K. (2002) Development of dynamic kinetic resolution processes for biocatalytic production of natural and nonnatural ʟ-amino acids, Org. Proc. Res. Dev. **6**, 452–457

- Drauz, K., Eils, S., Schwarm, M. (2002) Synthesis and production of enantiomerically pure amino acids, Chimica Oggi/Chemistry Today January/February, 15–21

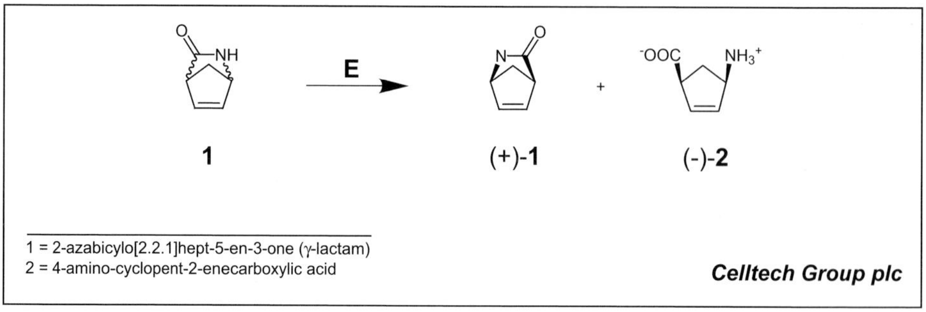

1 = 2-azabicylo[2.2.1]hept-5-en-3-one (γ-lactam)
2 = 4-amino-cyclopent-2-enecarboxylic acid

Celltech Group plc

Fig. 3.5.2.6 – 1

1) Reaction conditions

[1]:	1.83 M, 200 g · L⁻¹ [109.05 g · mol⁻¹]
T:	70 °C
medium:	aqueous
reaction type:	carboxylic acid amide hydrolysis
catalyst:	immobilized enzyme
enzyme:	β-lactamhydrolase (β-lactamase)
strain:	*Aureobacterium* sp.
CAS (enzyme):	[9001–74–5]

[1]: 1.83 M, 200 g · L^{-1} [109.05 g · mol^{-1}]
T: 70 °C
medium: aqueous
reaction type: carboxylic acid amide hydrolysis
catalyst: immobilized enzyme
enzyme: β-lactamhydrolase (β-lactamase)
strain: *Aureobacterium* sp.
CAS (enzyme): [9001–74–5]

2) Remarks

- The enzyme is purified (ammonium sulphate fractionation and anion-exchange chromatography) and immobilized on a glutaraldehyde-activated solid support.

- The biotransformation is operated in a batch reaction wherein an aqueous solution of the racemic lactam is circulated through the fixed bed of immobilized enzyme.

- The reaction is complete when the (–)-enantiomer is hydrolyzed completely (E-value > 7,000).

- The stability of the enzyme is improved by immobilization so that a nearly steady-state production can be achieved for more than 6 months.

- To separate the amino acid from the reaction mixture the solution has to be slurried with acetone. The (+)-lactam stays in solution while the amino acid crystallizes and can be removed by filtration.

- In comparison to the chemical resolution in which the lactam is converted to the salt of the amino acid, the biotransformation produces the neutral, zwitter-ionic form of the acid.

3) Flow scheme

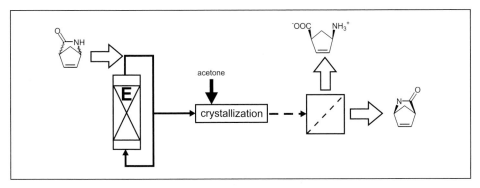

Fig. 3.5.2.6 – 2

4) Process parameters

reactor type: plug-flow reactor with circulation (batch)
down stream processing: extraction, filtration
company: Celltech Group plc, U.K.

5) Product application

- 4-Amino-cyclopent-2-enecarboxylic acid is used as a building block in the synthesis of carbocyclic nucleosides with the natural configuration.

6) Literature

- McCague, R., Taylor, S.J.C. (1997) Development of an immobilised lactamase resolution process for the (+)-γ-lactam and the ring-opened (–)-amino acid, in: Chirality In Industry II (Collins, A.N., Sheldrake, G.N., Crosby, J., eds.), pp. 187–188, John Wiley & Sons, New York

β-Lactamase
Pseudomonas solanacearum

EC 3.5.2.6

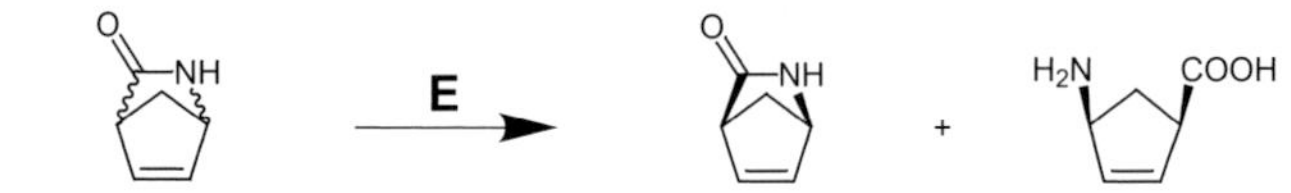

1 **(-)-1** **(+)-2**

1 = 2-azabicylo[2.2.1]hept-5-en-3-one (γ-lactam)
2 = 4-amino-cyclopent-2-enecarboxylic acid

Celltech Group plc

Fig. 3.5.2.6 – 1

1) Reaction conditions

[**1**]:	0.92 M, 100 g · L^{-1} [109.13 g · mol^{-1}]
pH:	7.0
T:	25 °C
medium:	aqueous
reaction type:	carboxylic acid amide hydrolysis
catalyst:	suspended whole cells
enzyme:	β-lactamhydrolase (β-lactamase)
strain:	*Pseudomonas solanacearum*
CAS (enzyme):	[9001–74–5]

2) Remarks

- The cells are grown by fermentation and the resulting cell mass is frozen and stored.
- The frozen cell mass is added in its crude form to the aqueous reaction solution. The advantage is that the biotransformation can be performed at different sites.
- The reaction rate is increased by some cell lysis, which is caused by the freeze-thaw process, liberating the enzyme.
- The whole cells are used since the isolated microbial enzyme is only of limited stability.
- Control of pH is not required, since the product 2-amino-cyclopent-2-ene carboxylic acid acts as its own buffer.
- The lactam is extracted with dichloromethane.
- The amino acid can be recovered from the aqueous phase as the hydrochloride by acidification (HCl) and evaporation of the water.

3) Flow scheme

Not published.

4) Process parameters

conversion:	55 %
yield:	45 %
selectivity:	82%
ee:	> 98 %
reactor type:	batch
capacity:	ton scale
residence time:	3 h
down stream processing:	harvesting cells, extraction of product with dichloromethane
production site:	Celltech Group plc, U.K.
company:	Celltech Group plc, U.K.

5) Product application

- The lactam can be converted to carbovir:

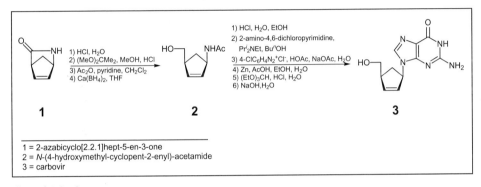

Fig. 3.5.2.6 – 2

- Carbovir is a potent and selective inhibitor of HIV-1.

- Its ability to inhibit infectivity and replication of the virus in human T-cell lines at concentrations 200–400 fold below toxic levels makes carbovir a promising candidate for development as a potential antiretroviral agent.

6) Literature

- Taylor, S.J.C., Sutherland, A.G., Lee, C., Wisdom, R., Thomas, S., Roberts, S.M., Evans, C. (1990) Chemoenzymatic synthesis of (–)-carbovir utilizing a whole cell catalyzed resolution of 2-azabicyclo[2.2.1]hept-5-en-3-one, J. Chem. Soc., Chem. Commun., 1120–1121

- Taylor, S.J.C., McCague, R. (1997) Resolution of the carbocyclic nucleoside synthon 2-azabicyclo[2.2.1]hept-5-en-3-one with lactamases, in: Chirality In Industry II (Collins, A.N., Sheldrake, G.N. and Crosby, J., eds.), pp. 184–190, John Wiley & Sons, New York

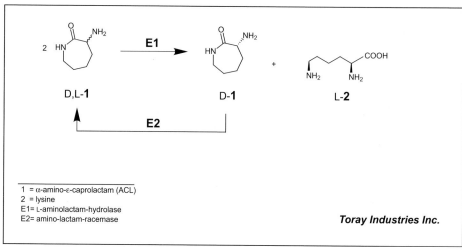

1 = α-amino-ε-caprolactam (ACL)
2 = lysine
E1= L-aminolactam-hydrolase
E2= amino-lactam-racemase

Toray Industries Inc.

Fig. 3.5.2.11 / 5.1.1.15 – 1

1) Reaction conditions

[1]:	0.78 M, 100 g · L^{-1} [128.09 g · mol^{-1}]
pH:	8.0–9.0
T:	40 °C
medium:	aqueous
reaction type:	carboxylic acid amide hydrolysis / racemization
catalyst:	suspended whole cells
enzyme:	E1 L-lysine-1,6-lactam hydrolase (L-lysine lactamase)/
	E2 2-aminohexano-6-lactam racemase (α-amino-ε-caprolactam racemase)
strain:	*Cryptococcus laurentii / Achromobacter obae*

2) Remarks

- The lactamase and racemase are fortunately active at the same pH, so that they can be used in one reactor.

- The combination of cells from *Candida humicola* and *Alcaligenes faecalis* is used alternatively.

- The reaction starts from cyclohexene leading to the oxime as intermediate. The Beckmann rearrangement gives the caprolactam for the enzymatic steps.

- This process has been totally replaced by highly effective fermentation methods.

3) Flow scheme

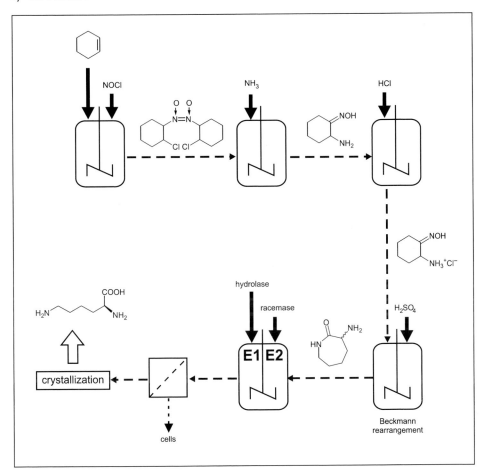

Fig. 3.5.2.11 / 5.1.1.15 – 2

4) Process parameters

ee:	> 99.5 %
reactor type:	batch
capacity:	4,000 t · a^{-1}
residence time:	25 h
down stream processing:	crystallization
start-up date:	1970
production site:	Japan
company:	Toray Industries Inc., Japan

5) Product application

- As nutrient and food supplement.

6) Literature

- Atkinson, B., Mavituna, F. (1991) Biochemical Engineering and Biotechnology Handbook, p. 1133 ff, Stockton Press, New York

- Crosby, J. (1991) Synthesis of optically active compounds: a large scale perspective, Tetrahedron **47**, 4789–4846

- Schmidt-Kastner, G., Egerer, P. (1984) Amino acids and peptides, in: Biotechnology, Vol. 6a, (Kieslich, K., ed.) pp. 387–419, Verlag Chemie, Weinheim

- Sheldon, R.A. (1993) Chirotechnology, p. 227 ff, Marcel Dekker Inc., New York

- Wiseman, A. (1995) Handbook of Enzyme and Biotechnology, p. 476 ff, Ellis Horwood, Chichester

Nitrilase
Acidovorax facilis

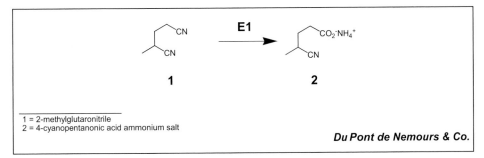

1 = 2-methylglutaronitrile
2 = 4-cyanopentanonic acid ammonium salt

Du Pont de Nemours & Co.

Fig. 3.5.5.1 – 1

1) Reaction conditions

[1]:	1.2 M, 129.8 g · L^{-1} [108.14 g · mol^{-1}]
[2]:	2 mM Ca acetate
pH:	7.0
T:	30 °C
reaction type:	nitrile hydrolysis
catalyst:	immobilized whole cell
enzyme:	nitrile aminohydrolase (nitrilase)
strain:	immobilized *Escherichia coli* transformant which expresses *Acidovorax facilis* 72 W nitrilase
CAS (enzyme):	[9024-90-2]

2) Remarks

- The volumetric productivity was increased by using immobilized *E. coli* transformant in alginate.

- The first commercial production of Xolvone (1,5-dimethyl-2-piperidone) employed direct hydrogenation of 2-methylglutaronitrile in the presence of methylamine and produced a mixture of 1,3- and 1,5-dimethyl-2-piperidones.

- A chemoenzymatic process is scheduled to replace the current chemical process.

- The chemoenzymatic process produces a single geometric isomer of dimethyl-2-piperidone with higher boiling point than the mixture of geometric isomers produced in the chemical process and in higher yield with less by-product formation.

- The productivity is 3500 g $_{product}$ · g $_{biocatalyst}^{-1}$

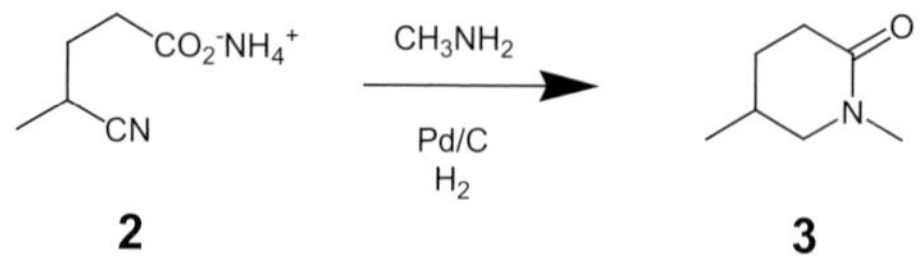

2 = 4-cyanopentanonic acid ammonium salt
3 = 1,5-dimethyl-2-piperidone (Xolvone)

Fig. 3.5.5.1 – 2

3) Flow scheme

Not published.

4) Process parameters

conversion:	100 %
yield:	98.7 %
selectivity:	98 %
reactor type:	consecutive stirred-batch
reactor volume:	500 L
space-time-yield:	$1896 \, g \cdot L^{-1}$
enzyme activity:	$3500 \, g_{\,product} \cdot g_{\,biocatalyst}^{-1}$
company:	Du Pont de Nemours and Co., USA

5) Product application

- Intermediate in the chemoenzymatic process for production of 1,5-dimethyl-2-piperidone (Xolvone), which is a precision cleaning solvent.

- Xolvone is used in a variety of industrial applications, including electronics cleaning, photoresist stripping, industrial degreasing, metal cleaning and resin cleanup; it can also be used in the formulation of inks and industrial adhesives and as a reaction solvent for the production of polymers and chemicals.

- Xolvone is not flammable, is completely miscible with water, has a good toxicological profile and is readily biodegradable.

6) Literature

- Hann, E.C., Sigmund, A.E., Hennessey, S.M., Gavagan, J.E., Short, D.R., Bassat, A.B., Chauhan, S., Fallon, R.D., Payne, M.S., DiCosimo, R. (2002) Optimization of an immobilized-cell biocatalyst for production of 4-cyanopentanoic acid, Org. Proc. Res. Dev. **6**, 492–496

- Cooling, F.B., Fager, S.K., Fallon, R.D., Folson, P.W., Gallagher, F.G., Gavagan, J.E., Hann, E.C., Herkes, F.E., Philips, R.J., Sigmund, A., Wagner, L.W., Wu, W., DiCosimo, R. (2001) Chemoenzymatic production of 1,5-dimethyl-2-piperidone, J. Mol. Catal. B: Enzymatic **11**, 295–306

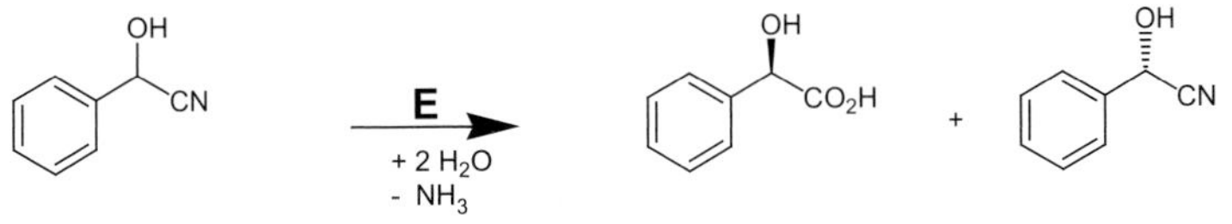

1 = (R,S)-cyanohydrine
2 = (R)-mandelic acid

BASF AG

Fig. 3.5.5.1 – 1

1) Reaction conditions

[1]:	[133.15 g · mol^{-1}]
pH:	7.2
T:	40 °C
medium:	aqueous
enzyme:	nitrile aminohydrolase (nitrilase)
strain:	*E. coli* JM (pDHE19.2)
CAS (enzyme):	[9024-90-2]

2) Flow scheme

Not published.

3) Process parameters

ee:	>99 %
reactor type:	batch
capacity:	multi-ton scale
company:	BASF AG, Germany

4) Literature

- Gröger, H. (2001) Enzymatic routes to enantiomerically pure aromatic α-hydroxy carboxylic acids: a further example for the diversity of biocatalysis, Adv. Synth. Catal. **343** (6-7), 547–558

- Ress-Löschke, M., Friedrich, T., Hauer, B., Mattes, R., Engels, D., (2000) Method for producing chiral carboxylic acids from nitriles with the assistance of a nitrilase or microorganisms which contain a gene for the nitrilase, WO 00/23577, 2000 (DE 19848129 A1, 2000)

Nitrilase / Hydroxylase
Agrobacterium sp.

EC 3.5.5.1 / 1.5.1.13

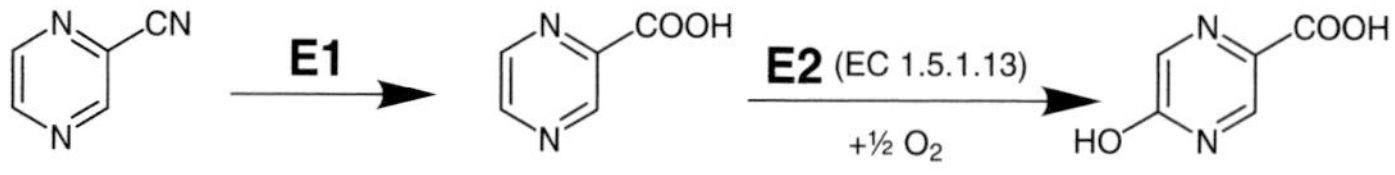

1 **2** **3**

1 = 2-cyanopyrazine
2 = pyrazine-2-carboxylic acid
3 = 5-hydroxypyrazine-2-carboxylic acid

Lonza AG

Fig. 3.5.5.1 / 1.5.1.13 – 1

1) Reaction conditions

[**1**]:	0.29 M, 30 g · L^{-1} [105.1 g · mol^{-1}]
pH:	6.0–8.0
T:	15 °C – 45 °C
medium:	aqueous
reaction type:	*N*-bond hydrolysis (nitrile hydrolysis)
catalyst:	suspended, living whole cells
enzyme:	**E1**: nitrilase (nitrile aminohydrolase) and **E2**: hydroxylase (nicotinate dehydrogenase) (EC 1.5.1.13)
strain:	*Agrobacterium* sp. DSM 6336 (**E1** + **E2**)
CAS (enzyme):	**E1**: [9024–90–2] / **E2**: [9059-03–4]

2) Remarks

- In contrast to the biotransformation the chemical synthesis of 5-substituted pyrazine-2-carboxylic acid leads to a mixture of 5- and 6-substituted pyrazinecarboxylic acids and requires multiple steps.
- Although the reaction sequence of growth and biotransformation are pretty similar, the cells grown on 2-cyanopyridine are much more active due to an optimized expression of the hydroxylase. Additonally, the degradation pathway in case of 2-cyanopyrazine can be stopped.

Nitrilase / Hydroxylase
Agrobacterium sp.

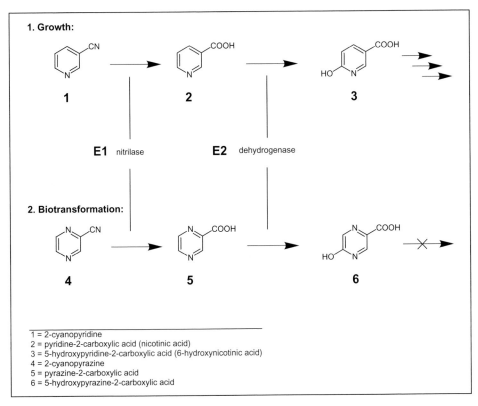

1. Growth:

1

2

3

E1 nitrilase   **E2** dehydrogenase

2. Biotransformation:

4

5

6

1 = 2-cyanopyridine
2 = pyridine-2-carboxylic acid (nicotinic acid)
3 = 5-hydroxypyridine-2-carboxylic acid (6-hydroxynicotinic acid)
4 = 2-cyanopyrazine
5 = pyrazine-2-carboxylic acid
6 = 5-hydroxypyrazine-2-carboxylic acid

Fig. 3.5.5.1 / 1.5.1.13 – 2

- Since a high pO_2 induces cell death, the oxygen partial pressure is limited to 90 mbar during cell growth. During the transformation it is reduced to 50 mbar.

- High oxygen partial pressure also causes the autooxidative dimerisation to di- and trihydroxylated pyridines.

- The biomass is separated by ultrafiltration (cutoff 10 kDa) after the biotransformation.

- 5-Hydroxypyrazine-2-carboxylic acid is precipitated from the permeate by acidification with sulfuric acid to pH 2.5.

- The lower practical yield of 80 % in comparison to the analytical yield of 95 % is due to repeated precipitation during product isolation.

3) Flow scheme

Not published.

4) Process parameters

yield:	95 % (analytical) / 80 % (isolated)
selectivity:	high
chemical purity:	> 99 % (isolated)
reactor type:	batch
reactor volume:	20 L
capacity:	multi kg
space-time-yield:	36 g · L^{-1} · d^{-1}
down stream processing:	ultrafiltration and precipitation
company:	Lonza AG, Switzerland

5) Product application

- Versatile building block for the synthesis of new antitubercular agents, e.g. 5-chloro-pyrazine-2-carboxylic acid esters:

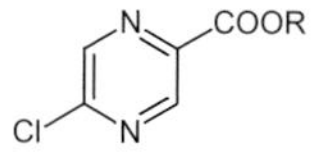

6) Literature

- Kiener, A. (1994) Mikrobiologisches Verfahren zur Herstellung von 5-Hydroxy-2-pyrazincarbonsäure, Lonza AG, EP 0578137 A1

- Kiener, A., Roduit, J.-P., Tschech, A., Tinschert, A., Heinzmann, K. (1994) Regiospecific enzymatic hydroxylations of pyrazinecarboxylic acid and a practical synthesis of 5-chloropyrazine-2-carboxylic acid, Synlett **10**, 814–816

- Roduit, J.-P. (1993) Verfahren zur Herstellung von Carbonsäurechloriden aromatischer Stickstoff-Heterocyclen, EP 0561421 A1

- Wieser, M., Heinzmann, K., Kiener, A. (1997) Bioconversion of 2-cyanopyrazine to 5-hydroxypyrazine-2-carboxylic acid with *Agrobacterium sp.* DSM 6336, Appl. Microbiol. Biotechnol. **48**, 174–180

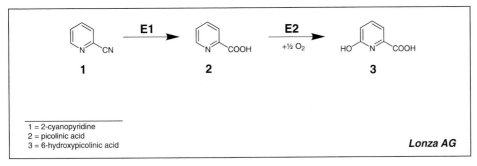

1 = 2-cyanopyridine
2 = picolinic acid
3 = 6-hydroxypicolinic acid

Lonza AG

Fig. 3.5.5.1 / 1.5.1.13 – 1

1) Reaction conditions

[2]:	> 0.016 M, > 2 g · L^{-1} [123.11 g · mol^{-1}]
[3]:	0.543 M, 75 g · L^{-1} [138 g · mol^{-1}]
pH:	7.0
T:	30 °C
medium:	aqueous
reaction type:	N-bond hydrolysis (nitrile hydrolysis)
catalyst:	suspended whole cells
enzyme:	**E1**: nitrile aminohydrolase (nitrilase), **E2**: hydroxylase (EC 1.5.1.13)
strain:	*Alcaligenes faecalis* DSM 6335 (**E1** + **E2**)
CAS (enzyme):	[9024–90–2]

2) Remarks

- The cells are grown on sodium fumarate (32 g · L^{-1}) as C-source up to an OD$_{650}$ of 16.

- Resting cells are employed as biocatalyst under aerobic conditions.

- Since 2-cyanopyridine is a solid at room temperature, the substrate solution is heated to 50 °C in order to add a liquid substrate.

- Since the intermediate picolinic acid is inhibiting the reaction to 6-hydroxypicolinic acid, the educt 2-cyanopyridine has to be maintained at a low concentration level. Therefore, the educt 2-cyanopyridine is continuously fed to the reaction solution. The feed rate is controlled by the on-line-analysis concentration of the intermediate picolinic acid.

- The reaction rate for the first step (**E1**) is 2.5 times faster than for the second step (**E2**).

- At the end of the biotransformation no intermediate can be found.

- To precipitate the product, the reaction solution is filtrated to remove the cells and the pH is adjusted to 2.5 using sulfuric acid at a temperature of 60 °C.

- Under strictly anaerobic conditions the hydroxylase activity is suppressed enabling the production of picolinic acid using similar conditions with a space-time yield of 138 g · L^{-1} · d^{-1} (chemical purity: 86 %). To prevent hydroxylation caused by oxygen, the reactor is aerated with nitrogen during the whole biotransformation.

- This enzyme acts on a wide range of aromatic nitriles and also on some aliphatic nitriles.

3) Flow scheme

Not published.

4) Process parameters

conversion:	100 %
yield:	87 %
reactor type:	fed-batch
capacity:	$1\ t \cdot a^{-1}$
residence time:	31 h
space-time-yield:	$58\ g \cdot L^{-1} \cdot d^{-1}$
down stream processing:	precipitation
company:	Lonza AG, Switzerland

5) Product application

- The product is used as an intermediate for pharmaceuticals, e.g. 2-oxypyrimidine, and herbicides.

6) Literature

- Fischer, E., Heß, K., Stahlschmidt, A. (1912) Verwandlung der Dihydrofurandicarbonsäure in Oxypyridincarbonsäure, Ber. Dtsch. Chem. Ges. **45**, 2456–2467

- Foster, C.J., Gilkerson, T., Stocker, R. (1991) Herbicidal carboxamide derivatives, Shell International Research Maatschappij B.V., EP 0447004 A2

- Glöckler, R., Roduit, J.-P. (1996) Industrial bioprocesses for the production of substituted aromatic heterocycles, Chimia **50**, 413–415

- Kiener, A. (1992) Mikrobiologisches Verfahren zur Herstellung von 6-Hydroxypicolinsäure, Lonza AG, EP 0504818 A2

- Kiener, A., Glöckler, R., Heinzmann, K. (1993) Preparation of 6-O-oxo-1,6-dihydropyridine-2-carboxylic acid by microbial hydroxylation of pyridine-2-carboxylic acid, J. Chem. Soc. Perkin Trans. I **11**, 1201–1202

- Kiener, A., Roduit, J.-P., Glöckler, R. (1999) Mikrobiologisches Verfahren zur Herstellung von heteroaromatischen Carbonsäuren mittels Mikroorganismen der Gattung *Alcaligenes*, Lonza AG, EP 0747486 B1

- Petersen, M., Kiener, A. (1999) Biocatalysis – preparation and functionalization of *N*-heterocycles, Green Chem. **2**, 99–106

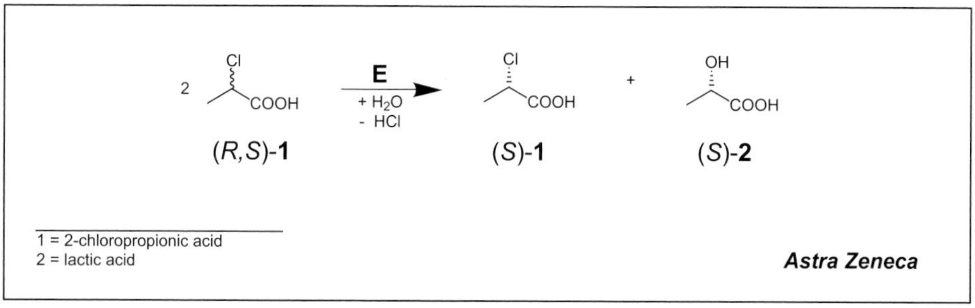

Dehalogenase

Pseudomonas putida

Fig. 3.8.1.2 – 1

1) Reaction conditions

[1]:	> 0.1 M, 10.9 g · L⁻¹ [108.52 g · mol⁻¹]

[1]: > 0.1 M, 10.9 g $\cdot$ L^{-1} [108.52 g $\cdot$ mol^{-1}]
pH: 7.4
T: 30 °C
medium: aqueous
reaction type: C-halide hydrolysis
catalyst: suspended whole cells
enzyme: 2-haloacid halidohydrolase (2-haloacid dehalogenase)
strain: *Pseudomonas putida*
CAS (enzyme): [37289–39–7]

2) Remarks

- *Pseudomonas putida* is a robust organism that grows rapidly on cheap carbon sources even in presence of high substrate concentrations.

- Since, at alkaline pH, 2-chloropropionic acid racemizes rapidly it is important that the dehalogenase retains its high specific activity at neutral pH.

- The K_M-value is very low so that the reaction rate remains high with typical end concentrations of the substrate of 1 g $\cdot$ L^{-1}.

- Since the stability of the cells is poor, the steps of fermentation and biotransformation are carried out separately.

- Usually an immobilization method would be used but in this case a special cell drying method is applied, where only 10 % enzyme activity is lost and the solid biocatalyst can be stored for over 12 months without deactivation.

- The main steps for the production of (S)-2-chloropropionic acid are:
  1. Continuous fermentation
  2. Biocatalyst preparation and drying
  3. Biocatalyst storage
  4. Fed-batch biotransformation
  5. Biocatalyst separation
  6. Solvent extraction of product

- The gene responsible for the intracellular enzyme production could be determined, cloned and overexpressed in an *E. coli* strain. Several manipulations of the strain increased the production level by about 20 times.

3) Flow scheme

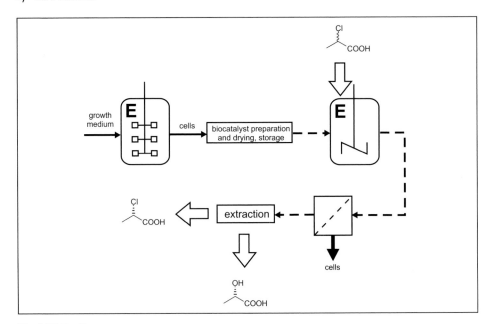

Fig. 3.8.1.2 – 2

4) Process parameters

yield:	> 42 %
ee:	99 %
chemical purity:	99 %
reactor type:	fed-batch
capacity:	2,000 t · a^{-1}
down stream processing:	solvent extraction / distillation
enzyme activity:	22 U · mg^{-1} (Protein)
enzyme supplier:	in-house manufacture
start-up date:	1988
production site:	Zeneca Life Science Molecules, U.K.
company:	Astra Zeneca, U.K.

5) Product application

- Applications include the synthesis of other optically active compounds such as pharmaceuticals.

- The main application of (S)-2-chloropropionic acid is in the synthesis of optically active phenoxypropionic acid herbicides:

dichlorprop

mecoprop

fluazifop

fenoxaprop

Fig. 3.8.1.2 – 3

- Many of these herbicides were sold as racemates but because of environmental considerations, the switch to the active enantiomer is preferred, thereby doubling capacities in existing plants and leading to cost reduction through raw material savings.

- Other chiral products synthesized by this dehalogenase technique are shown in the following figure:

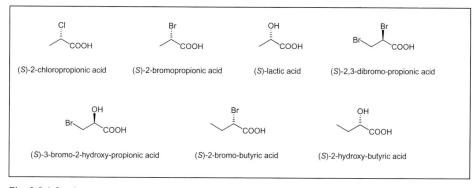

(S)-2-chloropropionic acid   (S)-2-bromopropionic acid   (S)-lactic acid   (S)-2,3-dibromo-propionic acid

(S)-3-bromo-2-hydroxy-propionic acid   (S)-2-bromo-butyric acid   (S)-2-hydroxy-butyric acid

Fig. 3.8.1.2 – 4

Dehalogenase
Pseudomonas putida

EC 3.8.1.2

6) Literature

- Barth, P.T., Bolton, L., Thomson, J.C. (1992) Cloning and partial sequencing of an operon encoding two *Pseudomonas putida* haloalkanoate dehalogenases of opposite stereospecificity, J. Bacteriol. **174**, 2612–2619

- Cheetham, P.S.J. (1994) Case studies in applied biocatalysis, in: Applied Biocatalysis (Cabral, J.M.S., Best, D., Boross, L., Tramper, J.; eds.) p. 70 ff., Harwood Academic Publishers, Chur, Switzerland,

- Liddell, J.M., Greer, W. (1990) Biocatalysts, Imperial Chemical Industries PLC, EP 366303

- Smith, J.M., Harrison, K., Colby, J. (1990) Purification and characterization of D-2-haloacid dehalogenase from *Pseudomonas putida* strain AJ1/23, J. Gen. Microbiol. **136**, 881–886

- Taylor, S.C. (1988) D-2-Haloalkanoic acid halidohydrolase, Imperial Chemical Industries PLC, US 4758518

- Taylor, S.C. (1997) (S)-2-Chloropropanoic acid: developments in its industrial manufacture, in: Chirality In Industry II (Collins, A.N., Sheldrake, G.N., Crosby, J., eds.) pp. 207–223, John Wiley & Sons, New York

Haloalkane dehalogenase

Alcaligenes sp. or *Pseudomonas* sp.

EC 3.8.1.5

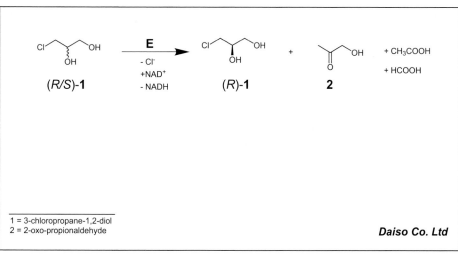

1 = 3-chloropropane-1,2-diol
2 = 2-oxo-propionaldehyde

Daiso Co. Ltd

Fig. 3.8.1.5 – 1

1) Reaction conditions

[1]:	< 0.9 M, < 100 g $\cdot$ L^{-1} [110.55 g $\cdot$ mol^{-1}]
pH:	6.8
T:	30 °C
medium:	aqueous
reaction type:	hydrolysis of halide bonds
catalyst:	immobilized whole cells
enzyme:	1-haloalkane halidohydrolasehalohydrin (haloalkane dehalogenase, HDDase,)
strain:	*Alcaligenes* sp. or *Pseudomonas* sp. depending on substrate
CAS (enzyme):	[95990–29–7]

2) Remarks

- The racemic starting materials are economically produced from propylene in the petrochemical industry. Therefore the limited yield of 50 % due to the kinetic resolution is economically tolerable.

- This microbial resolution can be carried out in an inorganic medium using bacteria that can assimilate (*R*)- or (*S*)- 2,3-dichloro-1-propanol and (*R*)- or (*S*)-3-chloro-1,2-propanediol as the sole source of carbon.

- Some related optically active halohydrins can also be resolved.

4-chloro-3-hydroxy-butyronitrile   4-chloro-3-hydroxy-butyroester   1,2-diol

- The enzyme shows a broad substrate specificity for alcohols, but not for acids.

439

- The whole cells are immobilized in calcium alginate.

- The fermenter is aerated with air at 20 L · min^{-1}.

- Glycidol can be easily synthesized from 3-chloropropane-1,2-diol:

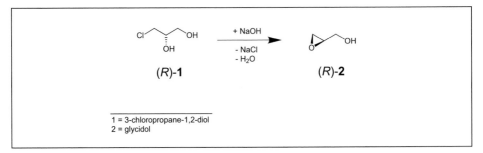

1 = 3-chloropropane-1,2-diol
2 = glycidol

Fig. 3.8.1.5 – 2

3) Flow scheme

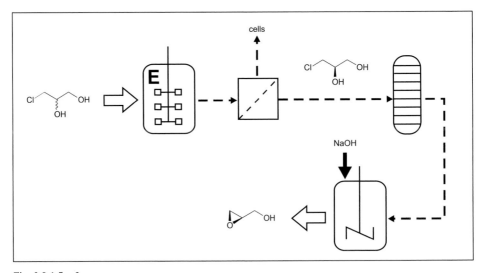

Fig. 3.8.1.5 – 3

4) Process parameters

conversion:	50%
yield:	max. 50%
ee:	99.5–100%
chemical purity:	> 98%
reactor type:	batch
reactor volume:	25,000 L
residence time:	48 h – 60 h
down stream processing:	extraction, distillation
enzyme activity:	HDDase: (in the presence of NAD$^+$ and O$_2$) 3.5 U · g^{-1} (*Alcanigenes* sp.) and 4.8 U · g^{-1} (*Pseudomonas* sp.)
start-up date:	1994
production site:	Daiso Co. Ltd, Japan
company:	Daiso Co. Ltd, Japan

5) Product application

- The products are chiral synthons for various chiral pharmaceuticals, agrochemicals and ferro-electroliquid crystals.

- Other possible products:

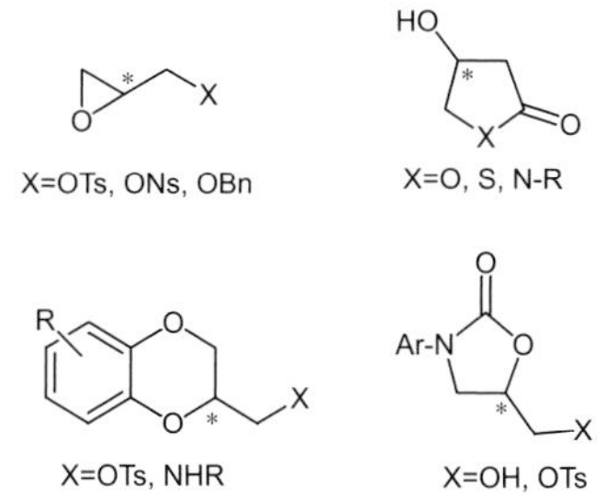

X=OTs, ONs, OBn   X=O, S, N-R

X=OTs, NHR   X=OH, OTs

Fig. 3.8.1.5 – 4

6) Literature

- Crosby, J., (1991) Synthesis of optically active compounds: A large scale perspective, Tetrahedron **47**, 4789–4846

- Kasai, N., Suzuki, T., Furukawa, Y. (1998) Chiral C3 epoxides and halohydrins: Their preparation and synthetic application, J. Mol. Catal. B: Enzymatic **4**, 237–252

- Kasai, N., Tsujimura, K., Unoura, K., Suzuki, T. (1992) Preparation of (*S*)-2,3-dichloro-1-propanol by *Pseudomonas* species and its use in the synthesis of (*R*)-epichlorohydrin, J. Indust. Microbiol. **9**, 97–101

- Suzuki, T., Kasai, N. (1991) A novel method for the generation of (*R*)-and (*S*)-3-chloro-1,2-propanediol by stereospecific dehalogenating bacteria and their use in the preparation of (*R*)-and (*S*)-glycidol, Bioorg. Med. Chem. Lett. **1**, 343–346

- Suzuki, T., Kasai, N., Yamamoto, R., Minamiura, N. (1992) Isolation of a bacterium assimilating (R)-3-chloro-1,2-propanediol and production of (S)-3-chloro-1,2-propanediol using microbial resolution, J. Ferment. Bioeng. **73**, 443–448

- Suzuki, T., Kasai, N., Yamamoto, R., Minamiura, N. (1993) Production of highly optically active (R)-3-chloro-1,2-propanediol using a bacterium assimilating the (S)-isomer, Appl. Microbiol. Biotechnol. **40**, 273–278

- Suzuki, T., Kasai, N., Yamamoto, R., Minamiura, N. (1994) A novel enzymatic dehalogenation of (R)-3-chloro-1,2-propanediol in *Alicaligenes sp.* DS-S-7G, Appl. Microbiol. Biotechnol. **42**, 270–279

- Suzuki, T., Kasai, N., Minamiura, N. (1994) A novel generation of optically active 1,2-diols from the racemates by using halohydrin dehydro-dehalogenase, Tetrahedron: Asymmetry **5**, 239–246

- Suzuki, T., Kasai, N., Yamamoto, R., Minamiura, N. (1994) Microbial production of optically active 1,2-diols using resting cells of *Alcaligenes sp.* DS-S-7G, J. Ferment. Bioeng. **78**, 194–196

- Suzuki, T., Idogaki, H., Kasai, N. (1996) A novel generation of optically active ethyl-4-chloro-3-hydroxybutyrate as a C4 chiral building unit using microbial dechlorination, Tetrahedron: Asymmetry **7**, 3109–3112

- Suzuki, T., Idogaki, H., Kasai, N. (1996) Production of (S)-4-chloro-3-hydroxybutyronitrile using microbial resolution, Bioinorg. Med. Chem. Lett. **6**, 2581–2584

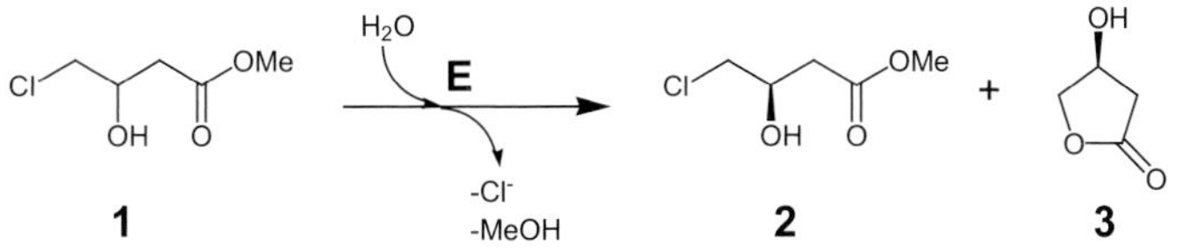

1 = methyl 4-chloro-3-hydroxybutyrate
2 = (*R*)-methyl 4-chloro-3-hydroxybutyrate
3 = (*S*)-3-hydroxy-γ-butyrolactone

Daiso Co. Ltd.

Fig. 3.8.1.5 – 1

1) Reaction conditions

[**1**]:	[152.58 g · mol^{-1}]
pH:	6.7
T:	30 °C
medium:	aqueous
reaction type:	ester resolution
catalyst:	whole cell
enzyme:	halohydrin dehydro-dehalogenase (dehalogenase)
strain:	*Enterobacter* sp. DS-S-75

2) Remarks

• Two chiral building blocks can be produced in a one-pot reaction.

• The reaction was carried out using resting whole cells. To harvest cells with a high specific activity it was necessary to keep the DO (dissolved oxygen) low during log phase.

• The base for controlling the pH had an important influence on the optical purity of (*S*)-3-hydroxy-γ-butyrolactone. Weak bases yield much higher optical purities than strong bases. 14 % (w/w) ammonium hydroxide solution gave the best optical purity and was finally used to control pH.

3) Flow scheme

Not published.

4) Process parameters

conversion:	50 %
ee:	>99 % (*R*)-methyl 4-chloro-3-hydroxybutyronitrile /
	95 % (*S*)-3-hydroxy-γ-butyrolactone
reactor type:	CSTR
reactor volume:	15,000 L
enzyme activity:	19.5 U · g$_{\text{wet cells}}^{-1}$

enzyme supplier: in-house production
company: Daiso Co. Ltd.

5) Product application

- (R)-Methyl 4-chloro-4-hydroxybutyrate can be used as building block for ʟ-carnitine and (R)-4-hydroxy-pyrrolidone, which is a moiety of several antibiotics. The synthesis of the (R)-enantiomer by chemical means using BINAP is difficult due to imperfect stereospecificity.

6) Literature

- Kasai, N., Suzuki, T. (2002) Industrialization of the microbial resolution of chiral C3 and C4 synthetic units: from a small beginning to a major operation, a personal account, Adv. Synth. Catal. **345** (4), 437–455

Fig. 3.8.X.X – 1

1 = ethyl-(S)-4-chloro-3-hydroxybutyrate
2 = ethyl-(R)-4-cyano-3-hydroxybutyrate

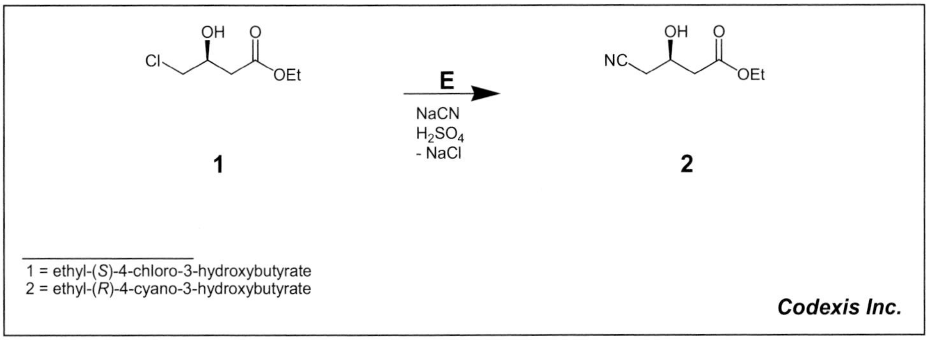

Codexis Inc.

1) Reaction conditions

[1]:	>0.577 M, >80 g · L^{-1} [138.55 g · mol^{-1}]
pH:	7.3
T:	40 °C
medium:	aqueous
reaction type:	cyanation
catalyst:	soluble recombinant enzymes
enzyme:	mutant halohydrin dehalogenase
enzyme:	mutant glucose dehydrogenase

2) Remarks

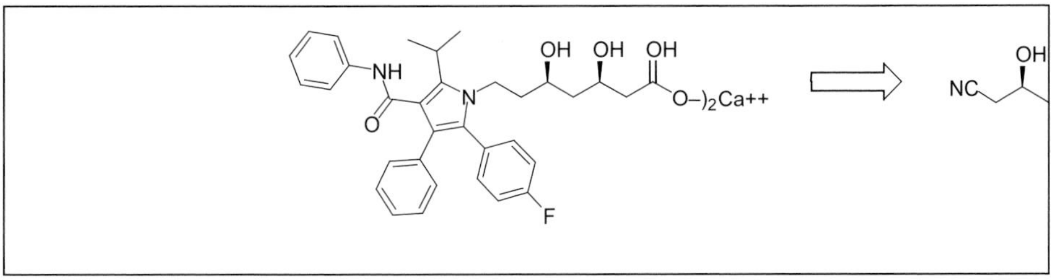

Fig. 3.8.X.X – 2

- First step of the process is the enantioselective reduction of ethyl-4-chloro-3-ketobutyrate (see page 191, EC 1.1.X.X)

3) Flow scheme

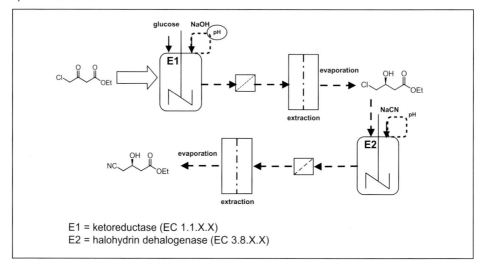

E1 = ketoreductase (EC 1.1.X.X)
E2 = halohydrin dehalogenase (EC 3.8.X.X)

Fig. 3.8.X.X – 3

4) Process parameters

conversion:	>99.5 %
yield:	>90 %
ee:	>99.9 % (retained)
selectivity:	extremely high
optical purity	extremely high
chemical purity:	>99.5 %
residence time:	reaction time <24 h
company:	Codexis Inc., USA

5) Product application

- Key chiral intermediate for atorvastatin

6) Literature

- Davis, C., Grate, J., Gray, D., Gruber, J., Huisman, G., Ma, S., Newman, L., Sheldon, R. (2005) Enzymatic process for the production of 4-substituted 3-hydroxybutyric acid derivatives, Codexis, Inc., WO04015132

- Davis, C., Jenne, S., Krebber, A., Huisman, G., Newman, L., (2005) Improved ketoreductase polypeptides and related polynucleotides, Codexis, Inc., WO05017135

- Davis, C., Fox, R., Gavrilovic, V., Huisman, G., Newman, L. (2005) Improved halohydrin dehalogenases and related polynucleotides, Codexis, Inc.,WO05017141

- Davis, C. et al. (2005) Enzymatic process for the production of 4-subsitituted 3-hydroxybutyric acid derivatives and vicinal cyano, hydroxyl substituted carboxylic acid esters, Codexis, Inc., WO05018579

446

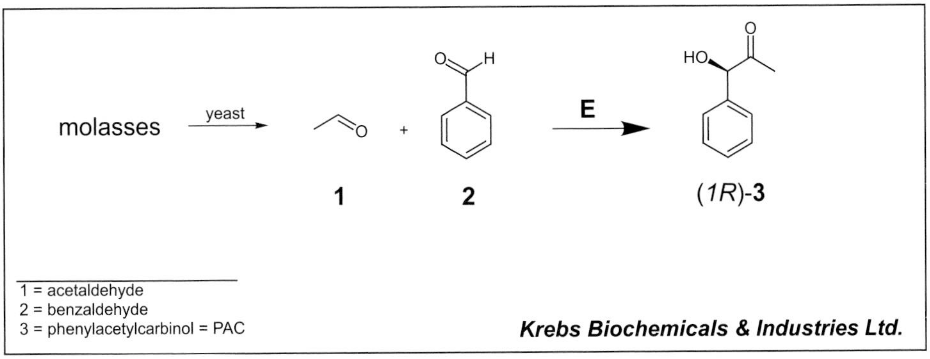

1 = acetaldehyde
2 = benzaldehyde
3 = phenylacetylcarbinol = PAC

Krebs Biochemicals & Industries Ltd.

Fig. 4.1.1.1 – 1

1) Reaction conditions

[1]:	0.022 M, 3.3 g · L^{-1} [150.17 g · mol^{-1}]
medium:	aqueous
reaction type:	carboligation
catalyst:	whole cells
enzyme:	2-oxo-acid carboxy-lyase (α-ketoacid carboxylase, pyruvate decarboxylase)
strain:	*Saccharomyces cerevisiae*
CAS (enzyme):	[9001-04-01]

2) Remarks

- Phenylacetylcarbinol production from benzaldehyde by yeast is also operated on a large scale by Knoll (BASF, Germany) and Malladi Drugs (India).

- Phenylacetylcarbinol is chemically converted in a two-step process to D-pseudoephedrine:

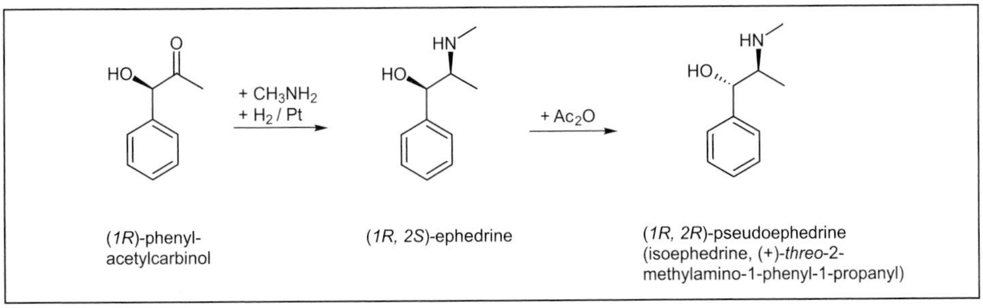

(1R)-phenyl-acetylcarbinol

(1R, 2S)-ephedrine

(1R, 2R)-pseudoephedrine
(isoephedrine, (+)-*threo*-2-methylamino-1-phenyl-1-propanyl)

Fig. 4.1.1.1 – 2

3) Flow scheme

Not published.

447

4) Process parameters

reactor type:	batch
reactor volume:	30,000 L
capacity:	$120 \ t \cdot a^{-1}$
enzyme consumption:	$6 \ g_{yeast} \cdot kg_{PAC}$
production site:	Hyderabad, India
company:	Krebs Biochemicals & Industries Ltd., India

5) Product application

- Ephedrine and pseudoephedrine are used for the treatment of asthma, hay fever and as a bronchodilating agent and decongestant.

6) Literature

- Dr. Ravi, R.T., Krebs Biochemicals Ltd., personal communication

- Cheetham, P.S.J. (1994) Case studies in applied biocatalysis – from ideas to products, in: Applied Biocatalysis (Cabral, J.M.S., Best, D., Boross, L., Tramper, J., eds.), pp. 87–89, Harwood Academic Publishers, USA

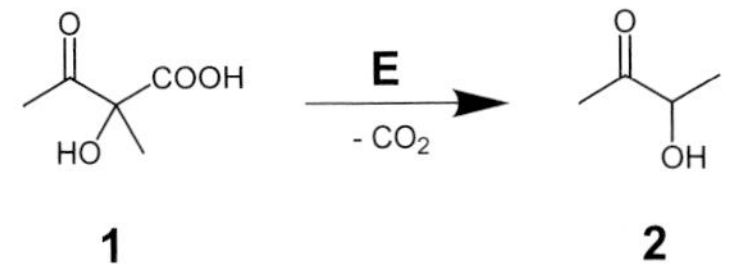

Acetolactate decarboxylase
Bacillus brevis

1 = α-acetolactate
2 = acetoin

Novo Nordisk

Fig. 4.1.1.5 – 1

1) Reaction conditions

[2]:	$< 1.13\ \mu M$, $< 0.1\ mg \cdot L^{-1}$ [88.11 g · mol^{-1}]
pH:	< 4.0
T:	13°C
medium:	aqueous
reaction type:	decarboxylation
catalyst:	solubilized enzyme
enzyme:	(S)-2-hydroxy-2-methyl-3-oxobutanoate carboxy-lyase (α-acetolactate decarboxylase)
strain:	*Bacillus brevis*
CAS (enzyme):	[9025-02–9]

2) Remarks

- During beer fermentation diacetyl is formed by a non-enzymatic oxidative decarboxylation of α-acetolactate.

- Diacetyl has a very low flavor threshhold, compared with acetoin.

- The problem is the slow reaction rate for the conversion of α-acetolactate to diacetyl with subsequent conversion of diacetyl to acetoin. The addition of acetolactate decarboxylase allows the bypassing of the slow oxidation step:

449

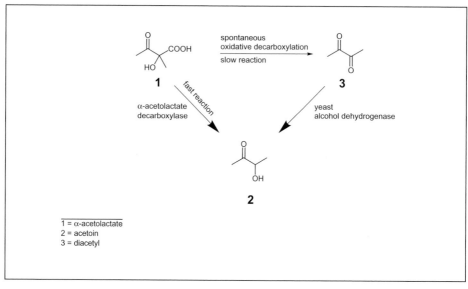

1 = α-acetolactate
2 = acetoin
3 = diacetyl

Fig. 4.1.1.5 – 2

- The yeast was harvested and reused for a total of five consecutive fermentations.

- The enzyme can be activated at low pH values by addition of glutaraldehyde (about 0.05 %) which intermolecularly cross-links the active dimer, which otherwise dissociates under acidic conditions.

3) Flow scheme

Not published.

4) Process parameters

residence time: > 6 d
company: Novo Nordisk, Denmark

5) Product application

- The process can be applied for beer fermentation procedures, in which diacetyl is produced as a by-product and has to be eliminated.

6) Literature

- Pedersen, S., Lange, N.K., Nissen, A.M. (1995) Novel industrial enzyme applications, Ann. N. Y. Acad. Sci. **750**, 376–390

Aspartate β-decarboxylase
Pseudomonas dacunhae

EC 4.1.1.12

Fig. 4.1.1.12 – 1

1) Reaction conditions

[1]:	2.5 M, 332.75 g · L^{-1} [133.1 g · mol^{-1}]
pH:	6.2
T:	37 °C
medium:	aqueous
reaction type:	decarboxylation
catalyst:	immobilized whole cells
enzyme:	L-aspartate 4-decarboxylase (L-aspartate β-decarboxylase)
strain:	*Pseudomonas dacunhae*
CAS (enzyme):	[9024–57–1]

2) Remarks

- L-Alanine is produced industrially by Tanabe Seiyaku, Japan, since 1965 via a batch process with L-aspartate β-decarboxylase from *Pseudomonas dacunhae*.

- To improve the productivity a continuous production was established in 1982. Here the formation of carbon dioxide was the main problem in comparison to the catalyst stability and the microbial enzyme activity. The production of CO_2 occurs stoichiometricaly (nearly 50 L of CO_2 for each liter of reaction mixture with 2 M aspartate). The consequences are difficulties in obtaining a plug-flow condition in fixed bed reactors and the pH shift that takes places due to formation of CO_2. Therefore a pressurized fixed bed reactor with 10 bar was designed.

- The enzyme stability is not affected by the elevated pressure.

- The main side reaction, the formation of L-malic acid, can be completely avoided.

- To improve the yield of L-alanine the alanine racemase and fumarase activities can be destroyed by acid treatment of the microorganisms (pH 4.75, 30 °C). The L-aspartate β-decarboxylase activity is stabilized by the addition of pyruvate and pyridoxal phosphate.

- The process is often combined with the aspartase catalyzed synthesis of L-aspartic acid from fumarate (see page 500) in a two step biotransformation (Fig. 4.1.1.12 – 4). The main reason for the separation in two reactors is the difference in pH optimum for the two enzymes (aspartase from *E. coli*: pH 8.5, L-aspartate β-decarboxylase: pH 6.0). This is the first commercialized system of a sequential enzyme reaction using two kinds of immobilized microbial cells:

- In this combination L-alanine can efficiently be produced by co-immobilization of *E. coli* and *Pseudomonas dacunhae* cells.

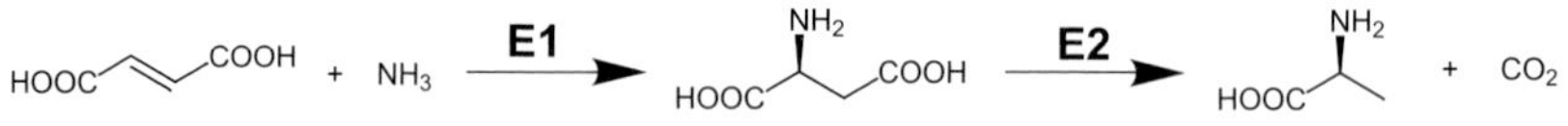

1 L-2 L-3

1 = fumaric acid
2 = aspartic acid
3 = alanine
E1 = L-aspartase ammonia-lyase (fumaric aminase)
E2 = aspartate β-decarboxylase (L-aspartate 4-decarboxylase)

Fig. 4.1.1.12 – 2

- If D,L-aspartic acid is used as a substrate for the reaction, L-aspartic acid is converted to L-alanine and D-aspartic acid remains unchanged in one resolution step. Both products can be separated after crystallization by addition of sulfuric acid. The continuous variant of the L-alanine and D-aspartic acid production is commercially in operation since 1988 (Fig. 4.1.1.12 – 5).

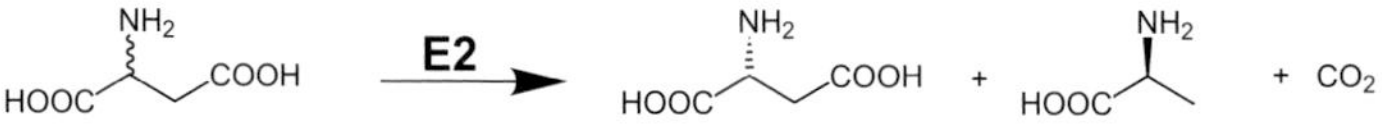

D,L-1 D-1 L-2

1 = aspartic acid
2 = alanine

Fig. 4.1.1.12 – 3

3) Flow scheme

- Production of L-aspartic acid from fumarate in a two step biotransformation:

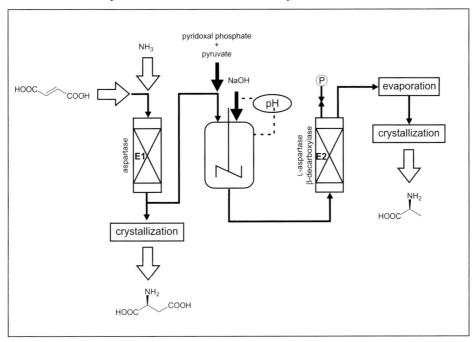

Fig. 4.1.1.12 – 4

- Production of L-alanine and D-aspartic acid:

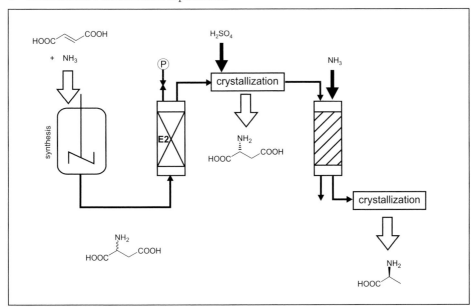

Fig. 4.1.1.12 – 5

4) Process parameters

conversion:	99 %
yield:	86 % after down stream processing
reactor type:	plug flow reactor
reactor volume:	1,000 L (each fixed bed column)
capacity:	114 t · a^{-1} D-aspartic acid; 61 t · a^{-1} L-alanine
residence time:	11 h
space-time-yield:	317 g · L^{-1} · d^{-1} D-aspartic acid; 170 g · L^{-1} · d^{-1} L-alanine
down stream processing:	crystallization
start-up date:	1982
production site:	Tanabe Seiyaku Co., Ltd., Japan
company:	Tanabe Seiyaku Co., Ltd., Japan

5) Product application

- The products are used in infusion solutions and as food additives.

- D-Aspartic acid is an intermediate for the semi-synthetic penicillin apoxycillin:

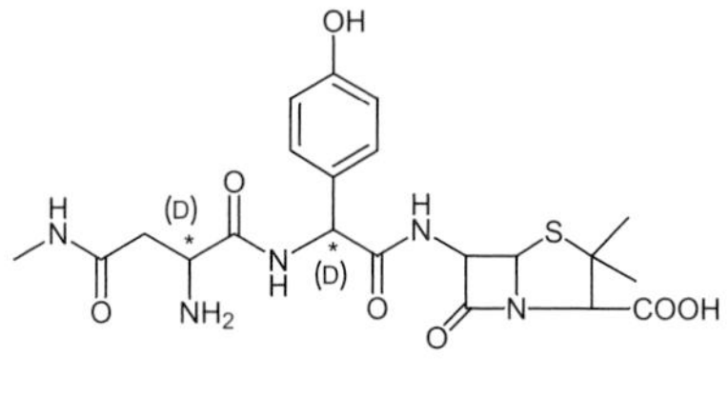

apoxycillin

Fig. 4.1.1.12 – 6

6) Literature

- Chibata, L., Tosa, T., Shibatani, T. (1992) The industrial production of optically active compounds by immobilized biocatalysts, in: Chirality in Industry (Collins, A.N., Sheldrake, G.N., Crosby, J., eds.) pp. 351–370, John Wiley & Sons, New York

- Furui, M., Yamashita, K. (1983) Pressurized reaction method for continuous production of L-alanine by immobilized *Pseudomonas dacunhae* cells, J. Ferment. Technol. **61**, 587–591

- Schmidt-Kastner, G., Egerer, P. (1984) Amino acids and peptides, in: Biotechnology, Vol. 6a, (Kieslich, K., ed.) pp. 387–419, Verlag Chemie, Weinheim

- Takamatsu, S., Umemura, I., Yamamoto, K., Sato, T., Tosa, T., Chibata, I. (1982) Production of L-alanine from ammonium fumarate using two immobilized microorganisms: elimination of side reactions, Eur. J. Appl. Microbiol. Biotechnol. **15**, 147–152

- Takamatsu, S., Tosa, T. (1993) Production of L-alanine, in: Industrial Application of Immobilized Biocatalysts, (Tanaka, A., Tosa, T., Kobayashi, T., eds.) Marcel Dekker Inc., New York

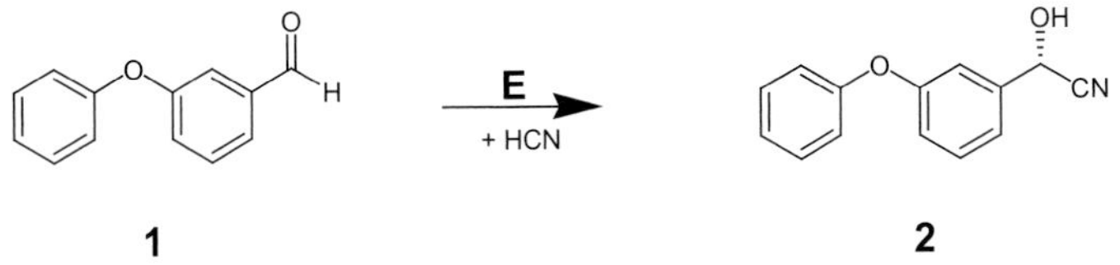

Oxynitrilase
Hevea brasiliensis

1 = *m*-phenoxybenzaldehyde
2 = (S)-*m*-phenoxybenzaldehyde cyanohydrin

DSM

Fig. 4.1.2.39 – 1

1) Reaction conditions

[1]:	[198.22 g · mol⁻¹]
medium:	biphasic solvent system: aqueous buffer / methyl *tert*-butyl ether
reaction type:	HCN addition
catalyst:	whole cells (*Hevea brasiliensis* cloned and overexpressed in a microbial host microorganism)
enzyme:	hydroxynitrilase (oxynitrilase)
strain:	*Hevea brasiliensis*
CAS (enzyme):	[202013-28-3]

2) Remarks

- The biphasic solvent system has also been shown to represent a suitable media for the enzymatic manufacture of (R)-mandelonitrile on an industrial scale.

- The enzyme can be reused more than 5 times with loss of activity.

3) Flow scheme

Not published.

4) Process parameters

yield:	98 %
ee:	98 %
space-time-yield:	2.1 mol · L⁻¹ · h⁻¹
company:	DSM, The Netherlands

5) Product application

- Intermediate for the manufacture of pyrethroids

6) Literature

- Gröger, H. (2001) Enzymatic routes to enantiomerically pure aromatic α-hydroxy carboxylic acids: a further example for the diversity of biocatalysis, Adv. Synth. Catal. **343** (6-7), 547–558

- Pöchlauer, P., Schmidt, M., Wirth, I., Neuhofer, R., Zabelinskaja-Mackova, A., Griengl, H., van den Broek, C., Reintjens, R., Wöries, J. (1999) EP 927766

- Griengl, H., Schwab, H., Fechter, M. (2000) The synthesis of chiral cyanohydrins by oxynitrilases, Opht. Gen. **18**, 252

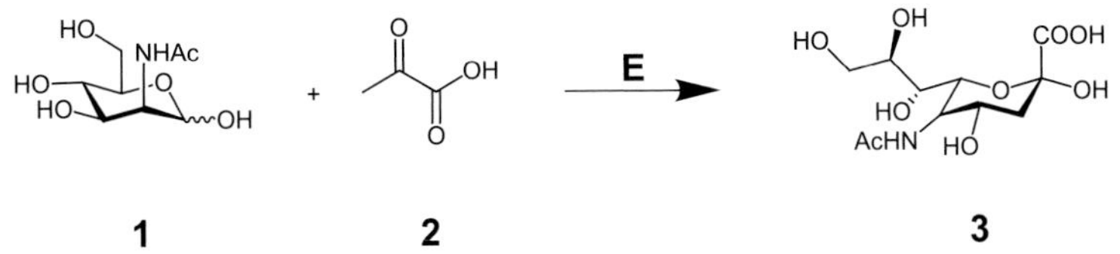

N-Acetyl-D-neuraminic acid aldolase
Escherichia coli

1 = N-acetyl-D-mannosamine
2 = pyruvic acid
3 = N-acetyl-D-neuraminic acid

GlaxoSmithKline

Fig. 4.1.3.3 – 1

1) Reaction conditions

[**1**]:	0.9 M, 200 g · L^{-1} [221.21 g · mol^{-1}]
pH:	7.5
T:	20°C
medium:	aqueous
reaction type:	C-C bond cleavage
catalyst:	immobilized enzyme
enzyme:	N-acetylneuraminate pyruvate-lyase (sialic aldolase)
strain:	Escherichia coli
CAS (enzyme):	[9027–60–5]

2) Remarks

- The chemical synthesis of Neu5Ac is costly since it requires complex protection and deprotection steps.

- The enzyme is covalently immobilized on Eupergit C.

- Since N-acetyl-D-mannosamine is very expensive, it is synthesized from N-acetyl-D-glucosamine by epimerization at C$_2$. The equilibrium of the epimerization is on the side of N-acetyl-D-glucosamine (GlcNAc:ManNAc = 4:1). After neutralization and addition of isopropanol GlcNAc precipitates. In the remaining solution a ratio of GlcNAc:ManNAc = 1:1 is reached. After evaporation to dryness and extraction with methanol, a ratio of GlcNAc:ManNAc = 1:4 is reached. In contrast to this chemical epimerization enzymatic epimerization using an epimerase is carried out by another company, see pages 459 and 506.

- N-Acetyl-D-glucosamine is not a substrate for the aldolase, but it is an inhibitor and limits by this way the applied maximal concentration of ManNAc.

- Non-converted GlcNAc can be recycled after down stream processing by epimerization to ManNAc.

- The natural direction of the aldolase-catalyzed reaction is the cleavage of Neu5Ac to pyruvate and ManNAc. The K_M for ManNAc is 700 mM. Therefore a very high ManNAc concentration of up to 20% w/V is used. By this means ManNAc itself drives the equilibrium. Pyruvate is used in a 1.5 molar ratio.

- Neu5Ac can be crystallized directly from the reaction mixture simply by the addition of acetic acid.

- In the repetitive batch mode the immobilized enzyme can be reused for at least nine cycles without any significant loss in activity.

3) Flow scheme

Not published.

4) Process parameters

yield:	60 % based on ManAc; 27 % based on GlcNAc
reactor type:	repetitive batch
capacity:	multi t
residence time:	5 h
down stream processing:	crystallization
enzyme supplier:	Röhm GmbH, Germany
company:	GlaxoSmithKline, U.K.

5) Product application

- Neu5Ac is the major representative of amino sugars (sialic acids) that are incorporated at the terminal positions of glycoproteins and glycolipids and play an important role in a wide range of biological recognition processes.

- The synthesis of Neu5Ac analogues and Neu5Ac-containing oligosaccharides is of interest in studies towards inhibitors of neuraminidase, hemagglutinin and selectin-mediated leucocyte adhesion.

6) Literature

- Dawson, M., Noble, D., Mahmoudian, M. (1994) Process for the preparation of *N*-acetyl-neuraminic acid, PCT WO 94/29476

- Mahmoudian, M., Noble, D., Drake, C.S., Middleton, R.F., Montgomery, D.S., Piercey, J.E., Ramlakhan, D., Todd, M., Dawson, M.J. (1997) An efficient process for production of *N*-acetyl-neuraminic acid using *N*-acetylneuraminic acid aldolase, Enzyme Microb. Tech. **20**, 393–400

- Zak, A., Dodds, D.R. (1997) Application of biocatalysis and biotransformations to the synthesis of pharmaceuticals, Drug Discovery Today **2**, 513–531

N-Acetyl-D-neuraminic acid aldolase
Escherichia coli

EC 4.1.3.3

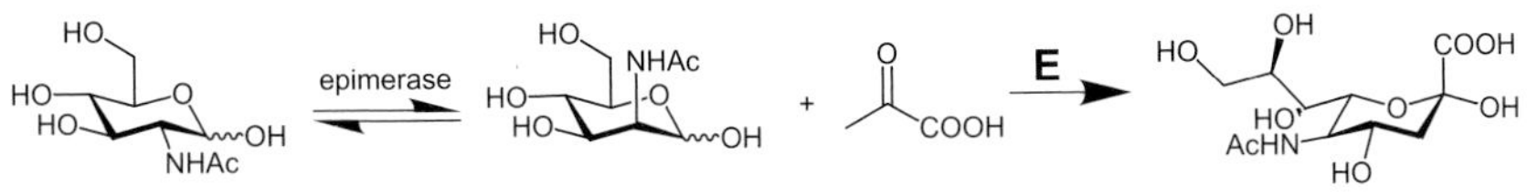

1 = N-acetyl-D-glucosamine (GlcNAc)
2 = N-acetyl-D-mannosamine (ManNAc)
3 = pyruvic acid
3 = N-acetyl-D-neuraminic acid (Neu5Ac)

Marukin Shoyu Company, Ltd.
Forschungszentrum Jülich, GmbH

Fig. 4.1.3.3 – 1

1) Reaction conditions

[1]:	0.8 M, 177 g · L^{-1} [221.21 g · mol^{-1}]
pH:	7.2
T:	30 °C
medium:	aqueous
reaction type:	C-C bond cleavage
catalyst:	solubilized enzyme
enzyme:	*N*-acetylneuraminate pyruvate-lyase (sialic aldolase)
strain:	*Escherichia coli*
CAS (enzyme):	[9027–60–5]

2) Remarks

- *N*-Acetyl-D-neuraminic acid aldolase from *E. coli* K-12 and *N*-acetyl-D-glucosamine epimerense from porcine kidney have been cloned and overexpressed in *E. coli*.

- The enzyme catalyzed aldol condensation to *N*-acetylneuraminic acid is combined in a one-vessel synthesis with the enzyme-catalyzed epimerization of *N*-acetyl-glucosamine (GlcNAc), see page 385.

- The production of Neu5Ac on a multi ton scale is carried out by Glaxo utilizing chemical epimerization (see page 338). In contrast to this synthesis here the native, non-immobilized enzyme is applied.

- Both enzymes (epimerase and aldolase) can be used in a pH range of 7.0 to 8.0. For the biotransformation pH 7.2 was chosen.

- Since excess amounts of pyruvate (educt for the aldolase) inhibit the epimerase, a fed batch in regard to pyruvate is performed. After the start of the reaction with a ratio of pyruvate to GlcNAc of 1:0.6, two times pyruvate is added twice up to a total amount of 251 mol (ratio of pyruvate to GlcNAc at start: 1:0.6; after first addition: 1:1.5; after second addition 1:2).

- Before the product is purified by crystallization (initiated by the addition of 5 volumes of glacial acetic acid), the enzymes are denaturated by heating to 80 °C for 5 minutes, afterwards the reaction solution is filtered.

- Another process layout is realized by the Research Center Jülich, Germany, that already established in 1991 the one-pot synthesis of Neu5Ac with the combined use of epimerase and aldolase. Here a continuously operated membrane reactor is used, where the enzymes are retained by an ultrafiltration membrane. By this technology kg quantities of N-acetylneuraminic acid were produced (GlcNAc = 300 mM; pyruvate = 600 mM; pH = 7.5; T = 25 °C; conversion = 78 %; residence time = 4 h; reactor volume = 0.44 L; space-time-yield = 655 g · L^{-1} · d^{-1}; enzyme consumption = 10,000 U · kg^{-1}). The advantage of this approach is a simplified downstream processing, since the catalysts are already separated. The product is additionally pyrogen free.

3) Flow scheme

Not published.

4) Process parameters

conversion:	77 %
reactor type:	fed batch
reactor volume:	200 L
capacity:	multi kg
down stream processing:	crystallization
enzyme supplier:	Marukin Shoyu Co., Ltd., Japan
company:	Marukin Shoyu Co., Ltd., Japan and Forschungszentrum Jülich GmbH, Germany

5) Product application

- Please see page 338.

6) Literature

- Ghisalba, O., Gygax, D., Kragl, U., Wandrey, C. (1991) Enzymatic method for N-acetylneuraminic acid production, Novartis , EP 0428947

- Kragl, U., Gygax, D., Ghisalba, O., Wandrey, C. (1991) Enzymatic process for prepaing N-acetylneuraminic acid, Angew. Chem. Int. Ed. Engl. **30**, 827–828

- Kragl, U., Kittelmann, M., Ghisalba, O., Wandrey, C. (1995) N-Acetylneuraminic acid: from a rare chemical isolated from natural sources to a multi-kilogram enzymatic synthesis for industrial application, Ann. N. Y. Acad. Sci. **750**, 300–305

- Maru, I., Ohnishi, J., Ohta, Y., Tsukada, Y. (1998) Simple and large-scale production of N-acetylneuraminic acid from N-acetyl-D-glucosamine and pyruvate using N-acyl-D-glucosamine 2-epimerase and N-acetylneuraminate lyase, Carbohydrate Res. **306**, 575–578

- Ohta, Y., Tsukada, Y., Sugimori, T., Murata, K., Kimura, A. (1989) Isolation of a constitutive N-acetylneuraminate lyase-producing mutant of *Escherichia coli* and its use for NPL production, Agric. Biol. Chem. **53**, 477–481

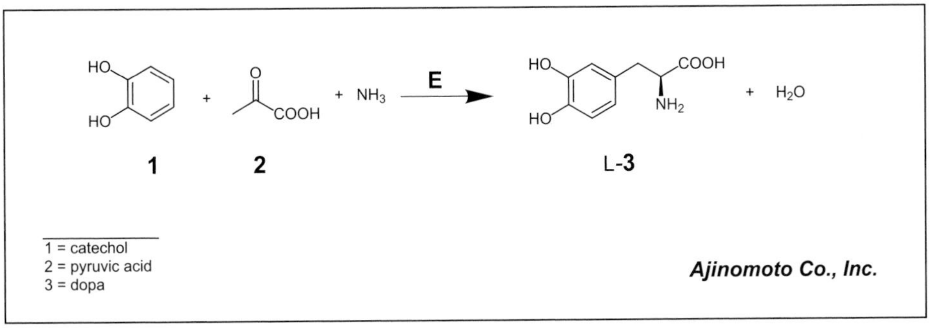

1 = catechol
2 = pyruvic acid
3 = dopa

Ajinomoto Co., Inc.

Fig. 4.1.99.2 – 1

1) Reaction conditions

[3]:   0.558 M, 110 g · L⁻¹ [197.19 g · mol⁻¹]<pH>
medium: aqueous
reaction type: α,β-elimination (reversed)
catalyst: suspended whole cells
enzyme: L-tyrosine phenol lyase (deaminating) (β-tyrosinase)
strain: *Erwinia herbicola*
CAS (enzyme):   [9059–31–8]

2) Remarks

- Monsanto has successfully scaled up the chemical synthesis of L-dopa:

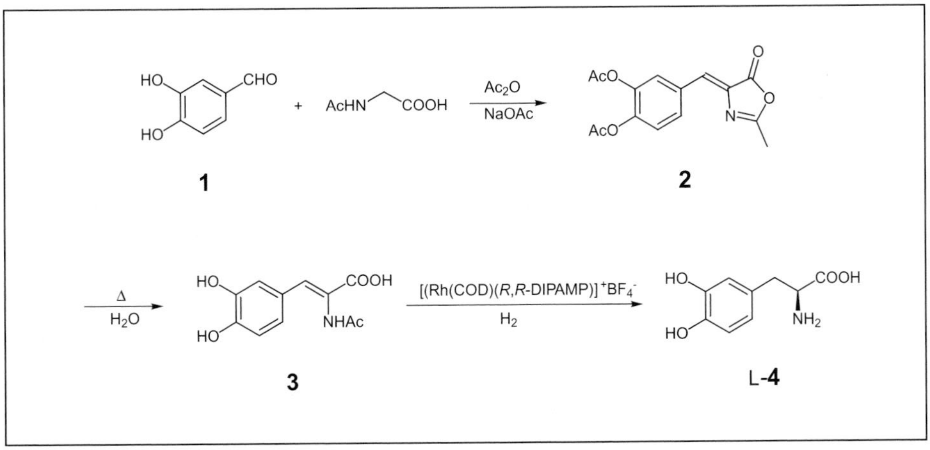

Fig. 4.1.99.2 – 2

3) Flow scheme

Not published.

4) Process parameters

reactor type: fed batch
reactor volume: 60,000 L
capacity: $250 \, t \cdot a^{-1}$
production site: Kawasaki, Kanagaua Prefecture, Japan
company: Ajinomoto Co., Ltd., Japan

5) Product application

- The product is applied for the treatment of Parkinsonism. Parkinsonism is caused by a lack of L-dopamine and its receptors in the brain. L-Dopamine is synthesized in organisms by decarboxylation of L-dopa. Since L-dopamine cannot pass the blood-brain barrier L-dopa is used in combination with dopadecarboxylase-inhibitors to avoid formation of L-dopamine outside the brain:

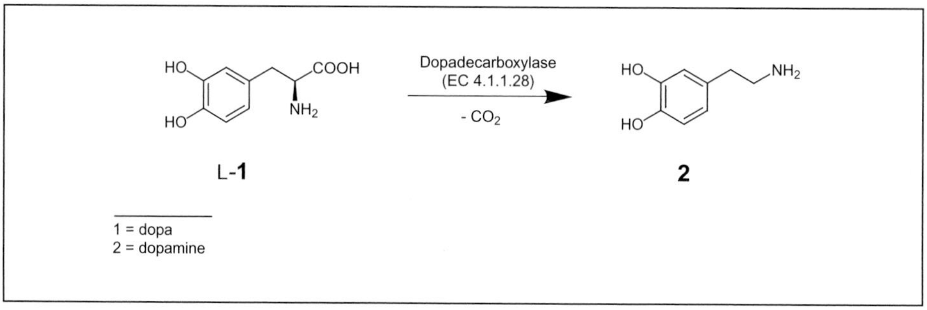

```
1 = dopa
2 = dopamine
```

Fig. 4.1.99.2 – 3

6) Literature

- Ager, D.J. (1999) Handbook of Chiral Chemicals, Marcel Dekker Inc., New York

- Tsuchida, T., Nishimoto, Y., Kotani, T., Iiizumi, K. (1993) Production of L-3,4-dihydroxyphenyl-alanine, Ajinomoto Co., Ltd., JP 5123177A

- Yamada, H. (1998) Screening of novel enzymes for the production of useful compounds, in: New Frontiers in Screening for Microbial Biocatalysis (Kieslich, K., van der Beek, C.P., de Bont, J.A.M., van den Tweel, W.J.J., eds.) pp. 13–17, Studies in Organic Chemistry **53**, Elsevier, Amsterdam

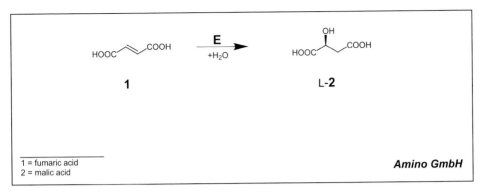

1 = fumaric acid
2 = malic acid

Amino GmbH

Fig. 4.2.1.2 – 1

1) Reaction conditions

[1]:	0.97 M, 150 g · L⁻¹ [154.14 g · mol⁻¹] (slurry of calcium salt)
[2]:	0.87 M, 150 g · L⁻¹ [172.16 g · mol⁻¹] (slurry of calcium salt)
pH:	8.0
T:	25 °C
medium:	aqueous
reaction type:	C-O bond formation
catalyst:	suspended whole cells
enzyme:	(*S*)-malate hydrolyase (fumarase, fumarate hydratase)
strain:	*Corynebacterium glutamicum*
CAS (enzyme):	[9032–88–6]

2) Remarks

- Only L-malate is produced, D-malate is not detectable.

- Microbial fumarases lead to a mixture of 85 % malate and 15 % fumarate.

- According to German drug regulations (DAB 10) the fumaric acid content of malic acid has to be less than 0.15 %.

- Fumaric acid separation is circumvented by forcing a quantitative transformation in a slurry reaction (solubility of calcium malate and calcium fumarate is approx. 1 %).

- The reaction is carried out in a slurry of crystalline calcium fumarate and crystalline calcium malate.

- The precipitation of the product shifts the equilibrium towards calcium malate (figure 4.2.1.2 – 2):

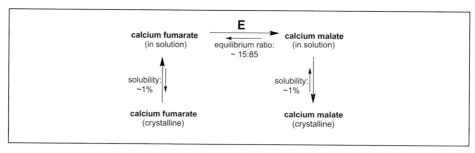

Fig. 4.2.1.2 – 2

- The biotransformation is carried out under non-sterile conditions, necessitating the addition of preservatives to prevent microbial growth.

- In this case p-hydroxybenzoic acid esters are used as preservatives because of their low effective concentrations and their biodegradability. They are permitted as food preservatives and therefore no special precautions need to be taken for their technical application.

- Enzyme stabilization is achieved by addition of soy bean protein or bovine serum albumin. The surplus of foreign protein can occupy interfaces at walls, stirrers and liquid surfaces and protect the enzyme from interface denaturation.

- Protein addition, temperature adjustment, pH adjustment and imidazole supplementation result in 25-fold increase of the (S)-malate hydrolyase productivity.

- Separation of the biocatalyst is performed easily by filtration of the slurry.

- Down-stream processing: Acidification with H_2SO_4 yields L-malic acid and gypsum. The latter is separated by filtration. L-Malic acid is purified by ion exchange chromatography.

- The same reaction is carried out by Tanabe Seiyaku Co., Japan, with an immobilized fumarase as catalyst (see page 466).

3) **Flow scheme**

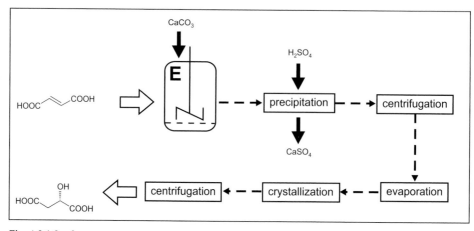

Fig. 4.2.1.2 – 3

4) Process parameters

yield:	85 %
chemical purity:	> 99 %
reactor type:	batch
capacity:	$2,000\ t \cdot a^{-1}$
down stream processing:	crystallization
enzyme consumption:	$3,500\ U \cdot kg^{-1}$
start-up date:	1988
production site:	Amino GmbH, Germany
company:	Amino GmbH, Germany

5) Product application

- About 40,000 t of malic acid are used worldwide annually as a supplement in the food, cosmetic and pharmaceutical industry.

- L-Malic acid is used as a component of amino acid infusions for parenteral nutrition.

6) Literature

- Daneel, H.J., Geiger, R., (1994) Verfahren zur Herstellung von L-Äpfelsäure aus Fumarsäure, AMINO GmbH, DE 4424664 C1

- Daneel, H.J., Geiger, R. (1994) Verfahren zur Abtrennung von Fumarsäure, Maleinsäure und/ oder Bernsteinsäure von einem Hauptbestandteil Äpfelsäure, AMINO GmbH, DE 4430010 C1

- Daneel, H.J., Busse, M., Faurie, R. (1995) Pharmaceutical grade L-malic acid from fumaric acid – development of an integrated biotransformation and product purification process; Med. Fac. Landbouww. Univ. Gent **60** / 4a, 2093–2096

- Daneel, H.J., Busse, M., Faurie, R. (1996) Fumarate hydratase from *Corynebacterium glutamicum* – process related optimization of enzyme productivity for biotechnical L-malic acid synthesis, Med. Fac. Landbouww. Univ. Gent **61** / 4a, 1333–1340

- Mattey, M. (1992) The production of organic acids, Crit. Rev. Biotechnol. **12**, 87–132

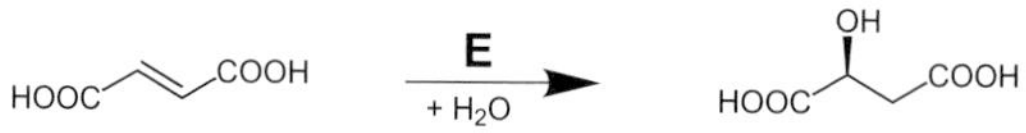

Fig. 4.2.1.2 – 1

1) Reaction conditions

[1]:	1.0 M, 116.1 g · L⁻¹ [116.1 g · mol⁻¹]

[1]:   1.0 M, $116.1 \text{ g} \cdot \text{L}^{-1}$ [$116.1 \text{ g} \cdot \text{mol}^{-1}$]
pH: 6.5–8.0
T: 37 °C
medium: aqueous
reaction type: C-O bond cleavage (elimination of H_2O)
catalyst: immobilized whole cells
enzyme: (S)-malate hydro-lyase (fumarate hydratase)
strain: *Brevibacterium flavum*
CAS (enzyme):   [9032–88–6]

2) Remarks

- The cells are immobilized on κ-carrageenan gel (160 kg wet cells in 1,000 L of 3.5 % gel).

- The side reaction (formation of succinic acid) can be eliminated by treatment of immobilized cells with bile extracts. Additionally, the activity and stability can be improved by immobilization in κ-carregeenan in the presence of Chinese gallotannin.

- The operational temperature of the immobilized cells is 10 °C higher than that of native cells.

- Earlier, the strain *Brevibacterium ammoniagenes* was used for the process. During optimization *Brevibacterium flavum* was discovered. The productivity with *B. flavum* is more than 9 times higher than with *B. ammoniagenes*.

- The cultural age of the cells also had a marked effect on the enzyme activity and the operational stability of immobilized cells.

- The same process is also employed by Amino GmbH, Germany, with the difference that they use the non-immobilized, native fumarase (see page 463).

3) Flow scheme

Not published.

4) Process parameters

conversion:	80 % (equilibrium conversion)
yield:	> 70 %
reactor type:	plug-flow reactor
reactor volume:	1,000 L
capacity:	468 t · a^{-1}
enzyme activity:	17 U · mL(gel)$^{-1}$ (37 °C); 28 U · mL(gel)$^{-1}$ (50 °C)
enzyme consumption:	$t_{1/2}$ = 243 d (37 °C); $t_{1/2}$ = 128 d (50 °C)
start-up date:	1974
company:	Tanabe Seiyaku Co., Ltd., Japan

5) Product application

- The product is used as an acidulant in fruit and vegetable juices, carbonated soft drinks, jams and candies, in amino acid infusions and for the treatment of hepatic malfunctioning.

6) Literature

- Tosa, T., Shibatani, T. (1995) Industrial applications of immobilized biocatalysts in Japan, Ann. N. Y. Acad. Sci. **750**, 364–375

- Tanaka, A., Tosa, T., Kobayashi, T. (1993) Industrial Application of Immobilized Biocatalysts, Marcel Dekker Inc., New York

- Lilly, M.D. (1994) Advances in biotransformation processes. Eighth P. V. Danckwerts memorial lecture presented at Glaziers' Hall, London, U.K. 13 May 1993, Chem. Eng. Sci. **49**, 151–159

- Wiseman, A. (1995) Handbook of Enzyme and Biotechnology, Ellis Horwood, Chichester

- Sheldon, R.A. (1993) Chirotechnology, Marcel Dekker Inc., New York

- Crosby, J. (1991) Synthesis of optically active compounds: a large scale perspective, Tetrahedron **47**, 4789–4846

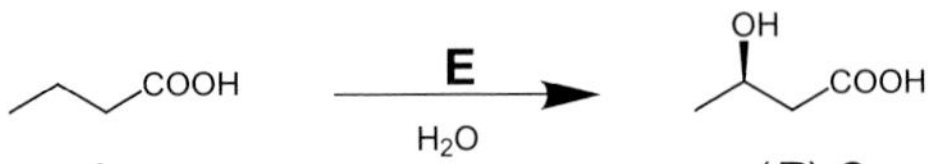

1 = butyric acid
2 = β-hydroxy-n-butyric acid

Kanegafuchi Chemical Industries Co., Ltd.

Fig. 4.2.1.17 – 1

1) Reaction conditions

pH:	7.2–7.5
T:	30 °C – 33 °C
medium:	aqueous
reaction type:	hydration (addition of H_2O to carbon-carbon double bond)
catalyst:	whole cells
enzyme:	(3 S)-3-hydroxyacyl-CoA hydro-lyase (β-hydroxyacid dehydrase, acyl-CoA dehydrase, enoyl-CoA hydratase)
strain:	*Candida rugosa* IFO 0750 M
CAS (enzyme):	[9027–13–8]

2) Remarks

- The biotransformation occurs in three steps. Initially the aliphatic acid is dehydrogenated to the α,β-unsaturated acid. In a subsequent step, enantioselective acylated and hydration takes place:

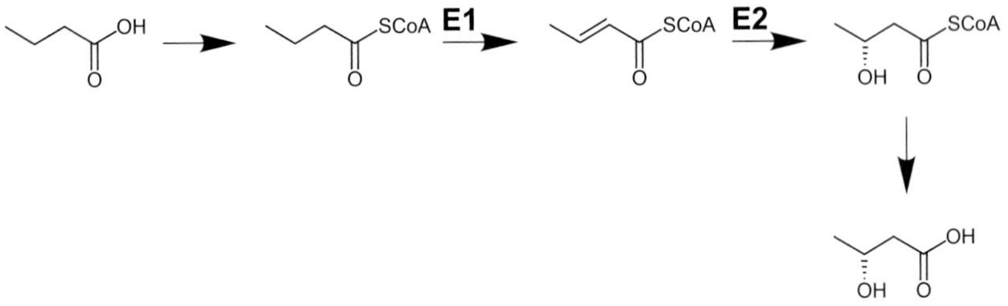

E1 = acyl-CoA dehydrogenase
E2 = enoyl-CoA hydratase

Fig. 4.2.1.17 – 2

3) Flow scheme

Not published.

4) Process parameters

ee: > 98 %
space-time-yield: 5–10 g · L^{-1} · d^{-1}
company: Kanegafuchi Chemical Industries Co., Ltd., Japan

5) Product application

- (R)-β-Hydroxy-*n*-butyric acid is used for the synthesis of a carbapenem intermediate:

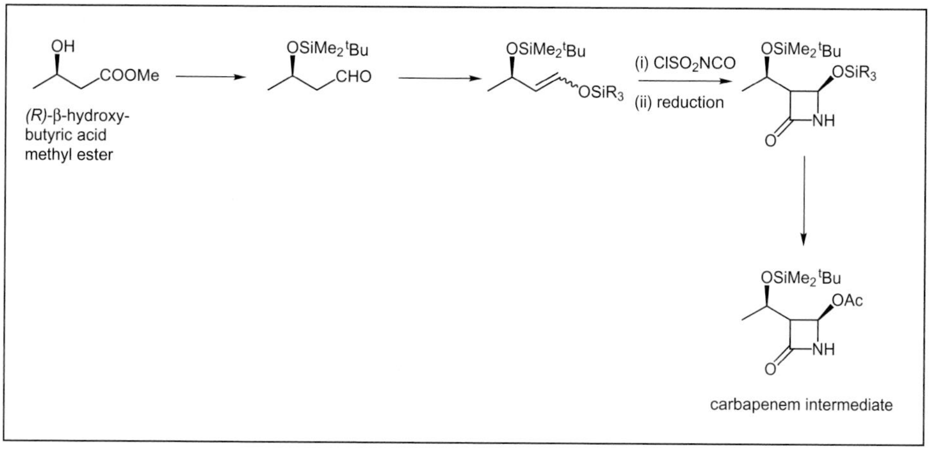

Fig. 4.2.1.17 – 3

6) Literature

- Sheldon, R.A. (1993) Chirotechnology, Marcel Dekker Inc., New York

- Kieslich, K., (1991), Biotransformations of industrial use, 5th Leipziger Biotechnology Symposium 1990, Acta Biotechnol. **11** (6), 559–570

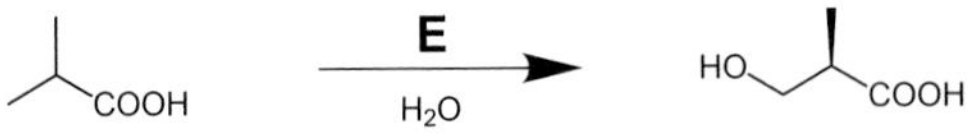

1 = isobutyric acid
2 = β-hydroxy-isobutyric acid

Kanegafuchi Chemical Industries Co., Ltd.

Fig. 4.2.1.17 – 1

1) Reaction conditions

pH: 7.2–7.5
T: 30 °C – 33 °C
medium: aqueous
reaction type: hydration (addition of H_2O to carbon-carbon double bond)
catalyst: whole cells
enzyme: (3S)-3-hydroxyacyl-CoA hydro-lyase (β-hydroxyacid dehydrase, acyl-CoA dehydrase, enoyl-CoA hydratase)
strain: *Candida rugosa* IFO 0750 M
CAS (enzyme): [9027–13–8]

2) Remarks

- The biotransformation occurs in three steps. Initially the aliphatic acid is dehydrogenated to the α,β-unsaturated acid. In a subsequent step, enantioselective acylated and hydration takes place:

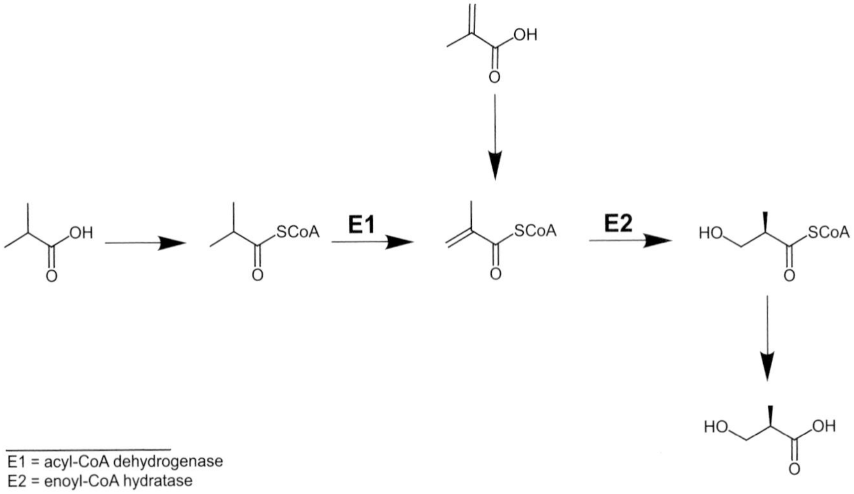

E1 = acyl-CoA dehydrogenase
E2 = enoyl-CoA hydratase

Fig. 4.2.1.17 – 2

470

3) **Flow scheme**

Not published.

4) **Process parameters**

yield:	98 %
ee:	97 %
space-time-yield:	5–10 g · L^{-1} · d^{-1}
company:	Kanegafuchi Chemical Industries Co., Ltd., Japan

5) **Product application**

- (R)-β-Hydroxy-isobutyric acid is used as a chiral synthon in the synthesis of captopril, an ACE-inhibitor (ACE = angiotensin converting enzyme):

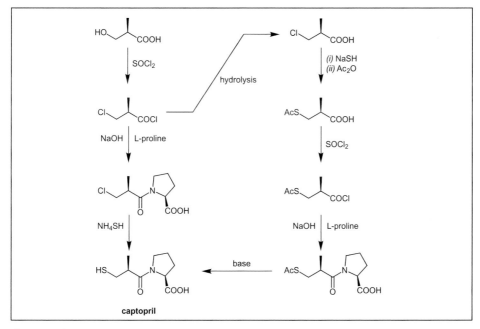

Fig. 4.2.1.17 – 3

6) **Literature**

- Crosby, J. (1991) Synthesis of optically active compounds: a large scale perspective, Tetrahedron **47**, 4789–4846

- Hasegawa, J., Ogura, M., Kanema, H., Noda, N., Kawaharada, H., Watanabe, K. (1982) Production of D-β-hydroxyisobutyric acid from isobutyric acid by *Candida rugosa* and its mutant, J. Ferment. Technol. **60**, 501–508

- Kieslich, K., (1991), Biotransformations of industrial use, 5th Leipziger Biotechnology Symposium 1990, Acta Biotechnol. **11** (6), 559–570

- Sheldon, R.A. (1993) Chirotechnology, Marcel Dekker Inc., New York

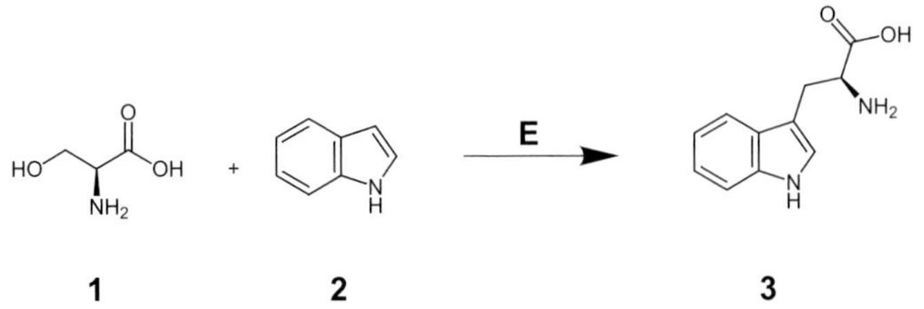

1 = L-serine
2 = indole
3 = L-tryptophan

Amino GmbH

Fig. 4.2.1.20 – 1

1) Reaction conditions

[1]:	0.1 M, 10.41 g · L^{-1} [104.1 g · mol^{-1}] (initial)
[2]:	0.01 M, 1.17 g · L^{-1} [117.16 g · mol^{-1}] (steady state)
[pyridoxal phosphate]:	> 5 · 10^{-5} M (cofactor)
[3]:	0.06 M, 12.25 g · L^{-1} [204.23 g · mol^{-1}](saturated solution)
pH:	8.0 – 9.0
T:	40 °C
medium:	aqueous
reaction type:	C-O bond cleavage
catalyst:	suspended whole cells
enzyme:	L-tryptophan synthase (L-serine hydrolyase)
strain:	*Escherichia coli*
CAS (enzyme):	[9014–52–2]

2) Remarks

- Pyridoxal phosphate is needed as cofactor.

- The enzyme works enantiospecifically for α-L-amino acid substrates of the type:

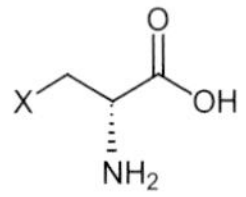

  where X indicates a small nucleophilic substituent like -OH, -Cl.

- The established process is dedicated to the production of L-tryptophan as a pharmaceutically active ingredient.

- The educt L-serine is separated from molasses. The best separation is performed with ion exchange chromatography (polystyrene resin) close to the isoelectric point of serine, pH 5.7. By concentration of the serine fraction to 35 % dry mass the main fraction of D-serine can be separated by filtration, leaving a L-serine stock solution.

- The fed batch is pH regulated and the indole dosage is directed via on-line HPLC analysis of the product/educt ratio.

472

- L-Tryptophan is produced in such high concentrations that it crystallizes instantly and it is isolated together with the cells at the end of the fed batch.

- The crude L-tryptophan is solubilized in hot water and the cells are separated after addition of charcoal.

3) Flow scheme

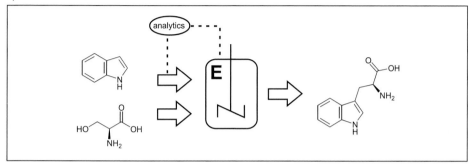

Fig. 4.2.1.20 – 2

4) Process parameters

yield:	> 95 % (based on indole)
reactor type:	fed batch
capacity:	$30 \, t \cdot a^{-1}$
residence time:	6 h / batch
space-time-yield:	$75 \, g \cdot L^{-1} \cdot d^{-1}$
down stream processing:	multiple fractional crystallizations from aqueous solution
enzyme activity:	$250 \, U \cdot L^{-1}$
enzyme consumption:	$4{,}500 \, U \cdot kg^{-1}$
enzyme supplier:	AMINO GmbH, Germany
start-up date:	1988/89
production site:	Amino GmbH, Germany
company:	Amino GmbH, Germany

5) Product application

- L-Tryptophan is used in parenteral nutrition (infusion solution) and as a pharmaceutical active ingredient in sedatives, neurolepticas, antidepressants and food additives.

- Intermediate for production of other pharmaceutical compounds.

6) Literature

- Bang, W.-G., Lang, S., Sahm, H., Wagner, F. (1983) Production of L-tryptophan by *Escherichia coli* cells, Biotech. Bioeng. **25**, 999–1011

- Plischke, H., Steinmetzer, W. (1988) Verfahren zur Herstellung von L-Tryptophan und D,L-Serin, AMINO GmbH, DE 3630878 C1

- Wagner, F., Klein, J., Bang, W.-G., Lang, S., Sahm, H. (1980) Verfahren zur mikrobiellen Herstellung von L-Tryptophan, DE 2841642 C2

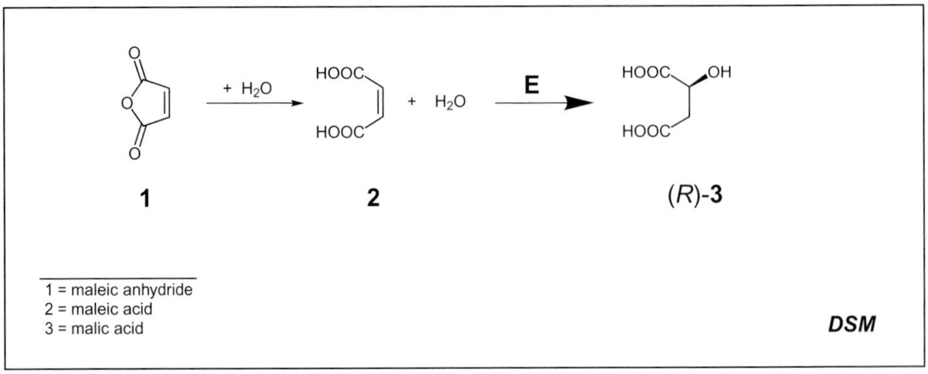

Fig. 4.2.1.31 – 1

1) Reaction conditions

[2]:	0.8 M, 92.9 g · L^{-1} [116.07 g · mol^{-1}]
pH:	7.0
T:	25 °C
medium:	aqueous
reaction type:	addition of H$_2$O
catalyst:	immobilized whole cells
enzyme:	(R)-malate hydro-lyase (maleate hydratase, malease)
strain:	*Pseudomonas pseudoalcaligenes*
CAS (enzyme):	[37290–71–4]

2) Remarks

- The cheaper maleic anhydride can be used instead of maleic acid which hydrolyses *in situ* to maleate.

- Two degradation pathways have been described for maleate as carbon and energy source:

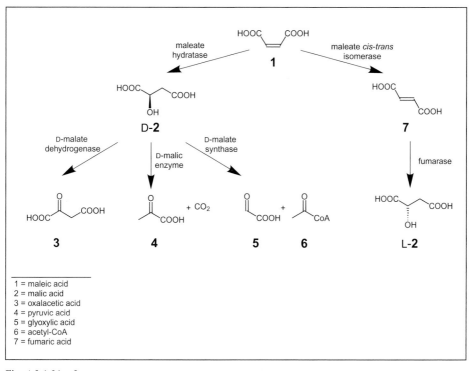

1 = maleic acid
2 = malic acid
3 = oxalacetic acid
4 = pyruvic acid
5 = glyoxylic acid
6 = acetyl-CoA
7 = fumaric acid

Fig. 4.2.1.31 – 2

- Fumarate accumulation during cell incubation with maleate under anaerobic conditions allows discrimination between the two degradation pathways and was used as selection criteria.

- 315 strains were screened.

- The chosen strain *Pseudomonas pseudoalcaligenes* is not able to grow on maleate as the sole source of carbon and energy. It is probably not capable of synthesizing a transport mechanism for maleate.

- To overcome transport problems of substrate and product across the cell membrane, the cells are permeabilized with Triton X-100.

- Cells have to be harvested before the substrate for growth is completely consumed, because otherwise malease activity drops rapidly. During growth the activity is constant in logarithmic phase. The presence of maleate does not influence the growth rate.

- The enzyme malease needs no cofactor.

- Although D-malate is a competitive inhibitor, it stabilizes the enzyme.

- The K_M-value is about 0.35 M.

- Malease can also catalyze the hydration of citraconate. The activity is only 56 % of the maleate hydration activity and the K_M-value is about 0.2 M:

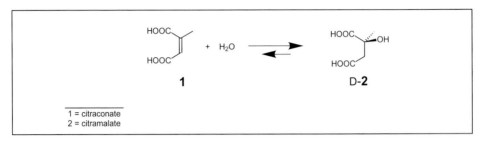

1   D-**2**

1 = citraconate
2 = citramalate

Fig. 4.2.1.31 – 3

3) Flow scheme

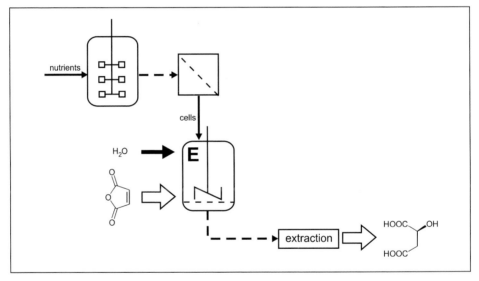

Fig. 4.2.1.31 – 4

4) Process parameters

yield:	> 99 %
ee:	> 99.99 %
enzyme activity:	2,000 U · g^{-1}$_{Protein}$ (permeabilized cells grown on 3-hydroxy-benzoate)
company:	DSM, The Netherlands

5) Product application

- D-Malate can be used as chiral synthon in organic chemistry or as resolving agent in resolution of racemic compounds.

6) Literature

- Chibata, I., Tosa, T., Shibatani, T. (1992) The industrial production of optically active compounds by immobilized biocatalysts, in: Chirality in Industry (Collins, A.N., Sheldrake, G., Crosby, J., eds.) pp. 351–370, John Wiley & Sons Ltd., New York

- Subramanian, S.S., Raghavendra Rao, M.R. (1968) Purification and properties of citraconase, J. Biol. Chem. **243**, 2367–2372

- van der Werf, M.J., van den Tweel, W.J.J., Hartmans, S. (1992) Screening for microorganisms producing D-malate from maleate, Appl. Environ. Microbiol. **58**, 2854–2860

- van der Werf, M.J., van den Tweel, W.J.J., Kamphuis, J., Hartmans, S., de Bont, J.A.M. (1994) D-Malate and D-citramalate production with maleate hydro-lyase from *Pseudomonas pseudoalcaligenes*, in: Proc. 6th Eur. Congr. Biotechnol. (Alberghina, L., Frontali, L., Sensi, P., eds.) pp. 471–474, Elsevier Science B.V., Amsterdam

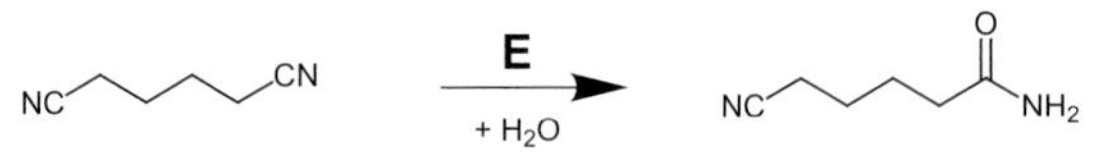

Nitrile hydratase
Pseudomonas chlororaphis

1 = adiponitrile
2 = 5-cyano-valeramide

DuPont

Fig. 4.2.1.84 – 1

1) Reaction conditions

[1]:	2.015 mol (organic phase), 218 kg [108.14 g · mol^{-1}]
pH:	7.0
T:	5 °C
medium:	two-phase: aqueous/organic
reaction type:	hydration of CN
catalyst:	immobilized whole cells
enzyme:	nitrile hydro-lyase (nitrile hydratase, acrylonitrile hydratase, NHase, L-NHase, H-NHase)
strain:	*Pseudomonas chlororaphis* B23
CAS (enzyme):	[82391–37–5]

2) Remarks

- The cells are immobilized in calcium alginate beads.

- For strain selection it is important that the cells do not show an amidase activity that would further hydrolyze the amide to the carboxylic acid.

- By this method 13.6 t are produced in 58 repetitive batch cycles, starting from 12.7 t of adiponitrile.

- As by-product adipodiamide is formed.

- This biotransformation is preferred to the chemical transformation due to a higher conversion and selectivity, production of more product per weight of biomass (3,150 kg · kg^{-1} (dry cell weight), and less waste:

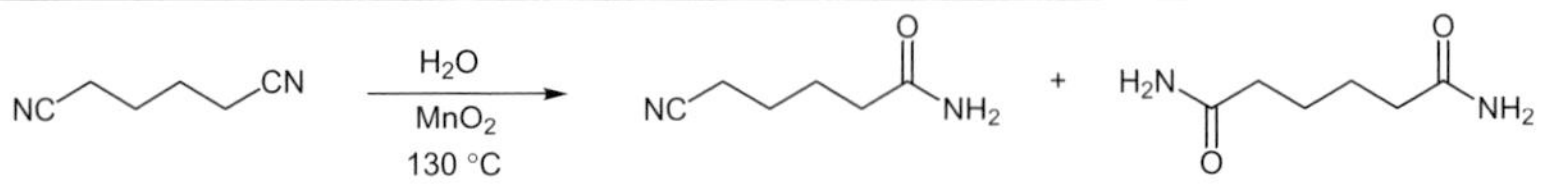

1 = adiponitrile
2 = 5-cyano-valeramide
3 = adipodiamide

Fig. 4.2.1.84 – 2

- Reactions with product concentrations higher than 0.45 M form two-phase systems.

- A reaction temperature of 5 °C is chosen, since the solubility of the by-product adipodiamide is only 37–42 mM in 1.0 to 1.5 M 5-cyanovaleramide.

- A batch reactor is preferred over a fixed-bed packed column reactor, because of the lower selectivity to 5-cyanovaleramide that is observed and the possibility of precipitation of adipamide and plugging of the column.

- After completion of the reaction 90 % of the product mixture is decanted and the reactor is recharged with 1,007 L of reaction buffer (23 mM sodium butyrate and 5 mM calcium chlorid pH 7 at 5 °C) and 218 kg (2,015 mol) adiponitrile.

- The catalyst productivity is 3,150 kg 5-cyanovaleramide per kg of dry cell weight.

- The catalyst consumption is 0.006 kg per kg product.

- Excess water is removed at the end of the reaction by distillation. The by-product adipamide as well as calcium and butyrate salts are precipitated by dissolution of the resulting oil in methanol at > 65 °C. The raw product solution is directly transferred to the next step in the herbicide synthesis.

3) Flow scheme

Not published.

4) Process parameters

conversion:	97 %
yield:	93 %
selectivity:	96 %
reactor type:	repetitive batch
reactor volume:	2,300 L
capacity:	several t · a^{-1}
residence time:	4 h
down stream processing:	distillation
company:	DuPont, USA

5) Product application

- 5-Cyanovaleramide is used as an intermediate for the synthesis of the DuPont herbicide azafenidine (Drauz).

6) Literature

- Hann, E.C., Eisenberg, A., Fager, S.K., Perkins, N.E., Gallagher, F.G., Cooper, S.M., Gavagan, J.E., Stieglitz, B., Hennesey, S.M., DiCosimo, R. (1999) 5-Cyanovaleramide production using immobilized *Pseudomonas chlororaphis* B23, Bioorg. Med. Chem. **7**, 2239–2245

- Yamada, H., Ryuno, K., Nagasawa, T., Enomoto, K., Watanabe, I. (1986), Optimum culture conditions for production of *Pseudomonas chlororaphis* B23 of nitrile hydratase, Agric. Biol. Chem. **50**, 2859–2865

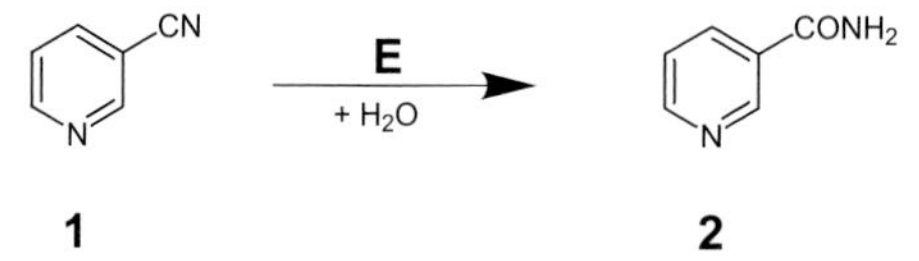

1 = 3-cyanopyridine
2 = nicotinamide = vitamin B3

Lonza AG

Fig. 4.2.1.84 – 1

1) Reaction conditions

[1]:	[$104.11 \text{ g} \cdot \text{mol}^{-1}$]
medium:	aqueous
reaction type:	C-O bond cleavage by elimination of water
catalyst:	immobilized whole cells
enzyme:	nitrile hydro-lyase (nitrile hydratase, acrylonitrile hydratase, NHase, L-NHase, H-NHase)
strain:	*Rhodococcus rhodochrous* J1
CAS (enzyme):	[82391–37–5]

2) Remarks

- In contrast to the chemical alkaline hydrolysis of 3-cyanopyridine with 4 % by-product of nicotinic acid (96 % yield) the biotransformation works with absolute selectivity.

- The same strain is used in the industrial production of acrylamide.

3) Flow scheme

Not published.

4) Process parameters

yield:	100 %
selectivity:	100 %
capacity:	$6{,}000 \text{ t} \cdot \text{a}^{-1}$
production site:	China
company:	Lonza AG, Switzerland

5) Product application

- Vitamin supplement for food and animal feed.

6) Literature

- Petersen, M., Kiener, A. (1999) Biocatalysis – Preparation and functionalization of N-heterocycles, Green Chem. 4, 99–106

480

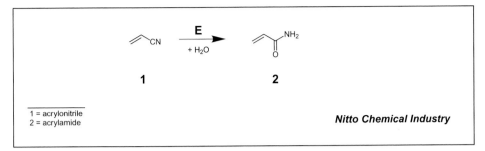

1 = acrylonitrile
2 = acrylamide

Nitto Chemical Industry

Fig. 4.2.1.84 – 1

1) Reaction conditions

[1]:	0.11 M, 6 g · L^{-1} [53.06 g · mol^{-1}] (fed batch)
[2]:	5.6 M, 400 g · L^{-1} [71.08 g · mol^{-1}]
pH:	7.0
T:	5 °C
medium:	aqueous
reaction type:	C-O bond cleavage by elimination of water
catalyst:	immobilized whole cells
enzyme:	nitrile hydro-lyase (nitrile hydratase, acrylonitrile hydratase, NHase, L-NHase, H-NHase)
strain:	*Rhodococcus rhodochrous* J1
CAS (enzyme):	[82391–37–5]

2) Remarks

- The chemical synthesis uses copper salt as catalyst for the hydration of acrylonitrile and has several disadvantages:

  1) The rate of acrylamide formation is lower than the acrylic acid formation,

  2) the double bond of educts and products causes by-product formations such as ethylene, cyanohydrin and nitrilotrispropionamide and

  3) at the double bond polymerization occurs.

- The biotransformation has the advantages that recovering the unreacted nitrile is not necessary since the conversion is 100 % and that no copper catalyst removal is needed.

- This biotransformation is the first example of an application in the petrochemical industry and the successful enzymatic manufacture of a bulk chemical.

- Although nitriles are generally toxic some microorganism can use nitriles as carbon / nitrogen source for growth.

- Since acrylonitrile is the most poisonous one among the nitriles, screening for microorganisms was conducted with low-molecular mass nitriles instead.

- More than 1,000 microbial strains were examined.

- Two degradation pathways of nitriles are known:

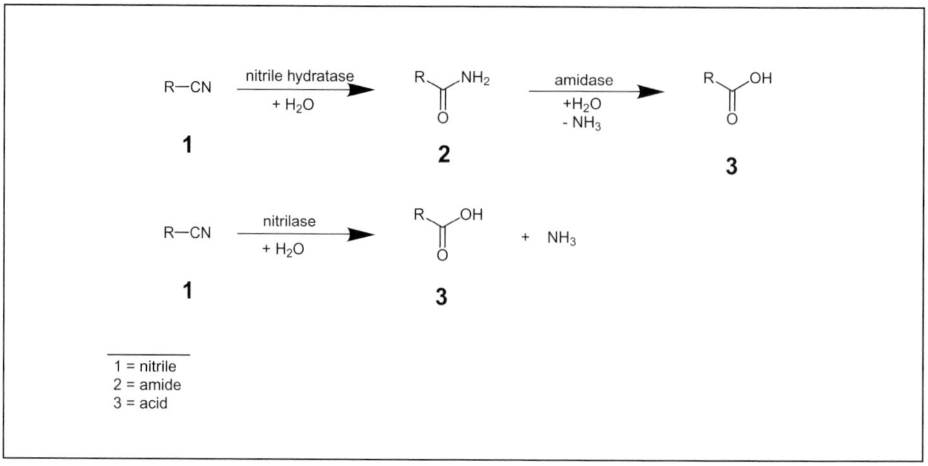

Fig. 4.2.1.84 – 2

- Microorganisms that produce amidases beside the nitrile hydratase are not suitable for the production of acrylamide without adding an amidase inhibitor.

- In the course of improvement of the biocatalyst for the production of acrylamide three main strains were used:

  1) *Rhodococcus* sp. N774

  2) *Pseudomonas chlororaphis*

  3) *Rhodococcus rhodochrous*

- The *Rhodococcus* sp. N774 strain was used for three years before the better *Pseudomonas chlororaphis* strain was found.

- The *Pseudomonas* strain cannot grow on acrylonitrile but grows on isobutyronitrile.

- The optimization of the *Pseudomonas* strain reveals that methacrylamide causes the greatest induction of nitrile hydratase. The addition of ferrous or ferric ions to the culture medium increases enzyme formation, no other ionic addition shows improvements, indicating that the nitrile hydratase contain Fe^{2+} ions as a cofactor.

- The growth medium can be optimized resulting in an amount of nitrile hydratase of 40 % of the total soluble protein formed in the cells.

- A problem during growth of *Pseudomonas chlororaphis* strain in the first optimized sucrose containing medium is the production of mucilaginous polysaccharides. These causes a high viscosity, resulting in difficulties during cell harvest.

- Using chemical mutagenesis methods (*N*-methyl-*N'*-nitro-*N*-nitrosoguanidine = MNNG) mucilage polysaccharide-non-producing mutants could be isolated. The following table shows the improvements (total activity increases 3,000-times) by optimizing the fermentation medium and by mutagenesis:

Nitrile hydratase
Rhodococcus rhodochrous

strain	specific activity (U/mg of dry cells)	total activity (U/mL)
parent (medium A*)	0.72	0.40
↓		
parent (medium R*)	66	363
↓ MNNG treatment		
Am 3	65	465
↓ MNNG treatment		
Am 324	125	952
↓ feeding of methacrylamide		
Am 324	141	1260

*medium A: dextrin, K_2HPO_4, NaCl, MgSO and isobutyronitrile
medium R: sucrose, K_2HPO_4, KH_2PO_4, $MgSO_4$, $FeSO_4$, soybean hydrolyzate and methacrylamide

Fig. 4.2.1.84 – 3

- The third generation of industrial strains is the *Rhodococcus rhodochrous* J1 that produces two kinds of nitrile-converting enzymes, the nitrilase and nitrile hydratase. The latter one was found after optimization of fermentation medium.

- Addition of cobalt ions greatly increases nitrilase hydratase activity in comparison to Fe ions for the *Pseudomonas chlororaphis* strain.

- The difference in metal-ion cofactors can be ascribed to a small number of amino acids at their ligand-binding sites, resulting in higher stability of *Rhodococcus rhodochrous* J1 strain against reducing and oxidizing agents. Although the association of 20 subunits depresses the flexibility of the protein, it increases the stability.

- The strains forms two kinds of nitrile hydratases with different molecular weights and characteristics. The following table compares the different enzymes and shows the advantages of the high-molecular mass hydratase:

| | *Pseudomonas chloraphis* B23 | *Rhodococcus rhodochrous* J1 | |
| | | low molecular weight (L-NHaseT) | high molecular weight (H-NHaseT) |
	(L-NHaseT)		
molecular weight	100,000	130,000	520,000
subunit molecular weight	α25,000	α26,000	α26,000
	β25,000	β29,000	β29,000
number of subunits	4	4-5	18-20
absorption maxima (nm)	280,720	280,415	280,415
(415/280) / (720/280)	0.014	0.031	0.016
optimum temperature (°C)	20	40	35
heat stability (°C)	20	30	50
optimum pH	7.5	8.8	6.5
pH stability	6.0-7.5	6.5-8.0	6.0-8.5
V_{max} at 20°C (U·mg^{-1} protein)	1,470		1,760
K_M at 20°C (mM)	34.6		1.89

Fig. 4.2.1.84 – 4

- As inducer urea is used, which is much cheaper than methacrylamide for the *Pseudomonas chlororaphis*. This allows an increase in the amount of L-NHase in the cell free extract to more than 50 % of the total soluble protein.

- The nitrile hydratases act also on other nitriles with yields of 100 %. The most impressive example is the conversion of 3-cyanopyridine to nicotinamide. The product concentration is about 1,465 g · L^{-1}. This conversion (1.17 g · L^{-1} dry cell mass) can be named 'pseudocrystal enzymation' since at the start of the reaction the educt is solid and with ongoing reaction it is solubilized. The same is valid for the product which crystallizes at higher conversions so that at the end of the reaction the medium is solid again (see also Lonza, page 480).

- The following table shows some examples and the end concentrations of possible products for *Rhodococcus rhodocrous* J1 induced by crotonamide:

product	concentration
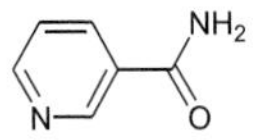	1,465 g·L^{-1}
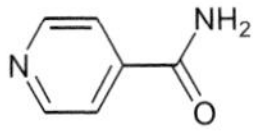	1,099 g·L^{-1}
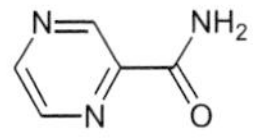	985 g·L^{-1}
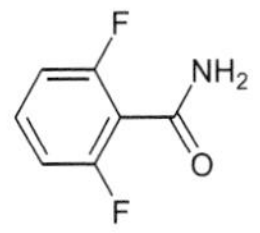	306 g·L^{-1}
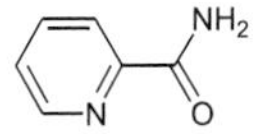	977 g·L^{-1}
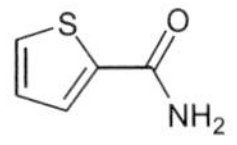	210 g·L^{-1}
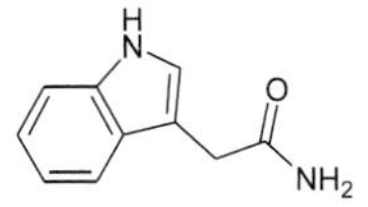	1,045 g·L^{-1}
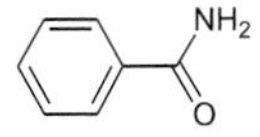	489 g·L^{-1}
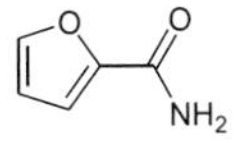	522 g·L^{-1}

Fig. 4.2.1.84 – 5

- Since acrylamide is unstable and polymerizes easily, the process is carried out at low temperatures (about 5 °C).

- Although the cells, which are immobilized on polyacrylamide gel, and the contained enzyme is very stable towards acrylonitrile, the educt has to be fed continuously to the reaction mixture due to inhibition effects at higher concentrations.

- The following table summarized important production data for the discussed strains:

	Rhodococcus sp. NT774	Pseudomonas chloraphis B23	Rhodococcus rhodochrous J1
tolarence to acrylamide (%)	27	40	50
arylic acid formation	very little	barely detected	barely detected
cultivation time (h)	48	45	72
activity of culture (U·mL^{-1})	900	1,400	2,100
specific activity (U·mg^{-1}cells)	60	85	76
cell yield (g·L^{-1})	15	17	28
acrylamide productivity (g·g^{-1}cells)	500	850	>7,000
total amount of production (t·a^{-1})	4,000	6,000	>30,000
final concentration of acrylamide (%)	20	27	40
first year of production scale	1985	1988	1991

Fig. 4.2.1.84 – 6

3) **Flow scheme**

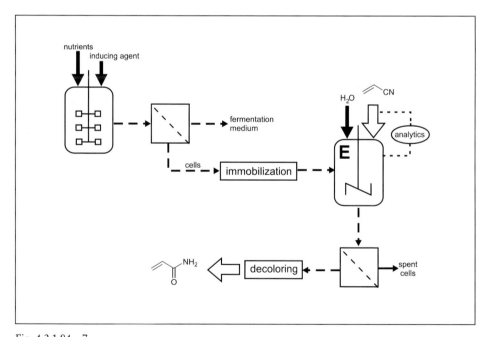

Fig. 4.2.1.84 – 7

4) Process parameters

conversion:	> 99.99 %
yield:	> 99.99 %
selectivity:	> 99.99 %
reactor type:	fed batch
capacity:	> 30,000 t $\cdot$ a^{-1}
residence time:	5 h
space-time-yield:	1,920 g $\cdot$ L^{-1} $\cdot$ d^{-1}
down stream processing:	filtration and decoloring
enzyme activity:	76,000 U $\cdot$ g$_{cells}$; 2,100,000 U $\cdot$ L^{-1}
start-up date:	1991
production site:	Nitto Chemical Industry Co., Ltd., Japan
company:	Nitto Chemical Industry Co., Ltd., Japan

5) Product application

- Acrylamide is an important bulk chemical used in coagulators, soil conditioners and stock additives for paper treatment and paper sizing, and for adhesives, paints and petroleum recovering agents.

6) Literature

- Nagasawa, T., Shimizu, H., Yamada, H. (1993) The superiority of the third-generation catalyst, *Rhodococcus rhodochrous* J1 nitrile hydratase, for industrial production of acrylamide, Appl. Microb. Biotechnol. **40**, 189–195

- Shimizu, H., Fujita, C., Endo, T., Watanabe, I. (1993) Process for preparing glycine from glycinonitrile, Nitto Chemical Industry Co., Ltd., US 5238827

- Shimizu, H., Ogawa, J., Kataoka, M., Kobayashi, M. (1997) Screening of novel microbial enzymes for the production of biologically and chemically usesful compounds, in: New Enzymes for Organic Synthesis; Adv. Biochem. Eng. Biotechnol. **58** (Ghose, T.K., Fiechter, A., Blakebrough, N., eds.), pp. 56–59

- Yamada, H., Tani, Y. (1982) Process for biologically producing amide, EP 093782 B1

- Yamada, H., Kobayashi, M. (1996) Nitrile hydratase and its application to industrial production of acrylamide, Biosci. Biotech. Biochem. **60**, 1391–1400

- Yamada, H., Tani, Y. (1987) Process for biological preparation of amides, Nitto Chemical Industry Co., Ltd., US 4637982

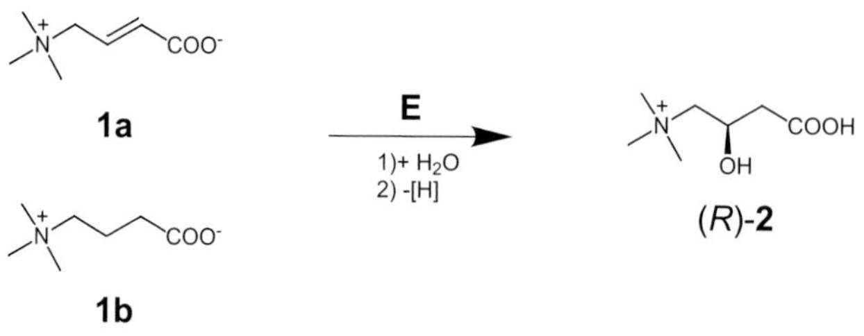

1a = crotonobetaine
1b = 4-butyrobetaine
2 = L-carnitine (3-hydroxy-4-(trimethylamino)butanoate)

Lonza AG

Fig. 4.2.1.89 – 1

1) Reaction conditions

[1a]:	0.69 M, 70 g · L^{-1} [102.11 g · mol^{-1}]
[1b]:	0.67 M, 70 g · L^{-1} [104.13 g · mol^{-1}]
[2]:	> 0.58 M, 70 g · L^{-1} [120.13 g · mol^{-1}]
pH:	7.0
T:	30 °C
medium:	aqueous
reaction type:	C-O bond cleavage by elimination of water
catalyst:	whole cell
enzyme:	carnitine: NAD$^+$ 3-oxidoreductase (L-carnitine 3-dehydrogenase, L-carnitine hydrolyase, carnitine dehydratase)
strain:	mutant strain of *Agrobacterium/Rhizobium* HK 1331-b, microorganisms DSM 3225 (HK1331-b), *Pseudomonas putida* T1
CAS (enzyme):	[104382–17–4]

2) Remarks

- *Agrobacterium* produces only L-carnitine.

- 4-Butyrobetaine is degraded via a reaction sequence that might be compared with the β-oxidation of fatty acids:

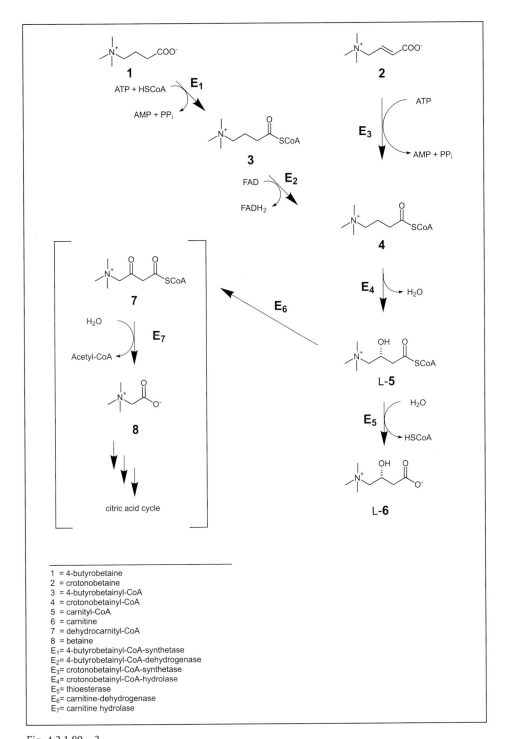

1 = 4-butyrobetaine
2 = crotonobetaine
3 = 4-butyrobetainyl-CoA
4 = crotonobetainyl-CoA
5 = carnityl-CoA
6 = carnitine
7 = dehydrocarnityl-CoA
8 = betaine
E_1= 4-butyrobetainyl-CoA-synthetase
E_2= 4-butyrobetainyl-CoA-dehydrogenase
E_3= crotonobetainyl-CoA-synthetase
E_4= crotonobetainyl-CoA-hydrolase
E_5= thioesterase
E_6= carnitine-dehydrogenase
E_7= carnitine hydrolase

Fig. 4.2.1.89 – 2

- The mutant strain blocks the ʟ-carnitine dehydrogenase **E6** and excretes the accumulated product.

- Only growing cells are active.

- The purified enzymes could not be used for the biotransformation due to their high instability.

- At high product concentration most tested strains show an inhibition and strains with lower inhibition are not so selective.

- The product tolerance and selectivity could be optimized in a strain development.

- Apart from usual batch fermentations, continuous production is also feasible since the cells go into a 'maintenance state' with high metabolic activity and low growth rate. The cells can be recycled after separation from the fermentation broth by filtration.

- In the continuous process only 92 % conversion is reached. For almost complete conversions, fed batch processes are used. As a result, product purification becomes easier.

- Duration of process development: 30 months.

- A chemical resolution process that was developed at Lonza was not competitive with the biotechnological route:

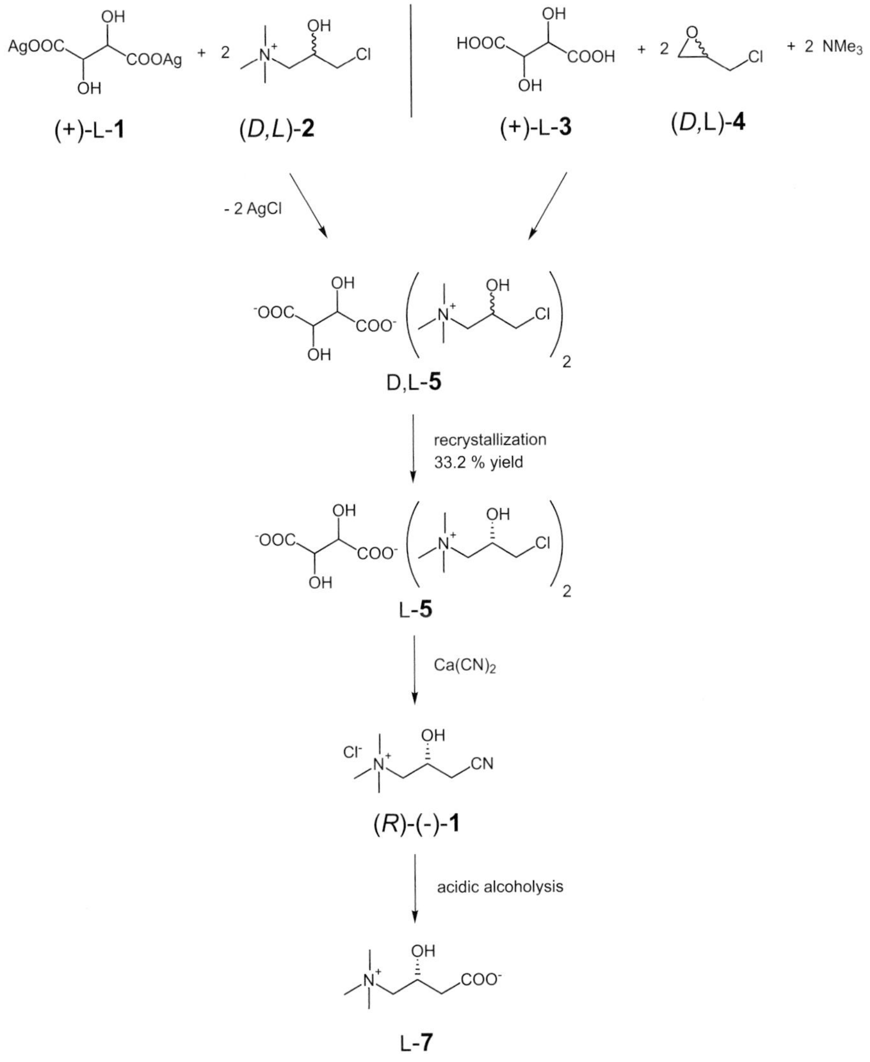

1 = silver tartrate
2 = (3-chloro-2-hydroxypropyl)-trimethylammonium salt
3 = tartaric acid
4 = epichlorohydrin
5 = tartrate
6 = (3-cyano-2-hydroxypropyl)-trimethylammonium salt
7 = carnitine

Fig. 4.2.1.89 – 3

3) Flow scheme

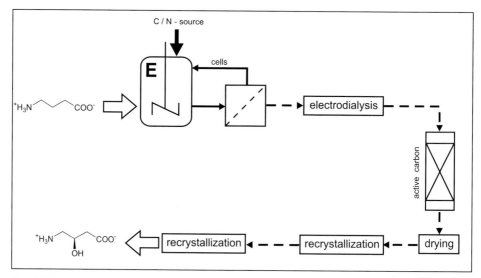

Fig. 4.2.1.89 – 4

4) Process parameters

conversion:	> 99.5 %
yield:	99.5 %
selectivity:	very high (only 3-dehydrocarnitine is accepted for transformation to L-carnitine)
ee:	99.9 %
chemical purity:	99.8 %
reactor type:	fed batch
reactor volume:	50,000 L
capacity:	300 t · a^{-1}
space-time-yield:	> 130 g · L^{-1} · d^{-1}
down stream processing:	cross-flow filtration, ion exchange, crystallization
start-up date:	1993
production site:	Czech Republic
company:	Lonza AG, Switzerland

5) Product application

- L-Carnitine is used as a thyroid inhibitor and in infant foods, health sport and geriatric nutrition.

- The physiological functions are:

  Transport of long-chain fatty acids through the mitochondrial membrane and their oxidation play an important role in energy metabolism.

  It is involved in regulating the level of blood lipids.

  L-Carnitine is utilized as a drug to increase cardiac output, improve myocardial function and treat cartinine deficiency (especially after hemodialysis).

6) Literature

- Hoeks, F.W.J.M.M. (1991) Verfahren zur diskontinuierlichen Herstellung von L-Carnitin auf mikrobiologischem Weg, Lonza AG, EP 0 410 430 A2

- Jung, W., Jung, K., Kleber, H.-P. (1993) Synthesis of L-carnitine by microorganisms and isolated enzymes, Adv. Biochem. Eng. Biotechnol. **50**, 21–44

- Kieslich, K. (1991), Biotransformations of industrial use, 5th Leipziger Biotechnology Symposium 1990, Acta Biotechnol. **11** (6), 559–570

- Kitamura, M., Ohkuma, T., Takaya, H., Noyori, R. (1988) A practical asymmetric synthesis of carnitine, Tetrahedron Lett. **29**, 1555–1556

- Kulla, H. (1991) Enzymatic hydroxylations in industrial application, CHIMICA **45**, 81–85

- Kulla, H., Lehky, P., Squaratti, A. (1991) Verfahren zur kontinuierlichen Herstellung von L-Carnitin, Lonza AG, EP 0 195 944 B1

- Kulla, H., Lehky, P. (1991) Verfahren zur Herstellung von L-Carnitin auf mikrobiologischem Wege, Lonza AG, EP 0 158 194 B1

- Macy, J., Kulla, H., Gottschalk, G. (1976) H_2-dependent anaerobic growth of *Escherichia coli* on L-malate: succinate formation, J. Bacteriol. **125**, 423–428

- Seim, H., Ezold, R., Kleber, H.-P., Strack, E. (1980) Stoffwechsel des L-Carnitins bei Enterobakterien Z. Allg. Mikrobiol. **20**, 591–594

- Vandecasteele, J.-P. (1980) Enzymatic synthesis of L-carnitine by reduction of an achiral precursor: the problem of reduced nicotinamide adenine dinucleotide recycling, Appl. Environ. Microbiol. **39**, 327–334

- Voeffray, R., Perlberger, J.-C., Tenud, L., Gosteli, J. (1987) L-Carnitine. Novel synthesis and determination of the optical purity, Helv. Chim. Acta. **70**, 2058–2064

- Zhou, B.-N., Gopalan, A.S., van Middlesworth, F., Shieh, W.-R., Sih, C.J. (1983) Stereochemical control of yeast reductions. 1. Asymmetric synthesis of L-carnitine, J. Am. Chem. Soc. **105**, 5925–5926

- Zimmermann, Th.P., Robins, K.T., Werlen, J., Hoeks, F.W.J.M.M. (1997) Bio-transformation in the production of L-carnitine, in: Chirality in Industry (Collins, A.N., Sheldrake, G.N., Crosby, J., eds.), pp. 287–305, John Wiley and Sons Ltd, New York

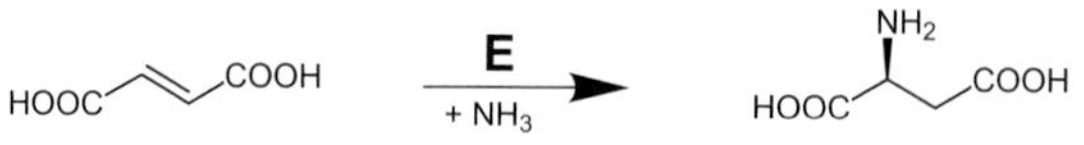

1 = fumaric acid
2 = aspartic acid

BioCatalytics Inc.

Fig. 4.3.1.1 – 1

1) Reaction conditions

[1]:   1.5 M, 174.1 g · L^{-1} [116.07 g · mol^{-1}]
[2]:   1.5 M, 199.5 g · L^{-1} [133.10 g · mol^{-1}]
pH: 8.5
T: inlet 27 °C; outlet 37 °C
medium: aqueous
reaction type: C-N-bond cleavage
catalyst: immobilized enzyme
enzyme: L-aspartate-ammonia lyase (fumaric aminase, aspartase)
strain: *Escherichia coli*
CAS (enzyme): [9027–30–9]

2) Remarks

- The presence of 1 mM MgCl$_2$ enhances the activity and stability of the enzyme.

- By using isolated, on silica-based support immobilized enzyme, a higher productivity was achieved than the comparable process using immobilized cells.

- Tanabe Seiyaku uses for the same synthesis an immobilized whole cell system since 1973 (see page 500), in contrast to Mitsubishi Petrochemical Co. Ltd., which uses suspended whole cells (see page 498). Kyowa Hakko Kogyo Co, Ltd., Japan, also uses an immobilized enzyme (see page 496).

- The product solution is acidified to pH 2.8, chilled and the precipitated product is filtered off.

3) Flow scheme

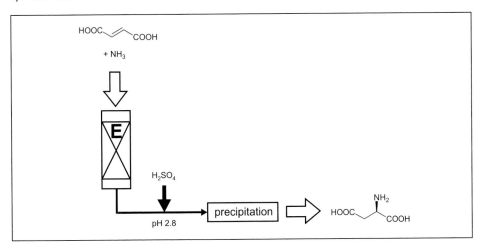

Fig. 4.3.1.1 – 2

4) Process parameters

conversion:	99 %
yield:	95 %
selectivity:	96 %
optical purity:	>99.9 %
chemical purity:	99 %
reactor type:	plug flow reactor
reactor volume:	75 L
residence time:	0.07 h
space-time-yield:	$3 \text{ kg} \cdot \text{L}^{-1} \cdot \text{h}^{-1}$
down stream processing:	precipitation
enzyme consumption:	$t_{1/2} = 6$ months
company:	BioCatalytics Inc., USA

5) Product application

- The product is used as a precursor for the synthesis of aspartame, an artificial sweetener (see page 373).

- It is also used as an acidulant, as a food additive, in parenteral nutrition and as a chiral synthon in organic synthesis.

6) Literature

- Rozzell, D. (1998) BioCatalytics, personal communication

Aspartase
Escherichia coli

1 = fumaric acid
2 = aspartic acid

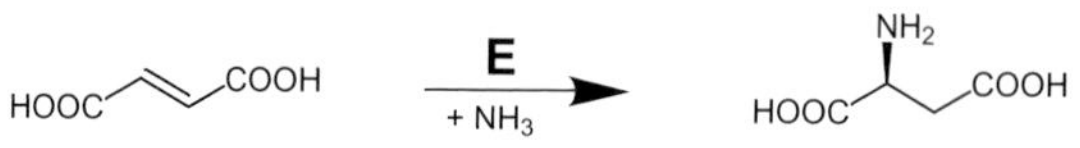

Kyowa Hakko Kogyo Co, Ltd.

Fig. 4.3.1.1 – 1

1) Reaction conditions

[1]:	2 M, 232.14 g · L⁻¹ [116.07 g · mol⁻¹]

[1]:   2 M, 232.14 g $\cdot$ L^{-1} [116.07 g $\cdot$ mol^{-1}]
T: 37 °C
medium: aqueous
reaction type: C-N bond cleavage
catalyst: immobilized enzyme
enzyme: L-aspartate ammonia-lyase (fumaric aminase, aspartase)
strain: *Escherichia coli*
CAS (enzyme): [9027–30–9]

2) Remarks

- 4 M NH$_4$OH is added as the amine source.

- Isolated enzyme is immobilized on Duolite A-7, a weakly basic anion-exchange resin.

- The column reactor is operated for over 3 months at over 99 % conversion.

- For the same syntheses, Tanabe Seiyaku has used an immobilized whole cell system since 1973 (see page 500), instead of suspended whole cells used by Mitsubishi Petrochemical Co. Ltd. (see page 498). Biocatalytics, Inc. uses an immobilized enzyme (see page 494).

3) Flow scheme

Not published.

4) Process parameters

conversion: > 99 %
ee: > 99.9 %
reactor type: plug flow reactor
residence time: 0.75 h
enzyme consumption: t$_{1/2}$ = 18 d
start-up date: 1974
company: Kyowa Hakko Kogyo Co, Ltd., Japan

496

5) Product application

- L-Aspartic acid is used as a precursor for the synthesis of aspartame, an artificial sweetener (see page 373).

- It is also used as an acidulant, as a food additive, in parentered nutrition and as a chiral synthon in organic synthesis.

6) Literature

- Tanaka, A., Tosa, T., Kobayashi, T. (1993) Industrial Application of Immobilized Biocatalysts, Marcel Dekker Inc., New York

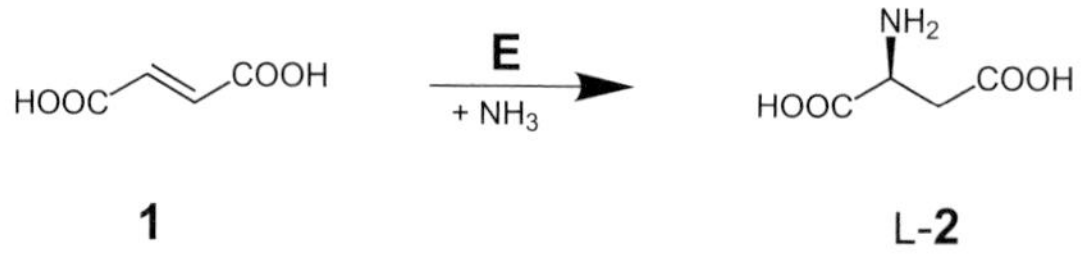

1 = fumaric acid
2 = aspartic acid

Mitsubishi Chemical Corporation

Fig. 4.3.1.1 – 1

1) Reaction conditions

[**1**]:	1.3 M, 150.9 g · L^{-1} [116.10 g · mol^{-1}]
pH:	9.0–10.0
T:	54 °C
medium:	aqueous
reaction type:	C-N bond cleavage
catalyst:	suspendend whole cells
enzyme:	L-aspartate ammonia-lyase (fumaric aminase, aspartase)
strain:	*Brevibacterium flavum*
CAS (enzyme):	[9027–30–9]

2) Remarks

- 4 M NH$_4$OH is added as the amine source.

- L-Malic acid is formed as a side product by an intracellular fumarase. But the fumarase can be inactivated by thermal effects. As optimal conditions for the suppression of fumarase activity the following parameters were established: Incubation of the cells (3 % w/v) at 45 °C for 5 hours in the presence of 2 M NH$_4$OH, 0.75 M L-aspartic acid, 0.0075 M CaCl$_2$ and 0.08 % (w/v) of the nonionic detergent Tween 20. L-Aspartic acid and CaCl$_2$ act as protectors against thermal inactivation. By the addition of Tween 20 the production of L-aspartic acid is increased by 40 %.

- L-Aspartic acid is produced from fumaric acid stoichiometrically.

- The bacterial cells are retained by ultrafiltration.

- The industrial production of L-aspartic acid has been carried out since 1953.

- Kyowa Hakko Kogyo Co. Ltd. and BioCatalytics use an immobilized enzyme (see page 496, 498) and Tanabe Seiyaku Co. use immobilized whole cells (see page 500).

3) Flow scheme

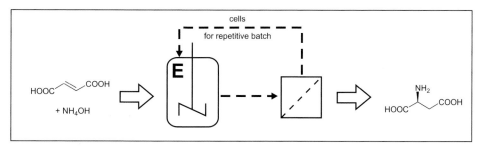

Fig. 4.3.1.1 – 2

4) Process parameters

yield:	> 99.99 %
selectivity:	> 99.99 %
reactor type:	repetitive batch
start-up date:	1986
company:	Mitsubishi Chemical Corporation, Japan

5) Product application

- L-Aspartic acid is used as a precursor for the synthesis of aspartame, an artificial sweetener (see page 343).

- It is also used as an acidulant, as a food additive, in parentered nutrition and as a chiral synthon in organic synthesis.

6) Literature

- Tanaka, A., Tosa, T., Kobayashi, T. (1993) Industrial Application of Immobilized Biocatalysts, Marcel Dekker Inc., New York

- Terasawa, M., Yukawa, H., Takayama, Y. (1985) Production of L-aspartic acid from *Brevibacterium* by cell re-using process, Proc. Biochem. **20**, 124–128

- Yamagata, H., Terasawa, M., Yukawa, H. (1994) A novel industrial process for L-aspartic acid production using an ultrafiltration membrane, Cat. Today **22**, 621–627

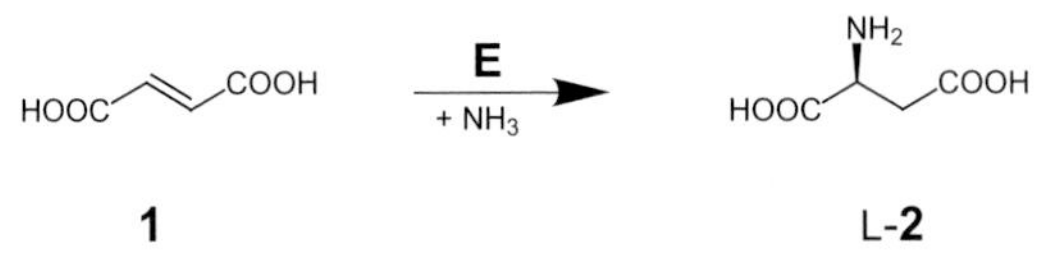

1 = fumaric acid
2 = aspartic acid

Tanabe Seiyaku Co., Ltd.

Fig. 4.3.1.1 – 1

1) Reaction conditions

[1]:	1 M, 150.14 g · L^{-1} [150.14 g · mol^{-1}] (ammonium fumarate)
pH:	8.5
T:	37 °C
medium:	aqueous
reaction type:	C-N bond cleavage
catalyst:	immobilized whole cells
enzyme:	L-aspartate ammonia-lyase (fumaric aminase, aspartase)
strain:	*Escherichia coli* B ATCC 11303
CAS (enzyme):	[9027–30–9]

2) Remarks

- L-Aspartic acid is produced batchwise via fermentation or enzymatic synthesis since 1953.

- The stability of isolated and immobilized aspartase is not satisfactory. Therefore the cells are immobilized on polyacrylamide or, preferably, on κ-carrageenan.

- This process is the first example of the application of immobilized whole cells.

- It is one of the rare examples where the synthesis of an amino acid via an enzymatic route is economically more attractive than the usual fermentation methods.

- The costs of the process are reduced to two thirds in comparison to batchwise operation.

- The activity of the cells is increased 10-fold by immobilization.

- The half-life of the cells is about 12 days. By addition of about 1 mM Mg^{2+}, Mn^{2+} or Ca^{2+} it can be increased to more than 120 days.

- A heat exchanger (multiple, small isothermic pipes) is used, because the reaction is exothermic. The use of a fixed bed reactor is advantageous.

- The product is isolated by titration to the isoelectric point (pH 2.8) with H$_2$SO$_4$ and filtration of the precipitate.

- This synthesis step is often combined with the synthesis of L-alanine in a two step biotransformation starting from fumaric acid (see page 451).

3) Flow scheme

See page 336

4) Process parameters

yield:	> 95 %
reactor type:	plug-flow reactor
reactor volume:	1,000 L
capacity:	700 t · a^{-1}
down stream processing:	isoelectric precipitation
enzyme activity:	3,200 U/g$_{cells}$
start-up date:	1973
company:	Tanabe Seiyaku Co., Ltd., Japan

5) Product application

- L-Aspartic acid is used as a precursor for the synthesis of aspartame, an artificial sweetener (see page 373).

- It is also used as an acidulant, as a food additive, in parentered nutrition and as a chiral synthon in organic synthesis.

6) Literature

- Adelberg, E.A., Mandel, M., Chen, G.C.C. (1965) Optimal conditions for mutagenesis by *N*-methyl-*N*-nitrosoguanidine in *Escherichia coli* K12, Biochem. Biophys. Res. Commun. **18**, 788–795

- Cheetham, P.S.J. (1994) Case studies in applied biocatalysis, in: Applied Biocatalysis (Cabral, J.M.S., Best, D., Boross, L., Tramper, J., eds.) pp. 77–78, Harwood Academic Publishers, Chur, Switzerland

- Chibata, I., Tosa, T., and Sato, T. (1974) Immobilized aspartase-containing microbial cells: preparation and enzymatic properties, Appl. Microbiol. **27**, 878–885

- Chibata, I. (1978) Immobilized Enzymes, Research and Development, John Wiley & Sons, New York

- Furui, M. (1985) Heat-exchange column with horizontal tubes for immobilized cell reactions with generation of heat, J. Ferment. Technol. **63**, 371–375

- Nishimura, N., Kisumi, M. (1984), Aspartase-hyperproducing mutants of *Escherichia coli* B, Appl. Environ. Microbiol. **48**, 1072–1075

- Sato, T., Tetsuya, T. (1993) Production of L-aspartic acid, in: Industrial Applications of Immobilized Biocatalysts (Tanaka, A., Tosa, T., Kobayashi, T., eds.), pp. 15–24, Marcel Dekker Inc., New York

- Takata, I., Tosa, T., and Chibata, I. (1977) Screening of matrix suitable for immobilization of microbial cells, J. Solid-Phase Biochem. **2**, 225–236

- Tosa, T., Shibatani, T. (1995) Industrial application of immobilized biocatalysts in Japan, Ann. N. Y. Acad. Sci. **750**, 364–375

L-Phenylalanine ammonia-lyase
Rhodotorula rubra

1 = *trans*-cinnamic acid
2 = ammonia
3 = L-phenylalanine

Genex Corporation

Fig. 4.3.1.5 – 1

1) Reaction conditions

[1]:	0.088 M, 13.02 g · L^{-1} [148.16 g · mol^{-1}]
[2]:	9.307 M, 158.5 g · L^{-1} [17.03 g · mol^{-1}]
[3]:	0.258 M, 42.7 g · L^{-1} [165.19 g · mol^{-1}]
pH:	10.6
T:	25 °C
medium:	aqueous
reaction type:	C-N bond formation
catalyst:	suspended whole cells
enzyme:	L-phenylalanine ammonia-lyase (PAL, L-phenylalanine deaminase, tyrase)
strain:	*Rhodococcus rubra* (Genex 1983), *Rhodotorula rubra* (Genex 1986), wild type
CAS (enzyme):	[9024–28–6]

2) Remarks

- The PAL-producing microorganisms are initially cultivated under aerobic, growth-promoting conditions.

- Due to the instability of the enzyme towards oxygen, the biotransformation is performed under anaerobic, static conditions.

- An aqueous solution of *trans*-cinnamic acid is mixed with 29 % aqueous ammonia and the pH is adjusted by addition of carbon dioxide.

- As biocatalyst 5.88 g · L^{-1} (dry weight) *Rhodotorula rubra* cells are added.

- The reaction is performed in fed batch mode with periodic addition of concentrated ammonium cinnamate solution.

- The enzyme is deactivated by oxygen and by agitation. Therefore the reaction medium is sparged with nitrogen before the addition of the cells. Instead of stirring, the bioreactor contents are mixed after each addition of substrate solution by sparging with nitrogen.

- Instead of starting from *trans*-cinnamic acid, the fermentation process now starts from glucose. The yields of this *de novo* process are high and up to 25 g · L^{-1} of L-phenylalanine are obtained.

502

- Prior to this and related processes, ʟ-phenylalanine was mainly obtained by extraction from human hair and other non-microbiological sources.

3) Flow scheme

Not published.

4) Process parameters

yield: 85.7 %
reactor type: fed batch
down stream processing: centrifugation, evaporation, crystallization
company: Genex Corporation, USA

5) Product application

- ʟ-Phenylalanine is used in the manufacture of the artificial sweetener aspartame and in parenteral nutrition.

- The product is used as a building block for the synthesis of the macrolide antibiotic rutamycin B:

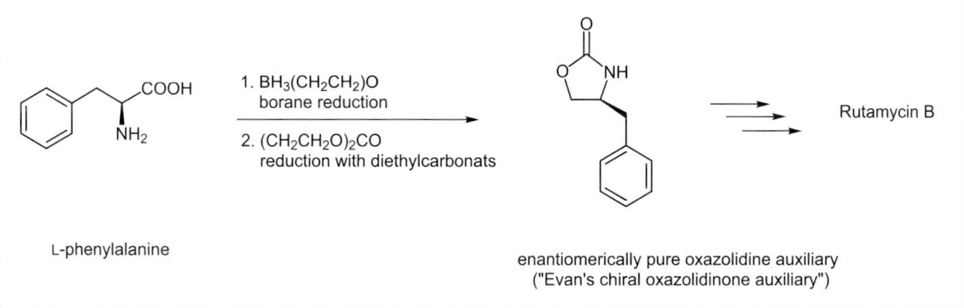

L-phenylalanine

1. BH₃(CH₂CH₂)O borane reduction
2. (CH₂CH₂O)₂CO reduction with diethylcarbonats

enantiomerically pure oxazolidine auxiliary
("Evan's chiral oxazolidinone auxiliary")

Rutamycin B

Fig. 4.3.1.5 – 2

- ʟ-Phenylalanine ammonia lyase is effective in the therapy of transplantable tumors in mice.

6) Literature

- Crosby, J. (1991) Synthesis of optically active compounds: A large-scale perspective, Tetrahedron **47**, 4789–4846

- Kirk-Othmer (1991–1997) Encyclopedia of Chemical Technology, John Wiley, New York

- Sheldon, R.A. (1993) Chirotechnology, Marcel Dekker-Verlag, New York

- Vollmer, P.J., Schruben, J.J., Montgomery, J.P., Yang, H.-H. (1986) Method for stabilizing the enzymatic activity of phenylalanine ammonia lyase during ʟ-phenylalanine production, Genex Corporation, US 4584269

- Wiseman, A. (1995) Handbook of Enzyme and Biotechnology, Ellis Horwood, Chichester

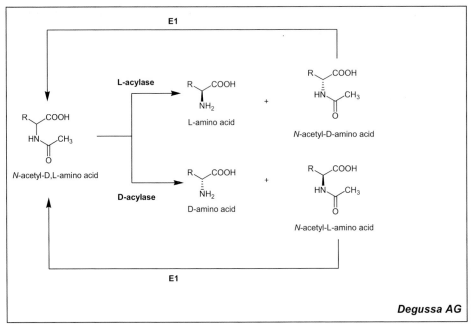

Fig. 5.1.1.10 – 1

1) Reaction conditions

medium:	aqueous
reaction type:	racemization
catalyst:	enzyme
enzyme:	L-acylase
enzyme:	N-acylamino acid racemase
strain:	*Amycolatopsis orientalis* subsp. *lurida*
CAS (enzyme):	[9068-61-5]

2) Remarks

- Several hundred tons of L-methionine are produced annually by this enzymatic conversion using an enzyme membrane reactor.

- The L-acylase process was optimized by introduction of novel racemases found by genetic screening.

- The novel racemase has a much lower substrate inhibition.

- Because of the dynamic resolution the novel process has a higher potential than the currently used classic resolution process.

- The N-acylamino acid racemase catalyzes the racemization of various industrially important aromatic as well as aliphatic N-acylamino acids. The K_m value and the V_{max} are 24 mM and 8 U mg^{-1} for N-acetyl-D-methionine and 35 mM and 13.7 U mg^{-1} for N-acetyl-L-methionine.

- The racemization reaction is subject to substrate inhibition with a K_i of 457 or 398 mM for *N*-acetyl-D- and -L-methionine, respectively. L-Methionine itself is not a substrate.

3) Flow scheme

Not published.

4) Process parameters

company: Degussa AG, Germany

5) Literature

- Drauz, K., Eils, S., Schwarm, M. (2002) Synthesis and production of enantiomerically pure amino acids, Chimica Oggi/Chemistry Today January/February, 15-21

- May, O., Verseck, S., Bommarius, A., Drauz, K. (2002) Development of dynamic kinetic resolution processes for biocatalytic production of natural and non-natural L-amino acids, Org. Proc. Res. Dev. **6**, 452–457

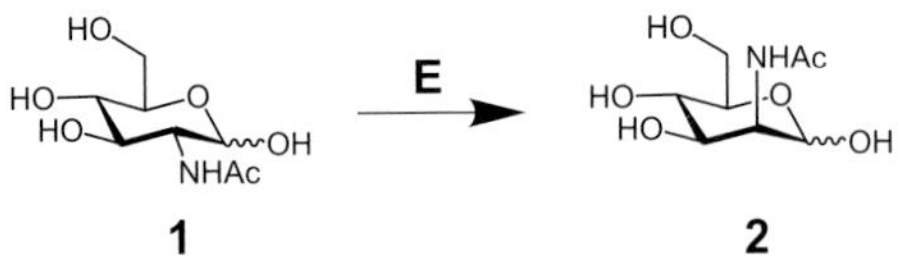

1 = *N*-acetyl-D-glucosamine
2 = *N*-acetyl-D-mannosamine

Marukin Shoyu Co., Ltd.

Fig. 5.1.3.8 – 1

1) Reaction conditions

[1]:	0.8 M, 177 g · L^{-1} [221.21 g · mol^{-1}]
pH:	7.2
T:	30 °C
medium:	aqueous
reaction type:	epimerization
catalyst:	solubilized enzyme
enzyme:	*N*-acyl-D-glucosamine 2-epimerase (*N*-acylglucosamine 2-epimerase)
strain:	*Escherichia coli*
CAS (enzyme):	[37318–34–6]

2) Remarks

- This biotransformation is integrated into the production of *N*-acetylneuraminic acid (Neu5Ac), see page 340.

- By application of the *N*-acylglucosamine 2-epimerase it is possible to start with the inexpensive *N*-acetyl-D-glucosamine instead of *N*-acetyl-D-mannosamine.

- The epimerase is used for the *in situ* synthesis of *N*-acetyl-D-mannosamine (ManNAc). Since the equilibrium is on the side of the educt, the reaction is driven by the subsequent transformation of ManNAc and pyruvate to Neu5Ac.

- The *N*-acylglucosamine 2-epimerase is cloned from porcine kidney and transformed and overexpressed in *Escherichia coli*.

- To reach maximal axctivitiy ATP and Mg^{2+} need to be added.

- Since the whole synthesis is reversible high GlcNAc concentrations are used.

- The chemical epimerization of GlcNAc is employed by Glaxo (page 459).

3) Flow scheme

Not published.

4) Process parameters

conversion:	84 %
reactor type:	batch
reactor volume:	200 L
capacity:	multi kg
down stream processing:	subsequent, *in situ* biotransformation to *N*-acetylneuraminic acid
enzyme supplier:	Marukin Shoyu Co., Ltd., Japan
company:	Marukin Shoyu Co., Ltd., Japan

5) Product application

- *N*-Acetyl-D-mannosamine serves as *in situ* generated substrate for the synthesis of *N*-acetyl-neuraminic acid.

6) Literature

- Kragl, U., Gygax, D., Ghisalba, O., Wandrey, C. (1991) Enzymatic process for preparing *N*-acetylneuraminic acid, Angew. Chem. Int. Ed. Engl. **30**, 827–828

- Maru, I., Ohnishi, J., Ohta, Y., Tsukada, Y. (1998) Simple and large-scale production of *N*-acetylneuraminic acid from *N*-acetyl-D-glucosamine and pyruvate using *N*-acyl-D-glucosamine 2-epimerase and *N*-acetylneuraminate lyase, Carbohydrate Res. **306**, 575–578

- Maru, I., Ohta, Y., Murata, K., Tsukada, Y. (1996) Molecular cloning and identification of *N*-acyl-D-glucosamine 2-epimerase from porcine kidney as a renin-binding protein, Biol. Chem. **271**, 16294–16299

Xylose isomerase

Bacillus coagulans/Streptomyces rubiginosus/Streptomyces phaechromogenes

EC 5.3.1.5

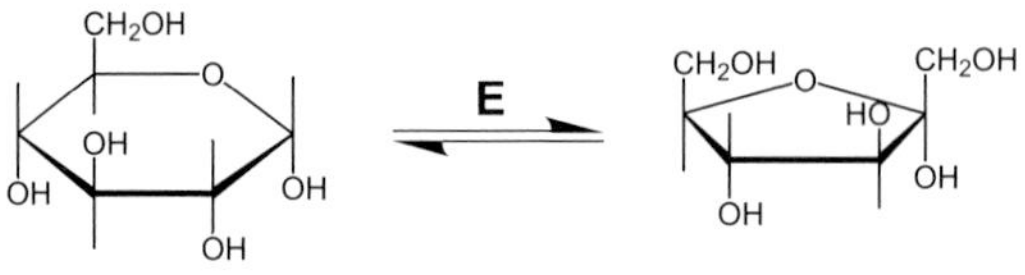

1 **2**

Novo Nordisk
DSM
Finnsugar
Nagase Company, Ltd.

1 = glucose
2 = fructose

Fig. 5.3.1.5 – 1

1) Reaction conditions

[**1**]:	> 95 % dry matter
pH:	7.5 – 8.0
T:	50 – 60 °C
medium:	aqueous
reaction type:	isomerization
catalyst:	immobilized whole cells or isolated enzyme
enzyme:	D-xylose ketol-isomerase (xylose-isomerase, glucose-isomerase)
strain:	several, see remarks
CAS (enzyme):	[9023–82–9]

2) Remarks

- Glucose isomerase is produced by several microorganisms as an intracellular enzyme. The following table shows some commercial examples:

Trade name	Microorganism	Company	Country
Sweetzyme	*Bacillus coagulans*	Novo-Nordisk	Denmark
Maxazyme	*Actinplanes missouriensis*	DSM	The Netherlands
Optisweet	*Streptomyces rubiginosus*	Nagase	Japan
Sweetase	*Streptomyces phaechromogenes*		

Fig. 5.3.1.5 – 2

- The commercially important varieties show superior affinity to xylose and are therefore classified as xylose-isomerases.
- Since the isolation of the intracellular enzyme is very expensive, whole cells are used instead. In almost all cases the enzymes or cells are immobilized using different techniques depending on strain and supplier.
- The substrate is purified glucose (dextrose) syrup from the saccharification stage.

Xylose isomerase
Bacillus coagulans/Streptomyces rubiginosus/Streptomyces phaechromogenes

- Since these isomerases belong to the group of metalloenzymes, Co^{2+} and Mg^{2+} are required.

- The reaction enthalpy is slightly endothermic and reversible. The equilibrium conversion is about 50 % at 55 °C.

- To limit byproduct formation, the reaction time must be minimized. This can be done economically only by using high concentrations of immobilized isomerase.

- Several reactors are operated in parallel and contain enzymes of different ages. The feed to a single reactor is controlled by the extent of conversion in this reactor.

- The feed syrup has to be highly purified (filtration, adsorption on charcoal, ion exchange) to prevent fast deactivation and clogging of the catalyst bed (for first part of process see page 334).

- Plants producing more than 1,000 t of HFCS (high fructose corn syrup) (based on dry matter) per day typically use at least 20 individual reactors.

- The product HFCS contains 42 % fructose (53 % glucose) or 55 % fructose (41 % glucose) (as dry matter).

- Glucose isomerases have half-lives of more than 100 days. To maintain the necessary activity the enzyme is replaced after deactivation to about 12.5 % of the initial value.

- The reaction temperature is normally above 55°C to prevent microbial infection although enzyme stability is lowered.

Xylose isomerase

Bacillus coagulans/Streptomyces rubiginosus/Streptomyces phaechromogenes

EC 5.3.1.5

3) Flow scheme

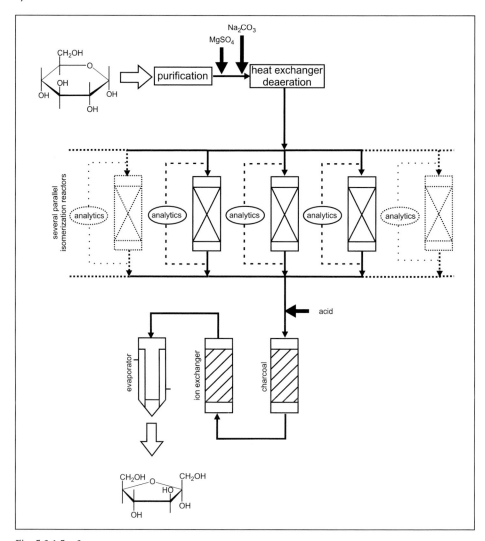

Fig. 5.3.1.5 – 3

4) Process parameters

reactor type:	continuous, fixed bed
reactor volume:	variable
capacity:	$> 7 \cdot 10^6 \, t \cdot a^{-1}$
residence time:	0.17 – 0.33 h
down stream processing:	55 % fructose: chromatography; 42 % fructose: no down stream processing
enzyme consumption:	see remarks
start-up date:	1967 by Clinton Corn Processing Co. (USA); 1974 with immobilized enzyme

510

Xylose isomerase

Bacillus coagulans/Streptomyces rubiginosus/Streptomyces phaechromogenes

EC 5.3.1.5

production site: numerous companies
company: Novo Nordisk, Denmark, DSM, The Netherlands, Finnsugar, Finland, Nagase & Co. Ltd., Japan

5) Product application

- The product is named high-fructose corn syrup (HFCS) or ISOSIRUP.

- It is an alternative sweetener to sucrose or invert sugar in the food and beverage industries.

- The chromatographically enriched form (55 % fructose) is used for sweetening alcoholic beverages.

- 42 % HFCS obtained directly by enzymatic isomerization is used mainly in the baking and dairy industries.

6) Literature

- Antrim, R.L., Colilla, W., Schnyder, B.J. and Hemmingesen, S.H., in: Applied Biochemistry and Bioengineering **2** (Wingard, L.B., Katchalski-Katzir, E., Goldstein, L., eds.), pp. 97–183, Academic Press, New York

- Blanchard, P.H., Geiger, E.O. (1984) Production of high-fructose corn syrup in the USA, Sugar Technol. Rev. **11**, 1–94

- Gerhartz, W. (1990) Enzymes in Industry: Production and Application, Verlag Chemie, Weinheim

- Hupkes, J.V., van Tilburg, R. (1976) Industrial applications of the catalytic power of enzymes, Neth. Chem. Weekbl. **69**, K14-K17

- Ishimatsu, Y., Shigesada, S., Kimura, S., (1975) Immobilized enzymes, Denki Kaguku Kogyo K.K. Japan, US 3915797

- Landis, B.H., Beery, K.E. (1984) Developments in soft drink technology, **3**, Elsevier Applied Science Publishers, London, pp. 85–120

- Oestergaard, J., Knudsen, S.L. (1976) Use of Sweetenzyme in industrial continuous isomerization. Various process alternatives and corresponding product types, Stärke **28**, 350–356

- Straatsma, J., Vellenga, K., de Wilt, H.G.J., Joosten, G.E.H. (1983) Isomerization of glucose to fructose. 1. The stability of a whole cell immobilized glucose isomerase catalyst, Ind. Eng. Chem. Process Des. Dev. **22**, 349–356

- Straatsma, J., Vellenga, K., de Wilt, H.G.J., Joosten, G.E.H. (1983) Isomerization of glucose to fructose. 2. Optimization of reaction conditions in the production of high fructose syrup by isomerization of glucose catalyzed by a whole cell immobilized glucose isomerase catalyst, Ind. Eng. Chem. Process Des. Dev. **22**, 356–361

- Tewari, Y.B., Goldberg, R.N. (1984) Thermodynamics of the conversion of aqueous glucose to fructose, J. Solution Chem., **13**, 523–547

- Weidenbach, G., Bonse, D., Richter, G. (1984) Glucose isomerase immobilized on silicon dioxide-carrier with high productivity, Starch/Stärke **36**, 412–416

- White, J.S., Parke, D.W. (1989) Fructose adds variety to breakfast, Cereal Foods World **34**, 392–398

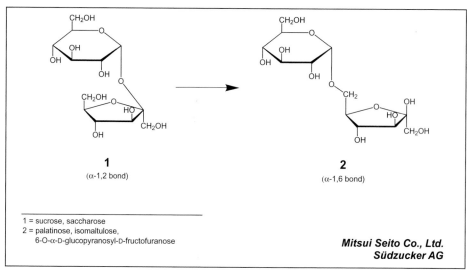

1
(α-1,2 bond)

2
(α-1,6 bond)

1 = sucrose, saccharose
2 = palatinose, isomaltulose,
    6-O-α-D-glucopyranosyl-D-fructofuranose

Mitsui Seito Co., Ltd.
Südzucker AG

Fig. 5.4.99.11 – 1

1) Reaction conditions

pH:	5.8
T:	40 °C
medium:	aqueous
reaction type:	isomerization
catalyst:	immobilized, whole cells
enzyme:	Sucrose glucosylmutase (isomaltulose synthetase, sucrose α-glucosyltransferase)
strain:	*Protaminobacter rubrum*
CAS (enzyme):	[159940–49–5]

2) Remarks

- Palatinose, a reducing disaccharide, occurs naturally in low amounts in honey and sugar-cane extract.

- The word palatinose is derived from palatin, the Latin name for the Pfalz region in Germany.

- α-Glucosyltransferase simultaneously produces isomaltulose and smaller amounts of trehalulose (1-O-α-D-glucopyranosyl-β-D-fructose) from sucrose.

- Palatinose (isomaltulose) is the kinetically preferred product, while trehalulose is the thermodynamically preferred product.

- The commercial product contains about 5 % water.

3) Process parameters

conversion:	> 99.5 %
yield:	85 %
selectivity:	85 %
capacity:	> 4,000 t · a^{-1}
start-up date:	1985
production site:	Mitsui Seito Co., Ltd., Japan and Südzucker AG, Germany
company:	Südzucker AG, Germany; Mitsui Seito Co. Ltd., Japan

4) Product application

- Palatinose and the hydrogenated product (palatinit, isomaltulit) used as sweeteners with a similar taste as sucrose but only 42 % of the sweetness of sucrose and with only half of the calorific value.

- It is used as a substitute for sucrose because of the low insulin stimulation.

- The advantage of palatinose or its hydrogenated derivativ is that it is decomposed only slowly by *Streptomyces* mutants and dental plaque suspensions resulting in a low cariogenic potential. The level of acid and glucan production is decreased compared to sucrose.

5) Literature

- Cabral, J., Best, D., Boross, L., Tramper, J. (1993) Applied Biocatalysis, Harwood Academic Publishers, Chur, Switzerland

- Cheetham, P.S.J., Imber, C.E., Isherwood, J. (1982) The formation of isomaltulose by immobilized *Erwinia rhapontici*, Nature **299**, 628–631

- McAllister, M., Kelly, C.T., Doyle, E., Fogarty (1990) The isomaltulose synthesizing enzyme of *Serratia plymuthica*, Biotechnol. Lett. **12**, 667–672

- Park, Y.K.; Uekane, R.T., Pupin, A.M. (1992) Conversion of sucrose to isomaltulose by microbial glucosyltransferase, Biotechnol. Lett. **14**, 547–551

- Takazoe, I., Frostell, G., Ohta, K., Topitsoglou, Sasaki, N. (1985) Palatinose – a sucrose substitute, Swed. Dent. J. **9**, 81–87

- Tosa, T., Shibatani, T. (1995) Industrial application of immobilized biocatalysts in Japan, Ann. N. Y. Acad. Sci. **750**, 365–375

7

Quantitative Analysis of Industrial Biotransformations

Adrie J. J. Straathof

The number of biotransformation processes that are carried out on an industrial scale is almost doubling every decade [1]. At present there are about 150 processes that are known to fulfill the following criteria:

- They describe a reaction or a set of simultaneous reactions in which a pre-formed precursor molecule is converted; as opposed to fermentation processes with *de novo* production from a source of carbon and energy, such as glucose, via the primary metabolism.
- They involve the use of enzymes and/or whole cells, or combinations thereof, either free or immobilized.
- They lead to the production of a fine-chemical or commodity product, which is usually recovered after the reaction.
- They have been reported to be operated on a commercial scale, or have been successfully scaled-up and announced to be commercialized at a scale of usually more than $100 \, \mathrm{kg \, a^{-1}}$.

In this chapter these processes will be quantitatively analyzed in order to obtain a better overview of the current status of industrial biotransformations. Most of the processes included in this analysis have been described in detail in the previous chapter of this book. However, some particular processes were left out of this analysis. These involve processes that have clearly only been performed on a pilot scale and also degradation processes, in which the biotransformation is used for removing an undesired compound rather than for making a desired product.

In addition, this analysis does include processes that have not been described in the previous chapter. This is due to the problem that public domain sources sometimes give incomplete or conflicting information. Moreover, some existing processes have not been described at all in public domain sources. Therefore, despite a number of additions and corrections since an earlier report [1], the numbers given in the quantitative information provided here cannot be taken as absolute numbers, but they will probably show the correct trends.

One aspect of the information that is only being reported in a modest number of cases is the operation scale of the processes. There are nine processes that are operated on a scale of more than $10000 \, \mathrm{t \, a^{-1}}$. These are mostly used for the production of carbohydrates, and applied in the food sector. Non-carbohydrate biotransformation products that are produced on a bulk scale are the monomer acrylamide and the pharmaceutical intermediate 6-amino-

Industrial Biotransformations. Andreas Liese, Karsten Seelbach, Christian Wandrey (Eds.)
Copyright © 2006 WILEY-VCH Verlag GmbH & Co. KGaA, Weinheim
ISBN: 3-527-31001-0

penicillanic acid (6-APA). For at least 40 biotransformation processes, the production scale is between 100 and 10 000 t a^{-1}, and for around the same number of processes the scale is even lower. Thus, most biotransformation products are fine chemicals. This does not imply that the processes involved will be less valuable than bulk processes. Fine-chemicals, in particular pharmaceutical intermediates, may have a very high value per kilogram.

Pharmaceuticals are the most important type of fine-chemical products, and this also applies to fine-chemical products that are produced by biotransformations, as shown in Figure 7.1. Some other types of fine chemicals, such as dyes, are never produced by biotransformations, but biotransformations are relatively important for food ingredients, and also seem to have found a place in the cosmetics industry. In the last two cases, the ease of selectively transforming natural ingredients by biocatalysis will play an important role.

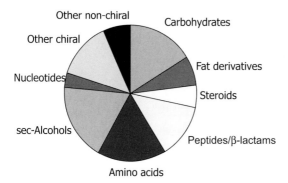

Fig. 7.1 Types of compounds produced in industrial biotransformations.

Natural compounds and their derivatives are indeed the prevailing classes of products produced by biotransformations (see Figure 7.2). The carbohydrate and fat derivatives are mainly used in the food sector, whereas the other types of compounds are mainly applied in the pharma sector. Only for the "other non-chiral" product class does this statement require refinement, because these products are applied in agrochemicals as much as in pharmaceuticals.

It is no surprise that chirality plays a key role in industrial biotransformations. Biocatalysts are chiral by nature, and chirality is a key issue at present in the production of pharmaceuticals. Moreover, carbohydrates and other natural compounds that are used as precursors for many biotransformations are usually chiral. For most of the enantiomerically pure products the enantiomeric configuration actually originates directly from the precursors (Figure 7.3); the biotransformation then usually leads to conservation of the stereochemistry. In about half of the biotransformations the optical activity of the product is due to the stereoselective conversion of a precursor that is not optically active. In this respect, kinetic resolution of racemic precursors occurs slightly more frequently than asymmetric synthesis using prochiral precursors. Asymmetric syntheses are carried out mostly by oxidoreductases and lyases, whereas kinetic resolutions involve almost exclusively hydrolases. In about a quarter of these kinetic resolutions, the hydrolases are applied in the synthetic rather than in the hydrolytic mode.

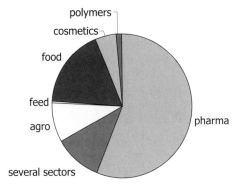

Fig. 7.2 Sectors in which the products of industrial biotransformations are applied.

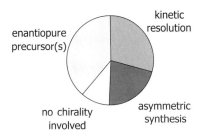

Fig. 7.3 Sources of chirality for the products of industrial biotransformations.

It is interesting to focus on the relatively few cases where no chirality is involved, where enantioselective catalysis is hence not an issue. These cases mainly involve regio-selective oxidations and nitrile hydrolysis or hydration. The latter lead to less side-products with biocatalysis than with the conventional chemistry. Thus, in all cases where biocatalysts are used in industrial biotransformations their selectivity seems to be exploited. Perhaps this statement should also be reversed: if biocatalysts provide no selectivity advantage, they will not be used for industrial biotransformations.

Hydrolases are the enzymes that are most frequently applied in industrial biotransformations (Figure 7.4). However, within the class of hydrolases, no single sub-class dominates. Lipases, proteases, amidases as well as glycosidases are important types of hydrolases. Obviously, lipases are often used for fat derivatives, proteases for peptides and glycosidases for carbohydrates. In addition, quite a variety of other hydrolases are being applied. Lyases and transferases are also applied in a reasonable number of cases. On an average, the lyase-catalyzed processes are more important than the transferase-catalyzed processes. The lyases are applied far more frequently in the synthetic mode (addition reactions) than in the reverse mode (elimination reactions), because the synthetic mode can be exploited for building up molecules and creating chiral centers. Ligases, which might also be synthetically valuable, have not yet been applied in industrial biotransformations. Their requirement for ATP or other cofactors can be more easily fulfilled in living cells. Isomerases are applied only to a modest extent, except for the use of xylose isomerase in the glucose–fructose isomerization. The other isomerases that are applied are mostly racemases.

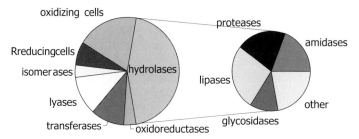

Fig. 7.4 Enzyme types used in industrial biotransformations.

Processes involving redox biotransformations are very important. Some of these involve two isolated oxidoreductases, one for the actual biotransformation and one for the regeneration of the cofactor NAD(P)H. However, in the majority of redox biotransformations, for example for all steroid biotransformations, metabolizing cells are applied. These cells can use their primary metabolism to regenerate redox cofactors. They often use glucose as the source of electrons, or oxygen as the sink for electrons.

Whole cells are also used in many non-redox biotransformations. In the absence of cofactor regeneration cycles, only the key biotransformation enzymes will be active. Cells can be used merely as crude enzyme preparations, in order to save on enzyme purification costs. Figure 7.5 indicates that whole cells are more popular than (partly) isolated enzymes. When whole cells are used, immobilization is less common than when isolated enzymes are used. For redox biotransformations with whole cells, immobilization will be avoided when oxygen is used. Owing to the very low aqueous solubility of oxygen, oxygen diffusion limitation in particles of several millimeters can not really be prevented at reasonable reaction rates. When redox biotransformations by whole cells are not taken into account, immobilization is performed in almost half of the industrial biotransformations, not only for enzymatic processes but also for the remaining whole cell processes. In a few cases ultrafiltration membranes are used to retain the biocatalysts, but binding on or in particles is much more common.

Retention of biocatalysts by these methods is a prerequisite for using continuous reactors in industrial biotransformations. On the other hand, in almost half of the cases

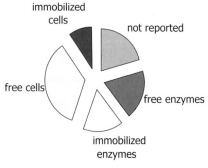

Fig. 7.5 Use of enzymes or whole cells in industrial biotransformations.

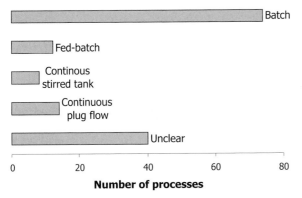

Fig. 7.6 Types of reactors in industrial biotransformations.

where immobilization is applied, the reactor is not continuous but the biocatalyst is reused in a batch or fed-batch reactor. Considering the abundance of fine chemicals as biotransformation products, it is not surprising that such batch reactors are more popular than continuous reactors (see Figure 7.6). Batch processing provides the flexibility that is usually required in fine chemistry.

Fed-batch reactors mainly involve processes with free whole-cells, where substrate toxicity can be a major issue. Continuous reactors are more common for enzymes than for whole cells, in particular when considering redox biotransformations. The origin of this may be the aforementioned diffusion limitations due to immobilization of living cells, but also the genetic instability of living cells.

In general, water is used as the solvent in industrial biotransformations. Figure 7.7 indicates that only a modest proportion of the industrial processes are carried out in organic solvents (without a separate water phase). These involve (trans)esterifications and amidations. More often, a water-miscible or water-immiscible organic liquid is added to an aqueous phase, leading a monophasic or biphasic liquid, respectively. The latter category is the more common one. Not only does it include processes where an organic solvent is used as a reservoir for the substrate and product, but also a significant number of processes where the organic phase is actually composed of just the liquid substrate or product. No organic solvent is added in such cases.

Fig. 7.7 Characteristics of solutions in industrial biotransformations.

When the substrate or product is not a liquid but a solid with a low solubility, the high product titers required for industrial processes may also lead to suspension reactions. Product precipitation occurs in at least 15 industrial processes, usually in aqueous solvents. In about half of these cases, a suspension of a substrate is converted into a suspension of a product.

The product titer has been found for a minority of the processes (see Figure 7.8), but this may be sufficient to draw some interesting conclusions. Only in rare cases is the product titer below 10 g L^{-1}. In most cases it is well above 100 g L^{-1}, with occasional values of about 500 g L^{-1}, irrespective of the product type. Thus, achievement of high product titers could be one of the strong points of biotransformation processes.

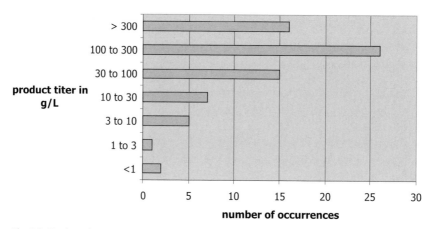

Fig. 7.8 Final product concentrations reported for industrial biotransformations.

With respect to the space–time yield, too few data are available to draw any statistically significant conclusion. The lowest space–time yield that was found is 1.7 g L^{-1} per day for human insulin production with carboxypeptidase B, whereas the highest value is 72 000 g L^{-1} per day for L-aspartate production with aspartase (see previous chapter). In the former case, the low space–time yield seems to be acceptable due to the high value of the product and the lack of suitable alternative catalysts. Also, too few data are available to draw general conclusions about biocatalyst consumption per amount of product. In relatively few cases, the biocatalyst concentration in the reactor and its lifetime can be estimated. When baker's yeast or bulk industrial enzymes such as subtilisin are used, the costs due to biocatalyst consumption will not be significant relative to those of the overall process, but in cases where dedicated biocatalysts have been developed, reduction of the biocatalyst consumption will be a major issue.

Reference

1 A. J. J. Straathof, S. Panke, A. Schmid 2002, (The production of fine-chemicals by bio-transformations), *Curr. Opin. Biotechnol.* 13, 548–556.

Index of enzyme name

enzyme name	EC number	strain	company	page
Acetolactate decarboxylase	4.1.1.5	*Bacillus brevis*	Novo Nordisk, Denmark	449
Alcohol dehydrogenase	1.1.1.1	*Acinetobacter calcoaceticus*	Bristol-Myers Squibb, USA	162
		Lactobacillus brevis	Wacker Chemie, Germany	170
		Lactobacillus kefiri	Jülich Fine Chemicals GmbH	167
		Neurospora crassa	Zeneca Life Science Molecules, U.K.	153
		Rhodococcus erythropolis	Forschungszentrum Jülich GmbH, Germany	157
Aldehyde reductase	1.1.1.2		Lonza AG, Switzerland	174
Amidase	3.5.1.4	*Comamonas acidovorans*	Lonza AG, Switzerland	377
		Klebsiella oxytoca	Lonza AG, Switzerland	382
		Klebsiella terrigena	Lonza AG, Switzerland	379
D-Amino acid oxidase	1.4.3.3	*Trigonopsis variabilis* ATCC 10679	Bristol-Myers Squibb, USA	211
		Trijonopsis variabilis	Sandoz AG, Switzerland	209
Amino acid racemase	5.1.1.10	*Amycolatopsis orientalis* subsp. *lurida*	Degussa AG, Germany	504
D-Amino acid transaminase	2.6.1.21	*Bacillus* sp.	NSC Technologies, USA	267
Aminoacylase	3.5.1.14	*Aspergillus niger*	Celltech Group plc, U.K.	403
		Aspergillus oryzae	Degussa AG, Germany	407
Aminopeptidase	3.4.11.1	*Pseudomonas putida*	DSM, The Netherlands	337
α-Amylase	3.2.1.1	*Bacillus licheniformis*	Several companies	332
Amyloglucosidase	3.2.1.3	*Aspergillus niger*	Several companies	332
Aryl alcohol dehydrogenase	1.1.1.90	*Pseudomonas putida* ATCC 33015	Pfizer Inc., USA	233
Aspartase	4.3.1.1	*Brevibacterium flavum*	Mitsubishi Chemical Corporation, Japan	498
		Escherichia coli	BioCatalytics Inc., USA	494
			Kyowa Hakko Kogyo Co., Ltd., Japan	496
L-Aspartase	4.3.1.1	*Escherichia coli*	Tanabe Seiyaku Co., Ltd., Japan	500
Aspartate β-decarboxylase	4.1.1.12	*Pseudomonas dacunhae*	Tanabe Seiyaku Co., Japan	451
Benzaldehyde dehydrogenase	1.2.1.28	*Pseudomonas putida* ATCC 33015	Pfizer Inc., USA	233
Benzoate dioxygenase	1.14.12.10	*Pseudomonas putida*	ICI, U.K.	225
Carbamoylase	3.5.1.77	*Pseudomonas* sp.	Dr. Vig Medicaments, India	415

Industrial Biotransformations. Andreas Liese, Karsten Seelbach, Christian Wandrey (Eds.)
Copyright © 2006 WILEY-VCH Verlag GmbH & Co. KGaA, Weinheim
ISBN: 3-527-31001-0

Index of strain

strain	enzyme name	EC number	company	page
Achromobacter obae	Racemase	5.1.1.15	Toray Industries Inc., Japan	424
Achromobacter xylosoxidans	Nicotinic acid hydroxylase	1.5.1.13	Lonza AG, Switzerland	213
Acidovorax facilis	Nitrilase	3.5.5.1	DuPont de Nemours & Co., USA	427
Acinetobacter calcoaceticus	Alcohol dehydrogenase	1.1.1.1	Bristol-Myers Squibb, USA	162
	Cyclohexanone monooxygenase	1.14.13.22	Sigma Aldrich, USA	227
Agrobacterium sp.	Hydroxylase	1.5.1.13	Lonza AG, Switzerland	430
	Nitrilase	3.5.5.1	Lonza AG, Switzerland	430
Alcaligenes faecalis	Hydroxylase	1.5.1.13	Lonza AG, Switzerland	433
	Nitrilase	3.5.5.1	Lonza AG, Switzerland	433
Alcaligenes sp. or *Pseudomonas* sp.	Haloalkane dehalogenase	3.8.1.5	Daiso Co. Ltd., Japan	439
Amycolatopsis orientalis	Amino acid racemase	5.1.1.10	Degussa AG, Germany	504
Arthrobacter sp.	Lipase	3.1.1.3	Sumitomo Chemical Co., Japan	304
	Oxygenase	1.13.11.1	Mitsubishi Chemical Corporation, Japan	219
Arthrobacter sp. DSM 9771	L-Hydantoinase	3.5.2.4	Degussa AG, Germany	417
Arthrobacter viscosus	Penicillin amidase	3.5.1.11	Dr. Vig Medicaments, India	386
Aspergillus niger	Aminoacylase	3.5.1.14	Celltech Group plc, U.K.	403
	Amyloglucosidase	3.2.1.3	Several companies	332
Aspergillus oryzae	Aminoacylase	3.5.1.14	Degussa AG, Germany	407
Aureobacterium sp.	β-Lactamase	3.5.2.6	Celltech Group plc, U.K.	420
Aureobasidium pullulans SC 13849	Reductase	1.1.X.X	Bristol-Myers Squibb, USA	197
Bacillus brevis	Acetolactate decarboxylase	4.1.1.5	Novo Nordisk, Denmark	449
	D-Hydantoinase	3.5.2.2	Kanegafuchi Chemical Industries Co., Ltd., Japan	411
Bacillus circulans	Cyclodextrin glycosyltransferase	2.4.1.19	Mercian Co., Ltd., Japan	264

Index of company

company	strain	enzyme name	EC number	page
Ajinomoto Co., Japan	*Erwinia herbicola*	Tyrosine phenol lyase	4.1.99.2	461
Amino GmbH, Germany	*Corynebacterium glutamicum*	Fumarase	4.2.1.2	463
	Escherichia coli	Tryptophan synthase	4.2.1.20	472
Asahi Kasei Chemical Corporation, Japan	*Bacillus megaterium*	Penicillin amidase	3.5.1.11	391
Asahi Kasei Chemical Industry Co., Ltd., Japan	*Pseudomonas* sp.	Glutaryl amidase	3.1.1.41	330
Asahi Kasei Chemicals Corporation, Japan	*Lactobacillus fermentum*	Urease	3.5.1.5	384
Astra Zeneca, U.K.	*Pseudomonas putida*	Dehalogenase	3.8.1.2	435
BASF AG, Germany	*Beauveria bassiana*	Oxidase	1.X.X.X	261
	Burkholderia plantarii	Lipase	3.1.1.3	273
	Escherichia coli	Nitrilase	3.5.5.1	429
Bayer AG, Germany	*Gluconobacter oxydans*	D-Sorbitol dehydrogenase	1.1.99.21	182
BioCatalytics Inc., USA	*Escherichia coli*	Aspartase	4.3.1.1	494
Bristol-Myers Squibb, USA	*Acinetobacter calcoaceticus*	Alcohol dehydrogenase	1.1.1.1	162
	Aureobasidium pullulans SC 13849	Reductase	1.1.X.X	197
	Bacillus megaterium	Glucose 1-dehydrogenase	1.1.1.118	201
		Glutamate dehydrogenase	1.4.1.3	201
	Beef liver	Glucose 1-dehydrogenase	1.1.1.118	201
	Candida boidinii	Formate dehydrogenase	1.2.1.2	207
		Phenylalanine dehydrogenase	1.4.1.20	207
	Geotrichum candidum	Dehydrogenase	1.1.X.X	187
	Nocardia salmonicolor SC 6310	Reductase	1.1.X.X	199
	Pichia methanolica	Reductase	1.1.X.X	195
	Pseudomonas cepacia	Lipase	3.1.1.3	277, 281, 311, 318
	Streptomyces sp. SC 1754	Monooxygenase	1.14.13.XX	241

company	strain	enzyme name	EC number	page
	Zygosaccharomyces rouxii	Dehydrogenase	1.1.1.1	164
Finnsugar, Finland	*Bacillus coagulans/ Streptomyces rubiginosus/ Streptomyces phaechromogenes*	Xylose isomerase	5.3.1.5	508
Forschungszentrum Jülich GmbH, Germany	*Escherichia coli*	N-Acetyl-D-neuraminic acid aldolase	4.1.3.3	459
	Rhodococcus erythropolis	Alcohol dehydrogenase	1.1.1.1	157
Fuji Chemicals Industries Co., Ltd., Japan	*Fusarium oxysporum*	Lactonase	3.1.1.25	321
Genencor International, Inc., USA	*Pseudomonas putida*	Naphthalene dioxygenase	1.13.11.11	221
Genex Corporation, USA	*Rhodotorula rubra*	L-Phenylalanine ammonia-lyase	4.3.1.5	502
GlaxoSmithKline, U.K.	*Candida antarctica*	Lipase	3.1.1.3	300, 302
	Escherichia coli	N-Acetyl-D-neuraminic acid aldolase	4.1.3.3	457
Hoffmann La-Roche AG, Switzerland	*Bacillus licheniformis*	Subtilisin	3.4.21.62	360, 364
	Bacillus sp.	Subtilisin	3.4.21.62	367
	Baker's yeast	Reductase	1.X.X.X	253
ICI, U.K.	*Pseudomonas putida*	Benzoate dioxygenase	1.14.12.10	225
International Bio Synthetics, Inc., The Netherlands	*Rhodococcus erythropolis*	Oxidase	1.X.X.X	255
Jülich Fine Chemicals GmbH	*Lactobacillus kefiri*	Alcohol dehydrogenase	1.1.1.1	167
Kanegafuchi Chemical Industries Co., Ltd., Japan	*Bacillus brevis*	D-Hydantoinase	3.5.2.2	411
	Candida rugosa	Enoyl-CoA hydratase	4.2.1.17	468, 470
Kaneka Corporation, Japan	*Escherichia coli*	Carbonyl reductase	1.1.1.1	172
Kao Corp., Japan	*Rhodococcus* sp.	Desaturase	1.X.X.X	257
Krebs Biochemicals & Industries Ltd., India	*Saccharomyces cerevisiae*	Pyruvate decarboxylase	4.1.1.1	447
Kyowa Hakko Kogyo Co., Ltd., Japan	*Escherichia coli*	Aspartase	4.3.1.1	496
Lonza AG, Switzerland	*Achromobacter xylosoxidans*	Nicotinic acid hydroxylase	1.5.1.13	213
	Agrobacterium sp.	Hydroxylase	1.5.1.13	430
		Nitrilase	3.5.5.1	430
	Alcaligenes faecalis	Hydroxylase	1.5.1.13	433
		Nitrilase	3.5.5.1	433
	Comamonas acidovorans	Amidase	3.5.1.4	377
	Escherichia coli	Aldehyde reductase	1.1.1.2	174
	Escherichia coli	Carnitine dehydratase	4.2.1.89	488
	Klebsiella oxytoca	Amidase	3.5.1.4	382
	Klebsiella terrigena	Amidase	3.5.1.4	379
	Pseudomonas putida	Monooxygenase	1.14.13.X	238

company	strain	enzyme name	EC number	page
	Rhodococcus rhodochrous	Nitrile hydratase	4.2.1.84	480
Marukin Shoyu Co., Ltd., Japan	*Escherichia coli*	GlcNAc 2-epimerase	5.1.3.8	506
Mercian Co., Ltd., Japan	*Bacillus circulans*	Cyclodextrin glycosyltransferase	2.4.1.19	264
Merck & Co., Inc., USA	*Candida sorbophila*	Dehydrogenase	1.1.X.X	191
Merck Sharp & Dohme, USA	*Nocardia autotropica*	Oxygenase	1.14.14.1	243
Mitsubishi Chemical Corporation, Japan	*Arthrobacter* sp.	Oxygenase	1.13.11.1	219
	Brevibacterium flavum	Aspartase	4.3.1.1	498
Mitsui Seito Co., Ltd., Japan	*Protaminobacter rubrum*	α-Glucosyl transferase	5.4.99.11	512
Nagase & Co., Ltd., Japan	*Bacillus coagulans/Streptomyces rubiginosus/Streptomyces phaechromogenes*	Xylose isomerase	5.3.1.5	508
Nippon Mining Holdings, Inc., Japan	*Nocardia corallina*	Monooxygenase	1.14.14.1	245
Nitto Chemical Industry Co., Ltd., Japan	*Rhodococcus rhodochrous*	Nitrile hydratase	4.2.1.84	481
Novartis, Switzerland	Microbial source	Catalase	1.11.1.6	217
Novo Nordisk, Denmark	*Bacillus brevis*	Acetolactate decarboxylase	4.1.1.5	449
	Bacillus coagulans/ Streptomyces rubiginosus/ Streptomyces phaechromogenes	Xylose isomerase	5.3.1.5	508
	Pig Pancreas	Trypsin	3.4.21.4	354
NSC Technologies, USA	*Bacillus* sp.	D-Amino acid transaminase	2.6.1.21	267
Pfizer Inc., USA	*Candida cylindracea*	Lipase	3.1.1.3	294
	Escherichia coli	Penicillin amidase	3.5.1.11	395
	Gluconobacter oxydans ATCC 621	D-Sorbitol dehydrogenase	1.1.99.21	185
	Leuconostoc mesenteroides	D-Lactate dehydrogenase	1.1.1.28	179
	Pseudomonas putida ATCC 33015	Aryl alcohol dehydrogenase	1.1.1.90	233
		Benzaldehyde dehydrogenase	1.2.1.28	233
		Monooxygenases	1.14.13.62	233
PGG Industries, USA	*Bacillus lentus*	Subtilisin	3.4.21.62	371
Sandoz AG, Switzerland	*Trijonopsis variabilis*	D-Aminoacid oxidase	1.4.3.3	209
Sanofi Aventis, France	*Escherichia coli*	Glutaryl amidase	3.1.1.41	326
	Pig Pancreas	Carboxypeptidase B	3.4.17.2	346
		Trypsin	3.4.21.4	352
Schering Plough, USA	*Candida antarctica*	Lipase	3.1.1.3	297
Several companies	*Aspergillus niger*	Amyloglucosidase	3.2.1.3	332
	Bacillus licheniformis	α-Amylase	3.2.1.1	332

company	strain	enzyme name	EC number	page
Sigma Aldrich, USA	*Acinetobacter calcoaceticus*	Cyclohexanone monooxygenase	1.14.13.22	227
	Escherichia coli	Oxygenase	1.14.13.44	230
Snow Brand Milk Products Co., Ltd., Japan	*Saccharomyces lactis*	β-Galactosidase	3.2.1.23	316
Südzucker AG, Germany	*Protaminobacter rubrum*	α-Glucosyl transferase	5.4.99.11	512
Sumitomo Chemical Co., Japan	*Arthrobacter* sp.	Lipase	3.1.1.3	304
	Saccharomyces lactis	β-Galactosidase	3.2.1.23	316
Tanabe Seiyaku Co., Japan	*Brevibacterium flavum*	Fumarase	4.2.1.2	466
	Pseudomonas dacunhae	Aspartate β-decarboxylase	4.1.1.12	451
	Escherichia coli	L-Aspartase	4.3.1.1	500
	Serratia marescens	Lipase	3.1.1.3	306
Toray Industries Inc., Japan	*Achromobacter obae*	Racemase	5.1.1.15	424
	Cryptococcus laurentii	Lactamase	3.5.2.11	424
UNICHEMA Chemie BV, The Netherlands	*Candida antarctica*	Lipase	3.1.1.3	313
Unifar, Turkey	*Escherichia coli*	Penicillin amidase	3.5.1.11	388
Wacker Chemie, Germany	*Lactobacillus brevis*	Alcohol dehydrogenase	1.1.1.1	170
Yamasa Corporation, Japan	*Erwinia carotovora*	Nucleosidase	3.2.2.1	335
		Phosphorylase	2.4.2.2	335
Zeneca Life Science Molecules, U.K.	*Neurospora crassa*	Alcohol dehydrogenase	1.1.1.1	153

Index of starting material

starting material	enzyme name	EC number	company	page
adiponitrile	Nitrile hydratase	4.2.1.84	DuPont, USA	478
alkene	Monooxygenase	1.14.14.1	Nippon Mining Holdings, Inc., Japan	245
amino acid; L-	Amino acid racemase	5.1.1.10	Degussa AG, Germany	504
	D-Amino acid transaminase	2.6.1.21	NSC Technologies, USA	267
D,L-amino acid	L-Hydantoinase	3.5.2.4	Degussa AG, Germany	417
amino acid amide; α-H or α-substituted	Aminopeptidase	3.4.11.1	DSM, The Netherlands	337
amino acid; α-H or α-substituted	Aminopeptidase	3.4.11.1	DSM, The Netherlands	337
amino methyl ester; α-H or α-substituted	Aminopeptidase	3.4.11.1	DSM, The Netherlands	337
amino-D-sorbitol; 1-	D-Sorbitol dehydrogenase	1.1.99.21	Bayer AG, Germany	182
ammonia	Aspartase	4.3.1.1	BioCatalytics Inc., USA	494
	L-Phenylalanine ammonia-lyase	4.3.1.5	Genex Corporation, USA	502
6-APA (= 6-amino-penicillanic acid)	Penicillin acylase	3.5.1.11	DSM, The Netherlands	397
arylallyl ether	Oxidase	1.14.X.X	DSM, The Netherlands	251
aspartic acid; D,L-	Aspartate β-decarboxylase	4.1.1.12	Tanabe Seiyaku Co., Japan	451
aspartic acid; L-	Aspartate β-decarboxylase	4.1.1.12	Tanabe Seiyaku Co., Japan	451
	Thermolysin	3.4.24.27	DSM, The Netherlands	373
azabicyclo[2.2.1]hept-5-en-3-one; 2-	β-Lactamase	3.5.2.6	Celltech Group plc, U.K.	420
azetidinone acetate; racemic cis-	Lipase	3.1.1.3	Bristol-Myers Squibb, USA	277
azetidinone; cis-[(2R,3S),(2S,3R)]-	Penicillin amidase	3.5.1.11	Eli Lilly, USA	393
azlactone of tert-leucine	Lipase	3.1.1.3	Celltech Group plc, U.K.	284
benzaldehyde	Pyruvate decarboxylase	4.1.1.1	Krebs Biochemicals & Industries Ltd., India	447
benzene; and derivatives thereof	Benzoate dioxygenase	1.14.12.10	ICI, U.K.	225
benzoic acid	Oxygenase	1.13.11.1	Mitsubishi Chemical Corporation, Japan	219
benzyl-3-[[1-methyl-1-((morpholino-4-yl) carbonyl)ethyl]sulfonyl] propionic acid ethyl ester; (R,S)-2-	Subtilisin	3.4.21.62	Hoffmann La-Roche AG, Switzerland	360
benzyl-3-(tert-butylsulfonyl)propionic acid ethyl ester; (R,S)-2-	Subtilisin	3.4.21.62	Hoffmann La-Roche AG, Switzerland	364
benzyloxy-3,5-dioxo-hexanoic acid ethyl ester; 6-	Alcohol dehydrogenase	1.1.1.1	Bristol-Myers Squibb, USA	162

starting material	enzyme name	EC number	company	page
4,5-dihydro-4-(4-methoxyphenyl)-6-(trifluoromethyl)-1*H*-1-benzazepine-2,3-dione	Reductase	1.1.X.X	Bristol-Myers Squibb, USA	199
dimethyl-1,3-dioxolane-4-methanol; (*R,S*)-2,2-	Oxidase	1.X.X.X	International Bio Synthetics, Inc., The Netherlands	255
dimethylcyclopropane-carboxamide; 2,2-	Amidase	3.5.1.4	Lonza AG, Switzerland	377
dimethylpyrazine; 2,5-	Monooxygenase	1.14.13.X	Lonza AG, Switzerland	238
5-(1,3-dioxolan-2-yl)-2-oxo-pentanoic acid (ketoacid acetal)	Phenylalanine dehydrogenase / Formate dehydrogenase	1.4.1.20 / 1.2.1.2	Bristol-Myers Squibb, USA	207
erythrono-γ-lactone; D,L-	Lactonase	3.1.1.25	Fuji Chemicals Industries Co., Ltd., Japan	321
ethyl 3-amino-5-(trimethylsilyl)-4-pentynoate	Penicillin amidase	3.5.1.11	Pfizer Inc., USA	395
ethyl-4,4,4-trifluoroacetoacetate	Aldehyde reductase	1.1.1.2	Lonza AG	174
ethyl 5-oxo-hexanoate	Reductase	1.1.X.X	Bristol-Myers Squibb, USA	195
ethyl acetoacetate	Alcohol dehydrogenase	1.1.1.1	Wacker Chemie, Germany	170
ethyl-4-chloro-3-hydroxybutyrate	Halohydrin dehalogenase	3.8.X.X	Codexis Inc., USA	445
ethyl-4-chloro-3-ketobutyrate	Ketoreductase	1.1.X.X	Codexis, USA	189
ethylmethoxyacetate	Lipase	3.1.1.3	BASF AG, Germany	273
(exo,exo)-7-oxabicyclo[2.2.1]heptane-2,3-dimethanol diacetate ester	Lipase	3.1.1.3	Bristol-Myers Squibb, USA	311
fumarate, calcium	Fumarase	4.2.1.2	Amino GmbH, Germany	463
fumaric acid			Mitsubishi Chemical Corporation, Japan	498
	Aspartase	4.3.1.1		498
			BioCatalytics Inc., USA	494
			Kyowa Hakko Kogyo Co., Ltd., Japan	496
	L-Aspartase	4.3.1.1	Tanabe Seiyaku Co., Ltd., Japan	500
	Fumarase	4.2.1.2	Amino GmbH, Germany	463
			Tanabe Seiyaku Co., Japan	466
galactono-γ-lactone; D,L-	Lactonase	3.1.1.25	Fuji Chemicals Industries Co., Ltd., Japan	321
GlcNAc = *N*-acetyl-D-glucosamine	*N*-Acetyl-D-neuraminic acid aldolase	4.1.3.3	GlaxoSmithKline, U.K.	457
	GlcNAc 2-epimerase	5.1.3.8	Marukin Shoyu Co., Ltd., Japan	506
	N-Acetyl-D-neuraminic acid aldolase	4.1.3.3	Forschungszentrum Jülich GmbH, Germany	459

starting material	enzyme name	EC number	company	page
			Marukin Shoyu Co., Ltd., Japan	459
glucono-δ-lactone; D,L-	Lactonase	3.1.1.25	Fuji Chemicals Industries Co., Ltd., Japan	321
glucooctanoic-γ-lactone; α,β-	Lactonase	3.1.1.25	Fuji Chemicals Industries Co., Ltd., Japan	321
glucose	Glutamate dehydrogenase / Glucose 1-dehydrogenase	1.4.1.3 / 1.1.1.118	Bristol-Myers Squibb, USA	201
	D-Sorbitol dehydrogenase	1.1.99.21	Bayer AG, Germany	182
	Xylose isomerase	5.3.1.5	DSM, The Netherlands	508
			Finnsugar, Finland	508
			Nagase & Co., Ltd., Japan	508
			Novo Nordisk, Denmark	508
glutaryl-7-aminocephalosporanic acid	Glutaryl amidase	3.1.1.41	Asahi Kasei Chemical Industry Co., Ltd., Japan	330
			Sanofi Aventis, France	326
			Toyo Jozo, Japan	330
glycero-D-gulo-heptono-γ-lactone; D,L-	Lactonase	3.1.1.25	Fuji Chemicals Industries Co., Ltd., Japan	321
glycero-D-manno-heptono-γ-lactone; D,L-	Lactonase	3.1.1.25	Fuji Chemicals Industries Co., Ltd., Japan	321
glycidate; (R,S)-	Lipase	3.1.1.3	DSM, The Netherlands	288
glyoxylic acid	D-Hydantoinase	3.5.2.2	Kanegafuchi Chemical Industries Co., Ltd., Japan	411
guanosine	Nucleosidase / Phosphorylase	3.2.2.1 / 2.4.2.2	Yamasa Corporation, Japan	335
gulono-γ-lactone; D,L-	Lactonase	3.1.1.25	Fuji Chemicals Industries Co., Ltd., Japan	321
H_2O_2	Catalase	1.11.1.6	Novartis, Switzerland	217
hexanedione; 2,5-	Alcohol dehydrogenase	1.1.1.1	Jülich Fine Chemicals GmbH	167
homogentisic acid lactone	Lactonase	3.1.1.25	Fuji Chemicals Industries Co., Ltd., Japan	321
hydroxy-2-methyl-3-oxobutanoate, 2-	Acetolactate decarboxylase	4.1.1.5	Novo Nordisk, Denmark	449
4-hydroxy-2-oxabicyclo[3.3.0]oct-7-en-3-one (ester)	Lipase	3.1.1.3	Celltech Group plc, U.K.	291
hydroxyphenylglycine-amide; D-(–)- (= HPGA)	Penicillin acylase	3.5.1.11	DSM, The Netherlands	397
hydroxyphenylglycineester; D-(–)- (= HPGM)	Penicillin acylase	3.5.1.11	DSM, The Netherlands	397
hydroxyphenyl)hydantoin; D,L-5-(p-	D-Hydantoinase	3.5.2.2	Kanegafuchi Chemical Industries Co., Ltd., Japan	411
hydroxyphenyl)hydantoin; D,L-5-(p-	Hydantoinase / Carbamoylase	3.5.2.4 / 3.5.1.77	Dr. Vig Medicaments, India	415

starting material	enzyme name	EC number	company	page
pantolactone; D,L-	Lactonase	3.1.1.25	Fuji Chemicals Industries Co., Ltd., Japan	321
penicillin-G	Penicillin amidase	3.5.1.11	Asahi Kasei Chemical Corporation, Japan	391
	Penicillin amidase	3.5.1.11	Dr. Vig Medicaments, India	386
	Penicillin amidase	3.5.1.11	Unifar, Turkey	388
phenol	D-Hydantoinase	3.5.2.2	Kanegafuchi Chemical Industries Co., Ltd., Japan	411
phenoxypropionic acid; (R)-2-(POPS)	Oxidase	1.X.X.X	BASF AG, Germany	261
phenyl acrylic acid; 3-	L-Phenylalanine ammonia-lyase	4.3.1.5	Genex Corporation, USA	502
phenylalanine-isopropylester; (R,S)-	Subtilisin	3.4.21.62	Coca-Cola, USA	357
phenylalanine methyl ester	Thermolysin	3.4.24.27	DSM, The Netherlands	373
phenyl-butyric acid; 2-	D-Amino acid transaminase	2.6.1.21	NSC Technologies, USA	267
phenylethylamine; 1-	Lipase	3.1.1.3	BASF AG, Germany	273
phenylglycineamide; D-(–)- = PGA)	Penicillin acylase	3.5.1.11	DSM, The Netherlands	397
phenylglycinemethylester; D-(–)- (= PGM)	Penicillin acylase	3.5.1.11	DSM, The Netherlands	397
phenylphenol; 2-	Oxygenase	1.14.13.44	Sigma Aldrich, USA	230
phenyl-2-propanone; 1-	Alcohol dehydrogenase	1.1.1.1	Forschungszentrum Jülich GmbH, Germany	157
picolinic acid	Nitrilase / Hydroxylase	3.5.5.1 / 1.5.1.13	Lonza AG, Switzerland	433
piperazine-2-carboxamide	Amidase	3.5.1.4	Lonza AG, Switzerland	379
propanol; 2-	Lipase	3.1.1.3	UNICHEMA Chemie BV, The Netherlands	313
pyrazine-2-carboxamide	Amidase	3.5.1.4	Lonza AG, Switzerland	379
pyrazine-2-carboxylic acid	Nitrilase / Hydroxylase	3.5.5.1 / 1.5.1.13	Lonza AG, Switzerland	430
pyridine-3-carboxylate	Nicotinic acid hydroxylase	1.5.1.13	Lonza AG, Switzerland	213
pyrocatechol	Tyrosine phenol lyase	4.1.99.2	Ajinomoto Co., Japan	461
pyruvic acid	N-Acetyl-D-neuraminic acid aldolase	4.1.3.3	Forschungszentrum Jülich GmbH, Germany	459
			GlaxoSmithKline, U.K.	457
			Marukin Shoyu Co., Ltd., Japan	459
	Tyrosine phenol lyase	4.1.99.2	Ajinomoto Co., Japan	461
racemic 6-hydroxynorleucine	D-Amino acid oxidase	1.4.3.3	Bristol-Myers Squibb, USA	211

starting material	enzyme name	EC number	company	page
methyl ester(2S)-7-(4,4'-bipiperidinylcarbonyl)-2,3,4,5,-tetrahydro-4-methyl-4-oxo-1H-14-benzodiazipine-2-acetic acid methyl ester – Lotrafiban; racemic	Lipase	3.1.1.3	GlaxoSmithKline, U.K.	300
ribono-γ-lactone; D,L-	Lactonase	3.1.1.25	Fuji Chemicals Industries Co., Ltd., Japan	321
ribose-1-phosphate	Nucleosidase / Phosphorylase	3.2.2.1/ 2.4.2.2	Yamasa Corporation, Japan	335
trifluoro-2-hydroxy-2-methylpropionamide; (R,S)-3,3,3-	Amidase	3.5.1.4	Lonza AG, Switzerland	382
R,S-2-carboalkoxy-3,6-dihydro-2H-pyran	Subtilisin	3.4.21.62	PGG Industries, USA	371
cyanohydrin; (R,S)-	Nitrilase	3.5.5.1	BASF AG, Germany	429
phenoxybenzaldehyde cyanohydrin; (R,S)-m-	Oxynitrilase	4.1.2.39	DSM, The Netherlands	455
(R,S)-N-(tert-butoxycarbonyl)-3-hydroxymethylpiperidine	Lipase	3.1.1.3	Bristol-Myers Squibb, USA	318
saccharose	α-Glucosyl transferase	5.4.99.11	Mitsui Seito Co., Ltd., Japan	512
			Südzucker AG, Germany	512
serine; L-	Tryptophan synthase	4.2.1.20	Amino GmbH, Germany	472
simvastatin	Oxygenase	1.14.14.1	Merck Sharp & Dohme, USA	243
sodium-3(-4-fluorophenyl)-oxopropanoate; 2-	D-Lactate dehydrogenase	1.1.1.28	Pfizer Inc., USA	179
starch	α-Amylase / Amyloglucosidase	3.2.1.1 / 3.2.1.3	Several companies	332
	Cyclodextrin glycosyltransferase	2.4.1.19	Mercian Co., Ltd., Japan	264
styrene	Styrene monooxygenase	1.14.13.69	DSM, The Netherlands	235
sucrose	α-Glucosyl transferase	5.4.99.11	Mitsui Seito Co., Ltd., Japan	512
			Südzucker AG, Germany	512
methoxycyclohexanol; (+/−)-trans-2-	Lipase	3.1.1.3	GlaxoSmithKline, U.K.	302
trimethylpyruvic acid	Leucine dehydrogenase	1.4.1.9	Degussa AG, Germany	203
tryptophan, L-	Naphthalene dioxygenase	1.13.11.11	Genencor International, Inc., USA	221
urea	D-Hydantoinase	3.5.2.2	Kanegafuchi Chemical Industries Co., Ltd., Japan	411
	Urease	3.5.1.5.	Asahi Kasei Chemicals Corporation, Japan	384
uridine	Nucleosidase / Phosphorylase	3.2.2.1/ 2.4.2.2	Yamasa Corporation, Japan	335

Index of product

product	enzyme name	EC number	company	page
	Penicillin amidase	3.5.1.11	Dr. Vig. Medicaments, India	386
	Penicillin amidase	3.5.1.11	Unifar, Turkey	388
arylglycidyl ether	Oxidase	1.14.X.X	DSM, The Netherlands	251
aspartame	Aminopeptidase	3.4.11.1	DSM, The Netherlands	337
aspartame (α-L-aspartyl-L-phenylalanine methyl ester, APM)	Thermolysin	3.4.24.27	DSM, The Netherlands	373
aspartic acid; L-	Aspartase	4.3.1.1	BioCatalytics Inc., USA	494
			Kyowa Hakko Kogyo Co., Ltd., Japan	496
			Mitsubishi Chemical Corporation, Japan	498
	L-Aspartase	4.3.1.1	Tanabe Seiyaku Co., Ltd., Japan	500
aspartic acid; D-	Aspartate β-decarboxylase	4.1.1.12	Tanabe Seiyaku Co., Japan	451
atenolol	Lipase	3.1.1.3	DSM, The Netherlands	288
	Oxidase	1.14.X.X	DSM, The Netherlands	251
azabicyclo[2.2.1]hept-5-en-3-one; (−)-2-		3.5.2.6	Celltech Group plc, U.K.	422
azabicyclo[2.2.1]hept-5-en-3-one; 2-	β-Lactamase	3.5.2.6	Celltech Group plc, U.K.	420
azetidinone; (4R,3R)-	Lipase	3.1.1.3	Bristol-Myers Squibb, USA	277
azetidinone acetate; (3R,4S)-	Lipase	3.1.1.3	Bristol-Myers Squibb, USA	277
azetidinone; cis-(2R,3S)	Penicillin amidase	3.5.1.11	Eli Lilly, USA	393
benzapril	Aminopeptidase	3.4.11.1	DSM, Netherlands	337
benzodiazepine	Dehydrogenase	1.1.1.1	Eli Lilly & Co., USA	164
benzyl-3-[[1-methyl-1-((morpholino-4-yl)carbonyl)ethyl]sulfonyl]propionic acid ethyl ester; (R)-2-	Subtilisin	3.4.21.62	Hoffmann La-Roche AG, Switzerland	360
benzyl-3-[[1-methyl-1-((morpholino-4-yl)carbonyl)ethyl]sulfonyl]propionic acid; (S)-2-	Subtilisin	3.4.21.62	Hoffmann La-Roche AG, Switzerland	360
benzyl-3-(tert-butylsulfonyl)propionic acid ethyl ester; (R)-2-	Subtilisin	3.4.21.62	Hoffmann La-Roche AG, Switzerland	364
benzyl-3-(tert-butylsulfonyl)propionic acid; (S)-2-	Subtilisin	3.4.21.62	Hoffmann La-Roche AG, Switzerland	364
benzyloxy-(3R,5S)-dihydroxy-hexanoic acid ethyl ester; 6-	Alcohol dehydrogenase	1.1.1.1	Bristol-Myers Squibb, USA	162
benzylserine; O-L-	Aminoacylase	3.5.1.14	Degussa AG, Germany	407

product	enzyme name	EC number	company	page
bis(4-fluorophenyl])-3-(1-methyl-1H-tetrazol-5-yl)-1,3-butadienyl]-tetra-hydro-4-hydroxy-2H-pyran-2-one; [4R-[4α,6β(E)]]-6-[4,4-	Lipase	3.1.1.3	Bristol-Myers Squibb, USA	281
boc-L-3-(4-thiazolyl)alanine; N- = Boc-Taz	Aminoacylase	3.5.1.14	Celltech Group plc, U.K.	403
Boc-Taz = N-boc-L-3-(4-thiazolyl)alanine	Aminoacylase	3.5.1.14	Celltech Group plc, U.K.	403
bromocatechol; 3-	Oxygenase	1.14.13.44	Sigma Aldrich, USA	230
butanedioic acid 4-ethyl ester; (R)-(2-methylpropyl)-	Subtilisin	3.4.21.62	Hoffmann La-Roche AG, Switzerland	367
butylcatechol; 3-sec-	Oxygenase	1.14.13.44	Sigma Aldrich, USA	230
carazolol	Lipase	3.1.1.3	DSM, The Netherlands	288
carbamoyl-D-hydroxyphenyl glycine; D-N-	Hydantoinase / Carbamoylase	3.5.2.4 / 3.5.1.77	Dr. Vig Medicaments, India	415
carbamoyl-D-phenylglycine; D-N-	D-Hydantoinase	3.5.2.2	Kanegafuchi Chemical Industries Co., Ltd., Japan	411
carnitine; L-	Carnitine dehydratase	4.2.1.89	Lonza AG, Switzerland	488
catechol	Oxygenase	1.13.11.1	Mitsubishi Chemical Corporation, Japan	219
cefaclor	Penicillin acylase	3.5.1.11	DSM, The Netherlands	397
cefadroxil	Penicillin acylase	3.5.1.11	DSM, The Netherlands	397
cefalexin	Penicillin acylase	3.5.1.11	DSM, The Netherlands	397
chloro-2,3-epoxypropane; 1-	Haloalkane dehalogenase	3.8.1.5	Daiso Co. Ltd., Japan	439
chloro-3-hydroxy-butanoate; (S)-4-	Carbonyl reductase	1.1.1.1	Kaneka Corporation, Japan	172
chloro-3-hydroxybutanoic acid methyl ester; (S)-4-	Dehydrogenase	1.1.X.X	Bristol-Myers Squibb, USA	187
chlorocatechol; 3-	Oxygenase	1.14.13.44	Sigma Aldrich, USA	230
chloropropionic acid; (S)-	Dehalogenase	3.8.1.2	Astra Zeneca, U.K.	435
cilazapril	Aminopeptidase	3.4.11.1	DSM, Netherlands	337
cyanophenylalanine, L-4-	Aminoacylase	3.5.1.14	Celltech Group plc, U.K.	403
cyanovaleramide; 5-	Nitrile hydratase	4.2.1.84	DuPont, USA	478
cyclodextrin; α-	Cyclodextrin glycosyltransferase	2.4.1.19	Mercian Co., Ltd., Japan	264
cyclodextrin; β-	Cyclodextrin glycosyltransferase	2.4.1.19	Mercian Co., Ltd., Japan	264
cyclodextrin; γ-	Cyclodextrin glycosyltransferase	2.4.1.19	Mercian Co., Ltd., Japan	264
delapril	Aminopeptidase	3.4.11.1	DSM, The Netherlands	337
deoxy-6-butylaminosorbose; 6-	D-Sorbitol dehydrogenase	1.1.99.21	Pfizer Inc., USA	185
desoxynojirimycin; 1-	D-Sorbitol dehydrogenase	1.1.99.21	Bayer AG, Germany	182
diepoxydecane; (R,R)-1,2,9,10-	Monooxygenase	1.14.14.1	Nippon Mining Holdings, Inc., Japan	245

product	enzyme name	EC number	company	page
ethylcatechol; 3-	Oxygenase	1.14.13.44	Sigma Aldrich, USA	230
ethyl-(*R*)-4-cyano-3-hydroxybutyrate	Halohydrin dehalogenase	3.8.X.X	Codexis Inc., USA	445
ethyl-(*S*)-4-chloro-3-hydroxybutyrate	Ketoreductase	1.1.X.X	Codexis, USA	189
ethyl-4,4,4-trifluoro-3-hydroxybutanoate; (*R*)-	Aldehyde reductase	1.1.1.2	Lonza AG	174
4-exo-hydroxy-2-oxybicyclo[3.3.0]oct-7-en-3-one butyrate ester	Lipase	3.1.1.3	Celltech Group plc, U.K.	291
fluorophenyl)-2-hydroxy propionic acid; (*R*)-3-(4-	D-Lactate dehydrogenase	1.1.1.28	Pfizer Inc., USA	179
fluvalinate	Aminopeptidase	3.4.11.1	DSM, The Netherlands	337
fructofuranose; 6-O-α-D-glucopyranosyl-D-	α-Glucosyl transferase	5.4.99.11	Mitsui Seito Co., Ltd., Japan	512
			Südzucker AG, Germany	512
fructose	Xylose isomerase	5.3.1.5	DSM, The Netherlands	508
			Finnsugar, Finland	508
			Nagase & Co., Ltd., Japan	508
			Novo Nordisk, Denmark	508
galactono-γ-lactone; D-	Lactonase	3.1.1.25	Fuji Chemicals Industries Co., Ltd., Japan	321
galactose	β-Galactosidase	3.2.1.23	Central del Latte, Italy	316
			Snow Brand Milk Products Co., Ltd., Japan	316
			Sumitomo Chemical Co., Japan	316
gluconic acid	Glutamate dehydrogenase / Glucose 1-dehydrogenase	1.4.1.3 / 1.1.1.118	Bristol-Myers Squibb, USA	201
glucono-δ-lactone; D-	Lactonase	3.1.1.25	Fuji Chemicals Industries Co., Ltd., Japan	321
glucooctanoic-γ-lactone; α,β-	Lactonase	3.1.1.25	Fuji Chemicals Industries Co., Ltd., Japan	321
glucose	α-Amylase	3.2.1.1	Several companies	332
	Amyloglucosidase	3.2.1.3	Several companies	332
	β-Galactosidase	3.2.1.23	Central del Latte, Italy	316
			Snow Brand Milk Products Co., Ltd., Japan	316
			Sumitomo Chemical Co., Japan	316
glutanic acid, D-	D-Amino acid transaminase	2.6.1.21	NSC Technologies, USA	267
glutaryl-7-amino-cephalosporanic acid	D-Aminoacid oxidase	1.4.3.3	Sandoz AG, Switzerland	209
glycero-D-gulo-heptono-γ-lactone; D-	Lactonase	3.1.1.25	Fuji Chemicals Industries Co., Ltd., Japan	321

product	enzyme name	EC number	company	page
glycero-D-manno-heptono-γ-lactone; D-	Lactonase	3.1.1.25	Fuji Chemicals Industries Co., Ltd., Japan	321
glycidate; (R)-	Lipase	3.1.1.3	DSM, The Netherlands	288
glycidol	Haloalkane dehalogenase	3.8.1.5	Daiso Co. Ltd., Japan	439
gulono-γ-lactone; D-	Lactonase	3.1.1.25	Fuji Chemicals Industries Co., Ltd., Japan	321
H$_2$O	Catalase	1.11.1.6	Novartis, Switzerland	217
hexanediol; (2R,5R)-	Alcohol dehydrogenase	1.1.1.1	Jülich Fine Chemicals GmbH	167
high fructose corn syrup (HFCS)	Xylose isomerase	5.3.1.5	DSM, The Netherlands	508
			Finnsugar, Finland	508
			Nagase & Co., Ltd., Japan	508
			Novo Nordisk, Denmark	508
homophenylalanine; L-	Aminoacylase	3.5.1.14	Degussa AG, Germany	407
HPBA = (R)-2-hydroxy-4-phenylbutyric acid	Lactate dehydrogenase	1.1.1.28	Ciba-Geigy, Switzerland	176
hydroxy-3-methyl-2-prop-2-ynyl-cyclopent-2-enone; (R)-4-	Lipase	3.1.1.3	Sumitomo Chemical Co., Japan	304
hydroxy-3-methyl-2-prop-2-ynyl-cyclopent-2-enone; (S)-4-	Lipase	3.1.1.3	Sumitomo Chemical Co., Japan	304
hydroxybutan-2-one, (R)-3-	Acetolactate decarboxylase	4.1.1.5	Novo Nordisk, Denmark	449
hydroxy-γ-butyrolactone; (S)-3-	Haloalkane dehalogenase	3.8.1.5	Daiso Co. Ltd., Japan	443
hydroxy-isobutyric acid; (R)-β-	Enoyl-CoA hydratase	4.2.1.17	Kanegafuchi Chemical Industries Co., Ltd., Japan	470
(1R,2R,3S,4S)-3-(hydroxymethyl)-7-oxa-bicyclo[2.2.1]heptan-2-yl)methyl acetate	Lipase	3.1.1.3	Bristol-Myers Squibb, USA	311
hydroxy-methyl-simvastatin; 6-β-	Oxygenase	1.14.14.1	Merck Sharp & Dohme, USA	243
hydroxymutilin; 8-hydroxy-, 7-hydroxy- and 2-	Monooxygenase	1.14.13.XX	Bristol-Myers Squibb, USA	241
hydroxy-n-butyric acid; (R)-β-	Enoyl-CoA hydratase	4.2.1.17	Kanegafuchi Chemical Industries Co., Ltd., Japan	468
hydroxynicotinate; 6-	Nicotinic acid hydroxylase	1.5.1.13	Lonza AG, Switzerland	213
hydroxynorleucine; L-6-	Glutamate dehydrogenase / Glucose 1-dehydrogenase	1.4.1.3 / 1.1.1.118	Bristol-Myers Squibb, USA	201
hydroxy-2-oxabicyclo[3.3.0]oct-7-en-3-one; 4- (endo)	Lipase	3.1.1.3	Celltech Group plc, U.K.	291
hydroxyphenoxy)propionic acid; (R)-2-(4′- (HPOPS)	Oxidase	1.X.X.X	BASF AG, Germany	261
hydroxy-4-phenylbutyric acid; (R)-2- (= HPBA)	Lactate dehydrogenase	1.1.1.28	Ciba-Geigy, Switzerland	176
hydroxyphenyl glycine; D-p-	D-Hydantoinase	3.5.2.2	Kanegafuchi Chemical Industries Co., Ltd., Japan	411

product	enzyme name	EC number	company	page
isopropylideneglycerol; (R)-	Oxidase	1.X.X.X	International Bio Synthetics, Inc., The Netherlands	255
isopropyl palmitate	Lipase	3.1.1.3	UNICHEMA Chemie BV, The Netherlands	313
isoprpoyl myristate	Lipase	3.1.1.3	UNICHEMA Chemie BV, The Netherlands	313
ketoadipinyl-7-aminocephalosporanic acid; α-	D-Aminoacid oxidase	1.4.3.3	Sandoz AG, Switzerland	209
2-keto-6-hydroxyhexanoic acid / L-6-hydroxynorleucine	D-Amino acid oxidase	1.4.3.3	Bristol-Myers Squibb, USA	211
lactic acid; (S)-	Dehalogenase	3.8.1.2	Astra Zeneca, U.K.	435
leucine; D-	D-Amino acid transaminase	2.6.1.21	NSC Technologies, USA	267
leucine; L-	Aminoacylase	3.5.1.14	Degussa AG, Germany	407
leucine; L-tert-	Leucine dehydrogenase	1.4.1.9	Degussa AG, Germany	203
lisinopril	Aminopeptidase	3.4.11.1	DSM, The Netherlands	337
low lactose milk	β-Galactosidase	3.2.1.23	Central del Latte, Italy	316
			Snow Brand Milk Products Co., Ltd., Japan	316
			Sumitomo Chemical Co., Japan	316
lysine; L-	Lactamase	3.5.2.11	Toray Industries Inc., Japan	424
	Racemase	5.1.1.15	Toray Industries Inc., Japan	424
malate, L-	Fumarase	4.2.1.2	Amino GmbH, Germany	463
malate, calcium	Fumarase	4.2.1.2	Amino GmbH, Germany	463
malic acid	Fumarase	4.2.1.2	Amino GmbH, Germany	463
malic acid; D-	Malease	4.2.1.31	DSM, The Netherlands	474
mailc acid; L-	Fumarase	4.2.1.2	Tanabe Seiyaku Co., Japan	466
malic acid; (R)-	Malease	4.2.1.31	DSM, The Netherlands	474
malic acid; (S)-	Fumarase	4.2.1.2	Tanabe Seiyaku Co., Japan	466
mandelic acid; (R)-	Nitrilase	3.5.5.1	BASF AG, Germany	429
ManNAc = N-acetyl-D-mannosamine	GlcNAc 2-epimerase	5.1.3.8	Marukin Shoyu Co., Ltd., Japan	506
mannono-γ-lactone; L-	Lactonase	3.1.1.25	Fuji Chemicals Industries Co., Ltd., Japan	321
methionine; L-	Aminoacylase	3.5.1.14	Degussa AG, Germany	407
methoxycyclohexanol; (+)-(1S,2S)-2-	Lipase	3.1.1.3	GlaxoSmithKline, U.K.	302
methoxyisopropylamine; (S)-	Transaminase	2.6.1.X	Celgene Corporation, USA	271
methoxyphenyl)glycidic acid; 3-(4-	Lipase	3.1.1.3	Tanabe Seiyaku Co. Ltd., Japan	306
methoxyphenyl)glycidic acid methyl ester; trans-(2R,3S)-(4-	Lipase	3.1.1.3	Tanabe Seiyaku Co. Ltd., Japan	306
methyl 4-chloro-4-hydroxybutyrate; (R)-	Haloalkane dehalogenase	3.8.1.5	Daiso Co. Ltd., Japan	443